The Microbiology of Safe Food

The Microbiology of Safe Food

Second edition

Stephen J. Forsythe

School of Science and Technology, Nottingham Trent University

⊛WILEY-BLACKWELL

A John Wiley & Sons, Ltd., Publication

This edition first published 2010
© 2010 by Blackwell Publishing Ltd

Blackwell Publishing was acquired by John Wiley & Sons in February 2007. Blackwell's publishing programme has been merged with Wiley's global Scientific, Technical, and Medical business to form Wiley-Blackwell.

Registered office
John Wiley & Sons Ltd, The Atrium, Southern Gate, Chichester, West Sussex, PO19 8SQ, United Kingdom

Editorial offices
9600 Garsington Road, Oxford, OX4 2DQ, United Kingdom
2121 State Avenue, Ames, Iowa 50014-8300, USA

For details of our global editorial offices, for customer services and for information about how to apply for permission to reuse the copyright material in this book please see our website at www.wiley.com/wiley-blackwell.

Library of Congress Cataloging-in-Publication Data

Forsythe, S. J. (Steve J.)
   The microbiology of safe food / Stephen J. Forsythe. – 2nd ed.
       p. ; cm.
   Includes bibliographical references and index.
   ISBN 978-1-4051-4005-8 (pbk. : alk. paper)   1. Food–Microbiology.   I. Title.
   [DNLM: 1. Food Microbiology.   2. Food Poisoning–microbiology.   3. Legislation, Food.
4. Risk Assessment.   5. Safety Management. QW 85 F735m 2010]
   QR115.F675 2010
   664.001'579–dc22   2009037331

A catalogue record for this book is available from the British Library.

Set in 10/12 pt Minion by Aptara® Inc., New Delhi, India
Printed and bound in Malaysia by KHL Printing Co Sdn Bhd

1   2010

Contents

A colour plate section is found facing page 222

See the supporting companion website for this book:
http://www.wiley.com/go/forsythe

Preface to second edition

Although I was pleased with the first edition of this book (MoSF), I nevertheless felt that it was not complete. This new edition tries to address this by including new sections on bioinformatics, biothreats and personnel, as well as updating many other sections. Since 2000, the topic of microbiological risk assessment has increased, and subsequently, I have incorporated parts of my other Blackwell's book *Microbiological Risk Assessment of Food* (2002) into Chapter 10 as it was a substantial improvement on the first MoSF edition's few pages. My appreciation is due to Simon Illingworth (LabM, Bury, UK) for reviewing Chapter 5 on detection methods for me.

A major change is the complementing websites at http://www.wiley.com/go/forsythe. This was available with the first edition, but was an afterthought and so unfortunately was not fully utilised. In fact, it was one of the first web-based supported books by Blackwells, and the listing of URL in the Appendix was considered a 'novelty'! How much has changed since 2000. I am using the web for two main purposes. Firstly to keep some chapters up-to-date, and secondly to offer various data exercises which are not in keeping with the book format. One aspect which I have been wanting to expand and encourage 'younger' readers to explore is the application of genomics, post-genomics and bioinformatics to food microbiology. Again, the first edition included microarrays but not the tools for one to investigate microbial genomes for oneself. In fact, 2000, the year MoSF was published, was also the year when the first version of the *Campylobacter jejuni* genome was released, and since the MoSF text was written in 1999, the whole topic of microbial genomics was not even on the radar. The fact that genomes are sequenced faster than they can be fully annotated means that one can quickly discover something which no one else has even known before, and I hope the bioinformatics aspects will enable and encourage readers to try *in silico* research. One topic which was gaining increasing public attention in 1999–2000 was BSE-vCJD. It appears in the intervening years that we have possibly passed the peak incidence, fortunately. However, over the same period, the spectre of bioterrorism has arisen and so this issue is addressed in a new section of this edition.

One thing which has not changed between these two editions is the unacceptable high incidence of foodborne disease. Even more alarming is that we still are only aware of the 'tip of the iceberg' with regard to its true incidence. When one considers that it has been estimated that in the United States, 3400 deaths are due to unknown foodborne agents (Frenzen *et al.* 2005), then there is evidently a considerable amount of research and investment still to be undertaken.

Confession time, it was my full intention to complete this new edition for publication in 2005. However, our intensive research into *Cronobacter* spp. (*Enterobacter sakazakii*) and related organisms have taken up more of my time than the hours in the day can permit. This emergent pathogen, sadly, can infect neonates causing severe illness, and even death. In order not to unbalance this book by excessive reference to this organism of my own personal interest, readers should consult the 2008 ASM Press book *Enterobacter sakazakii*, edited by myself and Jeff Farber (Health Canada) as well as my homepage (see http://www.wiley.com/go/forsythe).

As always, my thanks and appreciation go to Nigel Balmforth, David McDade and especially Katy Loftus at Wiley-Blackwell for their patience as the deadlines made a whooshing sound as they went by (frequently). Finally, a special thanks to my forever supportive wife Debbie, my children James and Rachel, and my parents – without whom none of this would have been possible.

Steve Forsythe
Professor of Microbiology
Nottingham Trent University

Preface to first edition

Throughout the world, food production has become more complex. Frequently, raw materials are sourced globally and the food is processed through an increasing variety of techniques. No longer does the local farm serve the local community through a local shop; nowadays, there are international corporations adhering to national and international regimes. Therefore, approaches to safe food production are being assessed on an expanding platform from national, European, trans-Atlantic and beyond. Against this backdrop, there have been numerous highly publicised food safety issues such as BSE and *E. coli* O157:H7 which have caused the general public to become more cautious of vociferous concerning food issues. The controversy in Europe over genetically modified foods is perceived by the general public within the context of 'food poisoning'.

This book aims to review the production of food and the level of micro-organisms which humans ingest. Certain circumstances require zero tolerances for pathogens, whereas more frequently, there are acceptable limits set, albeit with statistical accuracy or inaccuracy depending upon whether you subsequently suffer from food poisoning. Microbes are traditionally ingested in fermented foods and this has developed into the subject of pre- and probiotics with refuted health benefits. Whether engineered 'functional foods' will be able to attain consumer acceptance remains to be seen.

Food microbiology covers both food pathogens and food spoilage organisms. This book aims to cover the wide range of micro-organisms occurring in food, both as contaminants and deliberate inoculation. Due to the heightened public awareness over food poisoning, it is important that all companies in the food chain maintain high hygienic standards and assure the public of the safety of the produce. Obviously, over time, there are technological changes in production methods and methods of microbiological analysis. Therefore, the food microbiologist needs to know the effect of processing changes (pH, temperature, etc.) on the microbial load. To this end, this book reviews the dominant foodborne micro-organisms, the means of their detection, microbiological criteria as the numerical means of interpreting end-product testing, predictive microbiology as a tool to understand the consequences of processing changes, the role of 'Hazard Analysis Critical Control Point' (HACCP) and the objectives of Microbial Risk Assessment (MRA) and the setting of Food Safety Objectives which have recently become a focus of attention. In recent years, the web has become an invaluable source of information and to reflect this a range of useful food safety resource sites are given in the back to encourage the reader to boldly go and surf. Although primarily aimed for undergraduate and postgraduate courses, I hope the book will also be of use to those working in industry.

The majority of this book was written during the last months of 1999, a time when France was being taken to the European Court over its refusal to sell British beef due to BSE/nvCJD and there had been riots in Seattle concerning the World Trade Organisation. Whilst large organisations were wondering about the impact of the millennium bug, in the United Kingdom, the public were waiting to see the impact of the BSE 'bug' (a few hundred or a few thousand cases?).

As usual, no book can be achieved without assistance, and special thanks are due to Phil Vosey concerning MRA, Ming Lo for considerable help with the computer packages, Alison at Oxoid Ltd for the invaluable information on microbiological testing procedures around the world, Pete Silley and Andrew Pridmore at Don Whitley Scientific Ltd for the RABIT diagrams, and Garth Lang at Biotrace Ltd for the ATP bioluminescence data. Not forgetting of course Debbie and Cathy for reading through the draft copy, nevertheless all mistakes are the author's fault.

This book is especially dedicated to Debbie, James and Rachel, Mum and Dad for their patience whilst I have been burning the midnight oil.

Dr Steve Forsythe on 6 January, 2000

1 Foodborne infections and intoxications

This may seem a rather self-defeating opening sentence, but there is no universally accepted definition of 'safe food'. The reason is that we are dealing with a relative term, which is linked to determining the acceptable level of risk to a mixed population or maybe a specific subgroup. Our food supply involves international movement of ingredients and processed products. Our food is very diverse, and to ensure it is safe requires a systematic, proactive approach of minimising contamination from 'farm to fork'. Some procedures are well known to the general public such as refrigeration and canning. There is also the implementation of 'Hazard Analysis and Critical Control Point' (HACCP) in which the producer anticipates the likely hazards in the final product and ensures the processing reduces or eliminates them to an acceptable level. Unfortunately, illness due to foodborne contamination is still a major cause of morbidity and mortality. Foodborne illness can be defined as diseases commonly transmitted through food, and comprise a broad group of illnesses caused by microbial pathogens, parasites, chemical contaminants and biotoxins. An alternative phrase 'food poisoning' has often been used but nowadays is regarded as being too restrictive.

Dealing with food safety problems is challenging, in part because they are changing. We have changes in our economy, and therefore lifestyle, eating habits (ranges of food, eating at home or eating out), and an ageing population. The causative agents of foodborne illnesses are also changing, with the emergence of previously unrecognised emerging pathogens. Food producers, both industrially and domestically, need to be aware of these changes in order to improve the safety of our food. This first chapter will consider the magnitude of the foodborne illness, the diversity of sources and diseases, along with its economic consequences. These key topics will be covered in greater depth later in specific chapters. Definition of terms will be found in the glossary at the end of the book, where there is also a listing of useful hypertext links.

Food microbiology is a multidisciplinary topic, and there are rapid advances being made in a number of areas. In order to keep this book as up-to-date as reasonably possible, the reader should also refer to the supporting websites at http://www.wiley.com/go/forsythe, where additional information for specific chapters are given.

1.1 Origins of safe food production

The necessity of safe food must go back to early humans and developed following hunter–gatherer activities. The domestication of animals and cultivation of crops required cooking and storage activities. Barley production flourished in the Egyptian Nile Valley about 18 000 years ago. This necessitated the need to preserve the grain by keeping it dry to prevent fungal spoilage. Preventing the spoilage of more perishable foods by drying could easily have co-developed. Preservation by the addition of honey and olive oil were also early forms of food preparation. Once salt had been found to have a preservative capability, it became a major commodity. The word 'salary' originally meant 'soldier's allowance for the purchase of salt'.

Over time, humans learned to select edible animals and plants. We also learnt how to cultivate and farm, to harvest and organise our food resources according to the seasons, and habitat. Undoubtedly, there was considerable trial and error, but gradually the good habits became learnt and passed down through the generations. Numerous religious practices related to food have sound scientific basis for their time. This may include Jewish and Muslim faiths not eating pork, which we now know carries tapeworm *Trichinella spiralis*. The use of running water to bathe is more hygienic than using standing water.

The beginning of a more scientific approach to food preparation occurred with the development of food preservation by heat treatment. In 1795, the French government realised the strategic usefulness of preserved food for its troops and offered a large reward for anyone who could develop a new method of preserving food. The prize was won by Nicholas Appert, a Parisian confectioner by trade. His preservation method was putting the food into a wide-mouth glass bottle which was then sealed with a cork and put in boiling water for 6 hours. The use of tin cans instead of glass bottles was the idea of Durand in 1810 and is still the basis of the canning industry today. The thermal processing worked, but the rationale behind the procedures was not known until the work of Louis Pasteur and Robert Koch (Hartman 1997).

Although previous workers such as Antonie van Leeuwenhoek in 1677 had discovered 'little heat-sensitive animalcules', it was Louis Pasteur who started the science of microbiology. Due to his studies, between 1854 and 1864, he demonstrated that bacteria were the causative agents of food spoilage and disease. As a consequence, the French wine industry adopted the process of heating the wine to kill the spoilage organisms, before inoculating with favourable micro-organisms for the fermentation process. Later, the 'pasteurisation' process was applied to other foods such as milk. However, this later application is principally for the control of pathogenic micro-oganisms. Another founding figure in microbiology was the German Robert Koch who first developed a method of growing pure cultures of micro-organisms. In 1884, he was the first to isolate the bacterium *Vibrio cholerae*. From then on, the isolation and study of pure cultures have been a major activity for food microbiologists (Hartman 1997).

From those early days, medical, veterinary, environmental and food microbiology have each become established disciplines. Food microbiology itself encompasses a number of topics such as the detection of unwanted micro-organisms, and their products as well as the desirable use of microbial activity in the production of fermented foods such as beer, wine, cheese and bread. Or more simply put – 'the good, the bad and the ugly', for food production, food poisoning and food spoilage. Advances continue to be made and the development of HACCP (Section 8.5) was driven by the need for safe food for the US manned space programme, and was supported by the US army as well, a modern-day echo of Nicholas Appert's contribution to safe food.

1.2 Foodborne illness

Problems with food quality and safety have existed for many centuries, for example the adulteration of milk, beer, wine, tea leaves, and olive oil. Contaminated food causes one of the major health problems in the world, and leads to reduced economic productivity (Bettcher *et al.* 2000). Table 1.1 lists the pathogens which may be transmitted through contaminated food. Although foodborne illness is often regarded as referring to illness due to bacterial pathogens, Table 1.1 shows a wide range of organisms and chemicals that can cause foodborne illness. Some

Table 1.1 Hazards associated with food

Biological	Chemical	Physical
Macrobiological	Veterinary residues, antibiotics, growth stimulants	Glass
Microbiological	Plasticisers and packaging migration, vinyl chloride, bisphenol A	Metal
Viruses	Chemical residues, pesticides (DDT), cleaning fluids	Stones

Hepatitis A
  Norovirus
  Rotavirus		

Pathogenic bacteria	Allergens	Wood
Spore forming	Toxic metals, lead, cadmium, arsenic, tin, mercury	Plastic

Bacillus cereus
    Clostridium perfringens
    Cl. botulinum		

Non-spore forming	Food chemicals, preservatives, processing aids	Parts of pests

Campylobacter jejuni
    Pathogenic strains of *E. coli*
    Listeria monocytogenes
    Salmonella serovars		

Bacterial toxins	Radiochemicals, ^{131}I, ^{127}Cs	Insulation material
Staphylococcus aureus (source) *Bacillus cereus* (source)		
Shellfish toxins, domoic acid, okadaic acid	Dioxins, polychlorinated biphenyls (PCBs)	Bone
NSP, PSP	Prohibited substances	Fruit pits
Parasites and protozoa	Printing inks	

Cryptosporidium parvum
  Entamoeba histolytica
  Giardia lamblia
  Toxoplasma gondii
  Fasciola hepatica
  Taenia solium
  Anisakis spp.
  Trichinella spiralis		

Mycotoxins, ochratoxin, aflatoxins, fumonsins, patulin		

NSP, neurotoxic shellfish poison; PSP, paralytic shellfish poisoning.
Adapted from Snyder (1995) and Forsythe (2000).

compounds and organisms are external food contaminants, whereas some are intrinsic, for example oxalic acid in rhubarb and the alkaloid solanine in potatoes.

Microbial food poisoning is caused by a variety of micro-organisms with various incubation periods and duration of symptoms (Table 1.2). Organisms such as *Salmonella* and *Escherichia coli* O157:H7 are well known by the general public. But there are also viruses and fungal toxins which have relatively been poorly studied, and in the future, we may more fully recognise

Table 1.2 Common food poisoning micro-organisms

Micro-organism	Incubation period	Duration of illness
Aeromonas species	Unknown	1–7 days
C. jejuni	3–5 days	2–10 days
E. coli		
ETEC	16–72 hours	3–5 days
EPEC	16–48 hours	2–7 days
EIEC	16–48 hours	2–7 days
EHEC	72–120 hours	2–12 days
Hepatitis A	3–60 days	2–4 weeks
L. monocytogenes	3–70 days	Variable
Norovirus	24–48 hours	1–2 days
Rotavirus	24–72 hours	4–6 days
Salmonellas	16–72 hours	2–7 days
Shigellae	16–72 hours	2–7 days
Yersinia enterocolitica	3–7 days	1–3 weeks

their contribution to the general incidence of food poisoning. Micro-organisms causing food poisoning are found in a diverse range of foods including milk, meat and eggs. They have a wide range of virulence factors which can cause a range of adverse responses which may be acute, chronic or intermittent. Some bacterial pathogens such as salmonellas are invasive and can get through the intestinal wall into the bloodstream and cause generalised infections. Other pathogens produce toxins in the food prior to ingestion, or during infection and cause severe damage in susceptible organs such as the kidney, for example *E. coli* O157:H7. Complications can occur due to immune-mediated reactions (i.e. *Campylobacter* infection leading to reactive arthritis and Guillain–Barré syndrome) where the host immune response to the pathogen is also unfortunately directed against the host tissues. Therefore, foodborne illness can be much more severe than a short period of gastroenteritis, and instead may result in hospitalisation. The severity can be such that there may be residual (chronic) symptoms and the risk of death especially in the elderly and severely immuno-compromised. Hence, there is a considerable public health burden attributable to foodborne infections.

Because the consumer is unaware that there is a potential problem with the food, a significant amount of contaminated food is ingested and hence they become ill. Consequently, it is hard to trace which food was the original cause of food poisoning since the consumer will not recall noticing anything appropriate in their recent meals. They are likely to recall food which smelt 'off' or looked 'discoloured'. However, these attributes are related to food spoilage and not food poisoning.

Food poisoning micro-organisms are normally divided into two groups:
- Infections; *Salmonella* serotypes, *Campylobacter jejuni* and pathogenic *E. coli*
- Intoxications; *Bacillus cereus*, *Staphylococcus aureus*, *Clostridium botulinum*

The first group are the micro-organisms which multiply in the human intestinal tract, whereas the second group produce toxins either in the food or during passage in the intestinal tract. This division is very useful to help recognise the routes of food poisoning. The symptoms are also indicative of the infectious organism. As a generalisation, bacterial infections cause gastroenteritis, whereas ingestion of a toxin cause vomiting. Gastroenteritis with a fever could be due to Gram-negative infectious organism, as the host's immune system responds to the lipopolysaccharide causing the fever. Viral infections cause both vomiting and gastroenteritis.

Vegetative organisms are killed by heat treatment, whereas spores (produce by *Bacillus cereus* and *Clostridium perfringens*) may survive and hence germinate if the food is not kept sufficiently hot or cold after cooking.

An alternative grouping is according to severity of illness. This approach is useful in setting microbiological criteria (sampling plans) and risk analysis. The International Commission on Microbiological Specifications for Foods (ICMSF 1974, 1986 and 2002) divided the common foodborne pathogens into such groups in order to aid in the decision-making of sampling plans (Chapter 6). The ICMSF groupings are given later in Chapter 4. Detailed descriptions of certain foodborne pathogens are given in Chapter 4 and extensive details can be found in the numerous ICMSF publications listed in the References.

Despite an increasing awareness and understanding of food and waterborne micro-organisms, these diseases remain a significant problem and are an important cause of reduced economic productivity. While everyone is susceptible to foodborne diseases, there is a growing number of people who are more likely to experience such diseases and often with more severe consequences. These people include infants and young children, pregnant women, those who are immuno-compromised through medication or illness, and the elderly. There is evidence that the microbial causes of gastroenteritis vary with age and that viral agents are probably the major infectious agent in children under the age of 4 years (Figure 1.1). There is also a difference between the sexes (Figure 1.2), which is possibly due to differences in personal hygiene; that is males have a lower tendency to wash their hands after going to the toilet.

The production of food has increased by 145% since the 1960s. Of particular importance is the increase in developing countries such as Africa (140%), Latin America (200%) and Asia (280%). Food production has doubled in the United States, and by 68% in Western Europe. Yet hunger is still a worldwide problem. In this century, there are over 800 million people suffering from malnutrition. Over the same time period, the world's population has increased from 3 to 6 billion, and is expected to attain 9 billion by 2050. Subsequently, there will be ever increasing demands in food production.

The World Health Organisation (WHO) estimates that up to 30% of people in the developed countries are infected by food and water every year (WHO Food safety and foodborne illness. Fact sheet 237 http://www.who.int/mediacentre/factsheets/fs237/en/print.html). This number is supported by US, Canadian and Australian authorities (Majowicz *et al.* 2004, Mead *et al.* 1999, OzFoodNet Working Group 2003).

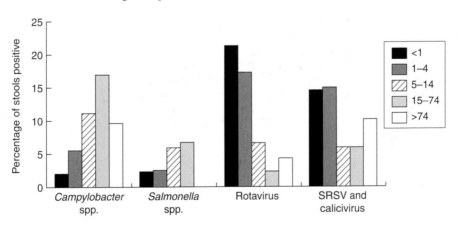

Figure 1.1 Variation in causative agent of gastroenteritis with age (Forsythe 2000).

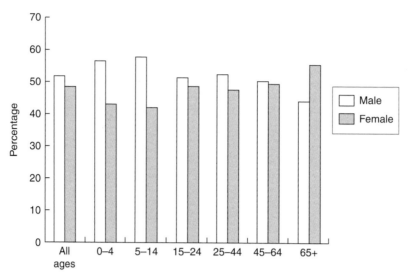

Figure 1.2 Variation in gastroenteritis with sex (Forsythe 2000).

In developing countries, foodborne diseases are a major (up to 70%) cause of diarrhoea in children under 5 years of age. They may suffer 2–3 episodes of diarrhoea per year, and for some, this may even be as many as 10. Pathogenic bacteria can contaminate weaning foods via the food or water supply, causing 25–30% of diarrhoeal infections. The repeated periods of diarrhoeal disease will have serious detrimental effects. The child's nutritional status and immune systems will be impaired due to reduced food intake, malabsorption of nutrients and vomiting. In addition, they will be more susceptible to other infections. Sadly, this cycle of infection results in about 13 million children under the age of 5 years dying each year. Rotavirus (principally transmitted via oral–faecal route) is one of the worst infectious organisms and kills up to 15 000–30 000 children/year in Bangladesh and one in 200–250 Indian children before they are 5 years old.

By the year 2025, more than 1000 million of the world population will be aged over 60, and more than two-thirds of them will live in developing countries. Growing old means increased risk from foodborne diseases. It is not surprising then that in some countries, one person in four is especially at risk of foodborne disease.

The exact number of annual foodborne illnesses can only be estimated. In many instances, only a small proportion of cases seek medical help and not all are investigated. Even where a country has a reporting infrastructure, only a small proportion of the foodborne-related cases are reported to the authorities. In the past, it has been assumed that in industrialised countries, less than 10% of the cases were reported, while in developing countries, reported cases probably account for less than 1% of the total. However, a better estimate is being achieved using sentinel studies as have been reported for the United States, the United Kingdom and the Netherlands. In the United States, it has been estimated that 76 million cases of foodborne diseases may occur each year, resulting in 325 000 hospitalisations and 5000 deaths (Mead *et al.* 1999). The UK study similarly estimates that the proportion of public experiencing gastroenteritis, due to foodborne pathogens, is 20% each year and maybe up to 20 per million die (Wheeler *et al.* 1999). The more recent sentinel study in the Netherlands estimated the number of microbial foodborne illnesses to be 79.7 per 10 000 person years (de Wit *et al.* 2001). Notermans and van der Giessen (1993) estimated the number to be 30% of the population per year.

Since food poisoning symptoms are generally mild and only last a few days, the person usually recovers without seeking medical care. Nevertheless, those at higher risk, such as the very young, pregnant and elderly, may suffer more serious, debilitating even life-threatening illness. These have previously been largely overlooked in determinations of the human burden of food poisoning.

1.3 Causes of foodborne illness

There are a number of factors which contribute to food being unsafe and causing illness (Table 1.3). The principal causes can be summarised as follows:
- Inadequate control of temperature during cooking, cooling and storage
- Personal hygiene
- Cross-contamination of raw and processed products
- Inadequate monitoring of processes

These contributing factors can be considerably reduced by adequate training of staff, implementation of HACCP combined with risk assessment (Chapters 8 and 9). As Chapter 8 will explain, it is now widely accepted that it is inadequate to rely on testing the end product for the presence of micro-organisms as a means of controlling the hygienic status of the process.

The key to the production of safe food is producing food which is microbiologically stable. In other words, any intrinsic microbes are unable to multiply to an infectious dose; ideally, they die off and that toxins are absent.

Table 1.3 Factors contributing to outbreaks of foodborne disease (various sources)

Contributing factors	Percentage[a]
Factors relating to microbial growth	
Storage at ambient (room) temperature	43
Improper cooling	32
Preparation too far in advance of serving	41
Improper warm holding	12
Use of leftovers	5
Improper thawing and subsequent storage	4
Extra large quantities prepared	22
Factors relating to microbial survival	
Improper reheating	17
Inadequate cooking	13
Factors relating to contamination	
Food workers	12
Contaminated processed non-canned foods	19
Contaminated raw foods	7
Cross-contamination	11
Inadequate cleaning of equipment	7
Unsafe source	5
Contaminated canned foods	2

[a]Percentages exceed a total of 100 since multiple factors often contribute to foodborne illness.
Adapted from Adams & Motarjemi (1999), with kind permission from World Health Organisation, Geneva.

Essentially, the cooking and cooling temperature profile should be aimed at the following:

1 Reducing the number of infectious organisms by 6 log orders (i.e. from 10^6 to 1 cells/g).
2 Not providing suitable conditions for the outgrowth of microbial spores which survive cooking.
3 Avoiding conditions which enable heat-stable toxins to be produced; by definition, these toxins are resistant to $100°C$ for 30 minutes and hence are not destroyed during cooking.

Cross-contamination causes post-processing (i.e. after the cooking step) contamination of the food. This can be avoided through the following:

1 Careful design of the factory layout
2 Control of personnel movement
3 Good personal hygiene habits

Foods which do not undergo a cooking process are normally acidified (i.e. fermented foods) and stored under chilled conditions. These practices rely upon the pH and temperature of the food stopping microbial growth. The growth range of the major food poisoning organisms has been documented (see ICMSF 1996a for details). Hence, it can be predicted by the pH and storage temperature of the food which will restrict the growth of foodborne pathogens.

1.4 Public perception of safe food

The increasing number and severity of food poisoning outbreaks worldwide have considerably increased public awareness about food safety. Public concern on the safety of the food they eat has been raised due to well-publicised issues such as food irradiation, BSE, *E. coli* O157:H7 and genetically modified foods.

'What is safe food?' invokes different answers depending upon who is asked. Essentially, the different definitions depend upon what is perceived as a significant risk. The general public might consider that 'safe food' means zero risk, whereas the food manufacturer would consider 'What is an acceptable risk?' The opinion expressed in this book is that **zero risk is not feasible** given the range of food products available, the complexity of the distribution chain and human nature. Nevertheless, the risks of food poisoning should be reduced during food manufacture to an 'acceptable risk'. Unfortunately, there is no public consensus on what constitutes an acceptable risk. After all, how can one compare the risk of hang gliding with eating rare beef? Hang gliding has known risks which can be evaluated and a decision 'glide or not to glide' taken. In contrast, the general public (rightly or wrongly) often feels it is not informed of relevant food risks. Table 1.4 shows the possible causes of death in the next 12 months. Many of these risks are acceptable to the general public; people continue to drive cars and cross the road. This table can be compared with Tables 1.5 and 1.6 which give the US and UK data on the chances of food poisoning. The US data indicate that each year, 0.1% of the population will be hospitalised due to food poisoning. Food scares cause public outcries and can give the industry an undeserved bad reputation. However, in fact, the majority of the food industry has a good safety record and is in the business to stay in business, not to go bankrupt due to adverse publicity.

Public concerns on food-related issues vary between countries and change with time (Sparks & Shepherd 1994). Prior to the 1990s, food additives were the focus of attention. Then in the 1990s, the public became more aware of foodborne pathogens such as *Salmonella*, and in the late 1990s, the cause for concern became the link between 'mad cow disease' and variant CJD, and also (particularly in Europe) food biotechnology. Hence, the public nowadays has a greater awareness of 'food poisoning' and have become critical and cautious of new

Table 1.4 Risk of death during the next 12 months

Cause	Mortality, chance of 1 in
Smoking 10 cigarettes a day	200
Natural causes, middle aged	850
Death through influenza	5 000
Dying in a road accident	8 000
Flying	20 000
Pedestrian struck by car	24 000
Food poisoning	25 000
Dying in a domestic accident	26 000
Being murdered	100 000
Death in a railway accident	500 000
Electrocution	200 000
Struck by lightning	10 000 000
Beef on the bone	1 000 000 000

production methods. The adverse public attention to the food industry leads to a distrust. This caution has been fed by numerous well-publicised instances of food poisoning or 'food scares' which have cost the food industry considerable sums of money; see Table 1.7.

Although the food industry needs to produce food that is safe to eat, it must also manufacture food that is of the quality expected by the consumer. However, as previously stated, 'safe food' is an arbitrary term, meaning different things to different people and does not necessarily equate with 'zero risk'. Hence, risk management and risk communication are important to the food industry and are defined in Chapter 9. The 'zero risk' approach is not feasible; for example the control of a foodborne pathogen requires the use of preservatives (with perceived toxicological risks) or heat treatment (with the possible production of carcinogens). The concept of a 'threshold' means that there exists a limit under which the risk is non-existent or negligible. In medical (and consequently food) microbiology, the 'minimal infectious dose' has been the threshold. However, this approach has been re-evaluated in microbiological risk assessment to estimate the probability of infection (P_i) by a single cell is calculated (Chapter 9; Vose 1998).

Novel foods and alternative methods of processing are being developed. Each new development needs to be considered for possible consequences, whether intentional or not, within the food chain. Some of these technologies will reduce or eliminate microbial hazards, whereas others might lead to the emergence of a new pathogen. In the application of HACCP to a novel process, the hazard analysis needs to be broad enough to consider the associated microbes, as well as the intrinsic and extrinsic conditions which affect microbial growth and toxin production. There is an increased market in 'organic' foods. This is partially linked to consumers perception of their greater nutritional content, and safer alternative to conventional produce which is not necessarily true. Cow manure can contain *Salmonella* and *E. coli* O157:H7, and there have been outbreaks linked with organically grown produce.

Industry and national regulators strive for production and processing systems which ensure that all food is 'safe and wholesome', although a complete freedom from risks is an unattainable goal. Safety and wholesomeness are related to a level of risk that society regards as reasonable in the context, and in comparison with other risks in everyday life. Putting food poisoning into context is not an easy task due to the high level of publicity which it receives in some countries. One can see in Table 1.4 that the risk of death due to food poisoning is nearly equivalent to the risk of being a pedestrian who is struck by a car. However, such tragic incidences occur to

Table 1.5 Risk of food poisoning in the United Kingdom

	Community		General practice		
	Number of cases[a]	Rate/1000 person years	Number of cases	Rate/1000 person years	Number of community cases/GP cases
Bacteria					
Aeromonas spp.	46	12.4	165	1.88	6.7
Bacillus spp. (>10^4/g)	0	0	4	0.05	—
Campylobacter spp.	32	8.7	354	4.14	2.1
Clostridium difficile cytotoxin	6	1.6	17	0.20	8.0
Cl. perfringens enterotoxin	9	2.4	114	1.30	1.9
E. coli O157	0	0	3	0.03	—
E. coli DNA probes					
Attaching and effacing	20	5.4	119	1.32	4.1
Diffusely adherent	23	6.2	103	1.18	5.3
Enteroaggregative	18	4.9	141	1.62	3.0
Enteroinvasive	0	0	0	0	—
Enteropathogenic	1	0.27	4	0.05	5.4
Enterotoxigenic	10	2.7	52	0.59	4.6
Verocytotoxigenic (non-O157)	3	0.82	6	0.06	13.4
Salmonella spp.	8	2.2	146	1.57	1.4
Shigella spp.	1	0.27	23	0.27	1.0
St. aureus (>10^6/g)	1	0.27	10	0.11	2.5
Vibrio spp.	0	0	1	0.01	—
Yersinia spp.	25	6.8	51	0.58	11.7
Protozoa					
Cryptosporidium parvum	3	0.81	39	0.43	1.9
Giardia intestinalis	2	0.54	28	0.28	1.9
Viruses					
Adenovirus group F	11	3.0	81	0.88	3.4
Astrovirus	14	3.8	77	0.86	4.4
Calicivirus	8	2.2	40	0.43	5.1
Rotavirus group A	26	7.1	208	2.30	3.1
Rotavirus group C	2	0.54	6	0.06	8.9
Norovirus	46	12.5	169	1.99	6.3
No organism identified	432	117.3	1305	14.82	7.9
Total	781	194	8770[b]	33.1	5.8

[a]Excluding cases where individual follow-up was not known.
[b]Total cases are greater than the sum of individual organisms due to cases for which a stool sample was not sent for testing. The general practice total includes cases from the enumeration arm, for which full stool testing was not carried out.
Reproduced from Wheeler *et al.* 1999, with permission from BMJ Publishing Group Ltd.

Table 1.6 Risk of food poisoning in the United States (Mead et al. 1999)

Disease or agent	Illnesses				Hospitalisations		Deaths	
	Total	Foodborne	% Foodborne transmission	Percentage of total foodborne	Foodborne	Percentage of total foodborne	Foodborne	Percentage of total foodborne
Bacterial								
B. cereus	27 360	27 360	100	0.2	8	0.0	0	0.0
Botulism, foodborne	58	58	100	0.0	46	0.1	4	0.2
Brucella spp.	1 554	777	50	0.0	61	0.1	6	0.3
Campylobacter spp.	2 453 926	1 963 141	80	14.2	10 539	17.3	99	5.5
Cl. perfringens	248 520	248 520	100	1.8	41	0.1	7	0.4
E. coli O157:H7	73 480	62 458	85	0.5	1 843	3.0	52	2.9
E. coli, non-O157 STEC	36 740	31 229	85	0.2	921	1.5	26	1.4
E. coli, enterotoxigenic	79 420	55 594	70	0.4	15	0.0	0	0.0
E. coli, other diarrhoegenic	79 420	23 826	30	0.2	6	0.0	0	0.0
L. monocytogenes	2 518	2 493	99	0.0	2 298	3.8	499	27.6
Salmonella Typhi	824	659	80	0.0	494	0.8	3	0.1
Salmonella, non-typhoidal	1 412 498	1 341 873	95	9.7	15 608	25.6	553	30.6
Shigella spp.	448 240	89 648	20	0.6	1 246	2.0	14	0.8
Staphylococcus food poisoning	185 060	185 060	100	1.3	1 753	2.9	2	0.1
Streptococcus, foodborne	50 920	50 920	100	0.4	358	0.6	0	0.0
Vibrio cholerae, toxigenic	54	49	90	0.0	17	0.0	0	0.0
V. vulnificus	94	47	50	0.0	43	0.1	18	1.0
Vibrio, other	7 880	5 122	65	0.0	65	0.1	13	0.7

(continued)

Table 1.6 (*Continued*)

Disease or agent	Illnesses				Hospitalisations		Deaths	
	Total	Foodborne	% Foodborne transmission	Percentage of total foodborne	Foodborne	Percentage of total foodborne	Foodborne	Percentage of total foodborne
Yersinia enterocolitica	96 368	86 731	90	0.6	1 105	1.8	2	0.1
Subtotal	5 204 934	4 175 565		30.2	36 466	59.9	1 297	71.7
Parasitic								
Cryptosporidium parvum	300 000	30 000	10	0.2	199	0.3	7	0.4
Cyclospora cayetanensis	16 264	14 638	90	0.1	15	0.0	0	0.0
G. lamblia	2 000 000	200 000	10	1.4	500	0.8	1	0.1
Toxoplasma gondii	225 000	112 500	50	0.8	2 500	4.1	375	20.7
Trichinella spiralis	52	52	100	0.0	4	0.0	0	0.0
Subtotal	2 541 316	357 190		2.6	3 219	5.3	383	21.2
Viral								
Norovirus	23 000 000	9 200 000	40	66.6	20 000	32.9	124	6.9
Rotavirus	3 900 000	39 000	1	0.3	500	0.8	0	0.0
Astrovirus	3 900 000	39 000	1	0.3	125	0.2	0	0.0
Hepatitis A	83 391	4 170	5	0.0	90	0.9	4	0.2
Subtotal	30 833 391	9 282 170		67.2	21 167	34.8	129	7.1
Grand total	38 629 641	13 814 924		100.0	60 854	100.0	1 809	100.0

Table 1.7 Examples of major outbreaks around the world due to contaminated food and water

Year	Location	Description
1964	Aberdeen, Scotland	*S.* Typhi in beef from Argentina; 507 cases, 3 deaths
1981	Spain	Toxic cooking oil with a toll of 800 dead and 20 000 injured (WHO 1984)
1985	United States	*S.* Typhimurium contamination of pasteurised milk leading to 170 000 infections (Ryan *et al.* 1987)
1986	Chernobyl, former USSR	Chernobyl accident led to a subsequent contamination of wide areas of Western Europe and subsequently food with radionuclides
1986	Birmingham, UK	*Salmonella* in tinned salmon, 2 died
1990	United States	Bottled sparkling water contaminated with benzene (up to 22 ppb). More than 160 million bottles of water were recalled worldwide, at an estimated cost of $263 million
1991	Shanghai, China	Hepatitis A infection of 300 000 people through contaminated clams (Halliday *et al.* 1991)
1992–1993	Washington, Idaho, California, and Nevada, US	>500 cases *E. coli* O157:H7
1993	United States	Cryptosporidium in drinking water; 403 000 cases
1994	41 States, United States	*S.* Enteritidis in ice cream leading to 224 000 cases (Hennesy *et al.* 1996)
1996	Japan	About 10 000 children infected (11 deaths) due to *E. coli* O157:H7 contamination of radish sprouts served in school lunches (Mermin & Griffin 1999)
1998	India	Toxic mustard seed oil led to a number of reported deaths
1999	Belgium	Dioxins that were found in poultry led to widespread trade disruption and losses of hundreds of millions of Euros
1999–2000	France	*L. monocytogenes* 4b pork tongue in jelly
2000	Japan	*St. aureus* in milk powder
	United Kingdom	BSE-vCJD has cost the United Kingdom more than $6 billion, not including lost jobs. UK beef product exports in 2000 were down by 99% from 1995. Continental Europe has seen a drop in meat sales
2005	Spain	2138 cases of gastroenteritis caused by *S.* Hadar due to a single brand of pre-cooked, vacuum-packed roast chicken product
2005	Denmark	Two Norovirus outbreaks due to raspberries imported frozen from Poland. The first outbreak involved 272 patients and employees at a hospital, and the other outbreak included at least 289 patients who received meals from the same food caterer in a home nursing scheme
2006	Seoul, South Korea	>1700 children infected with Norovirus through school meal
2006	United Kingdom	Contamination of chocolate in the United Kingdom with *S.* Montevideo. £20 million loss by company
2008	United States	>1000 *Salm.* St. Paul infections from jalapeño peppers and tomatoes
2008	Canada	28 listeriosis cases, with 1 death, of pregnant mothers and premature births linked to various cheeses. 12 deaths, out of >38 cases linked to meat products
2008	China	Melamine in milk products; >50 000 illness, >4 deaths

TVO7769

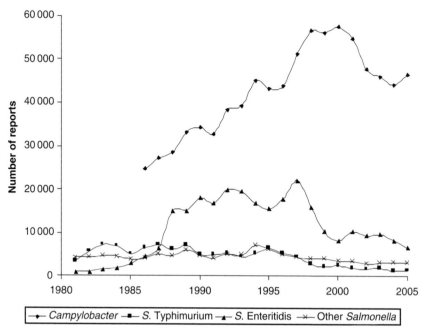

Figure 1.3 Twenty-year trend in *Campylobacter* and *Salmonella* reported cases for England and Wales (Data source HPA website).

individuals and only reach the local newspaper headlines, whereas food poisoning outbreaks involving a large number of people are more 'significant' to the media.

The reported increase in the number of 'food poisoning cases' is commonly cited by the media. However, these trends must be regarded with caution since they are the number of gastroenteritis cases which have been investigated, and the causative organism identified. Not all gastroenteritis cases may be due to food vectors, and an increase in public awareness can increase the number of general public seeking medical help. In addition, improved detection methods could 'increase' the number of identified cases over time. In fact, across Europe and the United States, the number of reported gastroenteritis cases has been decreasing (Anon. 1999b). In 2005, the Centers for Disease Control and Prevention (CDC) reported that the most common diseases associated with foodborne pathogens, parasites and other bacteria were on the decline or at least not increasing in incidence. *E. coli* O157:H7 infections decreased by 36% in just 1 year. Similarly, the number of *Campylobacter, Salmonella, Yersinia* and *Cryptosporidium* infections declined by 28%, 17%, 51% and 49%, respectively, in the 8-year surveillance period. However, this trend has not significantly changed in the following 3 years (FoodNet 2009). Figure 1.3 shows the 20-year trends in *Campylobacter* and *Salmonella* cases in England and Wales; note the decline in *Salmonella* cases since 1997. As given in Section 1.3, sentinel studies indicate the true incidence of gastroenteritis may be ~20%, though the mild nature means the illness is unnoticed and hence unreported. Possibly due to changes in eating habits, there is a marked seasonality in food poisoning incidences with pathogens such as *Salmonella* and *Campylobacter*. Figure 1.4 shows the peak incidence over the summer months. Typically, *Campylobacter* peaks a month or two before *Salmonella*.

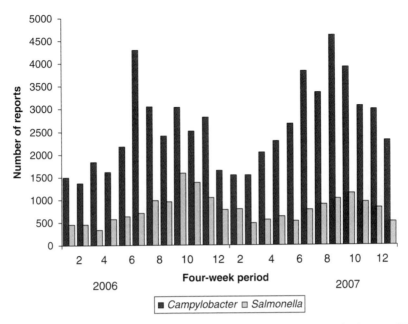

Figure 1.4 Seasonal trends in reported cases of *Salmonella* and *Campylobacter* enteric infections over 2006–2007 in England and Wales (Data source HPA site).

1.5 Host-related issues

The days of food being produced, processed, distributed and consumed locally have significantly decreased in recent decades. The regional, national and global food chain has required parallel changes in food science and technology, including preservation. At the same time, there have been social changes with an increasing number of meals being consumed outside of the home environment and also an ageing population. Public exposure to a foodborne pathogen may change due to changes in the processing (i.e. BSE exposure), changes in consumption patterns and the globalisation of the food supply chain. Many 'risk factors' influence the host susceptibility to infection (Chapter 9). These may be as follows:
- *Pathogen related*: ingested dose, virulence
- *Host related*: age, immune status, personal hygiene, genetic susceptibility
- *Diet related*: nutritional deficiencies, ingestion of fatty or highly buffered foods

The consumer demand for less processed and less additives means that food processors have less choice in their preservation method (Zink 1997). However, since the consumer also expects a long shelf-life, then this can led to problems with psychrotrophic pathogenic bacteria such as *Listeria*, *Yersinia* and *Aeromonas* (cf. *sous vide*). Hence, food processors are investigating new processing and preservation techniques: high pressure, ohmic eating and pulsed electric field (Chapter 3).

Eating away from home is a major trend of recent years. Many of these meals require extensive food handling and/or cold foods that are not cooked before consumption. Subsequently, this leads to an increase in the number of people handling food and hence the potential for an increase in the transmission of foodborne diseases from food handlers to consumers. It is plausible that since the individual spends less time in daily home food preparation, they have less knowledge regarding safe food preparation. Several studies have documented an increasing lack of knowledge about safe home food preparation or preservation practices, such as personal or utensils hygiene and correct temperature storage.

The immune system may be partially compromised or undeveloped in newborn babies, the very young, pregnant, those on medication or underlying illness (i.e. AIDS) and the elderly. The health effects of foodborne infections and case-fatality rates are up to 10 times greater in this part of the population. Young children are more likely than adults to develop illnesses from selected pathogens (Figure 1.1). Socio-economic factors affect vulnerability. For example, in developed countries, the case-fatality rate for typhoid is higher among individuals aged 55 years or older. In contrast, in developing countries, the higher risks of complication and death are for children from birth to 1 year of age and adults over 31 (Gerba *et al.* 1996).

Women during pregnancy have a suppressed immune system to reduce the rejection of the foetus. Subsequently, pregnant women have a greater risk of food poisoning due to *Listeria monocytogenes* than the general population. Infection of the foetus or the newborn babies from infected mothers can be extremely severe, resulting in abortion, stillbirth, or a critically ill baby that may present with an early onset (mortality rate 15–50%) or late onset (mortality rate 10–20%) form of listeriosis (Farber & Peterkin 1991, Farber *et al.* 1996). Neonates are uniquely susceptible to enterovirus infections such as coxsackie B and echovirus infections (Gerba *et al.* 1996).

Infections in the immuno-compromised (non-pregnant) host constitute a new and severe aspect of the food safety problem. Advances in medical treatment have resulted in an increased number of immuno-suppressed patients (e.g. cancer cases, organ transplants), and patients with serious underlying chronic diseases who may have an increased risk from foodborne infection and/or develop more severe illness. Obviously, enteric pathogens can easily cause persistent and generalised infection in the immuno-compromised host. The majority of AIDS patients (50–90%) suffer from chronic diarrhoeal diseases which can be fatal (Morris & Potter 1997).

The proportion of elderly people in the population is increasing. It has been projected that in the United States, one-fifth of the population will be above the age of 65 by the year 2030. The increased susceptibility of the elderly may be due to a number of physiological factors, such as the senescence of the gut-associated lymphoid tissue and/or the decrease in gastric acid secretion, thus reducing the natural barriers to gastrointestinal pathogens. Also, the immune system of the elderly is often weakened due to chronic illnesses. The incidence of salmonellosis, *Campylobacter* enteritis or *E. coli* O157:H7 appears to be higher in the elderly. Epidemiological studies have shown that the elderly experience higher case-fatality rates than the general population, for example 3.8% versus 0.1% for *Salmonella*, 11.8% versus 0.2% for *E. coli* O157:H7, 1% versus 0.01% for rotavirus (Gerba *et al.* 1996).

On a global scale, probably the leading cause of increased host susceptibility to foodborne infections is malnutrition. In developing countries, malnutrition affects about 800 million people. The region with the largest absolute number affected is Asia (524 million) and the region with the largest population affected is Africa (28%). In some individual countries, the percentage can be as high as 30%. Malnutrition increases host susceptibility to foodborne infections since it weakens and reduces the integrity of the intestinal epithelium and cell-mediated immunity. Malnutrition results in a 30-fold increase in the risk for diarrhoea-associated death (Morris & Potter 1997).

1.6 Hygiene hypothesis

Improvements in medical care, living conditions and hygiene have improved our quality of life and length of life compared with those 100 years ago. In the United Kingdom, less than 8 children in every 1000 die before their first birthday; yet in 1921, it was 80/1000. Also, life expectancy for males at birth is 75–80 years. However, in the 1850s, it was 18–50 depending upon your living conditions and wealth. There is a possibility that the reduction in exposure to bacteria has resulted in an increase over the past 20 years of disease such as allergies and asthma. The 'hygiene hypothesis' proposes that the immune systems need frequent challenges which do not occur if the environment is 'too clean'. While the rise in allergies and asthma over the past decades is real, alternative reasons could include improved diagnosis, and an increased exposure to diverse allergens. Currently, epidemiological studies on the hygiene hypothesis are contradictory. A refinement of the hypothesis is that it is the exposure to certain microbes, that is mycobacteria, during infancy, which have a protective effect against allergies.

If the hygiene hypothesis is true, then it will be important to ensure our immune system is appropriately challenged by exposure to non-pathogenic organisms and to avoid atopic disease, without reducing our protection against infectious diseases.

Therefore, hygiene measures will need to be focused to where and when they matter most.

1.7 The size of the foodborne illness problem

For several years, the WHO has been encouraging Member States to quantify the national burden and causes of foodborne disease. Although there are a number of foodborne disease burden estimates, they are mainly from developed countries. In large parts of the world, the data required to underpin such estimates are completely lacking. In 2007, the WHO established the Foodborne Disease Burden Epidemiology Reference Group (FERG) which will work towards estimating the burden of foodborne disease; see Section 11.2. It should be noted that across Europe, the numbers of food poisoning cases has been decreasing in recent years for all foodborne pathogens except *C. jejuni* (Anon. 1999b; FoodNet 2009).

Foodborne illness occurs when a person gets sick by eating food that has been contaminated with an unwanted micro-organism or toxin. The most common symptoms of foodborne illness include stomach cramps, nausea, vomiting, diarrhoea and fever. It is accepted that only a small proportion of cases of foodborne illness are brought to the attention of food inspection, control and health agencies. This is partially because many foodborne pathogens cause such mild symptoms, that the victim may not seek medical help. Hence, the notified number of cases is just the 'tip of the iceberg' with regard to true numbers of food poisoning cases (Figure 1.5). In the United States, England and the Netherlands, there have been studies to estimate the proportion of cases which are not recorded and hence obtain a more accurate figure of food poisoning numbers. Table 1.8 gives the proportion of under-reporting for the United States and the estimated number of foodborne infections for 1997.

Mead *et al.* (1999) reported that the total burden of foodborne illness per year caused approximately 76 million illnesses, 323 000 hospitalisations and 5000 deaths in the United States (Table 1.6). Together *Salmonella*, *Listeria* and *Toxoplasma* were responsible for 1500 deaths per year. Unknown agents caused 62 million illnesses, 265 000 hospitalisations and 3200 deaths. Using a population size of 270 299 000 for the same time period (US Census Bureau 1998), this equates to 28% of the population suffering from food poisoning each year and 0.1% being

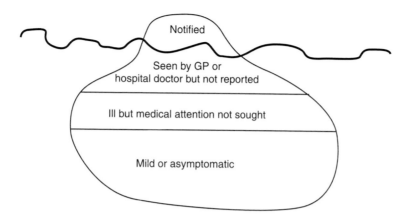

Figure 1.5 The reporting pyramid – 'tip of the iceberg'.

hospitalised due to foodborne illness. Since 1995, FoodNet in the United States has been gathering sentinel data from 10% of the population (Section 1.12.2). Specifically, for non-typhoidal *Salmonella* infections, Voetsch *et al.* (2004) reported that over a similar time period (1996–1999), there were 15 000 hospitalisations and 400 deaths annually.

The US studies can be compared with the sentinent study in England (Table 1.5; Wheeler *et al.* 1999, Sethi *et al.* 1999 and Tompkins *et al.* 1999). This was undertaken to determine a more accurate estimate of the incidence of food poisoning in England. The study estimated the overall extent of under-reporting, and that for every case detected by laboratory surveillance, there were 136 in the community (Figure 1.6). Hence, the scale of infectious intestinal disease in England was estimated at 9.4 million annual cases, of which 1.5 million cases are presented to general practitioners. Under-reporting for individual organisms varied, most likely due to the severity of the illness. The reporting of *Salmonella* (3.2:1) and *Campylobacter* (7.6:1) was higher than rotavirus (35.1:1) and Norovirus (approximately 1562:1), respectively. These under-reporting values differ from those of the United States (Table 1.8). Nevertheless, the total burden of foodborne diseases is that 20% of people in the general population of England are infected. This proportion is similar to the estimate of 28% in the United States. In England, the organisms most commonly detected in patients with infectious intestinal disease are *Campylobacter* spp. (12.2% of stools tested), rotavirus group A (7.7%) and small round structured virus (6.5%). No pathogen or toxin was detected in 45.1–63.1% of cases. Surprisingly, *Aeromonas* spp., *Yersinia* spp. and some enterovirulent groups of *E. coli* were detected as frequently from controls as from cases.

It is evident that causes of gastroenteritis vary with age (Figure 1.1). SRSV, caliciviruses and rotaviruses probably cause the majority of gastroenteritis in children under the age of 4 years, whereas bacteria (*Campylobacter* and *Salmonella* spp.) are the major causes of gastroenteritis in other age groups. Figure 1.2 indicates that males suffer from gastroenteritis more than females, except for one age group (>74, probably because of the lower ratio of men to women in this age group). A possible reason for part of this difference is that fewer men than women wash their hands after using the lavatory (33% compared with 60%, taking an average of 47 seconds, compared with 79 seconds).

Helms *et al.* (2003) compared the mortality rates of 48 857 cases of gastrointestinal infection with 487 138 controls from the general population. In their study, they considered mortality rates over longer periods of time than normal studies (<30 days), and also took coexisting

Table 1.8 Under-reporting of foodborne pathogens in the United States (Mead *et al.* 1999; Tauxe 2002)

Organism	Under-reporting factor	Estimated infections (1997)
Bacterial pathogens		
Campylobacter spp.	38	1 963 000
Salmonella, non-typhoid species	38	1 342 000
Cl. perfringens	38	249 000
St. aureus	38	185 000
E. coli	20	92 000
O157:H7		
STEC (VTEC) other than O157	Half as common as *E. coli* O157:H7 cases	Combined
Shigella spp.	20	90 000
Y. enterocolitica	38	87 000
E. coli enterotoxigenic (ETEC)	10	56 000
Streptococcus Group A	38	51 000
B. cereus	38	27 000
E. coli, other diarrhoegenic	Assumed to be as common as ETEC	23 000
Vibrio spp. other than those above	20	5 000
L. monocytogenes	2	2 000
Brucella spp.	14	777
S. Typhi	2	659
V. cholerae O1 or O139	2	49
V. vulnificus	2	47
Cl. botulinum	2	56
Parasitic pathogens		
G. lamblia	20	200 000
Tox. gondii	7	112 000
Cry. parvum	45	30 000
Cyc. cayetanensis	38	14 000
Tri. spiralis	2	52
Viral pathogens		
Norovirus	11% of all acute primary gastroenteritis	9 200 000
Rotavirus	Not given (number of cases taken as equal to birth cohort)	39 000
Astrovirus	Not given (number of cases taken as equal to birth cohort)	39 000
Hepatitis A	3	4 000

illnesses into account. This large study found that 2.2% of people with gastrointestinal infections died within 1 year after infection compared with 0.7% of controls. The relative mortality within 30 days of infection was high for *Salmonella, Campylobacter, Y. enterocolitica* and *Shigella*. These organisms were also associated with increased death rates after 30 days of infection.

1.8 Chronic sequelae following foodborne illness

The potential for chronic sequelae (secondary complications) has been recently recognised along with the variability of the human response (Table 1.9). It has been estimated that chronic

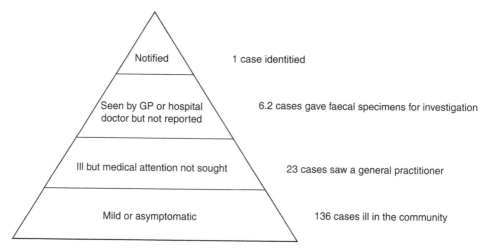

Figure 1.6 The under-reporting pyramid (Data collated from Wheeler *et al.* 1999).

sequelae occur in 2–3% of foodborne cases and may last in terms of weeks or even months. These sequelae may be more serious than the original illness and result in serious long-term disability or even death (Bunning *et al.* 1997, Lindsay 1997). The evidence that micro-organisms are implemented in chronic sequelae is not always certain because the chronic complications are unlikely to be epidemiologically linked to a foodborne illness. An appreciation of the chronic sequela is necessary to fully appreciate the economic burden of foodborne infections.

A major symptom of food poisoning is diarrhoea which may consequently lead to anorexia and malabsorption. Severe cases of diarrhoea may last for months or years and can be caused by *C. jejuni*, *Citrobacter*, *Enterobacter* or *Klebsiella* enteric infections. Subsequently, the intestinal

Table 1.9 Chronic sequelae following foodborne infection (Lindsay 1997; Mossel *et al.* 1995)

Disease	Associated complication
Brucellosis	Aotitis, orchitis, meningitis, pericarditis, spondylitis
Campylobacteriosis	Arthritis, carditis, cholecystitis, colitis, endocarditis, erythema, nodosum, Guillain–Barré syndrome, haemolytic uraemic syndrome, meningitis, pancreatitis, septicaemia, reactive arthritis, irritable bowel syndrome
E. coli (EPEC and EHEC types) infections	Erythema nodosum, haemolytic uraemic syndrome, seronegative arthopathy
Listeriosis	Meningitis, endocarditis, osteomyelitis, abortion and stillbirth, death
Salmonellosis	Aortitis, cholecystitis, colitis, endocarditis, orchitis, meningitis, myocarditis, osteomyelitis, pancreatitis, Reiter's syndrome, rheumatoid syndromes, septicaemia, splenic abscess, thyroiditis, irritable bowel syndrome
Shigellosis	Erythema nodosum, haemolytic uraemic syndrome, peripheral neuropathy, pneumonia, Reiter's syndrome, speticaemia, splenic abscess, synovitis
Taeniasis	Arthritis, epilepsy
Toxoplasmosis	Foetus malformation, congenital blindness
Yersiniosis	Arthitis, cholangitis, erythema nodosum, liver and splenic abcesses, lymphadenitis, pneumonia, pyomositis, Reiter's syndrome, septicaemia, spondylitis, Still's disease

wall permeability may be altered and absorb significant quantities of unwanted proteins which may induce atropy. Foodborne pathogens may interact with the immune system of the host to elude or alter the immunological process which may subsequently induce a chronic disease. In addition, genetic susceptibility of the host may predispose humans to some types of infection. Of particular importance are the following:

1 Guillain–Barré syndrome (*C. jejuni*, Section 4.3.1)
2 Reactive arthritis and Reiter's syndrome (*Salmonella* serotypes, Section 4.3.2)
3 Haemolytic uraemic syndrome (*E. coli* O157, Section 4.3.3)
   Less well-studied chronic sequelae are as follows:
1 Chronic gastritis due to *Helicobacter pylori*.
2 Crohn's disease and ulcerative colitis possibly caused by *Mycobacterium paratuberculosis*.
3 Long-term gastrointestinal and nutritional disturbances following infection by *C. jejuni*, *Citrobacter*, *Enterobacter* and *Klebsiella*.
4 Haemolytic anaemia due to *Campylobacter* and *Yersinia*.
5 Heart and vascular diseases caused by *E. coli*.
6 Atherosclerosis following *S.* Typhimurium infection.
7 Personality changes following toxoplasmosis.
8 Graves, disease (autoimmune disease) which is the result of autoantibodies to the thyrotropin receptor after *Y. enterocolitica* serotype O:3 infection.
9 Severe hypothyroidism due to *Giardia lamblia* infection.
10 Crohn's disease possibly due to *M. paratuberculosis* (causative agent of Johne's disease in ruminants) via pasteurised milk. Other causative bacteria may be *L. monocytogenes*, *E. coli* and *Streptococcus* species.
11 Viral induction of autoimmune disorders such as hepatitis A virus infection causing acute hepatitis with jaundice in adults. This is probably due to molecular mimicry.
12 Mycotoxins have a range of acute, subacute and chronic toxicities as some compounds are carcinogenic, mutagenic and even teratogenic (Section 4.7).

1.9 Changes in antibiotic resistance

Many bacterial species can develop resistance to the antibiotics they are exposed to in clinical and veterinary settings. This can occur by the following:
• Random mutation with subsequent positive selection of mutants resistant to the antibiotic, that is fluoroquinolone resistance in *C. jejuni*.
• Mobilisation, including horizontal gene transfer, of resistance genes, that is vancomycin resistance among the enterococci.
• Dissemination of strains with previously developed resistance, that is *S.* Typhimurium DT104.
   Resistance to medically important antibiotics has increased in *Salmonella* and *Campylobacter* to a disturbing degree. Fluoroquinolone resistance has also been reported in *Campylobacter* species, with a rise in ciprofloxacin (a medically important antibiotic) resistance. The source of antibiotic resistance is possibly the veterinary use of enrofloxacin (another fluoroquinolone antibiotic). This antibiotic may be used to reduce the carriage of *Salmonella* spp. It will also reduce the general gut flora, thereby increasing nutrient uptake due to the less 'inflamed' epithelial cell layer state, and act as a growth enhancer. The systemic use of this antibiotic may be imposing a selective pressure on the microbial flora and has resulted in the selection of DNA gyrase mutants (gyrA).

Table 1.10 Antibiotic resistance changes

Antibiotic	1990	1991	1992	1993	1994	1995	1996
Ampicillin	37[a]	50	72	85	88	90	95
Chloramphenicol	32	49	60	83	87	89	94
Streptomycin	38	52	75	85	92	97	97
Sulphonamides	37	53	76	86	93	90	97
Tetracyclines	36	50	74	83	88	90	97
Trimethoprim	0.4	3	3	2	13	30	24
Ciprofloxacin	0	0	0.2	0	1	7	14

[a]Percent resistant.

Molecular typing has revealed an association between resistant *C. jejuni* strains from chicken products and *C. jejuni* strains from human cases of campylobacteriosis (Jacobs-Reitsma *et al.* 1994, Smith *et al.* 1999). Another association is treatment with a fluoroquinolone antibiotic and foreign travel. A microbiological risk assessment of fluoroquinolone resistant *C. jejuni* has been published on the web, and is reviewed in Section 10.2.4.

A strain of *S.* Typhimurium DT104 resistant to ampicillin, chloramphenicol, streptomycin, sulphonamides and tetracyline (R-type ACSSuT) was first isolated in 1984, and appears to be highly clonal. It is likely that the genes were acquired through horizontal gene transfer, and afterwards, the strain was widely disseminated due to the veterinary use of antibiotics. This organism has been isolated from cattle, poultry, sheep, pigs and horses. Antimicrobial therapy is used extensively to combat *S.* Typhimurium infection in animals, and the evolution of a strain resistant to the commonly used antibiotics has made infections with *S.* Typhimurium in food animals difficult to control. The primary route by which humans acquire infection is by the consumption of a large range of contaminated foods of animal origin.

Table 1.10 shows the significant increase in antibiotic resistance in *S.* Typhimurium DT104 isolates between 1990 and 1996 in England and Wales. In addition to the antibiotic resistant type ACSSuT, many isolates are also resistant to trimethorpin and ciprofloxacin. Ciprofloxacin resistance has been most notable in *S.* Hadar (39.6%) compared to other *Salmonella* serovars.

Several countries are monitoring the occurrence of antimicrobial resistance and establishing guidelines (either through voluntary or regulatory means) to control their increase (Bower & Daeschel 1999). A number of international surveillance networks monitoring antibiotic resistance are covered in later, such as FoodNet and Global Salm-Surv (Sections 1.12.2 and 1.12.7). In Europe, there have been a number of concerted efforts to control microbial antibiotic resistance (Eurosurveillance 2008). In the 1960s, Denmark was the first European country to try and control methicillin-resistance *St. aureus* (MRSA), and have reduced antibiotic usage by 32% since 1999.

1.10 The cost of foodborne diseases

Several countries have estimated the economic consequences of foodborne illness. These costs include:
- loss of income by the affected individual,
- cost of health care,
- loss of productivity due to absenteeism,
- costs of investigation of an outbreak,

- loss of income due to closure of businesses,
- loss of sales when consumers avoid particular products.

It is clear that foodborne illness has considerable implications on a country's economy. The estimates, reviewed below, are for developed countries. There are no reliable estimates for developing countries, where the problem of diarrhoeal disease is far greater, and therefore the economic burden of foodborne illness must be even more severe.

In England and Wales, the 23 000 cases of salmonellosis were estimated to have resulted in an overall cost of £40–£50 million (Sockett 1991). Later, Roberts (1996) estimated that the medical and value of lives lost due to just five foodborne infections in England and Wales was £300–£700 million per year. In Australia, the cost of an estimated 11 500 food poisoning cases is AUD $2.6 billion per year (ANZFA 1999). Less recent estimations for Canada are $1.3 billion loss due to foodborne pathogens (Todd 1989a). The Swedish *Salmonella*-free poultry programme costs about $8 million per year, but saves an estimated $28 million per year in medical costs. The cost of *Campylobacter* infections in the United States has been estimated at $1.5–$8.0 billion. The total economical aspect of foodborne diseases is a loss of $5–$17 billion by the US Food and Drug Administration.

The Economic Research Service (ERS) is part of the US Department of Agriculture, and estimates the 'cost-of-foodborne-illness' for seven major foodborne pathogens (Table 1.11). These are based on estimated medical costs and productivity losses over the infected individual's lifetime. The medical costs include both acute illness and long-term chronic complications. Therefore, the amount of medical attention required is determined according to the severity of illness. This ranges from those who do not visit doctor, those who develop chronic complications, through to those who are hospitalised, and die prematurely. For each of these groupings, the medical costs are estimated for the number of days of treatment, average cost per treatment or service, and the number of patients (Buzby and Roberts 1997a). Productivity losses include changes in income and fringe benefits. High and low estimates of economic losses are calculated for cases where the person is unable to resume their normal job. This can be due to disability or death. The low estimate is based on loss of lifetime earnings and household production, whereas the high estimate is based on the 'risk premiums' in labour markets. The FDA uses the high estimate for productivity losses in its evaluation of food safety programmes, and adjusts this according to the person's age.

ERS estimates for the costs of acute infections and chronic complications due to foodborne infections are between $6.6 and $37.1 billion per year (Table 1.11; Buzby & Roberts 1997a). The cost of human illness due to only seven bacterial pathogens is $9.3–$12.9 billion annually. Of these costs, $2.9–$6.7 billion are attributable to the foodborne bacteria *Salmonella* serovars, *C. jejuni*, *E. coli* O157:H7, *L. monocytogenes*, *Staphylococcus aureus* and *Cl. perfringens*. A recent estimate for *E. coli* O157 costs in the United States (for 2003) based on 73 000 illnesses annually with 2000 hospitalisations and 60 deaths was $405 million/year (Frenzen *et al.* 2005). This was the total for premature deaths ($370 million), medical care ($30 million), and lost productivity ($5 million).

ERS has launched an online 'Foodborne illness cost calculator' for *Salmonella* and Shiga-toxin producing *E. coli* (STEC, also known as VTEC) with data from 1997 to 2006 (http://www.ers.usda.gov/data/foodborneillness) (see Plate 1). This allows users to change the assumptions made by ERS about the number of illnesses, medical costs, and lost productivity due to *Salmonella* or *E. coli* O157 infections, and then re-estimate the costs.

The Communicable Disease Center (US) estimates that 95% of *Salmonella* infections are foodborne in origin. Consequently, the FoodNet estimate of 1.4 million annual

Table 1.11 Medical costs and productivity losses estimated for selected human pathogens, 1993

Pathogen	Foodborne illness			Meat/poultry related			
	Cases	Deaths	Foodborne costs (billion $)	Per cent from meat/poultry (%)	Number of cases	Number of deaths	Total costs meat/poultry (billion $)
Bacteria							
C. jejuni or coli	1 375 000–1 750 000	110–511	0.6–1.0	75	1 031 250–1 312 500	83–383	0.5–0.8
Cl. perfringens	10 000	100	0.1	50	5 000	50	0.1
E. coli O157:H7	8 000–16 000	160–400	0.2–0.6	75	6 000–12 000	120–300	0.2–0.5
L. monocytogenes	1 526–1 767	378–485	0.2–0.3	50	763–884	189–243	0.1–0.2
Salmonella	696 000–3 840 000	696–3 840	0.6–3.5	50–75	348 000–2 880 000	348–2 610	0.3–2.6
St. aureus	1 513 000	1 210	1.2	50	756 500	605	0.6
Subtotal	3 603 526–7 130 767	2 654–6 546	2.9–6.7	N/A	2 147 513–4 966 884	1 395–4 191	1.8–4.8
Parasite							
T. gondii	3 056	41	2.7	100	2 056	41	2.7
Total	3 606 582–7 133 823	2 695–6 587	5.6–9.4	N/A	2 149 569–4 968 940	1 436–4 232	4.5–7.5

USDA, FSIS, Pathogen Reduction; Hazard Analysis and Critical Control Point (HACCP) Systems; Proposed Rule (13).

salmonellosis cases means that 1.3 million cases were due to consumption of foods contaminated by *Salmonella*. The estimated medical costs of *Salmonella* infections were based on the average medical care per case or each severity category, the estimated number of cases and the 1998 average US cost for each type of medical care. The value of a life ranges from $8.3 and $8.5 million at birth to $1.4 and $1.6 million at age 85 and above, for males and females, respectively. Since approximately two-thirds of deaths from *Salmonella* infections were aged 65 or older, the average forgone earnings per premature death were $4.1 million for males and $3.5 million for females. Medical care and lost productivity were calculated to be $0.5 billion. Time lost from work equated $0.9 to $12.8 billion. The annual cost of campylobacteriosis in the United States is approximately $0.8–$5.6 billion (Buzby & Roberts 1997b), and estimated total costs of *Campylobacter*-associated Guillain–Barré syndrome (GBS, an autoimmune reaction) are $0.2–$1.8 billion (Section 4.3.1). Hence, reducing the prevalence of *Campylobacter* in food could prevent up to $5.6 billion in costs annually. Therefore, it is not surprising that national regulatory authorities have set targets for the reduction in infections due to major foodborne pathogens.

The estimated productivity costs due to foodborne infections are likely to increase in the future as the calculation methods improve. One omission is that the current methods do not include estimates for how much consumers are willing to pay for improved food safety. Additionally, it is highly probable that more foodborne pathogens will be identified in the future along with increased recognition of chronic complications. Currently, chronic sequelae are estimated to occur in 2–3% of cases, and therefore, the long-term consequences to the individual and our economy may be more detrimental than the initial acute disease.

Given these economic costs, there is a considerable need, even in the developed countries, for more systematic and aggressive steps to be taken to significantly reduce the risk of microbiological foodborne diseases. Since the cost of 'gastroenteritis' in developed countries is determined to be in the billions of US$, how much more is the human cost of food and water disease in developing countries? WHO estimates that worldwide, almost 2 million children die every year from diarrhoea. A significant portion of deaths are caused by microbiologically contaminated food and water. Hence, this is a major challenge for the future, which requires a global response. Therefore, this book frequently refers to international organisation's activities such as Codex Alimentarius Commission and the WHO.

The microbial contamination of food has a major impact on food production. Worldwide, it is estimated that at least 10% of grain and legume production are lost, and this may even be as a high as 50% for vegetables and fruits. This contamination affects trade in two ways. Firstly, the contaminated food may not be accepted by the importing countries. Secondly, the loss in a country's reputation for food safety may result in loss in further trade. The socio-economic impact of unrelated issues such as BSE-vCJD, genetically engineered plants and dioxin contamination is still being experienced. One can only hope that the BSE-vCJD outbreak will catalyse improved, educated food safety systems that will prevent similar or even greater tragedies from occurring (Brown *et al.* 2001). An aspect that BSE-vCJD and publicised outbreaks of *E. coli* O157:H7 have emphasised is the need to control food safety 'from the farm to the fork'.

The cost benefits in preventing food poisoning through the assured food safety system HACCP (Section 8.7) have been estimated (Crutchfield *et al.* 1999). Due to the range of economic models used, the estimates varied considerably from $1.9 to $171.8 billion. Regardless of the exact figure, it can be predicted that implementing HACCP will on balance be less than the current medical costs due to foodborne infections, as well as decrease productivity losses.

In 2007, following a well-publicised outbreak of *Salmonella* with ~40 victims, a major UK manufacturer of chocolate was fined US$ 2 million. The company pleaded guilty to violating food and hygiene regulations, and also ordered to pay US$ 309 000 in legal costs. The cost of the product recall in 2006 was at least US$ 60 million, and the company is reported to have spent US$ 40 million on improvements, including changes to quality control procedures. The problem came to light after three people, including two young children, were taken to a hospital in a food poisoning alert linked to the manufacturer's chocolate. The victims' age ranged from babies to adults of 52 years, and most were children under the age of 4 years. Sales of the company's chocolate fell by 14% during the first 6 months following the outbreak. The crux of the loss of *Salmonella* control by the manufacturers was the change to using an MPN approach to detect the organism.

It should be noted that these values for clinical cases are where the agent has been isolated and identified. There are a significantly large number of cases for which the etiological agent is unidentified. Frenzen (2004) estimates that such foodborne organisms cause 3400 deaths annually in the United States. The most recent data from the WHO estimates that in 2005, 1.5 million people died from diarrhoeal disease, of which 70% were due to foodborne pathogens (Buzby & Roberts 2009).

1.11 Control of foodborne pathogens

Pathogenic microbes are present in soil and hence on crops and on farming livestock and fish. Hence, it is inevitable therefore that raw ingredients entering a process can contain pathogenic organisms. Therefore to control food poisoning, the pathogens associated with an ingredient and food product need to be recognised and controlled. Control programmes need to be in place and monitored to assess their efficacy, reviewed and modified, if necessary. As examples of control programmes in different foods, the control of *Salmonella* in Sweden and enteric pathogens in fresh produce are given in detail.

1.11.1 Example 1 – the control of *Salmonella* serovars in poultry

In large-scale poultry production, there is a breeding structure in which there is genetic selection at the top, which is amplified through the system down to breeder flocks, and finally to the production stage at the baseline. Consequently, the presence of *Salmonella* in the grandparent flocks will result in a corresponding increase in the prevalence of *Salmonella*-contaminated birds at the production stage.

Therefore, eradicating *Salmonella*-infected grandparent and parent birds will have a considerable effect in reducing the *Salmonella* occurrence at the production stage and human exposure. The European Food Safety Authority (EFSA) carried out a baseline study on the prevalence of *Salmonella* in holdings of laying hen flocks (EFSA 2007). This revealed a *Salmonella* spp. prevalence of 30.8%, ranging from 0% in Luxembourg and Sweden to 79.5% in Portugal.

In Denmark, it has been estimated that *Salmonella* control measures has saved the Danish society US$ 14.1 million in the year 2001 (Wegener *et al.* 2003). This was estimated by comparing *Salmonella* control costs in poultry production with the overall public health costs of salmonellosis. A Swedish cost benefit analysis from 1992 showed roughly the same figures.

In the 1950s, Sweden had a severe *Salmonella* food poisoning outbreak which caused about 9000 cases and 90 deaths. This resulted in the implementation of an effective *Salmonella* control programme, and the incidence of *Salmonella* in animal products for human consumption produced in the country is less that 1%. The programme has compulsory reporting of all cases of infection and subsequent elimination measures (Wierup *et al.* 1995). For example, whenever *Salmonella* is found in poultry feed, or in poultry, the *Salmonella*-contaminated feedstuffs and/or infected birds are removed from the food chain.

The importation of animal feeds is handled under licence. At least ten samples, or a 100 g sample for every 2 tonnes of feed, are sent to the Swedish National Food Administration for bacterial evaluation. If *Salmonella* is recovered, importation is not permitted. In domestic bonemeal and animal protein plants, at least five samples of 100 g are collected each day. These samples are pooled and examined for *Salmonella*. When *Salmonella* is recovered, an inspection is conducted. If needed, design modifications are recommended. The Swedish government has authority to investigate a feed operation if a farm it services is found to have *Salmonella* contamination. Samples of raw materials, dust and finished rations are cultured. Feed contaminated with *Salmonella* is subject to two sequences of heat treatment. If *Salmonella* is found in finished feed, the processing line is cleaned and fumigated. When necessary, the line is modified so that feed reaches 70–75°C. Because of *S.* Enteritidis, regulators now require heat treatment of feed and approved storage methods for grain to be used for laying hen operations.

Only day-old grandparent breeding chicks may be imported into Sweden. The birds are quarantined for 5 weeks. The liver, yolk sac and intestines of all chicks that die in transmit are examined for *Salmonella*. Cloacal swabs are taken from 100 birds and pooled to 20 swabs per sample at arrival. A total of 60 dead and culled birds are tested at 1–2, 4–6, 12–14 and 16–18 weeks. A total of 0.5% of the birds are tested, using dead and stunted birds, and composites of five faecal swabs if there is low mortality. The flock is destroyed if *Salmonella* is detected. Egg-type multiplier flocks are tested on a voluntary basis. Participating flocks are tested for *Salmonella* at 2, 6–10 and 14–18 weeks of age. During rearing, two composite faecal samples containing 30 droppings each and two composite necropsy samples containing liver and caecum from five birds each are cultured. After the multiplier flock is in production, two composite faecal samples and one composite necropsy sample are collected monthly. *Salmonella*-positive flocks are destroyed. Sanitation is the principal control for *Salmonella* in hatcheries. Eggs are disinfected before incubation and fumigated on day 3 of incubation. Eggs and chicks are separated by source. Hatcheries are bacteriologically monitored every 3 months.

Pullets are voluntarily monitored at 2–3 weeks of age and 2 weeks before placement. Three samples containing 30 faecal droppings each are collected for bacterial isolation; this procedure is repeated at 25 and 55 weeks of age. Producers that participate in voluntary monitoring are eligible for insurance to defray losses caused by the regulation of *Salmonella*. Egg-laying flocks with *S.* Enteritidis or *S.* Typhimurium are depopulated without processing or compensation. After flock depopulation, the poultry house must be cleaned and disinfected. Layer flocks with serotypes other than *S.* Enteritidis and *S.* Typhimurium are permitted to sanitary slaughter, followed by heat treatment of carcasses. Recovery of *S.* Enteritidis or *S.* Typhimurium from articles associated with a flock permits inspection of the farm. Articles may include diagnostic specimens, liquid or powdered egg. This authority may be used if a human isolate of *S.* Enteritidis implicates eggs.

These measures remained in place when Sweden joined the European Union in 1995. However, despite the strict control, reported *Salmonella* food poisoning has risen in Sweden. This is partially due to the increase in packaged holidays resulting in Swedish tourists bringing *Salmonella* back to the country. The total number of human salmonellosis cases in Sweden in 2006 was 4056 (44.9 per 100 000 population and per year) of which 1010 were domestically acquired (25%), and almost all the other *Salmonella* cases (2963, 73%) were imported cases. The relative small number of domestic cases supports the current Swedish *Salmonella* control measures which are financed by food producers and the state. It costs about $8 million per year, which is small compared to the cost of medical treatment of ~$22 million. The control programme costs less than 11 cents per kilo of chicken.

1.11.2 Example 2 – control of *E. coli* and *Salmonella* in fresh produce

Raw fruits and vegetables have been known to serve as vehicles of human disease for at least a century. Around 1900, typhoid infections were associated with eating celery, watercress and uncooked rhubarb. Some parasitic helminths (e.g. *Fasciola hepatica* and *Fasciolopsis buski*) require encystment on plants to complete their life cycle. Thus, the recognition of raw fruits and vegetables as potential vehicles for transmission of pathogenic micro-organisms known to cause human disease is not new. However, outbreaks of foodborne illness linked to the consumption of raw vegetables have recently emerged as a food safety issue and this needs to be addressed by both industry and regulatory authorities. In the 1980s, the consumption of fresh, uncooked vegetable sprouts increased and current figures for the United States give annual sales in the region of $12 billion. As the market continues to grow, the processors have to ensure the fresh, uncooked product is safe for consumption. Due to the sprouting conditions, vegetables can contain large numbers of micro-organisms. Microbial pathogens associated with the consumption of fresh fruits and vegetables include *Cyclospora cayetanensis*, *E. coli* O157:H7, hepatitis A virus, *L. monocytogenes*, Norovirus, *Salmonella* serovars, and *Shigella* spp. Since 1990, in North America, there have been over 500 foodborne outbreaks linked to fresh produce such as fresh fruits and vegetables. Between 1997–1999, there were at least seven outbreaks due to *Salmonella* and *E. coli* O157:H7 infections which were linked to the consumption of various raw vegetable sprouts. The outbreak in Japan was the largest ever outbreak of *E. coli* O157:H7 (Taormina *et al.* 1999). In 2006, in North America, there were outbreaks of pathogenic *E. coli* O157:H7 and *Salmonella* in fresh spinach, lettuce and tomatoes which caused over 700 illnesses and four deaths.

Unfortunately, the very nature of the fresh produce results in the high microbiological risk. Bacteria, viruses and other infectious organisms can contaminate raw fruits and vegetables through contact with faeces, polluted irrigation water or contaminated surface water. Fresh-cut produce has a high moisture and nutrient content which can support bacterial growth. Also, there is no heat step, or other lethal process to kill the pathogens; instead, the processing, storage, transportation and display temperatures may encourage bacterial growth. Fruits and vegetables can become contaminated while still on the plant in fields or orchards, during harvesting, transport, processing, distribution and marketing, or in the home. Pathogenic bacteria such as *Cl. botulinum*, *B. cereus* and *L. monocytogenes* are normal inhabitants of soil. In contrast, *Salmonella*, *Shigella*, *E. coli* and *Campylobacter* species are part of the intestinal flora of animals and humans. Spoilage organism will also be present as part of the surface flora. Fresh produce is neither cooked nor processed, and subsequently contamination may not be adequately removed. Contamination can also occur during the various post-harvest handling stages, preparation by retailers, food outlets and in the home.

It is normal to find micro-organisms on the surface of raw fruits and vegetables. These may originate from the soil or dust, or have grown using nutrients secreted from the plant. *Klebsiella* and *Enterobacter* are commonly found on plant vegetation and therefore the presence of lactose-fermenting *Enterobacteriaceae* (or 'coliforms') on raw fruits and vegetables does not necessarily indicate faecal contamination; see Section 4.2.1 for an explanation of the term 'coliform'. More recently, it has become evident that bacteria may be internalised within fresh produce (Ibarra-Sanchez *et al.* 2004).

During processing, fresh produce is cut. Since the natural physical barrier of the produce is broken, nutrients are released, and this increases the opportunity for bacteria to grow. Fresh-cut produce requires a lot of handling, which can provide opportunities for microbial contamination and spreading. In addition, inadequate cleaning procedures during processing will increase the risk of pathogen contamination.

Preventing the contamination of fruits and vegetables during the pre-harvest and post-harvest phases would be ideal. However, this is very hard to achieve. As previously stated, bacterial pathogens are normally present in the soil, on the produce surface, and possible contaminated irrigation water. However, reducing the extent of produce contamination can be achieved through improved industrial practices during harvesting, processing and distribution.

The simple practice of washing raw fruits and vegetables in hot water will reduce the microbial load. The use of disinfectants can result in 10- to 100-fold reduction in microbial load, although viruses and protozoan cysts have a greater resistance to disinfectants than bacteria or fungi. Chlorine has been used as a disinfectant in the raw fruit and vegetable industry. It is usually administered as elemental chlorine or one of the hypochlorites. The chlorine level is commonly about 50–200 ppm with a contact time of 1–2 minutes (cf. Section 7.4).

Food safety programmes such as the FDA's Guide to Industry *'Guide to Minimize Microbial Food Safety Hazards of Fresh-Cut Fruits and Vegetables'* (FDA online 2008) on the farm are very important. These guidance documents are aimed at improving the safety of fresh-cut produce by seed disinfection, good agricultural and manufacturing practices, and testing of spent irrigation water for pathogens.

1.12 Surveillance programmes

'A foodborne-disease surveillance program is an essential part of a food safety programme.'

Todd 1996b

Foodborne disease surveillance involves the collection, analysis and monitoring for trends of foodborne pathogens. Surveillance also enables the control of foodborne diseases. Hence, surveillance is an essential part of any food safety system. It is also important for surveillance systems to meet the obligations under the International Health Regulations (IHR 2005) and INFOSAN, as will be covered elsewhere. Unfortunately, only a few countries in the world have adequate surveillance programmes, and as a consequence, the real health impact and extent of foodborne diseases, particularly in developing countries, remain unknown. Attempts to obtain a global picture of foodborne disease are usually hindered by differences in national surveillance systems. This section reviews other networks which are being established to address this issue. The purpose of FERG in estimating the global food disease burden is covered in Section 11.2.

Foodborne disease surveillance consists of three principal activities:

1 Monitoring the incidence of specific pathogens
2 Identifying outbreaks
3 Determining the risk factors associated with sporadic cases

Surveillance refers to the systematic collection and use of epidemiological information for the planning, implementation and assessment of disease control. Until recently, with the exception of cholera, there was no obligation to report foodborne diseases internationally. This has now been addressed by the International Health Regulations (IHR 2005; Section 11.1).

Global surveillance of foodborne problems can play an important role in the early detection, early warning, rapid investigation and control of such problems. It could also help limit the

Table 1.12 Annual number of outbreaks by country

Country	B. cereus	Cl. perfringens	Salmonella	St. aureus	Total	Population in millions
Israel	0.0	3.0	4.4	4.0	11.4	4.4
Finland	3.8	8.8	7.8	5.6	26.0	5.0
Denmark	2.0	5.8	5.2	0.0	13.0	5.1
Scotland	3.0	6.6	152.0	2.4	164.0	5.2
Sweden	0.6	4.6	7.0	2.8	15.0	8.4
Hungary	5.2	5.0	131.2	16.0	157.4	10.6
Portugal	0.0	0.3	6.8	7.3	14.4	10.4
Cuba	4.8	13.4	6.8	60.8	85.8	10.4
Netherlands	4.8	2.4	8.0	0.0	15.2	14.8
Yugoslavia	0.4	2.8	46.0	13.0	62.2	24.0
Canada	14.0	18.0	49.4	18.8	100.2	26.0
Spain	0.6	5.0	467.6	0.0	473.2	39.3
England/Wales	28.3	53.3	450.0	9.5	541.1	50.4
France	0.0	17.6	177.0	12.8	207.4	56.0
Fed. Rep. of Germany	0.4	1.4	3.0	3.2	8.0	61.4
Japan	10.5	14.0	84.0	128.0	236.5	123.0
United States	3.2	4.8	68.4	9.4	85.8	247.4

[a]Number of outbreaks per year (mean 2–5 years).
Reprinted with permission from Todd 1996b.

extent and distribution of the problem, thereby preventing many cases of illness and minimising the negative impact on trade and the economics of individual countries. Due to the increase in world trade, the events within one country might also affect many other nations. International outbreaks can be recognised in two ways: one country recognising an outbreak, passing this information onto the network for information, and other countries recognising a similar occurrence in their country, or by analysing pooled, international databases for unusually high levels of infection by a specific pathogen.

The study by Todd (1996a) surveyed the incidence of food poisoning in 17 countries (Table 1.12). The survey demonstrated that there are considerable regional differences in the incidence of major foodborne diseases. For example, *St. aureus* was the major food poisoning organism in Cuba and Japan, whereas *Salmonella* dominated in most other countries. The most frequent vector of foodborne disease was meat in 13 countries and the most frequent place acquiring foodborne illness was in the home. The organisms included in a surveillance programme are not the same in every country. For example, *L. monocytogenes* has been the causative agent of listeriosis in the United States and Europe in the last decade. But foodborne outbreaks of listeriosis have not been recognised in Japan. Consequently, clinical laboratories do not examine food and faecal specimens for the organism.

1.12.1 International Food Safety Authorities Network (INFOSAN)

INFOSAN is operated by the WHO (Geneva) and includes 163 Member States. It enables food safety authorities and other relevant agencies to exchange important food safety information rapidly. Official national contact points are linked to deal with outbreaks and other international emergencies. This supports and complements the WHO Outbreak Alert and Response Network (GOARN). For more information, go to http://www.who.int/foodsafety to obtain INFOSAN information notes on numerous food safety issues including guidelines for investigation and control of foodborne disease outbreaks, and even nanotechnology.

1.12.2 FoodNet in the United States

The Foodborne Disease Active Surveillance Network (FoodNet) in the United States was established in 1995. In 1996, FoodNet began its surveillance for laboratory-confirmed cases of infection caused by *Campylobacter, Salmonella*, STEC O157, *Shigella, Listeria, Vibrio* and *Yersinia*. Cases of *Cryptosporidium* and *Cyclospora* infection were included in 1997 and STEC non-O157 infection in 2000. It is the principal foodborne disease component of the Emerging Infections Program (EIP) of the CDC of the United States. As a collaborative project, it combines the CDC, nine EIP state health department sites, the Food Safety and Inspection Service (FSIS), US Department of Agriculture (USDA) and the Food and Drug Administration (FDA). It covers areas of Minnesota, Oregon, Colorado, Tennessee, Georgia, California, Connecticut, Maryland and New York. The total population of this area is 33.1 million, which is 10% of the US population.

FoodNet is designed to:

- produce national estimates of the burden and sources of specific foodborne diseases in the United States through active surveillance and epidemiological studies;
- determine how much foodborne illness results from eating specific foods, such as meat, poultry and eggs;
- document the effectiveness of new food safety control measures, such as the USDA Pathogen Reduction and HACCP Rule, in decreasing the number of cases of major foodborne disease in the United States each year;
- describe the epidemiology of new and emerging bacterial, parasitic and viral foodborne pathogens;
- respond rapidly to new and emerging foodborne pathogens.

FoodNet has five activities:

- Active laboratory-based surveillance at over 300 clinical laboratories that test stool samples. Information is collected on every laboratory-diagnosed case of bacterial pathogens including *Salmonella, Shigella, Campylobacter* (including Guillain–Barré syndrome), *E. coli* O157 (including HUS), *L. monocytogenes, Y. enterocolitica* and *Vibrio* spp. The parasitic organisms *Cryptosporidium* and *Cyclospora* are also included.
- Survey of clinical laboratories to give a baseline information on which pathogens to include in routine bacterial stool cultures and to standardise procedures of sample collection and examination.
- Survey of physicians to obtain information on physician stool culturing practices to determine how often samples are requested and for what reason.
- Survey (by telephone questionnaire) of the population to determine the number of cases of diarrhoea in the general population and how often medical advice is sought.
- Epidemiological studies of *E. coli* O157, *Campylobacter* and *Salmonella* serogroups B and D, which cause 60% of *Salmonella* infections in the United States. The aim is to determine which foods or other exposures might be risk factors for these bacterial infections. Isolates are subjected to antibiotic resistance testing, phage typing and molecular subtyping (PFGE, see PulseNet in the next section). In the future sporadic outbreaks of *E. coli* O157, *Cryptosporidium* and *L. monocytogenes* will be included.

FoodNet has recorded the decrease in *Campylobacter, Salmonella* and *Cryptosporidium* infections and the decrease in meat and poultry products contaminated with *Salmonella*. It has reported high isolation rates of *Y. enterocolitica* in Georgina and *Campylobacter* in California. An outbreak of *Salmonella* infections was detected in Oregon which was due to alfalfa sprouts. It has also detected two *E. coli* O157:H7 outbreaks in Connecticut which were due to lettuce and apple cider. FoodNet has also assisted in the surveillance of vCJD. It has contributed to the investigations into multistate outbreaks of *Listeria* and *Cyclospora* (1997–1998). The latter outbreak was associated with raspberries from Guatemala and resulted in restrictions on the importation of raspberries into the United States.

To look for unusual cases clusters, CDC's Public Health Laboratory Information System (PHLIS) has an automated surveillance outbreak detection algorithm (SODA) that uses the 5-year mean number of cases from the same geographic area and week of the year. This is very effective at detecting case clusters of uncommon *Salmonella* serotypes, and the URL is given in the Web Resources section.

1.12.3 PulseNet: US *E. coli* O157:H7, *Salmonella* and *Shigella* detection network
Molecular typing schemes can improve epidemiological outbreak investigations by distinguishing unrelated sporadic cases from the main outbreak-associated strain. PulseNet is based on the pulsed-field gel electrophoresis (PFGE) technique which has been standardised in the US Department of Health and Human Services to identify distinctive DNA fingerprint patterns of *E. coli* O157:H7, *Salmonella*, *Shigella*, *Campylobacter*, *Vibrio cholera* and *Yersinia pestis*. The advantages of PFGE profiling are illustrated by the 1993 *E. coli* O157:H7 outbreak in the United States. PFGE revealed that the strain of *E. coli* O157:H7 found in patients had an indistinguishable profile from the strain found in hamburger patties served at a large chain of regional fast food restaurants. It is estimated that this prompt recognition may have prevented an estimated further 800 illnesses.

PFGE methodology is explained in detail in Section 5.5.1, but essentially, bacterial DNA is cut into shorter segments using random cutting DNA restriction enzymes. The DNA segments are separated using a pulsed electric field, especially designed to separate larger DNA fragments. The DNA fragments separate according to molecular size and generate a barcode-like appearance which is a 'fingerprint' of the organism. Using a computer network across 16 states, PFGE patterns from humans and suspected foods can be shared across the country. Hence, multistate foodborne disease outbreaks can be recognised rapidly and investigated. In the future, PulseNet will include other foodborne pathogens. Unfortunately, PulseNet's full potential is restricted as not all public health laboratories are part of the network, not all clinical laboratories routinely submit isolates to public health laboratories, and many states do not have the necessary resources to investigate individual cases or clusters of possible foodborne illnesses.

1.12.4 European Centre for Disease Prevention and Control (ECDC) and Enter-Net; European surveillance network for salmonellosis and shigatoxic *E. coli* (STEC)
The ECDC (URL: http://ecdc.europa.eu) started in October 2007, when it replaced Enter-Net as an international surveillance network for human gastrointestinal infections. Enter-Net was a continuation of the Salm-Net surveillance network which operated from 1994 to 1997. The current network involves 31 countries of the European Union, and European Fair Trade Association (EFTA). It conducts international surveillance of salmonellosis and shiga-toxin producing *E. coli* (STEC), including antimicrobial (antibiotic) resistance. The *Salmonella* database has been in existence since 1995, the STEC database since 2000 and the *Campylobacter* database since 2005. As given in Section 1.9, antibiotic resistance amongst *Salmonella* and *Campylobacter* species is of increasing concern.

The ECDC stated objectives are as follows:
1 To collect standardised data on the antimicrobial resistance patterns of salmonellas isolates.
2 To facilitate the study of resistance mechanisms and their genetic control by arranging the collection of representative strains of multiple drug-resistant salmonellas and coordinating

Table 1.13 International outbreaks recognised and investigated by Enter/Salm-Net

Outbreak	Number of cases	Countries with cases
S. Newport	100+	England & Wales and Finland
S. Livingstone	100+	Austria, Czech Republic, Denmark, England & Wales, Finland, France, Germany, Netherlands, Norway and Sweden
E. coli O157(HUS)	15	Denmark, England & Wales, Finland and Sweden
S. Anatum	19	England & Wales, Israel and United States
S. Agona	4000+	Canada, England & Wales, Israel and United States
S. Dublin	30+	France and Switzerland
S. Stanley	100+	Finland and United States
S. Tosamanga	28	Eire, England & Wales, France, Germany, Sweden and Switzerland
Sh. sonnei	100+	England & Wales, Germany, Norway, Scotland and Sweden

Enter-Net homepage (see Appendix for address).

the required research work between specialised centres, and where available, compare the resistances of animal isolates.

3 To extend the typing of STEC for surveillance purposes by:
  (i) extending the availability of phage typing for *E. coli* O157;
  (ii) using poly- and monovalent antisera to identify common non-O157 serogroups.
4 To pilot an international quality assessment scheme for laboratory methods used in the identification/typing of STEC.
5 To establish a core set of data items to accompany, where possible, each laboratory-typed STEC isolate.
6 To create an international database of STEC isolates which is updated regularly and is readily available to each participating team.
7 To detect clusters of STEC isolate types in time, place and person and to bring such clusters to the attention of collaborators rapidly.
8 To support the above objectives by continuing the Enter-Net surveillance system consisting of regular, frequent data exchange on salmonellas.
  Source: http://ecdc.europa.eu.

Prior to ECDC, a number of international outbreaks were recognised by the Salm-Net and subsequently by Enter-Net (Table 1.13).

1.12.5 Foodborne viruses in Europe network

As will be described later, the transmission of Norovirus and Hepatitis A virus (Sections 4.4.1 and 4.4.2) by contaminated food and water are being increasingly recognised. Such outbreaks can involve a large number of people over a wide geographical area. Therefore, a network of laboratories able to undertake standardised methods of molecular typing, electron microscopy and serotype is important. The 'Food-borne viruses in Europe' involving ten countries has been established for this purpose (Eurosurveillance 2002, http://www.Eufoodborneviruses.co.uk). The analytical methods were standardised in 2003, and the network is currently working on expanding the number of participants and procedures for investigating and controlling outbreaks. The network has reported the high number of Norovirus infections during the later period of 2007 in the United Kingdom, Ireland, Germany, the Netherlands and Sweden. The virus has evolved, probably due to selective pressure and antigenic variants, from the GII.4

strain. Three variants emerged in 1995, 2002 and 2004, and further subdivided. The recent GII.4-2006b strains have two mutants in Region A (part of the polymerase gene) and one in region C (part of the capsid gene). See 'http://hypocrates.rivm.nl' for background information on epidemiological and molecular typing. Unfortunately, there is a need for international standardised in the nomenclature of the variants types. The GII.4-2006b variant is known as Minerva in the United States, Kobe034 in Japan and v6 in the United Kingdom.

1.12.6 Rapid Alert System for Food and Feed (RASFF)

RASFF has been in place in the EU since 1979. Its purpose is to provide the EU authorities with an effective tool for exchange of information on measures taken to ensure food safety. It gives weekly overviews of border alerts, and recalls for products entering the EU, and being distributed within the EU. It also produces annual reports. See Table 1.14 for an example of notifications received in week 25 of 2006. This particular alert notification includes *S.* Montevideo in UK chocolate bars, which led to a multinational product withdrawal and a large financial loss to the company involved.

1.12.7 Global Salm-Surv (GSS)

The WHO aims to strengthen the capacities of its Member States in the surveillance and control of major foodborne diseases and to contribute to the global effort of containment of antimicrobial resistance in foodborne pathogens; see http://www.who.int/salmsurv/en/. In 1997, a survey of national reference laboratories revealed that only 66% routinely serotyped *Salmonella* for public health surveillance. Consequently, in 2000, the WHO established the GSS which now has almost 1000 members from 149 countries. This is a global network of laboratories that was initially concerned with the surveillance, isolation, identification and antimicrobial resistance testing of *Salmonella* serovars. When GSS started, its goals were to strengthen laboratory-based foodborne disease surveillance and to improve outbreak detection and response. It also acts as an international resource of expertise and training. Some countries may not submit data to Salm-Surv due to concerns regarding adverse publicity and adverse effects on trade. Other countries may not have the resources or training required for serotyping *Salmonella* isolates. As its name suggests, initially GSS was for the isolation and surveillance for *Salmonella*. However, it has expanded to include *E. coli* and *Campylobacter*.

The five main activities of GSS are:

1 International training courses
2 External quality assurance system
3 Focused national and regional projects
4 Electronic discussion group
5 WHO GSS country databank

Salm-Surv has shown that globally *S.* Enteritidis is the most common *Salmonella* serotype in humans, especially in Europe (85% cases), Asia (38%), Latin America and the Caribbean (31%); *S.* Typhimurium generally being the second most common (Galanis *et al.* 2006).

1.12.8 Surveillance of ready-to-eat foods in the United Kingdom

In the United Kingdom, there has been a programme of sampling ready-to-eat foods, and assessing their microbiological quality against the HPA (formerly PHLS) ready-to-eat guidelines (PHLS 2000). The details of these guidelines are covered in Section 6.11. Table 1.15 summarises the findings from surveys totalling more than 15 000 samples. In general, unsatisfactory and

Table 1.14 Rapid Alert System for Food and Feed (RASFF)

Date	Notified by	Ref.	Reason for notifying	Country of origin
19/06/2006	United Kingdom	2006.0382	Undeclared gluten and soya in snowballs and caramel log wafers	United Kingdom
19/06/2006	Denmark	2006.0385	Cadmium in crushed pineapple in its own juice	Kenya
20/06/2006	Germany	2006.0386	Migration of lead from ceramic plate	China via the Netherlands
20/06/2006	Sweden	2006.0387	*Salmonella* Senftenberg in soya bean meal	Brazil via the Netherlands
20/06/2006	Belgium	2006.0388	Migration of 4-4'-diaminodiphenylmethane from black nylon plastic kitchen utensils	Germany
21/06/2006	Belgium	2006.0389	Migration of formaldehyde from melamine (plastic) plates, cups, and bowls	via the Netherlands
21/06/2006	United Kingdom	2006.0390	Undeclared peanut in chicken satay	The Netherlands
21/06/2006	Italy	2006.0391	Food poisoning outbreak caused by defrosted tuna	Costa Rica via Spain
21/06/2006	Denmark	2006.0392	Ochratoxin A in rye flour	Denmark
22/06/2006	Denmark	2006.0393	Ochratoxin A in rye flour and rye	Denmark
22/06/2006	United Kingdom	2006.0394	Paralytic shellfish poisoning (PSP) toxins–saxitoxin in scallops	United Kingdom
22/06/2006	Italy	2006.0395	Residue level above MRL for tetracycline in pork legs	Austria
23/06/2006	Czech Republic	2006.0396	Ochratoxin A in dried raisins	Afghanistan
23/06/2006	Czech Republic	2006.0397	Fresh eggs unfit for human consumption and contaminated with mould	Poland
23/06/2006	The Slovak Republic	2006.0398	Benzo(a)pyrene in smoked winter sprats	Poland
23/06/2006	United Kingdom	2006.0399	*Salmonella* Montevideo in chocolate bars	United Kingdom
23/06/2006	Denmark	2006.0400	Yeast fermentation of raspberry jam	Denmark
23/06/2006	Italy	2006.0401	*Salmonella* spp. in frozen turkey meat mechanically separated (MSM)	France

Notifications in concern feed, all other notification concern food.

unacceptable levels were found in a small percentage of samples, and were frequently due to viable counts of *E. coli* $> 10^2$ cfu/g. Overall, these results indicate that production practices for ready-to-eat foods in the United Kingdom were good, and identified areas which warranted further attention.

Table 1.15 Survey of microbiological quality of ready-to-eat foods in the UK market

Food	Number of samples analysed	Satisfactory[a] (%)	Acceptable (%)	Unsatisfactory (%)	Unacceptable (%)	Cause of unsatisfactory or unacceptable categorisation
Butchery products	2192	87		16	0.3	*Salmonella* detected (2)[b] *St. aureus* $>10^4$ cfu/g *E. coli* $>10^4$ cfu/g (2)
Quiche	2513	93		6	<1	High aerobic colony counts *E. coli* $>10^4$ cfu/g (2)
Cooked rice	1972	94		0	1	*B. cereus* $>10^5$ cfu/g *E. coli* $>10^4$ cfu/g
Organic vegetables	3200	99.5		0.5	0	*E. coli* and *Listeria* spp. (not *L. monocytogenes*) $>10^2$ cfu/g
RTE foods to which spices have been added	1946	35		31	2	
Stuffing	147	76.3	15.6	8.2		High aerobic colony counts, *E. coli*, *Enterobacteriaceae*, *St. aureus*

[a]Categorisation is according to the HPA ready-to-eat guidelines (2000).
[b]Number of samples exceeding criteria are in parenthesis.

1.13 Outbreak investigations

An outbreak is an incident in which two or more persons have the same disease, similar symptoms or excrete the same pathogens and in which there is a time, place and/or person association between these persons. An outbreak may also be defined as a situation when the observed number of cases unaccountably exceeds the expected number. A foodborne or waterborne outbreak results from ingestion by those affected by food or water from the same contamination source or which has become contaminated in the same way.

Outbreaks can be small, such as within a household, or very large. One of the first well-documented large outbreaks was in Scotland in 1964 when 507 people became infected with S. Typhi and 3 died. The strain was phage type 34, which was rare in the United Kingdom but common in South America. The outbreak was traced to tinned corned beef, from a canning plant that had not chlorinated its cooling water for the previous 14 months and the organism most likely had entered through a seam. Contamination of the slicing machine led to cross-contamination of other meats and the larger number of cases. In June 2006, more than 1700 students at 25 schools in Seoul, Inchon and Kyonggi provinces in South Korea became ill with Norovirus after eating school meals supplied by a large manufacturer. The number of victims is the largest ever in school-lunch related illness in South Korea; see Table 1.7. Nevertheless, this is smaller than the 1996 outbreak of *E. coli* O157 in Japan through contaminated radish sprouts distributed through a centralised food production network. This led to ~10 000 infections in school children and 11 reported deaths.

A foodborne disease surveillance system must be able to detect and rapidly respond to potential outbreaks. Outbreaks caused by *Salmonella* and *Campylobacter* may be identified by detecting an increased incidence of a particular geno- or phenotype. Routine molecular typing of isolates by PFGE, and transmission of the patterns to a network of other laboratories (i.e. PulseNet and Enter-Net; Section 1.12) increases the potential for dispersed outbreaks to be recognised that would otherwise be overlooked by the surveillance of a single state, or even country. Subsequently, the source and mode of transmission can be identified. For example, in 1997, a *S.* Anatum outbreak amongst infants (aged 1–11 months) was traced back to a manufacturer of infant formula following a total of 18 salmonellosis cases in infants in Scotland, England, Wales, Belgium and France (Anon. 1997). The *S.* Anatum from 16 of the cases who had consumed a particular brand matched each other, and differed from *S.* Anatum strains from two infants who had not consumed that brand.

Outbreaks can also be identified when a group of people have similar illnesses. On these occasions, the outbreak is recognised before a causative agent has been identified. Therefore, these outbreaks must be investigated to identify the infectious organisms, as well as the source and vehicle of transmission. Table 1.16 summarises the incubation period and symptoms for the major foodborne pathogens. Therefore, when the causative agent has been identified, this can be compared with eating habits of the cases as collated from questionnaires.

There are various sources of information on outbreaks, and which can be accessed via the internet and are listed in the Section 'Food Safety Resources on the World Wide Web' at the end of the book

- The 2007 WHO publication '*Foodborne Disease Outbreaks: Guidelines for Investigation and Control*'; see Plate 2.
- CDC Foodborne Outbreak Response and Surveillance Unit (OutbreakNet) can be searched according to specific type of pathogen, year and State; Plate 3.
- Eurosurveillance reports on food-related outbreaks and surveillance of medically important organisms within Europe.
- National agencies; such the UK Health Protection Agency gives weekly total for common gastrointestinal infections, and description of general outbreaks of foodborne illness in England and Wales.

Outbreak investigations are often incomplete for several reasons. It could be because there was too long a delay between the outbreak event and the start of the investigation. Consequently, people involved in the outbreak were no longer available for further questioning or have forgotten the details. Also, the information gathered was limited because of language difficulties, poor employee communication skills or ineffective questioning by the investigators. Therefore, outbreak investigations require the prompt collaboration of environmental health agencies, public health laboratories and epidemiologists.

1.13.1 Preliminary outbreak investigation

An investigation and control of an outbreak will be in several phases. The preliminary phase will consist of several activities:

- Consider whether or not the cases have the same illness, and establish a tentative diagnosis.
- Determine if there is a real outbreak.

Table 1.16 Usual incubation and onset period for foodborne illness

Agent	Symptoms	Incubation period							Duration (days)	
		<2 hours	2–7 hours	8–14 hours	8–24 hours	1–2 days	1–7 days	Other	<1	>1
Chemical allergen										
Heavy metal: copper, tin, lead, zinc	Nausea, vomiting	■								
Fish toxins: PSP, ciguatera, and so on	Gastrointestinal and neurological symptoms	■								■
Monosodium glutamate	Burning sensation on body, tingling, dizziness, headache, nausea	■							■	
Food allergens: nuts, eggs, milk, wheat	Anaphylactic shocks. Respiratory failure, rashes, nausea, vomiting	■								
B. cereus, emetic	Nausea, vomiting, abdominal pain	■								
S. aureus	Nausea, vomiting, abdominal pain		■							
B. cereus, diarrhoeal	Abdominal pain, watery diarrhoea				■			■		
Clostridium perfringens	Abdominal pain, watery diarrhoea				■			■		
Salmonella	Abdominal pain, diarrhoea, chills fever, nausea, vomiting, loss of appetite				■	■		■		
Clostridium botulinum	Vertigo, blurred vision, difficulty in speaking, progressive nervous system failure and paralysis					■				■
Streptococcus Group A	Sore throat, fever, nausea, vomiting, rhinorroea, tonsilitis, maybe rash					■				■

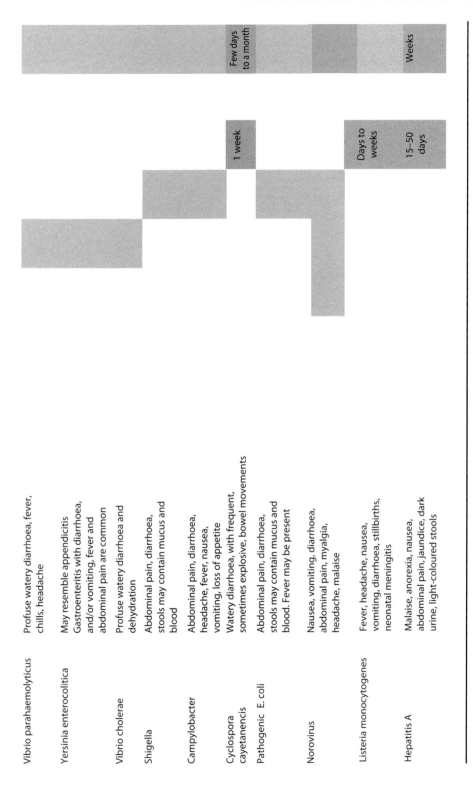

- Collect samples.
- Identify factors common to all or most cases.
- Conduct site investigation at implicated premises.
- Form preliminary hypothesis.
- Consider if there is a continuing public health risk.
- Initiate immediate control measure.
- Inform government agencies.

An inspection of the premises associated with the outbreak will depend upon the type of premises, but typical information sought will be as follows:

- Menus, food sources and suppliers
- Preparation, cooking and storage conditions, temperature measurements
- Details of HACCP records, pest control, cleaning regimes
- Staff details, health, hygiene training
- Flow diagram of food preparation area and processes

1.13.2 Case definition and data collection

The next phase would be to establish a case definition, and conduct an in-depth interview with initial cases using a standardised questionnaire. A 'case' is an epidemiological tool for counting cases. Its definition should be simple enough to apply, and include the following:

1 Clinical symptoms of the illness
2 A defined period of time
3 A defined location, for example attending a party, or living in a particular area
4 Limited 'persons', for example age group, or no recent international travel

The case definition should not only have high sensitivity to include all the true cases, but also have high specificity to exclude cases that are unrelated. This is the ideal situation, and in real life as each outbreak is different, the case definition needs to be appropriate. Further analysis of the data collected via questionnaires is required, and laboratory results are needed for confirmation. It is not uncommon for the case definition to change during an investigation as information comes to light. Also, during an investigation, there can be following cases:

- *Confirmed cases*: When the person has a positive laboratory result.
- *Probable cases*: When the person has the clinical symptoms of the illness but without laboratory confirmation.
- *Possible cases*: When the person does not have all the clinical symptoms.

Frequently, the cases that initiate an outbreak investigation are only a small portion of the total number of cases. In order to determine the size and severity of the outbreak, a proactive search for further cases is necessary. If the outbreak is likely to be a single source, such as a party, then contacting all the guests is appropriate. In contrast, for outbreaks with a continuous- or intermittent-common source with mild symptoms, finding a significant number of cases can be problematic. Subsequently, it may be necessary to contact the public via the media, as well as doctors and public health laboratories for additional cases. In some outbreaks, the national and international networks (such as PulseNet and Enter-Net, Section 1.12) may be used to link cases associated with a food product that is widely distributed.

The specific questionnaire would depend upon whether the outbreak was amongst a group who are already known to have eaten together, or individuals who are initially only linked by symptoms over similar time period. In general, the questionnaire would cover the following

points:
- Age, sex and occupation
- Date and time of first symptoms
- Nature, severity and duration of symptoms
- Information on any recent foreign travel
- Contact with ill people with similar symptoms
- Details, including location, of recent food and drink consumption
- Sources of domestic food, milk and water supplies
- Description of any samples provided

Examples of questionnaire designs are given in the WHO 'Foodborne Disease Outbreaks: Guidelines for Investigation and Control' document.

1.13.3 Data collation and interpretation

A response rate of at least 60% (preferably more than 80%) is achievable with proactive contacting of the cases and controls. Analysing the feedback information should lead to the calculation of incidence rate (proportion of those exposed to the infection who have developed illness), and time, place or person associations, and construct an epidemic curve. This may lead to a more detailed hypothesis as to the source, mode of spread of the disease and the need for microbiological sample collection.

The information from the initial questionnaire responses needs to be collated promptly to ascertain the distribution of clinical symptoms and other factors among cases. Table 1.17 shows an example of a case listing for an outbreak. Each column represents a variable, and each row is an individual case. Specific computer software is available for this, for example Epi Info[TM] and EpiData which are accessible from http://www.cdc.gov/epiinfo and http://www.epidata.dk, respectively.

Table 1.18 shows how the symptoms can be collated in decreasing order. This can indicate whether the cause is intoxication or an enteric infection. The symptoms of intoxications are typically vomiting without fever with a short incubation period (<8 hours), for example due to *B. cereus*, *Cl. perfringens* and *St. aureus*. In contrast, the general symptoms for an enteric infection are fever in the absence of vomiting, and an incubation period of more than 18 hours, for example, *Salmonella* and *Campylobacter*. See Table 1.16 for a fuller description of characteristic symptoms and organisms.

Time associations are when the onset of similar illnesses is within a few hours or days of each other. Place associations refer to eating or drinking at the same place, buying the food from the same retailer or living at the same address. Person associations refer to common experiences such as eating the same foods, being in the same age group, and so on.

To help identify the causative source of the outbreak, the frequency of those ill according to whether they ate each named food needs to be calculated. This is termed the 'attack rate'; examples are given in Table 1.19 where it can be seen that the attack rate for chocolate mousse is notably high compared with the others. Consequently, the method of chocolate mousse preparation would be investigated, and a likely scenario is that the mousse was prepared with raw eggs. In this example, the attack rate for ice cream is opposite to that for the chocolate mousse, as people were given a choice between the two desserts. Additionally, there is a large difference between the attack rates for eating peas; however, this is probably due to the small sample size of the 'Did not eat' category.

In Table 1.19, the attack rate varies for those who ate or did not eat the food between the food types. The 'relative risk' (or 'risk ratio') is the ratio of these attack rates. If the value is ~1, then

Table 1.17 Outbreak investigation – line listing

Identification code	Name	Age	Sex	Date and time of illness onset	Major signs and symptoms			Laboratory tests	
					Diarrhoea	Vomiting	Fever	Specimen	Results
1	GF	18	F	16.01.09 21:30		Y		ND	
2	SD	24	M	17.01.09 06:30	Y			Faeces	Salmonella
3	EH	32	M	17.01.09 07:30	Y			Faeces	Salmonella
4	AS	56	F	17.01.09 08:30	Y	Y		ND	
5	KG	6	M	18.01.09 15:00	Y	Y		Faeces	Pending
6	ML	35	M	19.01.09 11:00	Y	Y	Y	ND	

Cont.

Table 1.18 Frequency of symptoms amongst cases

Symptoms	Number of cases	Percentage (%)
Diarrhoea	71	91
Abdominal pain	35	45
Nausea	23	29
Fever	10	13
Vomiting	5	6
Total	78	

there is no association between the food and the illness. However, if the ratio is >1, then there is an increased risk of illness associated with eating the particular food. Again, this is clearly shown for the chocolate mousse. The risk ratio for the ice cream is considerably less than 1, due to the dessert choice, and therefore there is a decreased risk of illness. Table 1.20 differs from Table 1.19 as the number of people exposed to the risk is unknown. Instead of relative risk, the 'odds ratio' is calculated as the product of the number of cases exposed to the food type multiplied by the number of controls not exposed, which is divided by the product of the number of controls exposed multiplied by the number of cases not exposed. As for the 'relative risk' values, odds ratio greater than 1 indicate association between food type and illness. Odds ratios are normally used for outbreaks where the total number of exposed people is unknown, as in a case–control study (see later).

For most foodborne infections, the time course of an outbreak is usually shown as an epidemic curve. An epidemic curve is a graph which shows the distribution of the time of onset of symptoms of all cases which are associated with the outbreak; Figure 1.7. This is not always

Table 1.19 Specific attack rates for food types served at a wedding reception, a cohort study

Food type	Ate			Did not eat			Relative risk
	Ill	Not ill	Attack rate (%)	Ill	Not ill	Attack rate (%)	
Melon[a]	5	25	17	3	12	20	0.83
Prawn cocktail	3	8	27	10	25	29	0.95
Paté	2	3	40	15	25	38	1.07
Steak	12	23	34	4	6	40	0.86
Lamb	2	8	20	8	27	23	0.88
Chips	12	28	30	1	4	20	1.50
Potatoes	6	31	16	1	7	13	1.30
Peas	7	35	17	1	2	33	0.50
Ice cream	2	28	7	10	3	77	0.09
Chocolate mousse[b]	10	3	77	1	29	3	23.08
Orange juice	10	30	25	1	4	20	1.25
Tomato juice	2	3	40	25	15	63	0.64
Wine	9	32	22	1	3	25	0.88

[a]Attack rate calculation example, for those who ate melon is $(5/(5+25) \times 100) = 17\%$.
[b]Relative risk ratio calculation example, for chocolate mousse is $77/3 = 23.08$. Indicating a very strong association between the consumption of chocolate mousse and illness.

Table 1.20 Example of a case–control study, outbreak of *Salmonella* in the general population

Food type	Cases (n = 30)		Controls (n = 25)		Odds ratio
	Ate	Did not eat	Ate	Did not eat	
Green salad	15	10	12	8	1.00
Fresh fruit	7	3	10	6	1.40
Tomatoes	17	11	12	11	1.42
Mayonnaise	5	7	3	7	1.67
Chicken sandwiches[a]	26	2	5	10	26.00
Roast chicken meal	19	8	16	5	0.74

[a]Odds ratio calculation example, for chicken sandwiches $= (26 \times 10)/(2 \times 5) = 26.00$. Therefore, there is a strong association between the consumption of chicken sandwiches and *Salmonella* infection in the general population.

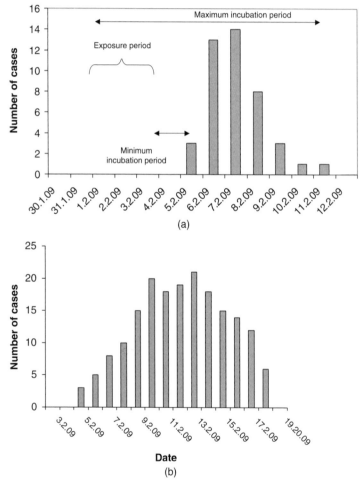

(a)

(b)

Figure 1.7 Four generalised types of epidemic curves: (a) point source (b) continuous-common source (c) intermittent-common source (d) person-to-person spread.

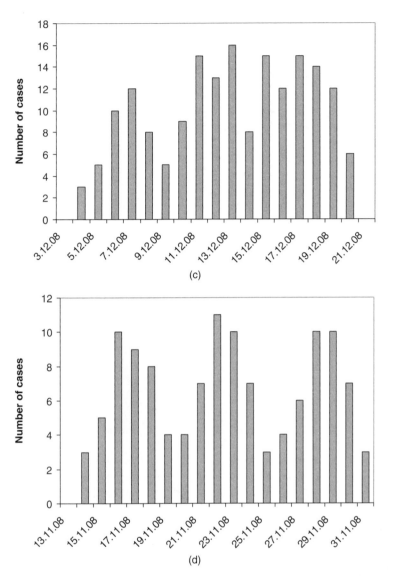

Figure 1.7 (*Continued*)

possible however with an organism with very prolonged/ill-defined incubation periods such as *L. monocytogenes*. The timescale varies depending on the duration of the outbreak.

The epidemic curve is very important as it can:
1 identify the mode of transmission;
2 determine the exposure period to the infection or toxin;
3 determine the incubation period of the infectious agent or toxin;
4 identify outlier cases – these may be invaluable in identifying the source of the outbreak.

Figure 1.7 shows four types of epidemic curves, which vary in their shapes. A point source outbreak shows a steep rise in number of cases, followed by a decrease (Figure 1.7a). An example could be a group of people eating one meal together, such as *Salmonella*-contaminated eggs used at a wedding reception for chocolate mousse preparation. The time difference between the first and last cases will be less than the incubation period. Additionally, if the causative agent has been identified, then the exposure period can be determined as the difference between the minimum and maximum incubation periods. A continuous-common source outbreak, where the cause has not been removed, will show a gradual increase followed by a plateau in the number of cases (Figure 1.7b). For example, a faulty milk pasteurisation unit will cause the number of cases to rise and eventually plateau until adequate control measures are put into place. An intermittent-common source outbreak will undulate according to the frequency of exposure (Figure 1.7c). A person-to-person transmitted outbreak will also undulate; however, the periodicity of the later will be close to the incubation period of the infectious organism (Figure 1.7d). Secondary cases, where a person is infected from a primary case and not the original source, can occur and these will complicate the interpretation of the epidemic curve.

When the people are linked by a place where they have eaten together (i.e. a wedding reception), a 'cohort study' is undertaken, otherwise a 'case–control study' is required. Cohort studies are more common than case–control studies. This is because a group of people with similar symptoms, who know each other, are more likely to report the incidence to health officials, and other authorities. A 'cohort' refers to a group of people who ate together, and therefore were potentially exposed to the foods being investigated. The food consumed by each person at the meal can be recorded via a questionnaire. The cohort study has the advantage over the case–control study in that there is no need to identify and select controls. This reduces the possibility of bias. Since the number of people exposed is known, the relative risk can be calculated (Table 1.19).

A case–control study is necessary when the population at risk cannot be identified, or when at risk, population is so large that it is not cost-effective to include everyone in the study. For example, a case–control study would be used if there was a sudden increase in the incidence of an uncommon *Salmonella* serovar, and the cases were not clustered together but from a wide regional area. Therefore, there is no single meal which the cases were exposed to. Subsequently, the cases need to be interviewed to determine which foods could be contaminated. The hypothesis is that the disease is linked to the consumption of a single or a small number of foods. A comparison is made with the diet of control people (matched by age, sex and location) who are well and have similar diets. Since the total number of persons exposed is unknown, an odds ratio is determined; Table 1.20.

The data gathered will need to be statistically analysed; usually the chi-square (χ^2) and Fisher's exact tests are used, the latter being more applicable for smaller sample sizes. The significance level is commonly set at 95%. This means that is there is a 1 in 20 likelihood that chance alone would account for the statistical difference between those who are ill, who ate a suspect food and those who are ill, who did not eat it. Sometimes confounding factors can arise because of foods being eaten together, such as fish dishes being served with sauces made with raw eggs.

Finally, control measures need to be put into place by the relevant authorities to:
- control the source, that is animal, human or environmental;
- control the mode of spread;
- protect those at risk;
- continue surveillance of the effectiveness of control measures.

Examples of a number of outbreak investigations are included with specific pathogens in Chapter 4.

1.14 Food terrorism and biocrimes

The majority of this book concerns accidental contamination of food by organisms that are well known to cause food poisoning. Such contamination can occur during harvesting, processing, retail distribution, preparation and consumption, in other words between 'farm to fork'. However, this section considers the deliberate contamination of food by individuals or organisations as a threat to human health. Deliberate contamination of food can range from tampering with individual food items to large-scale incidents involving bulk foods. Such acts could be carried out within a food company by a disgruntled employee for financial gain, personal vendetta or wider disruption by terrorists. Potentially, the malicious contamination of food within one manufacturer's production process could have serious repercussions on a global scale. Alternatively, food and water could be contaminated at a point of distribution such as a supermarket or a restaurant. The disruption being caused is not only by the number of those ill, but also through public fear and the loss of confidence in political and regulatory authorities. Many of the biocrime organisms are the same pathogens which cause outbreaks of foodborne illness due to accidental contamination. Such biocrimes could initially be interpreted as an accidental incidence and the evidence analysed by an unsuspecting hospital laboratory. Surveillance networks, both national and international, are considered elsewhere in greater detail; cf. Section 1.11. Such systems need to invoke rapid responses in the medical services and regulatory authorities. Even a small number of cases could represent a larger threat since the contaminated source could still be on the market, or in people's homes. In addition, surveillance systems are limited by the fact that only a proportion of people seek medical assistance when they are ill. This under-reporting is considered in Section 1.7.

Deliberate contamination of domestic water supplies with a biological agent is feasible. However, there are a number of barriers to its effectiveness, dilution, inactivation by chlorine treatment, or ozone treatment, filtration and the relatively small quantity of water that each person drinks. Nevertheless, the protozoan *Cryptosporidium parvum* can persist through these barriers (see Section 4.6.2). This is evident from the 1993 outbreak in Milwaukee where 403 000 people developed cryptosporidiosis. At least 54 people died, and 4400 were hospitalised. Obviously, poorly maintained water treatment plants pose a great threat to the risk of waterborne diseases through accidental and deliberate contamination. The contamination of food processors water supply could be even more detrimental through subsequent economic losses or product recall, as well as the number of illnesses. The international recall of 160 million bottles of sparkling bottled water due to possible benzene contamination in 1990 well illustrates this point.

Compared with accidental food contamination, the number of reported food sabotage cases is relatively few. Nevertheless, the potential impact of deliberate food contamination can be envisaged from the scale of many documented examples of unintentional outbreaks of foodborne disease (Table 1.7), some of which involved >100 000 cases of infections. The largest reported unintentional contamination of one food, hepatitis A in clams, led to 300 000 cases. Although most people recover from the primary symptoms of bacterial food poisoning, it should be remembered that secondary, chronic debilitating consequences can follow (Section 1.8). Given that a terrorist could use a more dangerous biological, chemical, or radionuclear agent to optimise the number of deaths, create panic, and economic loss, the scale of some outbreaks illustrate that a deliberate attack of food terrorism could be devastating.

The most frequently referred to biocrime cases are as follows:

- *1964–1968*: A research microbiologist deliberately contaminated food and beverages, causing several outbreaks of typhoid fever and dysentery in Japanese hospitals, family members and neighbours. The objective may have been to obtain clinical samples for a doctoral thesis.
- *1970*: A postgraduate student at McGill University contaminated his roommate's food with the parasite *Ascaris suum*, a large roundworm that infects pigs. Four victims became seriously ill, two with acute respiratory failure.
- *1978*: Mercury contamination of citrus fruit from Israel led to 12 children in Holland and West Germany being hospitalised.
- *1984*: S. Typhimurium contamination of salad bars in ten restaurants over a 2-week period in 1984. This was carried out by members of the Rajneesh religious group in the Dallas, Oregon, US, and caused 751 cases of salmonellosis and hospitalisation of 45 of the victims. It is generally believed that this was a trial run for a more extensive attack aimed at disrupting local elections. The cult also possessed S. Typhi, which causes more severe invasive illness than S. Typhimurium.
- *1996*: *Shigella dysenteriae* contamination of food. In 1996, a disgruntled laboratory worker at a Dallas hospital infected colleagues' food with a laboratory strain of *Sh. dysenteriae* type 2. In total, 12 people became ill, 9 of the victims were taken to hospital for treatment.
- *2001*: 120 people in China were poisoned when food was laced with rat poison by the owners of a noodle factory.
- *2002*: Similarly, in 2002, near Nanjing, China, about 40 people died, and more than 200 were hospitalised when the owner of a fast food outlet put rat poison in his competitor's breakfast foods.
- *2003*: A supermarket employee deliberately contaminated 200 pounds of ground beef with an insecticide containing nicotine. One hundred and eleven people, including approximately 40 children, became ill following consumption of the tainted meat.

The CDC have categorised bioweapon organisms according to their threat; Table 1.21. Category A, the high-priority biological agents, includes *Bacillus anthracis* (anthrax) and *Clostridium botulinum* (botulism). These are both well-known foodborne pathogens, which can be fatal. The majority of the foodborne pathogens are Category B agents. These are moderately easy to disseminate, cause moderate morbidity and low mortality.

Bacteria are not the only agents which can be used in food for a terrorist attack. Chemical weapons include pesticides, arsenic, mercury, lead, heavy metals, dioxins and polychlorinated biphenyls (PCBs). For example:

- In 1946, the group Nakam infiltrated a bakery that supplied bread to a prisoner-of-war camp outside Nuremberg. By spreading an arsenic-based poisoning, they killed hundreds of interned SS soldiers, and made thousands ill.
- In 1971–1972, in Iraq, more than 6500 people were hospitalised and 459 died after eating bread made from mercury-contaminated wheat.
- Deliberate mercury contamination of citrus fruit from Israel led to 12 children in Holland and West Germany being hospitalised in 1978.
- In 1981, in Spain, over 800 people died and about 20 000 were affected by toxic cooking oil.
- In 1985, in the United States, nearly 1400 people became ill after eating watermelons grown in soil containing the pesticide aldicarb.
- Alleged deliberate contamination of Chilean grapes laced with cyanide in 1989 led to the recall of all Chilean fruits from the United States.

Table 1.21 CDC categorisation of bioterror organisms

Category	Description	Organisms
A	Highest priority, easily disseminated or transmitted person to person, high mortality, potential for major health impact, causes social disruption and public panic, needs special action/intervention from public health services	Anthrax, *B. anthracis* Smallpox, *Variola major* Tularaemia, *Francisella tularensis* Plague, *Yersinia pestis* Botulinum, *Cl. botulinum* Viral haemorrhagic fevers (e.g. Ebola virus)
B	Second highest priority, moderately easily disseminated, moderate mortality, needs enhanced surveillance from CDC or other public services	Foodborne threats (*Salmonella* species, *E. coli* O157:H7) Waterborne threats (*Vibrio cholera*, *Cryptosporidium parvum*) Ricin toxin Q fever (*Coxiella burnetii*) Brucellosis Epsilon toxin of *Cl. perfringens* Staphylococcus enterotoxin B (SEB) Glanders Meliodies Psittacosis Typhus fever (*Rickettsia prowazekii*) Viral encephalitis (Venezuelan equine encephalitis, eastern equine encephalitis, western equine encephalitis)
C	Third highest priority, easy to produce and available and easy to disseminate, possible mass dissemination in future, can be high morbidity and/or mortality	Hantavirus Yellow fever Multidrug-resistant TB (tuberculosis) Tick-borne viruses (haemorrhagic, encephalitis) Nipah virus

- In 1995–1996, nine Russian soldiers and five civilians were killed by cyanide-laced champagne by the Tajik opposition at a New Year's celebration.

The economic losses associated with foodborne outbreaks can be considerable (see Section 1.10). The costs incurred include product recall, improving food safety and restoring consumer confidence. In 1998, following a *L. monocytogenes* contamination of frankfurters and luncheon meat, an American company spent in the order of $50–$70 million to recall 16,000 metric tonnes product, and a further $100 million improving food safety and regaining consumer's confidence. The *S*. Montevideo outbreak linked to chocolate from the United Kingdom in 2006, cost the manufacturers an estimate of $40 000 in product recall. The outbreak from *S*. Enteritidis contaminated ice cream in 1994 (Table 1.7) was estimated to have cost about $18.1 million in medical care and time lost from work. As expanded on in Section 1.10, the indirect costs of foodborne diseases are considerable. The USDA estimates that the most common pathogens cost the economy $6.9 billion annually. The economic implications of a deliberate contamination can be the intention of the perpetrator, such as the mercury sabotaging of Israeli citrus fruits in 1978.

Other agents that could be added to food deliberately are food toxicants and allergens. These could be added to foods which are not normally screened for such compounds. For example, foods which are not normally susceptible to aflatoxins or nut residue contamination. Numerous hoax threats of food and water contamination have been recorded. This can have considerable effect of political pressure, a public sense of security and wasted response by health services.

Countries import large quantities of food; more than 75% fresh fruits and vegetables, and 60% of seafood are imported into the United States. Because of the globalisation of the food chain, food that has been contaminated in one country can have significant effect upon another country albeit on the other side of the world. For example:

- *1989*: Cantaloupes imported from Mexico were contaminated with S. Chester and led to 25 000 infections in 30 states in the United States.
- *1991*: Three cases of cholera in Maryland, US, were associated with contaminated frozen coconut milk imported from Thailand.
- *1994–1995*: An outbreak of S. Agona in Israel involving >2000 cases was traced to a contaminated children's kosher savory snack, and also caused some cases in the United States, France, England and Wales.
- *1996–1997*: Cyclospora-contaminated raspberries from Guatemala caused 2500 infections in 21 states in the United States and two Canadian provinces.
- *1997*: Raspberries from Slovenia contaminated with Norovirus caused ~300 illnesses in Canada.
- *1997*: Also in 1997, 151 cases of hepatitis A in four schools were linked to frozen strawberries from Mexico.

The disruption to international trade was demonstrated during the Belgium dioxin contamination which necessitated the international recall of food products, and more acutely with the link between bovine spongiform encephalopathy (BSE) (also known as 'mad cow disease') and variant Crutzfeld–Jacob (vCJD) disease in the United Kingdom in the 1990s; see Section 4.8.1 for more details. The transference of the agent from cattle to humans was initially denied, but later accepted in 1996. This resulted in widespread public anxiety, import bans, slaughtering of millions of cattle, crippling the UK beef industry, loss of trust in regulatory authorities as well as human cost. The UK government (following a change in the leading political party) responded with the creation of a new food regulatory authority, the 'Food Standards Agency'.

1.15 Food safety following natural disasters, and conflict

Following natural disasters, it is essential that safe water and food are available. The victims may be suffering from injuries, shock, disorientation and fear. The risk of further illness will be high due to the inability to cook combined with poor hygiene and the lack of toilet facilities. Hence, food and water can quickly become contaminated with microbial pathogens causing hepatitis A, typhoid fever, cholera and dysentery. To help governments in their planning and response to natural disasters, the WHO has developed the guide '*Ensuring Food Safety in the Aftermath of Natural Disasters*'; see Web Resources section for URL. It offers specific advice to those involved in food storage, handling and preparation during disaster situations. In summary:

1 *Preventive food safety measures in the aftermath of natural disasters*: All water should be treated as contaminated with surface water unless it is boiled or otherwise made safe. Only after treatment can it be consumed or used in preparing food. However, the food may still contain chemical hazards. Agricultural areas should be assessed for where food can still be harvested or where food has been safely stored after harvesting.

2 *Inspecting and salvaging food*: All food stocks should be inspected and assessed for their safety. Canned foods with broken seams, serious dents, or leaks, and jars with cracks should be discarded. Food deemed as safe should be segregated from other foods. In areas of flooding, food should be stored in a dry area. Mould growth (and possible toxin production) is more likely to occur in vegetables, fruits and cereals which are stored under moist conditions, or have become wet. Refrigerated food, in particular meat, fish, poultry and milk, which can no longer be kept cold, should be consumed before it has been in the 'danger zone' (5–60°C) for more than 2 hours. Other foods, normally refrigerated, can be kept for longer than 2 hours, but should be discarded if showing signs of spoilage.

3 *Provision of food after a natural disaster*: When cooking facilities become re-established, it is usual to distribute dried food. Therefore, instructions on food preparation, especially the crucial need for safe water not only for reconstitution, but also for washing hands and utensils, will be needed.

2 Basic aspects

2.1 The microbial world

The world of microbiology covers a diverse range of life. Using the loose definition that microbiologists study life forms that are not clearly visible to the naked eye means that it includes complex organisms such as protozoa and fungi as well as simpler bacteria and viruses. The major micro-organisms studied to date have been the bacteria. This is because of their medical importance, and because they are easier to cultivate than other organisms such as viruses, and the more recently recognised 'prions', which are uncultivatable infectious organisms. Despite the predominance of bacteria in our understanding of microscopic life, there are numerous important organisms in other microbial categories.

The cell structure reveals whether an organism is 'eucaryotic' (also spelt 'eukaryotic') or 'procaryotic' (also spelt 'prokaryotic'). Eucaryotes have cellular organisation with mitochondria, endoplasmic reticulum and a defined nucleus, whereas procaryotes have no obvious organelle differentiation and are in fact similar in size to the organelles of eucaryotes. Analysis of the genetic information encoding for part of the ribosome has revealed a plausible relationship and evolution of life from procaryotes to eucaryotes through intracellular symbiotic relationships (see 16S rDNA analysis in Section 2.9.2).

Until relatively recently, the classification of micro-organisms has been based on morphology and phenotypical (biochemical) properties. Morphological analysis extends from whether the bacterium is spherical (coccoid), filamentous, curved or rod shaped to whether the bacterium has a Gram-negative or Gram-positive type cell wall (Section 2.2.2). The possession of certain enzymes (e.g. α-galactosidase for lactose fermentation) has helped in identifying and defining a general group of bacteria associated with faecal contamination, the 'coliforms'. In general, these characteristics enable the categorisation of bacteria into broad groups. In the past, for more precise classification, biochemical and enzyme activities were used. More recently, advancements in DNA sequencing have enabled a fuller understanding of the genetic capabilities of micro-organisms. Certain genes, termed 'housekeeping genes', have undergone limited variation during evolution and can be used as a means of constructing a family tree of relationships between organisms. This is called 'phylogenetics'. A commonly used gene for this purpose encodes for one of the rRNA molecules, which is essential for the ribosome to function for protein synthesis. This molecule is called the 16S rRNA gene in prokaryotes, and 18S rRNA gene in eukaryotes (Section 2.9.2). A phylogenetic tree of organisms which cause foodborne disease is given in Figure 2.1. Viruses do not possess this gene and therefore do not appear on this 'tree of life'. Being able to sequence genes in organisms of interest and compare the order of DNA nucleotides is just one example of 'bioinformatics' (Section 2.9) which enables us to better understand the microbial world.

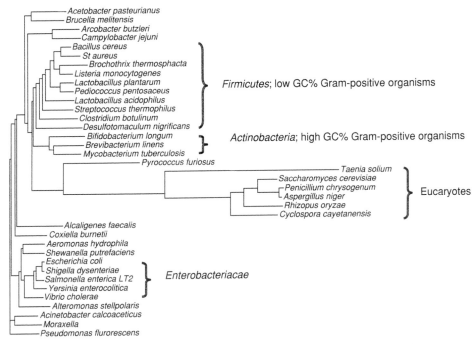

Figure 2.1 Phylogenetic tree of organisms associated with food microbiology and the Archaea organism *Pyrococcus furiosus*, based upon the DNA sequence ribosomal small subunit, constructed using ClustalW (freeware). Details of sequence accession numbers and tree construction can be obtained from http://www.wiley.com/go/forsythe, and Section 2.9.2.

Although the 'species' concept is conventionally applied to bacteria, there are difficulties. In higher organisms, a species can be defined as an interbreeding population which is reproductively isolated. The genome of each individual is derived most equally from its two parents, and that as species diverge, hybrids become less viable. In contrast, bacteria multiply by binary fission. The acquisition of additional genes by bacteria is limited to the transfer of a few genes between related strains, and sometimes even different species (horizontal gene transfer). Whole genome sequence analysis can reveal regions of DNA which differ from the majority of the genome due to its G:C ratio or codon bias. These regions can be from bacteriophages, and less closely related bacteria. Whereas in multilocus sequence analysis (Section 5.5.4), housekeeping genes are sequenced and form a core of genes which help to define the species. Therefore, although there are differences between 'species' concept in bacteria and higher organisms, in both cases, individual 'species' can be recognised by reference to a gene pool. Inevitably, as our knowledge of organisms at the genome level improves, previous species may be redefined. Some bacterial species were primarily defined according to biochemical (phenotyping) traits, which represent only a small portion of the total coding of an organism.

A brief survey of food poisoning micro-organisms can start with eucaryotic organisms such as helminths (see Table 2.1 and Figure 2.1; cf. Section 4.6). These include the cestode worms that are responsible for taeniasis: *Taenia solium*, the pork tapeworm; and *T. sainata*, the beef tapeworm. Both have worldwide distribution. Infection results from the ingestion of undercooked or raw meats containing the cysts. The mature worms can infect the eye, heart,

Table 2.1 Microbiological contaminants

	Where found	Sources
Viruses		
A wide range which cause diseases, including hepatitis A	Most common in shellfish, raw fruits and vegetables	Associated with poor hygiene and cultivation in areas contaminated with untreated sewage and animal and plant refuse
Bacteria		
Includes *Bacillus* spp. *Campylobacter*, *Clostridium*, *E. coli*, *Salmonella*, *Shigella*, *Staphylococcus* and *Vibrio*	Raw and processed foods: cereal, fish and seafood, vegetables, dried food and raw food of animal origin (including dairy products)	Associated with poor hygiene and unclean conditions generally carried by animals such as rodents and birds, and human secretions
Moulds		
Aspergillus flavis and related fungi	Nuts and cereals	Products stored in high humidity and temperature
Protozoa		
Amoebae and Sporidia	Vegetables, fruits and raw milk	Contaminated production areas and water supplies
Helminths		
A group of internal parasites including *Ascaris*, *Fasciola*, *Opisthorchis*, *Taenia*, *Trichinella* and *Trichuris*	Vegetables and uncooked or undercooked meat and raw fish	Contaminated soil and water in production areas

liver, lungs and brain. A third tapeworm is *Diphyllobothrium latum* which is found in a variety of freshwater fish including trout, perch and pike. Contaminated water supplies can carry infectious organisms including pathogenic protozoa such as *Cyclospora* and *Cryptosporidium*.

Fungi (a term which includes yeast and moulds) are also eucaryotic. Well-known examples of moulds are *Penicillium chrysogenum* and *Aspergillus niger*. As shown in Figure 2.1, they form a unique branch of life, and despite their superficial appearance, they are not plants. They are not differentiated into the usual roots, stems and leaves system. They also do not produce the green photosynthetic pigment chlorophyll. Their morphology can be a branching mycelium with differentiating cells producing hyphae to disperse spores or exist as more single cell forms commonly referred to as yeast. Yeasts such as *Saccharomyces cerevisiae* are very important in food microbiology. The brewing and bakery industries are dependent upon yeast metabolism of sugars to generate ethanol (and other alcohols) for beer and wine production, and carbon dioxide for bread manufacture.

Mycotoxicoses are caused by the ingestion of poisonous metabolites (mycotoxins) which are produced by fungi growing in food (cf. Section 4.7). Aflatoxins are produced by the fungi *Aspergillus flavus* and *A. parasiticus*. There are four main aflatoxins designated B1, B2, G1 and G2 by the blue (B) or green (G) fluorescence when viewed under a ultraviolet (UV) lamp. Ochratoxins are produced by *A. ochraceus* and *Penicillium viridicatum*. Ochratoxin A is the most potent of these toxins.

Bacteria are procaryotes and are divided into the 'Eubacteria' (true bacteria) and 'Archaea' organisms (old term 'Archaebacteria') according to 16S rDNA analysis, detailed cell composition and metabolism studies. There are no *Archaea* organisms of importance in the food industry. *Pyrococcus furiosus* (found in hydrothermal vents, growing at 100°C) has been in-cluded in Figure 2.1 to illustrate the separation of the *Archaea*. As a generalisation, the size of a rod-shaped bacterium is about 2 μm × 1 μm × 1 μm. Although they are very small, even 500 cells of *Listeria monocytogenes* can be an infectious dose to a pregnant woman resulting in a stillbirth. Familiar foodborne pathogens such as *Salmonella* spp., *Escherichia coli* and *C. jejuni* are eubacteria which are able to grow at body temperatures (37°C) and produce various toxins which damage human cells resulting in 'food poisoning' symptoms such as diarrhoea and vomiting (see Section 4.3 for more information on (eu)bacteria causing food poisoning).

Viruses, such as Noroviruses, are much smaller than bacteria. The large ones, such as the cowpox virus, are about 0.3 μm; the smaller ones, such as the foot and mouth disease virus, are about 0.1 μm (see Section 4.4 for more information). Because of their small size, viruses pass through bacteriological filters and are invisible under the light microscope. Bacterial viruses are termed bacteriophages (or 'phages'). They can be used to 'fingerprint' bacterial isolates which are necessary in epidemiological studies; this is termed 'phage-typing'.

Prions (short for 'proteinaceous infectious particles') have a very long incubation period (months, or even years) and resistance to high temperature, formaldehyde and UV irradiation; see Section 4.8.1 for more details. With regard to sheep and cattle, the isomer of the normal cellular protein PrP^C, termed PrP^{SC}, accumulates in the brain causing holes or plaques. This leads to the symptoms of scrapie in sheep and BSE in cattle. The equivalent disease in humans, variant Creutzfeldt–Jakob disease, is probably due to ingestion of some infectious agent from cattle.

2.2 Bacterial cell structure

2.2.1 Morphology
Bacteria are characteristically unicellular organisms. Essentially, they are bags of self-replicating information within a controlled micro-environment that is semi-protected from extremes of variation in the external environment by a semi-permeable, cytoplasmic membrane. Bacterial morphology can be straight or curved rods, cocci or filaments depending upon the organism concerned. The morphology of the organism is consistent with the type of organism, and to a lesser extent, with the growth conditions. Bacteria in the genera *Bacillus*, *Clostridium*, *Desulphotomaculum*, *Sporolactobacillus* and *Sporosarcina* form spores in the cytoplasm under certain environmental stress-related conditions. The spore is more resistant to heat, drying, pH, and so on, than the vegetative cell. Hence, it enables the organism to persist until more favourable conditions prevail when the spore can germinate and grow into a vegetative cell.

2.2.2 Cell membrane structure and the Gram stain
The structural integrity of the cytoplasmic membrane is maintained by the peptidoglycan layer that allows the cell to cope with extremes in osmolarity in the external environment. There is, however, a price to pay for maintaining homeostasis within the cytoplasm. This is in the energy costs of transporting nutrients into the cell and harmful materials out of the cell. There are many specialised transport proteins within the cell, which are from five separate super-families. The super-families contain both permeases that transport materials into the cell, and efflux pumps that transport materials out of the cell. The energy for transportation is derived either from hydrolysis of adenosine triphosphate, ion gradients across the cytoplasmic membrane or from hydrolysis of phosphoenol pyruvate.

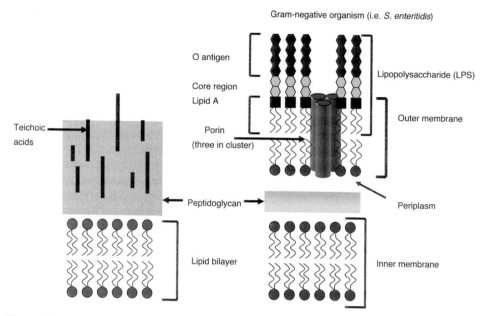

Figure 2.2 General structure of the bacterial cell wall.

Gram-negative organisms have an additional structure in the cell wall called the outer membrane that acts as a barrier to limit diffusion of harmful materials into the cell. The outer membrane, together with efflux pumps, enables Gram-negative bacteria to survive and thrive in environments that are harmful to other types of micro-organisms. The outer membrane would also limit diffusion of nutrients into the cell, but there are channels in the outer membrane called porins; see Figure 2.2. These allow passive diffusion of nutrients into the periplasmic space, but porins have size exclusion limits and some have specificities for the charge on molecules. These proteins have evolved to allow some Gram-negative organisms to survive in the gut, where there are detergents present at around 20 mM. After nutrients have crossed into the periplasmic space, they are then taken up by substrate-specific permeases and transported across the cytoplasmic membrane. Some of the permeases are highly specific for their substrates, particularly those of the phosphotransferase system (PTS). Others, particularly those of the proton symport type, are much less discriminatory.

Christian Gram was a Danish microbiologist who wanted to visualise bacteria in muscle tissue. He tried a range of histological staining protocols and eventually came up with the, nowadays ubiquitous, 'Gram stain' procedure. Christian Gram noted that the bacteria either stained dark blue or red. This was due to the precipitation of crystal violet with Lugol's iodine in the cell cytoplasm which could not be extracted using solvents such as ethanol or acetone from certain organisms (Gram-positives) but was extracted from others (Gram-negatives). Since the latter cells are no longer stained by the crystal violet–iodine complex, counterstaining is required with safranine or basic fuschin.

The reason for the differential extraction of the crystal violet–iodine complex is due to differences in the cell wall structure (Figure 2.2). Gram-positive organisms have a thick cell

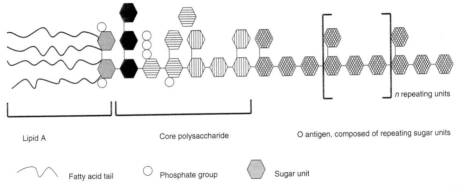

Lipid A   Core polysaccharide   O antigen, composed of repeating sugar units

⌇ Fatty acid tail   ○ Phosphate group   ⬡ Sugar unit

Figure 2.3 The structure of the lipopolysaccharide molecule of Gram-negative organisms, such as *Salmonella*.

wall surrounding the cytoplasmic membrane. This is composed of peptidoglycan (also known as murein) and teichoic acids, whereas the Gram-negative organisms have a thinner cell wall which is surrounded by an outer membrane. Hence, Gram-negative organisms have two membranes. This outer membrane differs from the inner membrane and contains the molecule known as lipopolysaccharide (LPS). Since peptidoglycan is the site of action for the original penicillin antibiotic (penicillin G), the difference in the cell wall structures explains why the antibiotic was initially so effective against streptococci and staphylococci (both Gram-positives) rather than Gram-negatives, such as *E. coli*. Consequently, the semi-synthetic penicillins were developed to widen the range of sensitive organisms to penicillin.

2.2.3 Lipopolysaccharide, O antigen

The outer membrane of Gram-negative organisms contains the molecule LPS. It is composed of three regions: lipid A, core and O antigen (Figure 2.3). Lipid A anchors the molecule in the outer membrane and is toxic to human cells and is also known as 'endotoxin'. During an infection, it causes the fever symptoms and is referred to as 'pyrogenic'. This is a virulence factor for Gram-negative organisms such as *Salmonella* and *Chlamydia*.

The LPS core region is composed of sugar molecules, the sequence of which reflects the organism's identity. The O region is more variable than the core region. In some organisms, the O region may only contain a few sugar residues, whereas in others, there are repeating units of sugars. Isolates of a single species vary in the amount of the O region present and hence gives rise to the terms 'smooth' and 'rough' variants. The antigens are on the body of the organism and are known as the 'somatic' or 'O antigens'. 'O' is for 'Ohne', the German for 'without' which originally referred to non-swarming or non-flagellated forms. The LPS structure is resistant to boiling for 30 minutes and is therefore also referred to as a 'heat-stable' antigen. One of the factors causing the fatality of Gram-negative septicaemia is that the presence of LPS causes the overproduction of tumour necrosis factor (TNF), which leads to overstimulation of nitric oxide synthase.

2.2.4 Flagella (H antigen)

Most rod-shaped (and a few coccoid) bacteria are motile in liquid media. Motility occurs due to the rhythmic movement of thin filamentous structures called flagella. The cell may have a single flagellum (monotrichus) or a tuft of flagella (lophotrichus) at one or both poles, or many flagella (peritrichus) over the entire surface. The proteinaceous nature of the flagella gives rise to its antigenicity. The flagellar antigen is called the H antigen. 'H' is from the German word 'Hauch' meaning 'breath' which originated from describing the appearance of *Proteus* swarming on moist agar plates being similar to the light mist caused by breathing on cold glass. Flagella are denatured by heat (100°C, 20 minutes), and therefore the H antigen is referred to as 'heat labile' (LT) in contrast to the heat-stable LPS antigen. Flagella are also denatured by acid and alcohol.

Salmonella spp. may express two flagellar antigens: Phase 1 which is possessed by only a few other serotypes of *Salmonella* and Phase 2 which is less specific. Phase 1 antigen is represented by letters, Phase 2 antigen by numbers. A culture may be entirely expressing one phase (monophasic culture) which will give rise to mutants in the other phase (diphasic culture), especially if the culture is incubated for more than 24 hours. The serotyping of *Salmonella* serovars is covered in more detail in Section 4.3.2. *E. coli* has a total of 173 different O antigens and 56 different H antigens. *E. coli* O157:H7 is a very pathogenic strain of *E. coli*, which is recognised by serotyping the strain's O and H antigens.

2.2.5 Capsule (Vi antigen)

Some bacteria secrete a slimy polymeric material composed of polysaccharides, polypeptides or polynucleotides. If the layer is very dense, then it can be visualised as a capsule. The possession of the capsule endows a resistance to white blood cell engulfment. The capsule's antigenicity is called the Vi antigen and was originally thought to be responsible for the virulence of S. Typhi. See Section 4.3.2 for more information on *Salmonella* serotyping.

2.3 Bacterial toxins and other virulence determinants

The ability of micro-organisms to cause disease is dependent upon the possession of various virulence determinants. These include the previously covered LPS (pyrogenic), capsule (engulfment evasion), flagella (motility towards target cells) as well as adhesins, invasins and toxins. To infect the host, most pathogens must adhere to the intestinal surface to overcome peristalsis. Adhesins are molecules on the cell surface composed of glycoproteins or glycolipids which bind to specific host surface structure. Fimbriae, also known as pili, are hair-like structures on the bacterial surface. Bacterial pathogens may have several different adhesins which result in host and tissue specificity. Invasins are molecules that enable the pathogen to invade a host cell where they can persist intracellularly. *Salmonella* serovars, *Shigella* species and *L. monocytogenes* invade the intestinal epithelial layer (cf. Section 4.3). *Listeria* has a surface protein, 'internalin', to facilitate invasion. Some pathogens have Type III secretory systems that pump bacterial proteins into the host cell. These cause changes on the host cell surface such as membrane ruffling, and macropinocytosis leading to engulfment and internalisation.

2.3.1 Bacterial toxins

There are a number of different bacterial toxin definitions, some of which overlap:

1 *Exotoxins*: These are essentially the same as 'bacterial protein toxins', and are also called 'secreted toxins'. The term 'exotoxin' emphasises the nature of the substance, being extracellular, excretory, heat labile and antigenic. They exert their biological activity (often lethal) in minute doses and are released during the death phase of batch culture. These are considered in more detail below.

2 *Enterotoxins*: These are exotoxins which result in extremely watery diarrhoea. They are further categorised by their mode of action as follows:

(i) Exotoxins which induce by direct action on gut tissue, biochemical and/or structural lesions which lead to diarrhoea.

(ii) Exotoxins which cause specific action (i.e. net fluid loss) in the gut.

(iii) Exotoxins responsible for elevating cAMP levels, which subsequently causes ion flux changes and excess fluid secretion.

3 *Cytotonic enterotoxins*: This is the term applied to those toxins, such as cholera toxin, that induce net fluid secretion by interfering with biochemical regulatory mechanisms without causing overt histological damage.

4 *Cytotoxic enterotoxins*: These enterotoxins induce actual damage to intestinal cells as a necessary prelude to onset of net fluid secretion. Cytotoxins can be a protein or LPS (endotoxin) and may also be called 'cell-associated' toxins and 'cytolysins'. They are invasive and kill target cells. Their mode of action is either intracellularly or by formation of pores within cells. Intracellular cytotoxins inhibit cellular protein synthesis and/or actin filament formation. Cytotoxins which cause pore formation in target cells can also be detected by their lytic activity upon erythrocytes, and hence are also known as 'haemolysins'. The result of cytotoxin invasion is inflammatory diarrhoea which often contains blood and leukocytes.

5 *Endotoxins*: Endotoxins are heat-stable, cell-associated, complex LPS structures. The LPS is part of the outer membrane of Gram-negative bacteria (Figure 2.2). Lipid A is the part of the LPS which is responsible for toxicity. They cause toxic shock, inflammation and fever as occurs with *Salmonella* spp. infections. They elicit cytotoxic activity upon cells.

6 *Cytolethal distending toxins (CLDTs)*: These toxins affect host cell-cycle regulation. These have been found in a variety of unrelated organisms: *C. jejuni, E. coli, Sh. dysenteriae, Haemophilus ducreyi* and *Actinobacillus actinomycetemcomitans* (Picket & Whitehouse 1999). The toxin blocks cells in G_2. CLDT inhibits the dephosphorylation of the protein kinase cdc2 and this prevents the cells entering mitosis.

Exotoxins are toxic bacterial proteins. The term is derived from the early observation that many bacterial exotoxins were excreted into the medium during growth. This differentiated them from endotoxins (LPS). This differentiation is not exact since some exotoxins are localised in the cytoplasm or periplasm and are released upon cell lysis. The naming of exotoxins is not systematic. Some toxins are named to indicate the type of host cell. Exotoxins that attack a variety of different cell types are called cytotoxins, whereas exotoxins that attack specific cell types can be designated by the cell type or organ affected, such as neurotoxin, leukotoxin, hepatotoxin and cardiotoxin. Exotoxins can also be named after their mode of action, that is lecithinase produced by *Cl. perfringens*. However, exotoxins can also be named after the producer organism, or the disease which they cause. Two examples are as follows: (1) cholera toxin causes cholera and is produced by *Vibrio cholerae* and (2) Shiga-toxin causing bacterial dysentery is produced by *Shigella* spp. Some toxins have more than one name, such as the Shiga-like toxin of *E. coli* O157:H7 which is also called a verotoxin since it affects the mammalian Vero cell culture line. This confusion is reflected by the name changes for VTEC and STEC pathogenic *E. coli* (Section 4.2.2).

Exotoxins can be divided into four groups (Henderson *et al.* 1999):

1 Those that act at the cell membrane.

2 Those that attack the membrane.

3 Those that penetrate the membrane to act inside the cell.

4 Those that are directly transported from the bacterium into the target eucaryotic cell by a secretion system.

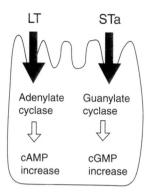

Figure 2.4 Mode of action of LT and ST toxins produced by pathogenic strains of *E. coli*.

Toxins that act at the cell membrane

Although proteins are normally sensitive to high temperatures (e.g. 100°C; LT toxins), there are a few small exotoxins that are heat stable (ST toxins). The heat-stable toxin of *E. coli* is a peptide of 18–19 amino acids with three disulphide bonds. This structure is responsible for the toxin's resistance to heat denaturation. The toxin causes diarrhoea since it binds to the endogenous ligand guanylin, a peptide hormone that regulates salt and water homeostasis in the gut and kidney (Figure 2.4). The *B. cereus* emetic toxin (also called cereulide) is a ring of three repeats of four amino and/or oxy acids '-D-O-leucine-D-alanine-O-valine-L-valine-'. It is plausible that this toxin binds to the 5-HT$_3$ receptor to stimulate the vagus afferent nerve, leading to vomiting; see Section 4.3.10 for more detail.

Superantigens are toxins which act by stimulating T cells to release cytokines and so induce an inappropriate immune response. Superantigens are produced by *St. aureus* and *Strep. pyogenes*, and lack an AB subunit structure. One of the most studied is TSST-1 (~22 kDa) from *St. aureus* which causes toxic shock syndrome (associated with the use of tampons). *St. aureus* also produces the enteroxins A–E (~27 kDa) which are responsible for food poisoning (see Section 4.3.7 for more detail).

Membrane-damaging toxins

These toxins have A and B subunits which do not separate and act by disrupting the host cell membranes, that is listeriolysin O. This group is subdivided into channel-forming (pore-forming) toxins which allow cytoplasmic contents to leak out of the host cell. The thiol-activated cholesterol-binding cytolysins (52–60 kDa) bind to the cholesterol of the host cell to form pores of about 30–40 nm. Examples include listeriolysin O (*L. monocytogenes*), perfringolysin O (*Cl. perfringens*) and cereolysin O (*B. cereus*). These may sometimes be called 'haemolysins' due to the common use of blood cells to detect the presence of a toxin. In the past, they have also been termed 'oxygen labile' and 'sulphydryl activated' toxins, though these terms are now recognised as misleading. The second type is phospholipase (PLC) which removes the charged head group from the lipid portion of phospholipids. A number of both Gram-positive and Gram-negative bacteria produce PLCs. *L. monocytogenes* produces a phophoinositol-specific PLC which enables the organism to escape the vacuole after engulfment and a broad-range PLC involved in cell-to-cell spread; see Section 4.3.5 for more detail. *Cl. perfringens* produces phospholipase C also known as α-toxin which has necrotic and cytolytic activity (Titbull *et al.* 1999). *St. aureus* produces a β-haemolysin.

Membrane penetrating toxins

The membrane penetrating toxins have a so-called AB structure, where A is the catalytic subunit and the B subunit binds to the host cell receptor. There are two common types of AB toxins:

1 A and B are joined in one molecule, though the structure can be disrupted to yield two subunits joined by a disulphide bridge. Examples include botulinum toxins types A–F and tetanus toxin (Figure 2.5). These are large (>150 kDa) single-chain toxins that are proteolytically cleaved during activation and cellular entry. The botulinum toxin blocks the action of peripheral nerves (Figure 2.6). See Section 4.3.9 for more detail.

2 AB$_5$ where five B subunits (pentamer) form a doughnut-like ring structure and possibly the A subunit enters the cell through the central hole. Examples include cholera toxin, *E. coli* heat-labile toxin and the Shiga toxins (Figure 2.7). Cholera and Shiga toxins bind to glycolipid ganglioside receptors on the host cell. The Shiga toxins attack 28S rRNA to depurinate adenine 4324 which is involved in elongation factor 1 mediated binding of tRNA to the ribosomal complex and subsequently inhibits protein synthesis. The A subunit of Shiga toxin has sequence and structural homology with the ricin family of plant toxins which have an identical mode of action. The A subunit is activated by proteolytic cleavage on cell entry to yield two fragments joined by a disulphide bridge which is subsequently reduced. Cholera toxin and *E. coli* heat-labile (LT) toxin ADP-ribosylate G$_3$, a heterotrimeric G protein involved in stimulation of adenylate cyclase (Figure 2.8). The A subunit of G$_s$ is modified at Arg201 which inhibits the GTPase activity. This keeps the G$_s$ protein in the 'on' position, leading to permanent activation of adenylate cyclase. The resultant high concentration of cAMP in the gut epithelial cells causes massive fluid accumulation in the gut lumen and watery diarrhoea which can be fatal. Cholera toxin is chromosomally encoded on a phage, whereas *E. coli* LT and ST toxins are plasmid encoded, and Shiga toxin is chromosomally encoded.

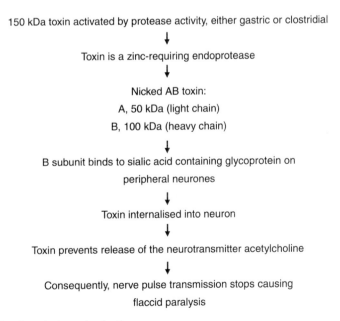

150 kDa toxin activated by protease activity, either gastric or clostridial

↓

Toxin is a zinc-requiring endoprotease

↓

Nicked AB toxin:

A, 50 kDa (light chain)

B, 100 kDa (heavy chain)

↓

B subunit binds to sialic acid containing glycoprotein on

peripheral neurones

↓

Toxin internalised into neuron

↓

Toxin prevents release of the neurotransmitter acetylcholine

↓

Consequently, nerve pulse transmission stops causing

flaccid paralysis

Figure 2.5 *Cl. botulinum* toxin mode of action.

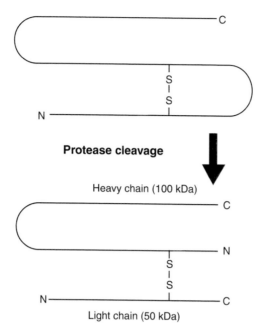

Figure 2.6 Structure and activation of *Cl. botulinum* toxin.

Transported toxins

Toxins in this group do not display toxic activity when purified from bacteria. They are transported directly from the bacterial cytoplasm to the eucaryotic cytoplasm by a complex array of proteins that bridge the membranes of the two cell types. An example is the Type III secretion system of *Salmonella* and *E. coli* O157. This secretion system is often encoded on pathogenicity islands, described in the next section. Bacteria known to produce such toxins include *Salmonella, E. coli, Shigella* and *Yersinia* spp.

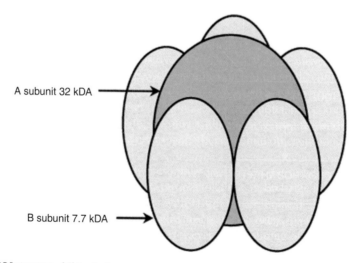

Figure 2.7 AB5 structure of Shiga toxin.

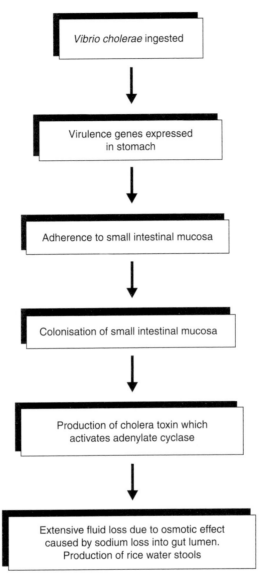

Figure 2.8 Mode of action of cholera toxin.

2.3.2 Pathogenicity islands

Pathogenicity islands (PAIs) are large (>30 kb) distinct chromosomal elements encoding virulence-associated genes (Table 2.2). Because the %GC content of PAIs is often distinct from the rest of the bacterial DNA %GC content, it has been proposed that they may have been acquired in the past by horizontal gene transfer from other bacterial species. They are often found at tRNA loci. In addition to PAIs, there are shorter sequences of DNA, termed 'pathogenicity islets' which are transferred between bacterial pathogens. PAIs constitute major routes in the evolution of bacterial pathogens since in one acquisition they can transform a benign organism into a virulent one.

Table 2.2 Pathogenicity islands of three important foodborne pathogens

Pathogen	PAI designation	Size (kb)	G+C ratio (PAI/host)	Phenotype
E. coli	Pai I	70	40/51	Haemolysin production
	Pai II	190	40/51	Haemolysin, P-fimbriae production
	LEE (Pai III)	35	39/51	Induction of attaching and effacing lesions on enterocytes
S. Typhimurium	SPI-1	40	42/52	Invasion of non-phagocytic cells
	SPI-2	40	45/52	Survival in macrophages
	SPI-3	17	Unknown	Survival in macrophages
V. cholerae	VPI	39.5	35/46	Colonisation, expression of phage CTXΦ receptor

PAI, pathogenicity island.
Adapted from Henderson *et al.* (1999).

Both enterohaemorrhagic *E. coli* (EHEC) and enteropathogenic *E. coli* (EPEC) contain the Locus of Enterocyte Effacement (LEE) island. This encodes for a Type III secretion system and other virulence factors which are essential for disease. LEE is a PAI (35 kb) of EPEC strains. It induces the attaching and effacing lesions on enterocytes and encodes the secretion system for toxin transfer from the *E. coli* cell to the host cell. This results in cytoskeletal rearrangements and the formation of a pedestal on which the *E. coli* cell is located (cf. Section 4.3.3). Uropathogenic *E. coli* cause urinary tract infections. These have a completely different PAI inserted in exactly the same site as EPEC strains. The PAI encodes for P fimbriae (an adhesin) and haemolysin (a toxin), which together are virulence factors required for urinary tract colonisation.

Although *Yersinia* and *Shigella* species have Type III systems on their plasmids, they have other virulence attributes that are on the chromosome and not localised in islands.

V. cholerae produces cholera toxin encoded by the *ctx*A and *ctx*B genes that are encoded by the filamentous phage CTX. The bacterial receptor for phage infection is the toxin co-regulated pilus (TCP) that is also an important adherence determinant. TCP is found on the 39.5 kb PAI called VPI in *V. cholerae*. It is believed that the acquisition of VPI enables aquatic *V. cholerae* strains to colonise the human intestine and the subsequent generation of epidemic and pandemic *V. cholerae* strains. A possible Gram-positive PAI is found in pathogenic strains of *L. monocytogenes*. The 10 kb element encodes for genes required for listeriolysin O (*hly*), *act*A and *plc*B (responsible for intra- and intercellular movement).

PAIs are covered in more detail with the specific pathogens *E. coli* (Section 4.3.3) and *Salmonella* (Section 4.3.2). Locating PAI regions on bacterial genomes using bioinformatics is explained in Section 2.9.4.

2.3.3 Bacterial toxins encoded in bacteriophages

Bacteriophages are involved in the transfer of virulence factors between pathogens. For example, the Shiga toxin of *Sh. dysenteriae* is encoded on a bacteriophage which is integrated in the chromosome. *E. coli* O157:H7 causes haemorrhagic colitis and haemolytic uraemic syndrome and contains the Shiga-toxin genes. It is plausible that a bacteriophage encoding the Shiga toxin from *Shigella dysenteriae* was transferred to an EPEC strain and this led to the generation of a new pathogen, EHEC. Under certain circumstances, the phage encoding the Shiga toxin becomes lytic, multiples and subsequently releases the toxin. Another example of

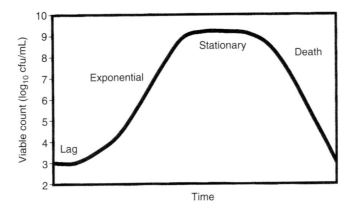

Figure 2.9 The microbial growth curve.

bacteriophage-encoded toxin is in *V. cholerae* which produces cholera toxin encoded by the *ctx*A and *ctx*B genes. These are encoded by the filamentous phage CTX. The bacterial receptor for phage infection is the TCP that is also an important adherence determinant. Consequently, the phage only infects bacteria that already possess an essential adhesin, thus ensuring virulence.

2.4 Microbial growth cycle

Microbial growth cycle is composed of six phases (see Figure 2.9):

1 *Lag phase*: Cells are not multiplying, but are synthesising enzymes appropriate for the environment. Lag times are more variable than growth rate possibly due to the effects of the physiological history of the cell and the environment.

2 *Acceleration phase*: An increasing proportion of the cells are multiplying.

3 *Exponential (or log) phase*: The cell population is multiplying by binary fission (1-2-4-8-16-32-64, etc.). The cell numbers are increasing at such a rate that to graphically represent them it is best to use exponential values (logarithms). This results in a straight line, the slope of which represents the μ_{max} (rate of maximum growth) and the doubling time 't_d' (time required for the cell mass to increase twofold) can be determined.

4 *Deceleration phase*: An increasing proportion of cells are no longer multiplying.

5 *Stationary phase*: The rate of growth equals the rate of death, resulting in equal numbers of cells at any given time. Death is due to the exhaustion of nutrients, the accumulation of toxic end products and/or other changes in the environment, such as pH changes.

   The length of the stationary phase is dependent upon a number of factors such as the organism and environmental conditions (temperature, etc.). Spore-forming organisms will develop spores due to the stress conditions.

6 *Death phase*: The number of cells dying is greater than the number of cells growing. Cells which form spores will survive longer than non-spore formers.

   The length of each phase is dependent upon the organism and the growth environment, temperature, pH, water activity, and so on. The growth cycle can be modelled using sophisticated computer programmes and leads to the area of microbial modelling and predictive microbiology; see Section 2.8.

2.5 Death kinetics

2.5.1 Expressions
There are a number of expressions used to describe microbial death:
- *D value*: Decimal reduction time (*D* value) is defined as the time at any given temperature for a 90% reduction (= 1 log value) in viability to be effected.
- *Z value*: It is defined as the temperature increase required to increase the death rate tenfold, or in other words reduce the *D* value tenfold.
- *P value*: It refers to the heating period at 70°C. Cooking for 2 minutes at 70°C will kill almost all vegetative bacteria. For a shelf-life of 3 months, the *P* value should be 30–60 minutes according to other risk factors.
- *F value*: This value is the equivalent time, in minutes at 250°F (121°C), of all heat considered, with respect to its capacity to destroy spores or vegetative cells of a particular organism.

Since these values are mathematically derived, they can be used in predictive microbiology (Section 2.7) and microbiological risk assessment (Chapter 9).

2.5.2 Decimal reduction times (*D* values) and *Z* values
To design an effective temperature treatment regime, it is imperative to have an understanding of the effects of heat on micro-organisms. The thermal destruction of micro-organisms (death kinetics of vegetative cells and spores) can be expressed logarithmically. In other words, for any specific organism, in a specific substrate and at a specific temperature, there is a certain time required to destroy 90% (= 1 log reduction) of the organism. This is the decimal reduction time (*D* value). Plotting the survival numbers (as $\log_{10}$ cfu/mL) for an organism against time generally gives a straight-line relationship which is more precisely known as a log-linear relationship; Figure 2.10.

The rate of death depends upon the organism, including the ability to form spores, and the environment (Table 2.3). Free (or planktonic) vegetative cells are more sensitive to detergents than fixed cells (i.e. biofilms). The heat sensitivity of an organism at any given temperature varies according to the suspending medium. For example, the presence of acids and nitrite will increase the death rate, whereas the presence of fat may decrease it. The *D* value is also dependent upon the inoculum preparation and the enumeration conditions. This has been demonstrated for *E. coli* O157:H7 and is summarised in Figure 2.11 (Stringer *et al.* 2000). Hence, *D* values quoted in books and journals cannot be taken as fixed values and directly applied to processes.

Although most frequently applied to thermal death rates, *D* values can also be used to express the rate of death due to other lethal effects such as acid and irradiation.

Note that it is questionable whether the first-order death kinetics (log-linear) with temperature dependence is always appropriate for survival curve calculations. Although the log-linear relationship has been the standard approach to thermal susceptibility determination, the straight-line relationship does not always occur when plotting experimental data. Complications arise because the microbial cell can become non-viable due to various reasons: disruption of the cell envelope, protein denaturation or nucleic acid damage (Oliver 2005). Some people have reported shoulders and tails when plotting log (N/N_0) survival curves (Geeraerd *et al.* 2005). This may be due to reasons such as cell clumping, mixed populations with varying thermal sensitivities, changes in resistance during treatment or inactivation of a number of essential loci. Various mathematical models are available to address this issue, the derivation of which is outside the scope of this book. However, a useful Excel™ add-on called 'GInaFIT' is available as freeware (see Web Resources section in the Appendix section) which enables the

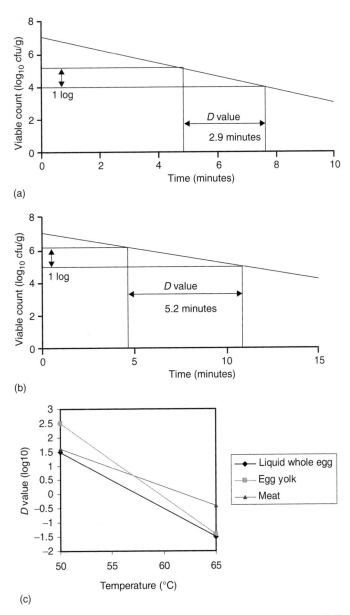

Figure 2.10 (a) Death rate of *E. coli* O157:H7 in beef at 60°C. (b) Death rate of *St. aureus* in liquid egg at 60°C. (c) *D* value for *S.* Enteritidis in eggs and meat (FSIS 1998, Fazil *et al.* 2000).

user to try nine different mathematical fits to their data from within Excel (Geeraerd *et al.* 2005). The models are (i) classical log-linear curves, (ii) curves with a shoulder, (iii) curves showing tailing, (iv) survival curves with both shoulder and tailing, (v) concave curves, (vi) convex curves, (vii) convex/concave curves followed by tailing, (viii) biphasic inactivation kinetics and (ix) biphasic inactivation kinetics preceded by a shoulder.

Table 2.3 Variation in microbial heat resistance of micro-organisms according to test conditions

Organism	Medium	pH	Temperature (°C)[a]	D value (min)	Z value
A. hydrophila	Saline		51.0	8.08–122.8	5.22–7.69
Brucella spp.	—	—	65.5	0.1–0.2	—
B. cereus	—	—	100	5.0	6.9
(spores)			100	2.7–3.1	6.1
(toxin destruction; diarrhoeal/emetic)			56.1/121	5/stable	
B. coagulans	Buffer	4.5	110	0.064–1.46	—
	Red pepper	4.5	110	5.5	—
B. licheniformis	Buffer	7.0	110	0.27	—
	Buffer	4.0	110	0.12	—
B. stearothermophilus			120	4.0–5.0	10
B. subtilis	—	—	100	11.0	
C. jejuni	Buffer	7.0	50	0.88–1.63	6.0–6.4
	Beef	—	50	5.9–6.3	
Cl. butyricum	—	—	100	0.1–0.5	
Cl. perfringens	—	—	100	0.3–20.0	3.8
(spores)			98.9	26–31 minutes	7.2
(toxin)	—	—	90	4	5.5
Cl. botulinum					
(type A and B proteolytic strain's spores	—	—	121.1	0.21	10
type E and non-proteolytic types B and F)			82	0.49–0.74	5.6–10.7
(toxin destruction)			85	2.0	4.0–6.2
Cl. thermosaccharolyticum	—	—	120	3–4	7.2–10
D. nigrificans	—	—	120	2–3	
E. coli O157:H7	Beef		62.8	0.47	4.65
	Apple juice				
		3.6	58	1.0	4.8
		4.5	58	2.5	4.8
	Growth, 23°C		58	1.6	4.8
	Growth, 37°C		58	5.0	4.8
L. monocytogenes	Beef		62.0	2.9–4.2	5.98
S. Enteritidis	Liquid whole egg		62.8	0.06	3.30
S. Senftenberg	Beef		62.0	2.65	5.91
St. aureus	—	—	65.5	0.2–2.0	4.8–5.4
(toxin destruction)			98.9	>2 hours	(approximately 27.8)
Streptococcus Group D	Cured meat		70	2.95	10
Y. enterocolitica	Saline		60.0	0.4–0.51	4.0–5.2
Sac. cerevisiae	Saline	4.5	60	22.5[b]	5.5
Z. bailii	Saline	4.5	60	0.4[c]	3.9
	Saline	4.5	60	14.2[b]	—

[a]To convert to °F, use the equation °F = (9/5)°C + 32. As a guidance: 0°C = 32°F, 4.4°C = 40°F, 60°C = 140°F.
[b]Ascospores.
[c]Vegetative cells.
Various sources, including Mortimore & Wallace 1994; Borche *et al.* 1996, and ICMSF 1996a.

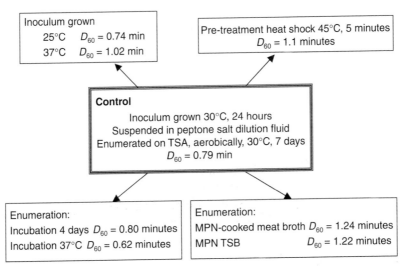

Figure 2.11 Changes in *E. coli* O157:H7 *D* value at 60°C with inoculum and recovery conditions (adapted from Stringer *et al.* 2000).

Plotting the *D* value against temperature can be used to determine the change in temperature required to obtain a tenfold increase (or decrease) in the *D* value. This coefficient is called the *Z* value (Figure 2.12). The integrated lethal value of heat received by all points in a container during processing is designated as F_s or F_o. This represents a measure of the capacity of a heat process to reduce the number of spores or vegetative cells of a given organism per container. When we assume instant heating and cooling throughout the container of spores, vegetative cells or food, F_o may be derived as follows:

$$F_o = D_r(\log a - \log b)$$

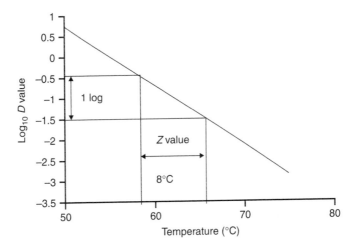

Figure 2.12 *C. jejuni Z* value in lamb cubes.

Table 2.4 Effect of cooking temperature on the survival of E. coli O157:H7

Temperature (°C)	Time for viable count to be reduced by 6 log cycles[a] (min)
58.2	28.2
62.8	2.8
67.5	0.28

[a] For example 10^7 cfu/g to 10 cfu/g.
Data taken from Table 2.3; $D_{62.8} = 0.4$, $Z = 4.65$.

where $a =$ number of cells in the initial population, and $b =$ number of cells in the final population.

See Table 2.4 for an example of the effect of cooking temperature on the survival of E. coli O157:H7 in beef using D and Z values.

12-D concept

The 12-D concept refers to the process lethality requirement which has been used for a long time in the canning industry. It implies that the minimum heat process should reduce the probability of survival of the most resistant Cl. botulinum spores to 10^{-12}. Since Cl. botulinum spores do not germinate and produce toxin below pH 4.6, this concept is observed only for foods above this pH value.

$$F_o = D_r(\log a - \log b)$$
$$F_o = 0.21(\log 1 - \log 10^{-12})$$
$$F_o = 0.21 \times 12 = 2.52$$

Processing for 2.52 minutes at 250°F (121°C) should reduce the Cl. botulinum spores to one spore in 1 million containers (10^{12}). When it is considered that some flat-sour spores have D_r values of about 4.0 and some canned foods receive F_o treatments of 6.0–8.0, the potential number of Cl. botulinum spores is reduced even more (see Table 2.3).

Application of D value concept in pasteurisation

Pasteurisation treatment aims to reduce the number of pathogenic-spoilage organisms by a set amount (frequently a 6 log reduction) and to ensure that the product formulation and storage conditions inhibit the growth of any surviving cells during the intended shelf-life of the product. For more detail, see Section 3.6.1. For example, if the D value of a target organism is 0.5 minutes at 70°C and the Z value was 5°C, a process of 3 minutes at 70°C would give a 6 log reduction. At 75°C, the organism would die ten times faster; hence, 0.3 minutes (20 seconds) at 75°C is an equivalent treatment giving a 6 log reduction.

Since the determination of actual Z values for each organism and product would require considerable work, assumed appropriate values are often used. Although this may be satisfactory for general quality assurance purposes, it is important to establish accurate values for the most appropriate micro-organism present in the product.

Table 2.5 Methods of food preservation (ICMSF 1988)

Operation	Intended effect
Cleaning, washing	Reduces microbial load
Cold storage (below 8°C)	Prevents the growth of most pathogenic bacteria; slows the growth of spoilage microbes
Freezing (below −10°C)	Prevents growth of all microbes
Pasteurising (60–80°C)	Kills most non-sporing bacteria, yeast and moulds
Blanching (95–110°C)	Kills surface vegetative bacteria, yeast and moulds
Canning (above 100°C)	'Commercially sterilises' food; kills all pathogenic bacteria
Drying	Stops growth of all microbes when $a_w = 0.60$
Salting	Stops growth of most microbes at $ca.$ 10% salt
Syruping (sugars)	Halts growth when $a_w = 0.70$
Acidifying	Halts growth of most bacteria (effects depend on acid type)

a_w denotes water activity.

2.6 Factors affecting microbial growth

Traditional ways to control microbial spoilage and safety hazards in foods are summarised in Table 2.5. How these are used in methods of food preservation are considered in more detail in Section 3.5. In order to design adequate treatment processes, an understanding of the factors affecting microbial growth is necessary. To complement this, Table 2.6 lists the common faults during processing which result in the growth of foodborne pathogens or toxin production.

2.6.1 Intrinsic and extrinsic factors affecting microbial growth

Food is a chemically complex matrix, and predicting whether, or how fast, micro-organisms will grow in any given food is difficult. Most foods contain sufficient nutrients to support microbial growth. Several factors encourage, prevent or limit the growth of micro-organisms in foods; the most important are a_w, pH and temperature.

A useful acronym is 'FATTOM', defined as follows:

Food refers to the nutritional content

Acidity is the pH inhibitory to microbial growth

Time and

Temperature are together and can refer, for example, to cooking time and temperature

Table 2.6 Most common faults in handling food which allow growth or formation of toxins by micro-organisms

Temperature abuse	Other process parameters
No or not sufficient chilling	Not properly controlled
Inadequate heat processing	a_w
temperature too low	pH
process time too short	NO_2^-, NO_3^- and concentration of
holding time too long at $\leq 65°C$	other preservatives not adequate

Reprinted with permission from Sinell (1995) International Journal of Food Microbiology, Vol. 25, Issue 3, p. 209, Copyright 1995, with permission from Elsevier.

Table 2.7 Intrinsic and extrinsic parameters affecting microbial growth

Intrinsic parameters	Extrinsic parameters
Water activity, humectant identity	Temperature
Oxygen availability	Relative humidity
pH, acidity, acidulant identity	Atmosphere composition
Buffering capacity	Packaging
Available nutrients	
Natural antimicrobial substances	
Presence and identity of natural microbial flora	
Colloidal form	

Oxygen presence and the retardation of microbial growth under modified atmosphere and vacuum packaging.

Moisture refers to the amount of available water.

Factors affecting microbial growth are divided into two groups: intrinsic and extrinsic parameters (Table 2.7).

2.6.2 Water activity

When other substances (solutes) are added to water, water molecules orient themselves on the surface of the solute and the properties of the solution change dramatically. The microbial cell must compete with solute molecules for free water molecules. Except for *St. aureus*, bacteria are rather poor competitors for free water, whereas moulds are excellent competitors (Chirife & del Pilarbuera 1996).

Water activity (a_w) is a measure of the available water in a sample. The a_w is the ratio of the water vapour pressure of the sample to that of pure water at the same temperature:

$$a_w = \frac{\text{Water vapour pressure of sample}}{\text{Pure water vapour pressure}}$$

A solution of pure water has an a_w of 1.00. The addition of solute decreases the a_w to less than 1.00. a_w varies very little with temperature over the range of temperatures that support microbial growth.

The a_w of a solution may dramatically affect the ability of heat to kill a bacterium at a given temperature. For example, a population of S. Typhimurium is reduced tenfold in 0.18 minutes at 60°C if the a_w of the suspending medium is 0.995. If the a_w is lowered to 0.94, 4.3 minutes are required at 60°C to cause the same tenfold reduction.

An a_w value stated for a bacterium is generally the minimum a_w which supports growth. At the minimum a_w, growth is usually minimal, increasing as the a_w increases. At a_w values below the minimum for growth, bacteria do not necessarily die. The bacteria may remain dormant, but infectious. Most importantly, a_w is only one factor, and the other factors (e.g. pH and temperature) of the food must be considered. It is the interplay between factors that ultimately determines whether a bacterium will grow or not. The a_w of a food may not be a fixed value; it may change over time, or may vary considerably between similar foods from different sources. The minimum water activity supporting growth for various micro-organisms is given in Table 2.8.

Table 2.8 Limits of microbial growth; water activity (a_w), pH and temperature

Organism	Minimal water activity (a_w)	pH range	Temperature range (°C)[a]	Growth rate[b] (t_d)
A. hydrophila	0.970	(7.2 optimum)	−0.1–42	12 hours, 4°C
B. cereus	0.930	4.3–9.3	4–52	4 hours/generation, 8°C
B. stearothermophilus	—	5.2–9.2	28–72	
C. jejuni	0.990	4.9–9.5	30–45	6 hours/generation, 32°C
Cl. botulinum type A and proteolytic B and F	0.935	4.6–9.0	10–48	(8 days, 10°C)[c]
Cl. botulinum type E and non-proteolytic B and F	0.965	5.0–9.0	3.3–45	(8 days, 10°C)[c]
Cl. perfringens	0.945	5.0–9.0	10–52	12 hours, 12°C
E. coli	0.935	4.0–9.0	7–49.4	25 hours/generation, 8°C
Lactobacillus spp.	0.930	3.8–7.2	5–45	
L. monocytogenes	0.920	4.4–9.4	−0.4–45	I day, 4.4°C
Salmonella spp.	0.940	3.7–9.5	5–46	(60 hours),[d] 10 hours, 10°C
Shigella spp.	0.960	4.8–9.3	6.1–47.1	(3.6 days, 8°C)[e]
St. aureus	0.830	4.0–10	7–50	(2.8 days),[d] I day, 10°C
(toxin production)	0.850	4.0–9.8	10–48	
V. cholerae	0.970	5.0–10.0	10–43	(4 hours),[d] 98 minutes, 20°C
V. parahaemolyticus	0.936	4.8–11	5–44	60 minutes, 18°C
V. vulnificus	0.960	5.0–10	8–43	
Y. enterocolitica	0.945	4.2–10	−1.3–45	17 hours, 5°C
Saccharomyces spp.	0.85	2.1–9.0	—	
Asp. oryzae	0.77	1.6–13.0	10–43	
F. miniliforme	0.87	<2.5–<10.6	2.5–37	
Pen. verrucosum	0.79	<2.1–<10.0	0–31	

[a]To convert to °F, use the equation: $°F = (9/5)°C + 32$. As a guidance: $0°C = 32°F$, $4.4°C = 40°F$, $60°C = 140°F$.

[b]These are only examples of doubling time (td). The values will vary according to food composition.

[c]Time to toxin production.

[d]Lag time. Average time to turbidity (inoculation from 1:1 dilution of 5 hours, 37°C culture).

[e] Various sources used (principally ICMSF 1996a, Corlett 1998, Mortimore & Wallace 1994, where data have differed between sources and the wider growth range has been quoted).

The water activity has been widely used as the preservation factor by the addition of salt and sugar. Sugar has been traditionally used in the preservation of fruit products (jams and preserves). In contrast, salt has been used for the preservation of meat and fish. The water activity of a range of foods is given in Table 2.9.

2.6.3 pH

The pH range of a micro-organism is defined by a minimum value (at the acidic end of the scale) and a maximum value (at the basic end of the scale). There is a pH optimum for each micro-organism at which growth is maximal (Table 2.8). Moving away from the pH optimum in either direction slows microbial growth. A range of food pH values is given in Table 2.10. Shifts in pH of a food with time may reflect microbial activity, and foods that are poorly buffered (i.e. do not resist changes in pH), such as vegetables, may shift pH values considerably. For meats, the pH of muscle from a rested animal may differ from that of a fatigued animal.

Table 2.9 Water activity of various foods

Water activity (a_w) range	Foods
1.00–0.95	Highly perishable foods: meat, vegetables, fish, milk Fresh and canned fruit Cooked sausages and breads Foods containing up to 40% sucrose or 7% NaCl
0.95–0.91	Some cheeses (e.g. Cheddar) Cured meat (e.g. ham) Some fruit juice concentrates Foods containing 55% sucrose or 12% NaCl
0.91–0.87	Fermented sausages (salami), sponge cakes, dry cheeses, margarine Foods containing 65% sucrose or 15% NaCl
0.87–0.80	Most fruit juice concentrates, sweetened condensed milk, chocolate syrup, maple and fruit syrups, flour, rice, pulses containing 15–17% moisture, fruit cake
0.80–0.75	Jam, marmalade, marzipan, glace fruits
0.75–0.65	Rolled oats, fudge, jelly, raw cane sugar, nuts
0.65–0.60	Dried fruits, some toffees and caramels, honey
0.60–0.50	Noodles, spaghetti
0.50–0.40	Whole egg powder
0.40–0.30	Cookies, crackers
0.30–0.20	Whole milk powder, dried vegetables, cornflakes

Table 2.10 pH values of various foods (ICMSF 1988)

pH range	Food	pH
Low acid (pH 7.0–5.5)	Whole eggs	7.1–7.9
	Frozen eggs	8.5–9.5
	Milk	6.3–8.5
	Camembert cheese	7.44
	Cheddar cheese	5.9
	Roquefort cheese	5.5–5.9
	Bacon	6.6–5.6
	Carcass meat	7.0–5.4
	Red meat	6.2–5.4
	Ham	5.9–6.1
	Canned vegetables	6.4–5.4
	Poultry	5.6–6.4
	Fish	6.6–6.8
	Crustaceans	6.8–7.0
	Milk	6.3–6.5
	Butter	6.1–6.4
	Potatoes	5.6–6.2
	Rice	6.0–6.7
	Bread	5.3–5.8
Medium acid (pH 5.3–4.5)	Fermented vegetables	5.1–3.9
	Cottage cheese	4.5
	Bananas	4.5–5.2
	Green beans	4.6–5.5
Acid (pH 4.5–3.7)	Mayonnaise	4.1–3.0
	Tomatoes	4.0
High acid (<pH 3.7)	Canned pickles and fruit juice	3.9–3.5
	Sauerkraut	3.3–3.1
	Citrus fruits	3.5–3.0
	Apples	2.9–3.3

Foods in parentheses are outside the pH range, but are included for comparison.

Table 2.11 Grouping of micro-organisms according to temperature growth range

Group	Minimum (°C)	Optimum (°C)	Maximum (°C)
Psychrophiles	−5	12–15	20
Mesophiles	5	30–45	47
Thermophiles	40	55–57	60–90

Note: To convert to °F, use the equation: $°F = (9/5)°C + 32$. As a guidance: $0°C = 32°F$, $4.4°C = 40°F$, $60°C = 140°F$.

A food may start with a pH which precludes bacterial growth, but as a result of the metabolism of other microbes (yeasts or moulds), pH shifts may occur and permit bacterial growth.

2.6.4 Temperature

Temperature values for microbial growth, such as pH values, have a minimum and maximum range with an optimum temperature for maximal growth. The optimum growth temperature determines its classification as a thermophile, mesophile or psychrophile (Table 2.11). A thermophile cannot grow at room temperature and therefore canned foods can be stored at room temperature even though they may contain thermophilic micro-organisms which survive the high processing temperature.

2.6.5 Interplay of factors affecting microbial growth in foods

Although each of the major factors listed above plays an important role, the interplay between the factors ultimately determines whether a micro-organism will grow in a given food. Often, the results of such interplay are unpredictable, as poorly understood synergism or antagonism may occur. Advantage is taken of this interplay with regard to preventing the outgrowth of *Cl. botulinum*. Food with a pH of 5.0 (within the range for *Cl. botulinum*) and an a_w of 0.935 (above the minimum for *Cl. botulinum*) may not support the growth of this bacterium. Certain processed cheese spreads take advantage of this fact and are therefore shelf stable at room temperature even though each individual factor would permit the outgrowth of *Cl. botulinum*.

Therefore, predictions about whether a particular micro-organism will grow in a food can, in general, only be made through experimentation. Also, many micro-organisms do not need to multiply in food to cause disease.

2.7 Microbial response to stress

The common means of food preservation (Table 2.5) are being expanded by new techniques such as:
- Mild heating
- Modified atmosphere and vacuum packaging
- Inclusion of natural antimicrobial agents
- High hydrostatic pressure
- Pulse electric field and high-intensity laser

The development of these techniques is partially due to consumer demand for less processed, less heavily preserved yet of higher quality foods; see Section 3.6. They are 'milder' than the traditional methods and commonly rely on storage and distribution at

Table 2.12 Food poisoning micro-organisms of concern in minimally processed foods

Minimum growth temperature (°C)	Heat resistance	
	Low[a]	High[b]
0–5	L. monocytogenes Y. enterocolitica A. hydrophila	Cl. botulinum type E and non-proteolytic type B B. cereus B. subtilis B. licheniformis
5–10	Salmonella spp. V. parahaemolyticus Pathogenic strains of E. coli St. aureus	
10–15		Cl. botulinum type A and proteolytic type B Cl. perfringens

[a]Organisms undergo a 6 log kill following heat treatment at 70°C for 2 minutes.
[b]Organisms require heat treatment at 90°C or above to destroy spores.
Reprinted from Abee & Wouters (1999), with permission from Elsevier.
Note: To convert to °F, use the equation °F = (9/5)°C + 32. As a guidance: 0°C = 32°F, 4.4°C = 40°F, 60°C = 140°F.

refrigeration temperatures for their preservation. These are commonly called 'minimal processed foods'.

The micro-organisms of concern in minimal processed foods are psychrotrophic and mesophilic micro-organisms. The psychrotrophic organisms, by definition, can grow at refrigeration temperatures, whereas mesophilic pathogens can survive under refrigeration and grow during periods of temperature abuse; see Table 2.12.

Micro-organisms are able, within limits, to adapt to stress conditions such as thermal stress, pH stress, osmotic shock, oxidative stress and nutrient limitation (Table 2.13). The mechanism of adaptation is by signal transduction systems which control the coordinated expression of genes involved in cellular defence mechanisms (Huisman & Kolter 1994, Kleerebezem et al. 1997). Microbial cells are able to adapt to many processes used to retard microbial growth starvation. These include cold shock, heat shock, (weak) acids, high osmolarity and high hydrostatic pressure. The easiest mechanism of survival to recognise is in Bacillus and Clostridium spp. These organisms form spores under stress conditions which can germinate under favourable conditions later. Other organisms, such as E. coli, undergo significant physiological changes to enable the cell to survive environmental stresses such as starvation, near-UV radiation, hydrogen peroxide, heat and high salt. As already shown, due to adaptation, E. coli O157:H7 D_{60} value is dependent upon the inoculum preparation and the enumeration conditions (Figure 2.11).

The regulatory mechanism involves the modification of sigma (δ) factors whose primary role is to bind to core RNA polymerase conferring promoter specificity. The sigma factors of B. subtilis and E. coli have been extensively studied. The main sigma factor is responsible for the transcription from the majority of the promoters. Alternative sigma factors have different promoter specificities, directing expression of specific regulons involved in heat-shock response, the chemotactic response, sporulation and general stress response (Abee & Wouters 1999, Haldenwang 1995).

Table 2.13 Response mechanisms in micro-organisms

Environmental stress	Stress response reaction
Low nutrient levels	Nutrient scavenging, oligotrophy, generation of viable non-culturable forms
Low pH, presence of weak organic acids	Extrusion of hydrogen ions, maintenance of cytoplasmic pH and membrane pH gradient
Reduced water activity	Osmoregulation, avoidance of water loss, maintenance of membrane turgor
Low temperature – growth	Membrane lipid changes, cold-shock response
High temperature – growth	Membrane lipid changes, heat-shock response
High oxygen levels	Enzymic protection from oxygen-derived free radicals
Biocides and preservatives	Phenotypic adaptation and development of resistance
Ultraviolet radiation	Excision of thymine dimers and repair of DNA
Ionising radiation	Repair of DNA single-strand breaks
High temperature – survival	Low water content in the spore protoplast
High hydrostatic pressure – survival	Low spore protoplast water content
High-voltage electric discharge	Low conductivity of spore protoplast
Ultrasonication	Structural rigidity of cell wall
High levels of biocides	Impermeable outer layers of cells
Competition	Formation of biofilms, aggregates with some degree of symbiosis

Reprinted from Gould (1996), with permission from Elsevier.

2.7.1 General stress response (GSR)

Bacterial responses to stress can be either general or specific. The specific responses to acid, heat, cold and osmotic stress will be covered in the next sections. The GSR regulon is a group of genes concerned with growth and survival under various stress conditions. In *E. coli*, there are more than 50 genes involved in the GSR, and these are coordinately regulated by the sigma factor σ^s which is encoded by the *rpoS* gene. This is an alternative sigma subunit of RNA polymerase which accumulates (due to decreased degradation) in the cell during stress exposure, and activates the genes which express the GSR. This response regulator controls the stability of the protein and hence its degradation rate. In *B. subtilis*, the GSR is modulated principally by an alternative sigma factor known as σ^B. When *B. subtilis* cells are stressed, an anti-sigma factor is produced which binds to σ^B making it no longer accessible. Several genes in the *E. coli* rpoS-mediated stress protection system have homologues in the *B. subtilis* σ^{32} stress protection system regulon. These are likely to encode traits which help protect the organisms during stress conditions in the soil and intestinal environments.

The GSR for one stress may induce cross-protection, whereby one stress causes the organism to be less sensitive to a second stress, for example heat exposure and acid resistance. When *S.* Enteritidis is heat shocked by shifting the incubation temperature from 20 to 45°C, there is a threefold increase in the *D* value at pH 2.6. This acid tolerance is increased tenfold, when the incubation temperature is raised to 56°C.

Since bacteria are constantly mutating, pathogens will continue to emerge, which may have new characteristics enabling them to cause infections. This can include activating certain virulence determinants to survive in an otherwise incompatible environment. Therefore, inducing these responses during food processing and preparation could increase their infectivity. Examples include desiccation inducing acid tolerance, and the GSR regulatory protein *rpoS* of enabling the survival of *Salmonella* inside phagosomal vacuoles.

2.7.2 pH stress

Acidification is a commonly used method of preserving food, such as dairy products. This method is very effective since the main foodborne pathogens grow best at neutral pH values. These organisms, however, are able to tolerate and adapt to weak acid stresses, and this possibly enables them to survive passage through the acidic human stomach. Additionally, acid resistance can cross-protect against heat treatment (cf. *E. coli* in Table 2.3; Buchanan & Edelson 1999).

Acid stress is the combined effect of low pH and weak (organic) acids such as acetate, propionate and lactate. Weak acids in their unprotonated form can diffuse into the cell and dissociate (see Chapter 3.5.1). Consequently, they lower the intracellular pH (pH_{in}) resulting in the inhibition of various essential cytoplasmic enzymes. In response, micro-organisms have an inducible acid survival strategies. Regulatory features include an alternative sigma factor σ^s, an acid shock protein, two-component signal transduction systems (e.g. PhoP and PhoQ) and the major iron regulatory protein FUR (ferric uptake regulator, Bearson *et al.* 1997).

The important aspect of the acid tolerance response (ATR) is the induction of cross-protection to a variety of stresses (heat, osmolarity, membrane active compounds) in ex-ponentially grown (log phase) cells. Acid-adapted cells are those that have been exposed to a gradual decrease in environmental pH, whereas acid-shocked cells are those which have been exposed to an abrupt shift from high pH to low pH. It is important to differentiate between these two conditions since acid-adapted, but not acid-shocked, *E. coli* O157:H7 cells in acidi-fied tryptone soya broth and low-pH fruit juices have enhanced heat tolerance in TSB at 52 and 54°C, and in apple cider and in orange juice at 52°C. Acid-induced general stress resistance may reduce the efficiency of hurdle technologies which are dependent upon multiple stress factors; see Section 3.4.

Acid resistance can be induced by factors other than acid exposure (Baik *et al.* 1996, Kwon & Ricke 1998). *S.* Typhimurium acid tolerance can be induced by exposure to short-chain fatty acids, used as food preservatives and which also occur in the intestinal tract; see Section 4.1. Subsequently, the virulence of *S.* Typhimurium may be enhanced by increasing acid resistance upon exposure to short-chain fatty acids, such as propionate and further enhanced by anaerobiosis and low pH.

2.7.3 Heat shock

Food preservation by heat is a common food preservation method, that is blanching, pas-teurisation and sterilisation (Table 2.5). The targets for heat inactivation of microbes are the intrinsic stability of macromolecules, that is ribosomes, nucleic acids, enzymes and intracel-lular proteins and the membrane. The exact primary cause for cell death due to heat exposure is not fully understood (Earnshaw *et al.* 1995). Bacterial thermotolerance increases upon ex-posure to sublethal heating temperatures, phage infection and chemical compounds such as ethanol and streptomycin.

Microbes can adapt to mild heat treatment in a variety of ways:

- Cell membrane composition changes by increasing the saturation and the length of the fatty acids in order to maintain the optimal membrane fluidity and the activity of intrinsic proteins.
- Accumulation of osmolytes that may enhance protein stability and protect enzymes against heat activation.
- *Bacillus* and *Clostridium* species produce spores.
- Heat-shock proteins (HSPs) are produced.

When bacteria cells are exposed to higher temperatures, a set of HSPs is rapidly induced. The primary structure of most HSPs appears to be highly conserved in a wide variety of micro-organisms. HSPs involve both chaperones and proteases which act together to maintain quality control of cellular proteins. Both types of enzymes have as their substrates a variety of misfolded and partially folded proteins that arise from slow rates of folding or assembly, chemical or thermal stress, intrinsic structural instability and biosynthetic errors. The primary function of classical chaperones, such as the *E. coli* DnaK (Hsp70) and its co-chaperones, DnaJ and GrpE, and GroEL (Hsp60) and its co-chaperone, GroES, is to modulate protein folding pathways, thereby preventing misfolding and aggregation and promoting refolding and proper assembly. HSPs are induced by several stress situations, for example heat, acid, oxidative stress and macrophage survival, which suggests that HSPs contribute to bacterial survival during infection. In addition, HSPs may enhance the survival of foodborne pathogens in foods during exposure to high temperatures. It is important to note that micro-organisms develop a complicated, tightly regulated response upon an upshift in temperature. Different stressors can activate (parts of) this stress regulon by which they can induce an increased heat tolerance. The process of adaptation and initiation of defence against elevated temperature is an important target when considering food preservation and the use of hurdle technology; see Section 3.4.

The GSR becomes more stable at higher temperatures due to decreased degradation, as the heat-shock response involves the synthesis of an alternative sigma factor, σ^{32}. This results in an increased rate of heat-shock promoters being transcribed by the σ^{32} RNA polymerase. The heat-shock regulon encodes ~30 proteins. In *E. coli*, there is a second heat-shock system which is controlled by σ^E (σ^{24}) that includes a mechanism for sensing and coordinating responses to thermal stress in the periplasm.

2.7.4 Cold shock

Cold adaptation by micro-organisms is of particular importance due to the increased use of frozen and chilled foods and the increased popularity of fresh or minimally processed food, with little or no preservatives. The relevant growth ranges of *L. monocytogenes*, *Y. enterocolitica*, *B. cereus* and *Cl. botulinum* are of particular importance; see Table 2.8.

Mechanisms of cold adaptation are as follows:
- Membrane composition modifications, to maintain membrane fluidity for nutrient uptake.
- Structural integrity of proteins and ribosomes.
- Production of cold-shock proteins (CSPs).
- Uptake of compatible solutes (i.e. betaine, praline and carnitine).

To maintain membrane fluidity and function at low temperatures, micro-organisms increase the proportion of shorter and/or unsaturated fatty acids in the lipids. In *E. coli*, the proportion of *cis*-vaccenic acid (C18:1) increases at low temperature at the expense of palmitic acid (C16:0). The increase in average chain length has the opposite effect on membrane fluidity, but is outweighed by the greater fluidity of increased unsaturation. In *L. monocytogenes*, the fatty acid composition is altered in response to low temperature by an increase of a C15:0 fatty acid and a decrease of a C17:0 fatty acid. Compatible solutes (e.g. betaine, proline and carnitine) may play a role in osmoprotection and in cold adaptation. At 7°C, *L. monocytogenes* increases its uptake of betaine 15-fold compared to growth at 30°C. CSPs are small (7 kDa) proteins which are synthesised when bacteria are subjected to a sudden decrease in temperature. CSPs are involved in protein synthesis and mRNA folding.

Different cold-shock treatments prior to freezing result in differences in microbial survival of bacterial after freezing. This might result in a high survival rate of bacteria in frozen food products. Furthermore, low temperature adapted bacteria may be relevant to food quality and safety.

2.7.5 Osmotic shock

Lowering water activity (a_w) is one of the common ways of preserving food; see Tables 2.5 and 2.8. This works by increasing the osmotic pressure on the microbial cell. This is achieved by the addition of high amounts of salts and sugar (osmotically active compounds) or desiccation. The internal osmotic pressure in bacterial cells is higher than that of the surrounding medium. This results in an outward pressure called 'turgor pressure'. This pressure is necessary for cell elongation.

The microbial response to loss of turgor pressure is the cytoplasmic accumulation of 'compatible solutes' that do not interfere too seriously with cellular functions (Booth *et al.* 1994). These compounds are small organic molecules, which share a number of common properties:
* Soluble to high concentrations
* Can be accumulated to very high levels in the cytoplasm
* Neutral or zwitterionic molecules
* Specific transport mechanism present in the cytoplasmic membrane
* Do not alter enzyme activity
* May protect enzymes from denaturation by salts or protect them against freezing and drying

The adaptation of *E. coli* O157:H7, *S.* Typhimurium, *B. subtilis*, *L. monocytogenes* and *St. aureus* to osmotic stress is mainly through the accumulation of betaine (*N,N,N*-trimethylglycine) via specific transporters. Other compatible solutes include carnitine, trehalose, glycerol, sucrose, proline, mannitol, glucitol, ectoine and small peptides.

2.8 Predictive modelling

The major factors affecting microbial growth are as follows:
* pH
* water activity
* atmosphere
* temperature
* presence of certain organic acids, such as lactate

However, our ability to predict accurately the subsequent growth of microbes in food with regard to the safety and shelf-life of food is very limited.

Predictive food microbiology combines elements of microbiology, mathematics and statistics to develop models that describe and predict the growth and decline of microbes under prescribed (including varying) environmental conditions (Whiting 1995). Obviously, if food is subjected to temperature fluctuations during distribution and storage, then the rate of microbial growth will be affected. There are 'kinetic' models which model the extent and rate of microbial growth and 'probability' models which predict the likelihood of a given event occurring, such as sporulation (Ross & McMeekin 1994).

The main objective is to describe mathematically the growth of micro-organisms in food under prescribed growth conditions. Initially, the data for models are collected using a range of bacterial strains to represent the variation of the target organism present in the commercial situation. Ideally, this will include strains associated with outbreaks, the fastest growing strains and the most frequently isolated strain. Although there is a considerable wealth of knowledge from the growth of microbes in bioreactors (fermenters), this is not directly applicable to the food industry. In food, the environmental factors are more varied and fluctuate and one is often dealing with a mixed population. In the past 20 years, predictive microbiology has become established as a scientific discipline. A detailed study of predictive microbiology is given by McMeekin *et al.* (1993) and McDonald & Sun (1999).

2.8.1 Predicting modelling development

Predictive models have various applications (Whiting 1995):

- *Prediction of risk*: The model can estimate the likelihood of pathogen survival during storage.
- *Quality control*: Predicting the effect of environmental factors can assist deciding on critical control points in a HACCP plan (see Chapter 8).
- *Product development*: Microbial survival can be predicted for changes in processing and new product formulation without the need for extensive laboratory analysis.
- *Education*: The implications of changes in temperature, pH, and so on, can be visualised during a staff training programme.

The origin of predictive microbiology is in the canning industry and the *D* values used to describe the rate of microbial death (Section 2.5.2). However, the ability to routinely solve complex mathematical equations required the revolution in computing power and this, in turn, has greatly facilitated the development of predictive modelling. Models developed to predict microbial survival and growth may become an integral tool to evaluate, control, document and even defend the safety designed into a food product (Baker 1995). The application of microbial models in Hazard Analysis Critical Control Point and microbial risk assessment are covered in Chapters 8 and 9. A form of predictive microbiology is the development of expert systems which can quantify the safety risks of food products without the necessity of extensive laboratory work (Schellekens *et al.* 1994, Wijtzes *et al.* 1998).

Predictive models can be at the following three levels:

1. Primary level models describe changes in microbial numbers (or equivalent) with time.
2. Secondary level models show how the parameters of the primary model vary with environmental conditions.
3. Tertiary level combines the first two types of models with user-friendly application software or expert systems that calculate microbial behaviour under changing environmental conditions.

2.8.2 Primary models and the Gompertz and Baranyi equations

Primary models quantify the increase in microbial biomass as colony forming units per millilitre or absorbance at a specific environmental condition, that is temperature, pH and water activity. Alternatively, it could quantify changes in media composition, such as metabolic end products, conductivity and toxin production.

According to Whiting (1995), the first primary models were simple equations such as growth versus no-growth conditions, that is acetic acid and sugar concentrations preventing the growth of spoilage yeast. Subsequent models described the time between inoculation and growth (or equivalent). This approach was used to model the time required for toxin production by *Cl. botulinum*. Primary models describing growth parameters started by plotting the growth curve and determining the rate of growth from the exponential phase:

$$N_t = N_0 e^{kt/\ln 2}$$

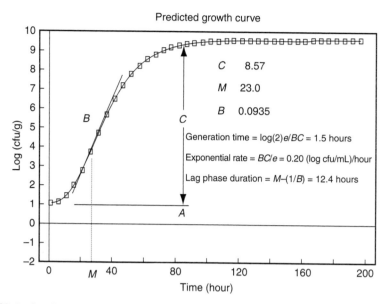

Figure 2.13 Predicted microbial growth curve using the Gompertz function.

where

N_t = population size (logarithm) at a specified time
N_0 = initial population size (logarithm) at time 0
k = slope
t = time

This simple modelling was applied to the growth of *L. monocytogenes* in milk in the presence of pseudomonads (Marshall & Schmidt 1988) and in various meats (Grau & Vanderlinde 1992, Yeh *et al.* 1991). Further models were developed which included the lag phase.

Gompertz model
The Gompertz function has become the most widely used primary model (Figure 2.13, Pruitt & Kamau 1993).
The Gompertz function can be represented as follows:
where

N_t = population density (cfu/mL) at time t (hours)
A = initial population density [log(cfu/mL)]
C = difference in initial and maximum population densities [log(cfu/mL)]
M = time of maximum growth rate (hours)
B = relative maximum growth rate at M [log(cfu/mL)/hour]

This Gompertz function produces a sigmoidal curve that consists of four phases comparable to the phases of microbiological growth: lag, acceleration, deceleration and stationary.

Baranyi model
Baranyi and co-workers developed an alternative equation based on the basic growth model which incorporated the lag, exponential and stationary phases and the specific growth rate:

$$N_t = N_{max} - \ln[1 + (\exp(N_{max} - N_0) - 1 \exp(-\mu_{max} A(t))]$$

where

N_t = population size (logarithm)
N_o = initial population size (logarithm)
N_{max} = maximum population (logarithm)
μ_{max} = maximum specific growth rate
$A(t)$ = integral of the adjustment function

The model fitted experimental data better than the Gompertz function with regard to predicting the lag time and exponential growth phase.

The Baranyi growth model (Baranyi & Roberts 1994) is used as the basis (with modification) for the MicroFit software. This is a freeware programme (see Web Resources section in the Appendix section) developed by the UK Food Standards Agency (formerly MAFF) and four other partners. The programme enables microbiological growth data to be easily analysed to determine the following:

- μ_{max}, doubling time, lag time, initial cell count and final cell count
- estimate confidence intervals on the above parameters
- simultaneously analyse two data sets and compare them graphically
- perform statistical testing on difference between two data sets

An example of *C. jejuni* is given in Figure 2.14.

The model is very easy to use, but cannot analyse the death phase as this is not described by the Baranyi growth model.

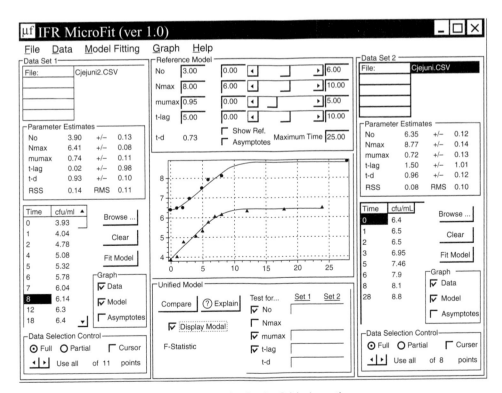

Figure 2.14 An example of the MicroFit screen shot for the *C. jejuni* growth curve.

2.8.3 Secondary models
Commonly used secondary models describe the responses to changes in an environmental factor, that is temperature, pH and a_w. Other models determine the time required for a tenfold log reduction in viability or the length of the lag phase in response to changes in pH, temperature, and so on.

There are three types of models: second-order response surface equation, the square root model (Belehardek) and Arrhenius relationships. The Arrhenius equation is applicable if the growth rate is determined by a single rate-limiting enzymic reaction. Broughall *et al.* (1983) used the Arrhenius equation to describe the lag and generation time of *St. aureus* and *S.* Typhimurium. This model was later modified to take into account the pH value (Broughall & Brown 1984). Initially, the response surface equation was used when a number of factors affected the primary model. The square root model (Ross 1993) is based on the linear relationship between the square root of growth rate and temperature. An important feature is that the equation includes the concept of a 'biological zero' which is the temperature when growth rate is zero. The square root model has been applied to the growth of *E. coli*, *Bacillus* sp., *Y. enterocolitica* and *L. monocytogenes* (Adams *et al.* 1991, Gill & Phillips 1985, Heitzier *et al.* 1991, Wimptheimar *et al.* 1990). In order to have sufficient data for model fitting, a large number of data points must be collected. Growth can be measured by a number of methods such as turbidity and plate counts. Methods of automatic data collection are of obvious benefit since they are less laborious and the data are digitised. The Bioscreen (Labsystems) automatically records turbidity in a large number of samples at any given time. Alternative measurements include changes in the media conductivity (Borch & Wallentin 1993). The growth rate of *Y. enterocolitica* in pork was modelled using conductance microbiology and the data closely fitted to the Gompertz function. A listing of selected modelling papers is given in Table 2.14.

2.8.4 Tertiary models
Tertiary models use the primary and secondary models to generate models to calculate the microbial response to changing conditions and to compare the effects of different conditions. A

Table 2.14 Examples of predictive modelling of microbial growth and toxin production

Organism	Comments	Reference
B. cereus		Zwietering *et al.* (1996)
Brochothrix thermosphacta		McClure *et al.* (1993)
Cl. botulinum types A and B	Growth and toxin production	Lund *et al.* (1990), Roberts & Gibson (1986), Robinson *et al.* (1982)
Cl. botulinum	Toxin production	Lindroth & Genigeorgis (1986), Baker & Genigeorgis (1990), Hauschild *et al.* (1982), Meng & Genigeorgis (1993, 1994)
E. coli O157:H7		Sutherland *et al.* (1995)
L. monocytogenes		McClure *et al.* (1997)
St. aureus		Broughall *et al.* (1983); Sutherland *et al.* (1994)
S. Typhimurium		Broughall *et al.* (1983)
Y. enterocolitica		Sutherland & Bayliss (1994)

number of models have been developed and are accessible via the internet. See Web Resources section in the Appendix section for the downloading addresses.

The 'Pathogen Modelling Programme' and the 'FoodMicro' model programme are two easily available tertiary models. The Pathogen Modelling Programme was developed by the USDA Food Safety group as a spreadsheet software-based system. It includes models for the effect of temperature, pH, water activity, nitrite concentration and atmospheric composition on the growth and lag responses of major food pathogens, as well as survival curves following heat treatment and gamma irradiation. Figures 2.15 and 2.16 show examples of screen prints. The Pathogen Modelling Programme currently covers the following organisms: *Aeromonas hydrophila*, *Bacillus cereus*, *Clostridium perfringens*, *E. coli O157:H7*, *Listeria monocytogenes*, *Salmonella* serovars, *Shigella flexneri*, *Staphylococcus aureus* and *Yersinia enterocolitica*. The 'Food MicroModel' developed in the United Kingdom (the Campden Food and Drink Research Association, UK) includes environmental parameters similar to the Pathogen Modelling Programme. It has predictive equations for growth, survival and death of pathogens. The Food MicroModel can be used for risk assessment within the development of a Hazard Analysis Critical Control Point (HACCP) plan, and this is described in Section 8.4. 'Growth Predictor' (see Web Resources section in the Appendix section for the downloading address) has a series of models for predicting the growth of various pathogenic and spoilage bacteria and yeasts at a range of temperatures, pH and water activities. The 'Seafood Spoilage Predictor' was developed to predict the shelf-life of seafood under both constant and fluctuating temperature storage conditions. The software can be linked to temperature loggers and therefore evaluate the effect of fluctuating temperatures on the shelf-life. The Sym'Previus software is a combined simulation tool and a database for *L. monocytogenes*, *Salmonella*, *B. cereus* and *E. coli*. Strains can be selected according to their source (i.e. dairy, seafood and meat products).

2.8.5 Application of predictive microbial modelling
Predictive microbial modelling has many applications:
1 Product research and development
2 Shelf-life studies (Section 3.2)

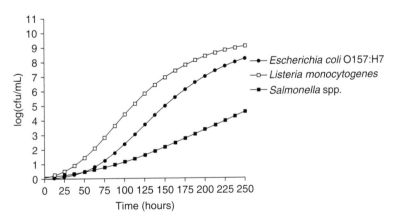

Figure 2.15 Pathogen Modelling Programme simulation for the growth of *E. coli* O157:H7, *L. monocytogenes* and *Salmonella* spp. at 10°C, pH 6.5 and 0.9% NaCl.

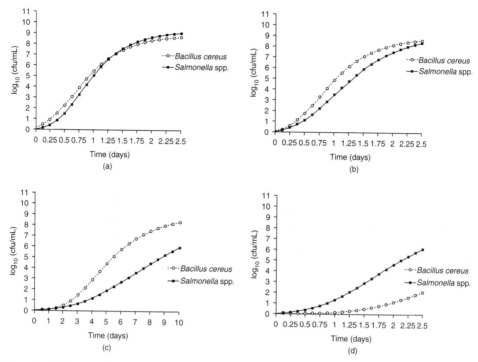

Figure 2.16 The growth of *Salmonella* species and *Bacillus cereus* under different growth conditions as predicted using the Pathogen Modelling Programme. (a) Temperature 20°C, pH 6.8, a_w 0.997. (b) As per (a) except pH 5.6. (c) As per (a) except temperature is 10°C (d) As per (a) except a_w 0.974. Note change in scale for (c).

3 Education and training (Section 7.3.1)

4 HACCP (Section 8.5)

5 Risk assessment studies (Chapter 9)

Growth models have been generated for most food pathogens. Gompertz equation and non-linear regression can be used to predict the survival curve shape and response to heat of *L. monocytogenes* under many environmental conditions.

Zwietering *et al.* (1992) described an expert system which modelled bacterial growth and spoilage in food production and distribution chains. The system combined two databases of physical parameters of the food (temperature, a_w, pH and oxygen level) with the physiological data (growth kinetics) of spoilage organisms. This enables predictions of possible spoilage types and kinetics of deterioration to be made.

Predictive models are available for use in HACCP and microbiological risk analysis (Ross and McMeekin 2003, van Gerwen & Zwietering 1998). Models can be used to assess the risk or probability and determine the consequence of a microbiological hazard in food (see Chapter 10 for detailed examples). By using predictive models, ranges and combinations of process parameters can be established as critical limits for critical control points in HACCP implementation. Predictive microbiological models are tools to aid in the decision-making processes of risk assessment and in describing process parameters necessary to achieve an acceptable level of risk. Using a combination of risk assessment and predictive modelling, it is possible to determine the effect of processing modifications on food safety. Figure 2.17 shows

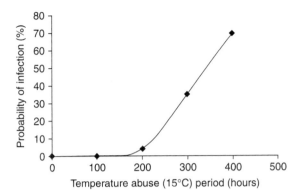

Figure 2.17 The predicted effect of temperature abuse on the probability of infection for *Sh. flexneri*. (Buchanan & Whiting 1996. Reprinted with permission from the Journal of Food Protection. Copyright held by the International Association for Food Protection, Des Moines, Iowa, US).

the predicted effect of temperature abuse on the probability of infection (see Section 9.5.7) for *Sh. flexneri*. The Pathogen Modelling Programme was used to predict the rate of *Sh. flexneri* growth under defined conditions (pH, salt and temperature) for increasing time periods and the subsequent likelihood of infection (Buchanan & Whiting 1996).

2.9 Bioinformatic studies

2.9.1 Bioinformatics and genomes

Our understanding of the microbial world has been revolutionised in recent years due to ever-increasing rate of sequencing the total DNA (chromosomal and plasmid) in bacteria, termed 'genomics'. These sequences are deposited in databases, and can be studied using internet-based tools ('bioinformatics'). Alongside this, there has been the expansion in protein analysis ('proteomics') and microarrays ('transcriptomics') to study the expression of the DNA sequence. This is a very large topic, and can only be covered in brief here. However, the reader can access the necessary software, example exercises and internet sites via the website supporting this book at http://www.wiley.com/go/forsythe.

As stated above, bacterial, viral and fungal genomes are being sequenced and annotated at an increasing rate. The first bacterium sequenced was *Haemophilus influenzae* by TIGR in 1995. It has a genome 1.9 Mb in size, where 1 mega base = 100 000 base pairs in length. In part, this was to demonstrate the reliability of the shotgun sequencing technique as an alternative method for the human genome project. To sequence a bacterial genome (chromosome and plasmid), the DNA is first fragmented and cloned into pUC (or similar) plasmids. The plasmids are then transformed into *E. coli* and grown on a medium to generate clones of plasmids containing a DNA insert from the fragmented target organism. The inserts are amplified by the PCR reaction and labelled using nucleotide base-specific fluorescent markers and sequenced using capillary electrophoresis. This is known as the 'Sanger sequencing method'. Although each sequence is only ~900 bases long, which is a small fraction of the total genome, since so many fragments of the original genome are cloned, there is a good chance that overlapping regions would be identified and used to form longer DNA sequences called 'contigs'. It is not possible to construct the whole genome from these contigs as there are often gaps due to uncloned regions. Eventually, through further directed sequencing, these gaps are closed, the genome is circularised. Specific software identifies the open reading frames (ORFs) in the genome. The ORFs are then compared with databases for homologous regions which have previously been identified in other organisms with experimental evidence. This later step is called 'annotation'. Since the genome is a prediction based on overlapping regions, and not a step-by-step sequencing, some errors do occur. Some of the earlier genomes, such as *C. jejuni* ATCC 11168, have been revised as researchers have investigated the genome in further detail. Table 2.15 summarises the genomes for a number of key food-related organisms.

The earlier Sanger method of sequencing could sequence 8 million bases per week from bacterial colonies of cloned fragments. However, the new Roche's 454 and Illumina's sequencers do not require fragment cloning, colony picking or capillary electrophoresis. Instead, they use sequencing-by-synthesis (SbS) approaches which detect the addition of each base of the newly synthesised fragment in real time. Consequently, these technologies are 1000 times faster, and less expensive than the Sanger method. Each machine can sequence 2–8 billion bases per week; each read length being 500–600 bases for Roche's 454 and 35–100 bases for Illumina. Despite this high throughput, their disadvantage is that they cannot accurately determine repetitive regions. Nevertheless, it is now feasible to sequence a bacterial genome in 1 day. The major consequence of this is that many bacterial genomes are released to the public databases (i.e. Genbank) having been annotated automatically with varying levels of accuracy and require manual investigation by interested groups independent of the sequencing facility for more precise annotation.

As well as sequencing whole organisms, the new science of 'metagenomics' has arisen in which whole microbial communities are sequenced. There are two forms of metagenomics. The first is sequence based in which the total DNA of a community is extracted and cloned without any cultivation or pre-selection. The DNA fragments are sequenced and then compared with database to determine the microflora. A second form of metagenomics is expression driven. In this format, the DNA fragments are cloned into expression vectors and then screened for novel enzyme activities. In 2007, the US National Research Council (2007) proposed a global metagenomic initiative which equals the human genome project for scale. Complementing this, the 'human microbiome project' is sequencing the microbial flora of the human body; URLs are http://nihroadmap.nih.gov/hmp and http://www.hmpdacc.org/bacterial_strains.php. Currently, it is aiming to sequence 1000 reference genomes of organisms that live on or are in the human body, and also to use the metagenomics approach to sequence samples from five (nasal, oral, skin, gastrointestinal and urogenital) body sites of 375 healthy and ill individuals to show evidence of clinical correlation. Similarly, the MetaHIT project (http://www.metahit.eu) aims to use metagenomics to link the human intestinal flora with obesity and inflammatory bowel disease. Hattori & Taylor (2009) have reviewed the subject and give extensive information on techniques and data.

There are a number of large international resources available over the internet for genome information. Two of the most popular ones are National Center for Bioinformatics and Information (NCBI) and the J. Craig Venter Institute (formerly TIGR). When DNA sequences are deposited into Genbank, they are given an accession number which is a unique identifier. These can be searched using the Basic Local Alignment Search Tool (BLAST). There are essentially six forms of BLAST, whereby nucleotide and amino sequences can be used to search nucleotide or protein databases. The preferable BLAST will depend upon the source of the unknown sequence, and the question being asked. For example, comparing DNA sequences for a structural protein in *Bacillus* and *Staphylococcus* would be TBLASTX. Although the amino acid sequence will have been conserved during evolution for a protein, each bacterial genus will

Table 2.15 Genome sequencing of major food-related organisms

Organism	Genome accession number[a]	Publication year[b]	Size of genome (Mb)	Description
A. hydrophila subsp. *hydrophila* ATCC 7966	CP000462	2006	4.7	Gammaproteobacteria
Arcobacter butzleri RM4018	CP000361	2007	2.3	Epsilonproteobacteria
B. cereus ATCC 14579	AE016877	2003	5.42	Firmicutes
Bifidobacterium longum NCC2705	AE014295	2002	2.26	Actinobacteria
Brucella melitensis biovar Abortus 2308	AM040264	2005	3.28	Alphaproteobacteria
Brucella suis ATCC 23445	CP000911	2002	3.3	Alphaproteobacteria
C. jejuni RM1221	CP000025	2005	1.8	Epsilonproteobacteria
C. jejuni subsp. *jejuni* NCTC 11168	AL111168	2000	1.6	Epsilonproteobacteria
C. jejuni subsp. *doylei* 269.97	CP000768	2007	1.8	Epsilonproteobacteria
C. jejuni subsp. *jejuni* 81-176	CP000538	2007	1.68	Epsilonproteobacteria
C. jejuni subsp. *jejuni* 81116	CP000814	2007	1.6	Epsilonproteobacteria
Clostridium botulinum A str. ATCC 3502	AM412317	2007	3.92	Firmicutes
Clostridium perfringens ATCC 13124	CP000312	2006	3.3	Firmicutes
Coxiella burnetii RSA 493	AE016828	2003	2.03	Gammaproteobacteria
Cronobacter sakazakii ATCC BAA-894	CP000783	2007	4.56	Gammaproteobacteria
E. coli str. K-12 substr. MG1655	U00096	1997	4.6	Gammaproteobacteria
E. coli O157:H7 EDL933	AE005174	2001	5.59	Gammaproteobacteria
Hepatitis A	NC_001489	1987	0.008	Picornaviridae
Lactobacillus acidophilus NCFM	CP000033	2005	2	Firmicutes
Lactobacillus brevis ATCC 367	CP000416	2006	2.35	Firmicutes
Lactobacillus delbrueckii subsp. *bulgaricus* ATCC 11842	CR954253	2006	1.86	Firmicutes
Lactobacillus plantarum WCFS1	AL935263	2003	3.34	Firmicutes
Lactococcus lactis subsp. *cremoris* SK11	CP000425	2006	2.56	Firmicutes
L. monocytogenes str. 4b F2365	AE017262	2004	2.91	Firmicutes
Mycobacterium tuberculosis CDC1551	AE000516	2002	4.4	Actinobacteria
Oenococcus oeni PSU-1	CP000411	2006	1.8	Firmicutes
Pediococcus pentosaceus ATCC 25745	CP000422	2006	1.8	Firmicutes
Pseudomonas fluorescens	CP000094	2008	6.44	Gammaproteobacteria
Salmonella enterica subsp. *enterica* serovar Choleraesuis str. SC-B67	AE017220	2005	4.99	Gammaproteobacteria
Salmonella Typhimurium LT2	AE006468	2001	4.99	Gammaproteobacteria
Shewanella putrefaciens CN-32	CP000681	2007	4.7	Gammaproteobacteria
Shigella dysenteriae Sd197	CP000034	2005	4.56	Gammaproteobacteria
St. aureus	AJ938182	2008	2.7	Firmicutes
Streptococcus thermophilus CNRZ1066	CP000024	2004	1.8	Firmicutes
Vibrio cholerae O395	CP000626	2007	4.1	Gammaproteobacteria
Vibrio parahaemolyticus RIMD 2210633	BA000031	2000	5.17	Gammaproteobacteria
Vibrio vulnificus CMCP6	AE016795	2003	5.1	Gammaproteobacteria
Norovirus	M87661	2002	0.008	Caliciviridae
Y. enterocolitica subsp. *enterocolitica* 8081	AM286415	2007	4.67	Gammaproteobacteria

[a] Go to http://www.ncbi.nlm.nih.gov/genomes/lproks.cgi.
[b] Not necessarily the last modification.

have preferred codon usage for certain amino acids. In contrast, comparing *Shigella* and *E. coli* would use BLASTN as the organisms are so closely related. To further appreciate the impact of bioinformatics in microbiology and learn some of the key skills, go to the accompanying website (http://www.wiley.com/go/forsythe) for this book. Most of the diagrams in this section have been generated using free software tools which are covered in the online exercises, and the source files have been made available.

Genomic data on pathogenic bacteria have been highly informative. For example, even with the extremely well-studied bacterium *E. coli*, a large proportion (~40%) of identified ORFs have no known function. Also, considerably more horizontal gene transfer has occurred than previously thought, including the sharing of virulence-related genes (including PAIs; Section 2.3.2) and genes encoding for antibiotic resistance (cf. Section 1.9). Comparative genomics may also lead to improvements in phylogenetic classification. Bergey's Manual of Systematic Bacteriology (2001 onwards) is now based on this approach.

The total expression of genes by an organism can be studied using DNA microarrays. These enable expression profiling of the organism under different environmental conditions (Schwartz 2000). Microarrays (Section 5.4.6) can also be applied for the rapid and miniaturised diagnosis and detection of pathogenic and spoilage bacteria. Complementing advances in genomics are also advances in proteomics, the study of the complete protein complement of an organism.

The following section has examples of where DNA sequence analysis has helped us to further investigate and understand a number of important foodborne pathogens. This list is hardly exhaustive, and the reader is encouraged to use bioinformatics tools on topics of interest to complement laboratory studies. In the future, the microbial flora of the food chain will be better understood due to the application of 'metagenomics' as described above. In the future, advances in DNA sequencing should enable the monitoring of viruses, bacteria and fungi simultaneously from 'farm to fork'. This will include changes in microbial composition of food during processing and in parallel will enable the rapid detection of foodborne pathogens.

2.9.2 16S rRNA gene sequence and denaturing gradient gel electrophoresis (DGGE)

Advances in molecular biology, sequence databases and phylogenetic analysis have revolutionised the analysis of microbial communities including the human intestinal tract, cheese manufacturing and wine production (Amann *et al.* 1995, Ogier *et al.* 2004, Renouf *et al.* 2006). It is culture independent, and instead relies upon phylogenetic analysis of the DNA sequences. Figure 2.1 was constructed using an online freeware programme called ClustalW (http://www.ebi.ac.uk/Tools/clustalw2/index.html), although more sophisticated programmes are available for purchase. The tree is based on comparison of DNA sequences encoding for the RNA in the ribosomal small unit (rRNA). In bacteria, this is called the 16S rDNA gene (~1500 nucleotides in length), and in eukaryotes, this is 18S rDNA gene due to the difference in subunit sizes. Despite the considerable differences between organisms, the gene has been highly conserved during evolution. These sequences are readily accessible from downloading from specific internet sites such as the Ribosome Database Project II (http://rdp.cme.msu.edu).

Since the technique does not require any laboratory-based cultivation of organisms, 16S rDNA gene sequencing has enabled the study of microbial communities (metagenomics) and uncultivated organisms. The most widely used method is to amplify the 16S rDNA gene directly from total community DNA using specific primers to rDNA genes. The PCR amplification results in PCR products which are of the same length, but contain sequence differences for each species present in the community. These can be separated using DGGE in which the polyacrylamide gel contains a linear gradient of urea and formamide which are DNA denaturing agents. As discussed above, the 16S rDNA genes are amplified using specific primers, but one has a 'G+C clamp' of approximately 39 nucleotides attached to the 5′ end. This 'G+C clamp' prevents the two DNA strands from dissociating completely, even under highly denaturing conditions. The equal-length PCR products will stop migrating at different distances when the DNA denatures, which will depend upon their G+C content. This method therefore enables the separation of individual sequences from a single sample of mixed species.

The bands in the profile will represent the dominant microbial populations, and changes in banding profile between samples will reflect changes in the microbial diversity. Although the technique does not enumerate the identified organisms, the intensity of each band may give a relative estimation of the target sequence in the sample.

The PCR-DGGE method may not show the full diversity of the microbial flora present in the sample due to failing in cell lysis and low prevalence. Also, the 16S rDNA gene exists in many copies in some bacteria, and sequence variation may hinder interpretation of the results (Coeny & Vandamme 2003). The RNA polymerase beta subunit gene *rpoB* has been used as an alternative, though databases of its sequence are not as extensive as that of the 16S rRNA gene (Renouf *et al.* 2006).

2.9.3 *Campylobacter jejuni* and *Campylobacter coli* genome sequence

From Figure 2.1, it can be seen that the Gram-negative organisms split into various groups, with *Campylobacter*, *Helicobacter* and *Arcobacter* being more closely related to each other, than to the *Enterobacteriaceae* (*Salmonella*, *E. coli* and *Shigella* spp.) and *Pseudomonas* spp. *C. jejuni* is recognised as the predominant cause of bacterial gastroenteritis, and knowledge of its genome could help in developing further control measures. Parkhill *et al.* (2000) published the first *C. jejuni* sequence. This was of strain NCTC 11168, a human clinical isolate. The genome is 1.6 Mb in length, and since ~94% encodes for proteins, it is one of the densest that has been sequenced (Table 2.16). There were a number of discoveries following the genome sequencing, including the revelation of a capsular polysaccharide, considerable capacity for phase variable gene expression and also lipo-oligosaccharide structural phase variation. In general, the *Campylobacter* genome has a high %GC content, but sequencing revealed two regions of low %GC (25%) and were within the lipo-oligosaccharide (LOS) and extracellular

Table 2.16 Comparison of genome features for five *Campylobacter* species

Trait	C. jejuni NCTC 11168	C. jejuni RM1221	C. coli RM2228	C. lari RM2100	C. upsaliensis RM3195
Chromosome size (Mb)	1.64	1.78	~1.68	~1.5	~1.66
%GC	30.55	30.31	31.37	29.64	34.54
Open reading frames (ORFs)	1634	1835	1764	1554	1782
% ORF assigned function	—	61	74	73	68
Phage/genomic island regions	0	4	0	1	1
Plasmids (kb)	0	0	1 (~178)	1 (~46)	2 (~3.1, 110)
Two-component systems	15	15	15	13	11
Outer membrane proteins	14	16	12	10	11
Transmembrane proteins	87	92	94	74	118

Adapted from Fouts *et al.* (2005).

polysaccharide (EP) biosynthesis gene clusters. Genes encoding for CLDT, haemolysin and phospolipase were identified, and these may have a role in pathogenesis. No cholera-like toxin or Type III secretion systems were found; however, plasmid-encoded components of a putative Type IV secretion system were identified.

Apart from the LOS, EP and flagellar modification, there is little gene organisation into operons or clusters. Most of the hypervariable sequences are in localised regions of the genome, close to the clusters of the LOS and EP biosynthesis genes and flagellar modification genes. Genome sequencing also revealed variations in the length of polyG:C tracts, which in some organisms is responsible for phase variation of surface antigenicity. It appears that *C. jejuni* does not possess many of the adaptive responses, such as RpoS homologues present in other foodborne pathogens and so does not show GSR; see Section 2.7.1. It also lacks CSPs; though its minimum growth temperature is only ∼30°C. However, it shows considerable genetic diversity, and considerable strain–strain variation in virulence and tolerance to stresses (Park 2005, Scott *et al.* 2007). The genome encodes for five iron-acquisition systems, including enterochelin, siderophore receptor protein and haemin uptake operon. The sequences of five *C. jejuni* genomes are aligned in Plate 4. There are considerable sequence differences between *C. jejuni* subsp. *dolyei* and *C. jejuni* subsp. *jejuni* which account for the rearrangements shown. The major difference between the two *C. jejuni* genomes is the presence of four large integrated elements or regions. Three of these encode for phage, or phage-related genes, and the fourth may be an integrated plasmid.

The genome sequences can be compared between different *Campylobacter* species. Comparing the genome of *C. jejuni* with *C. coli*, *C. lari* and *C. upsaliensis* shows that there are considerable genomic structural differences between the organisms. These are associated with the insertion of phage- and plasmid-like regions, as well as differences in the LOS complex (Fouts *et al.* 2005). Table 2.16 summarises the results.

2.9.4 *Salmonella* evolution and PAIs

Although an organism is identified according to key features, the comparison of bacterial genomes has revealed that considerable diversity in genes exists within a bacterial 'species'. Within the *Salmonella* species, there are many clones that are specialised according to their host and also the disease that they cause. This also occurs in the *E. coli* species as described in the next section. The variation in genes between clones causes the major genetic diversity within a bacterial species. It is recognised that there can be up to 20% difference in the DNA between two strains of the same species. Genes found in most strains can be regarded as the core set of genes, which characterise the species. In contrast to the gain in genes through horizontal (or lateral) gene transfer, genes that become deleterious or are no longer of benefit will be lost following an accumulation of mutations and deletions.

As discussed in more detail in Section 4.3.2, different *Salmonella* serovars can have very different host ranges and the diseases that they cause. Therefore, it is of interest to compare the complete genomes of *Salmonella* serovars. Using WebACT to visualise an alignment of S. Typhimurium, S. Typhi and S. Paratypi genomes, one can see large regions that have been inverted, as indicated by the blue triangles in Plate 5. They are all *Salmonella*, and possess nearly the same genes, but they are in a different order in different serovars. At a greater resolution, one can determine the regions where prophage DNA remnants can be located, as well as antibiotic resistance factors. A fuller treatment of bioinformatics investigation of *Salmonella* can be accessed at the URL supporting this book (http://www.wiley.com/go/forsythe).

Salmonella spp. have ∼100 genes required for virulence and contain at least 5 PAIs (Table 2.2, Section 2.3.2). PAIs have been a major area of pathogenicity research and in *Salmonella* are particularly of interest in respect to the virulence and adaptation of the species. The ability of *Salmonella* to cause disease is attributed to specific virulence factors. The genes encoding these virulence factors are often found, clustered together at specific loci on the chromosome or plasmids called PAIs. These are large (>30 kb) distinct chromosomal elements encoding virulence-associated genes (Table 2.2). Because the G+C ratio of PAIs is often distinct from the rest of the bacterial DNA G+C ratio, it has been proposed that they may have been acquired in the past by horizontal gene transfer from other bacterial species. In *Salmonella*, there are five PAIs and their distribution between serovars suggests that their acquisition by horizontal gene transfer was essential in the organism becoming a pathogen. *Salmonella* PAI (SPI)-1 is associated with the toxin secretion system and encodes for ∼25 genes (Table 2.2). It contains genes required for the invasion of epithelial cells and multiplying in the lymphoid tissue. As this is absent in *E. coli*, it could have enabled *Salmonella* to occupy different niches to its close relative. Further gene acquisition, such as SPI-2, enabled *Salmonella enterica* serovars to survive macrophage uptake to spread to the bloodstream. SP1-2 is found in *S. enterica* serovars and not in *S. bongori*. The acquisition of SPI-3 and SPI-4 resulted in the ability to multiply within macrophages. SPI-3 also contains ORFs similar to the ToxR regulatory protein of *V. cholerae*, and another that is similar to an adhesion of EPEC. SPI-5 encodes the genes for enteropathogenicity. This illustrates an application of genomics to the study of bacterial virulence and pathogen evolution. These genetic regions usually have a different %GC content of DNA, often have repetitive ends, and are often inserted into or near tRNA genes. They can be located on bacterial genomes by looking at a plot of the %GC content, and noting regions of atypical values. Further instructions and examples of locating PAIs can be found in the supporting website for this book.

2.9.5 *E. coli* O157:H7 genome sequence

As outlined above with *Salmonella*, genomics has greatly facilitated our understanding of pathogen evolution. This can also be demonstrated by studying *E. coli* O157:H7 which is thought to have evolved from ancestral *E. coli* by stepwise acquisition or loss of virulence and phenotypic traits (Feng *et al.* 1998). Comparing the sequenced genomes of *E. coli* O157:H7 to a non-pathogenic *E. coli* reveals some surprising information. As would be expected, both strains share a backbone of about 4.1 million similarly arranged base pairs. However, *E. coli* O157:H7 genome contains an additional 1387 new genes. The Venn diagram, Figure 2.18, shows the distribution of genes amongst *E. coli* K12 (non-pathogenic), *E. coli* O157:H7 and uropathogenic *E. coli*. Furthermore, the non-pathogenic *E. coli* K12 has 528 genes that are not found in *E. coli* O157:H7. Possibly, the most significant discovery is that the additional DNA in the pathogenic *E. coli* is distributed among 177 different regions, each of which is likely to have been inherited through independent events. Also, about 18% of its current genes were obtained by horizontal gene transfer from other species. It has been concluded that these two *E. coli* strains diverged about 4.5 million years ago. This followed divergence from *Salmonella* about 100 million years ago.

As shown in Figure 2.19, it is believed that *E. coli* O55:H7 is ancestral to *E. coli* O157:H7. 'Primative' O157 fermented sorbitol, and was β-glucuronidase positive, and had the Shiga-toxin 2 gene (*stx*₂). This divided into two lineages. First lineage has the loss of sorbitol metabolism, gain of *stx*₁ gene, and later loss of β-glucuronidase. The second lineage is composed of SOR+, GUD+, and non-motile strains. The common virulence genes shared by EHEC serotypes O157:H7 with O26:H11 and O111:H8 show that the acquisition of the same virulence gene elements has occurred on multiple occasions. A fuller description of pathogenic *E. coli* strains is given in Section 4.3.2.

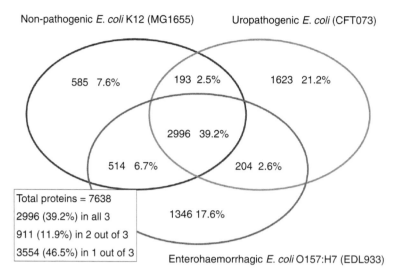

Non-pathogenic *E. coli* K12 (MG1655)　　Uropathogenic *E. coli* (CFT073)

585　7.6%　　　193　2.5%　　　1623　21.2%

2996　39.2%

514　6.7%　　　　　204 2.6%

Total proteins = 7638
2996 (39.2%) in all 3
911 (11.9%) in 2 out of 3
3554 (46.5%) in 1 out of 3

1346 17.6%

Enterohaemorrhagic *E. coli* O157:H7 (EDL933)

Figure 2.18　Venn diagram of *E. coli* genomes.

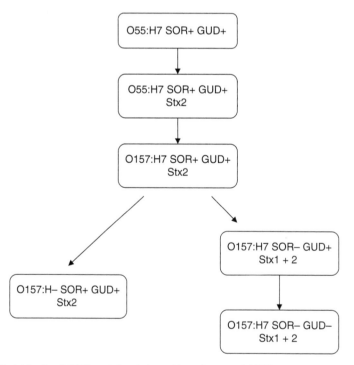

O55:H7 SOR+ GUD+

O55:H7 SOR+ GUD+
Stx2

O157:H7 SOR+ GUD+
Stx2

O157:H– SOR+ GUD+
Stx2

O157:H7 SOR– GUD+
Stx1 + 2

O157:H7 SOR– GUD–
Stx1 + 2

Figure 2.19　Model for *E. coli* O157 evolution (adapted from Feng *et al.* 1998).

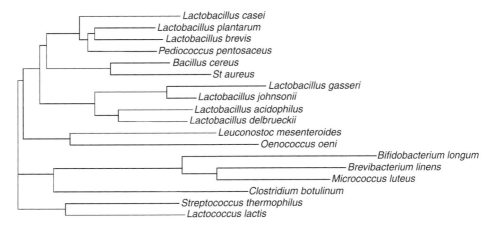

Figure 2.20 Phylogenetic tree of lactic acid bacteria and other Gram-positive organisms.

2.9.6 The diversity of the lactic acid bacteria and bifidobacteria

The lactic acid bacteria are a heterogenous family of bacteria that are grouped together on the basis of being Gram-positive rods or cocci, production of lactic acid from sugars and catalase negative. The bifidobacteria are also Gram-positive rods, which show a bificating morphology and are generally grouped with the lactic acid bacteria. However, this phenotypical basis of grouping does not fully reflect the diversity of the group as shown by 16S rDNA sequence analysis (Figure 2.20). Figure 2.1 shows that the Gram-positive organisms are split into two groups: the *Firmicutes* (low %GC content) and the *Actinobacteria* (high %GC content). The lactic acid bacteria, as well as the *Staphylococcus* and *Listeria* genera, are in the *Firmicutes*, whereas the bifidobacteria are in the *Actinobacteria* and therefore are genetically very dissimilar in origin.

Initially, genome sequencing of the lactic acid bacteria was primarily focused on pathogenic species from the *Streptococcus* genera. This has now broadened to include a large number of non-pathogenic, even probiotic associated, human intestinal species. Table 2.17 summarises the genomes of selected lactic acid bacteria, brevibacteria and bifidobacteria species which have been sequenced to date. The *Lactobacillales* (phylum *Firmicutes*, class *Bacilli*, sister taxon to the order *Bacillales*) have relatively small genomes (~2 Mb), with evidence of gene inactivation, acquisition and duplication. The large number of genomes sequenced (>20 to date) from multiple strains within the same species, and related species will enable insights into genome evolution and structure unrivalled by other bacterial groups (Makarova & Koonin 2007). There appears to be three lineages within the order *Lactobacillales*:

1 *Leuconostoc group: Leu. mesenteroides and Oenococcus oeni.*
2 *Lactobacillus casei-Pediococcus group: Lb plantarum, Lb. casei, Ped. pentosaceus and Lb. brevis.*
3 *Lb. delbrueckii group: Lb. delbrueckii, Lb. gasseri and Lb. johnsonii.*

The *Streptococcus* and *Lactococcus* species form a separate branch. The availability of these genome sequences has improved the phylogenetic analysis of this considerably important group of bacteria. The common ancestor of the *Lactobacillales* had at least 2100–2200 genes, and after divergence from the *Bacilli* ancestor, it lost 600–1200 genes and gained <100 genes. Many of these changes reflect an evolution to growth in nutritionally rich environments, such as milk and intestinal contents. Horizontal gene transfer has resulted in the acquisition of at least 84 genes from different sources. Interestingly, during the evolution of the *Lactobacillales*, many

Table 2.17 Genomes of selected lactic acid bacteria, brevibacteria and bifidobacteria

Genus	Species	Strain	Size (Mb)	%GC	Accession number
Brevibacterium	linens	BL2/ATCC9174	4.4	62.9	NZ_AAGP00000000
Bifidobacterium	adolescentis	ATCC 15703	2.1 2.3	59.3	AP009256
	longum	NCC2705		60.1	AE014295
Enterococcus	faecalis	V583	3.36	37.2	AE016830
Lactobacillus	acidophilus	NCFM	2.0	34.7	CP000033
	brevis	ATCC 367	2.35	46.1	CP000416
	casei	ATCC 334	2.93	46.1	CP000423
	delbrueckii subsp. bulgaricus	ATCC BAA365	1.9	49.7	CP000412
	delbrueckii subsp. bulgaricus	ATCC 11842	1.86	49.7	CR954253
	gasseri	ATCC 333323	1.9	35.3	CP000413
	helveticus	DPC 4571	2.1	37.1	CP000517
	johnsonii	NCC533	2.0	34.6	AE017198
	plantarum	WCFS1	3.3	44.4	AL935263
	reuteri	F275	2.0	38.9	CP000705
	sakei subsp. sakei	23K	1.9	41.3	CR936503
	salivarius subsp. salivarius	UCC118	2.1	33.0	CP000233
Lactococcus	lactis spp. cremoris	SK11	2.6	35.8	CP000425
	lactis spp. cremoris	MG1363	2.5	35.7	AM406671
	lactics spp. lactis	IL1403	2.4	35.3	AE005176
Leuconostoc	mesenteroides	ATCC 8293	2.0	37.7	CP000414
Oenococcus	oeni	PSU-1	1.8	37.9	CP000411
Pediococcus	pentosaceus	ATCC 25745	1.8	37.4	CP000422
Proprionibacterium	fredenreichii subsp. shermanii	CIP 103027	Incomplete	67	Incomplete
Streptococcus	agalactiae	2603V/R	2.2	35.6	AE009948
	agalactiae	A909	2.1	35.6	CP000114
	mutans	UA159	2.0	36.8	AE014133
	pneumoniae	TIGR4	2.2	39.7	AE005672
	pyogenes	M1 GAS	1.9	38.5	AE004092
	thermophilus	LMD-9	1.9	39.1	CP000419
	thermophilus	LMG18311	1.8	39.1	CP000023
	thermophilus	CNRZ1066	1.8	39.1	CP000024

genes involved in sugar metabolism and transport were duplicated. These included several PTSs, β-galactosidase and enolase.

The genome scale analysis has revealed their broad saccharolytic ability as would be expected from their wide range of environmental niches. There is a noted predominance of PTS transporters which enable the organisms to metabolise various carbohydrates. As well as sugar uptake, 13–17% of genes in many genomes encode for protein transport, and amino acid uptake systems predominate over sugar and peptide uptake systems.

There is evidence that the lactic acid bacteria have evolved to nutritionally complex environments such as milk through gene decay with the loss of dispensable functions, and secondly gene acquisition via horizontal gene transfer along with gene duplication. Apart from *L. lactis*, amino acid biosynthetic pathways are incomplete in most other lactic acid bacteria reflecting their fastidious nature. For example, *L. plantarum* is deficient in a few pathways such as those for the synthesis of branched chain amino acids, whereas species of the 'L. acidophilus complex' (*L. acidophilus*, *L. gasseri* and *L. johnsonii*) are largely deficient in amino acid biosynthetic capacity. To compensate for this deficiency, the organisms have a large number of peptidases, amino acid permeases and multiple oligo-peptide transporters. In *Strep. thermophilus*, 10% of genes are pseudogenes and non-functional. This genome decay is particularly noted for genes involved in carbohydrate metabolism, uptake and fermentation, and may be an adaptation to nutritionally rich environment of milk and human intestinal tract. In fact, a specific lactose transported is found in *Strep. thermophilus* which is absent in other pathogenic streptococci. There is also some evidence of horizontal gene transfer from *Lb. bulgaricus* and *Lb. lactis* for methionine biosynthesis, which is a rare amino acid in milk. In turn, *Lb. lactis* has acquired plasmid DNA encoding important genes for growth and survival in milk, lactose metabolism, proteolytic activity, exopolysaccharide production, bacteriocin production and bacteriophage resistance.

Whole genome comparison of *Lb. plantarum* shows a lack of synteny (co-localisation of genes) with the three species of the *Lb. acidophilus* complex: *Lb. acidophilus*, *Lb. johnsonii* and *Lb. grasseri*. The later three show extensive conservation of gene content and gene order, except for two apparent chromosomal inversions in *Lb. gasseri* (Plate 6). The high degree of similarity between the three species explains the difficulty normally experienced when attempting to differentiate these organisms using phenotyping or molecular techniques. It is possible to find genes unique to one species by (electronically) subtracting one genome from another, and thereby design species-specific DNA probes. Due to variation in genomes between strains in the same species, these DNA probes would need to be further validated using a strain collection for each species.

Bifidobacterium species are considered as the lactic acid bacteria. They are Gram-positive, heterofermentative, non-motile, non-spore forming rods. However, unlike the lactic acid bacteria considered above, bifidobacteria are found in the *Actinobacteria* branch of the Gram-positive organisms, with high %GC content (42–67%; Figure 2.1). The majority of *Bifidobacterium* species inhabit the mammalian gastrointestinal tract. They can display a range of cell morphologies including rods and branched (bificating) shapes. The genome of *Bif. longum* has been sequenced and therefore direct genome comparison with other lactic acid bacteria is possible (Table 2.17). The genome size is $\sim$2.3 Mb and is of similar size to that of other lactic acid bacteria. Despite living in the same habitat as several *Lactobacillales*, there are only seven genes in common between *Bif. longum* and other lactic acid bacteria that are not found in other *Actinobacteria*. Nevertheless, there are similar trends in the loss and gain of genes. Losses in both *Bif. longum* and *Lactobacillales* include oxidases, catalase, heme and fatty acid biosynthesis. Both groups have gained additional amino acid metabolism by the acquisition of dipeptidase activity. Bifidobacteria are part of the early flora that develops in breastfed infants and are regarded as beneficial to the infants' health (Section 4.1.3). Therefore, it is interesting to learn that recent genomic analysis has revealed that *Bif. longum* subsp. *infantis* has adapted to metabolise human milk by the presence of a 43 kb cluster of catabolic genes, and permeases for the uptake of milk oligosaccharides (Sela *et al.* 2008). Five of these oligosaccharides are not degraded by host enzymes and apparently are of no nutritional benefit to the infant, but are preferred substrates for *Bif. longum* subsp. *infantis*.

Bacteriocins such as nisin and pediocin are short proteins, often including rare amino acids, and have highly divergent sequences. They can show bactericidal or bacteriostatic activity towards closely related organisms. Nisin is already used commercially in food preservation and could be of further use in the design of novel antibiotics (cf. Section 3.7.1). Using genome comparisons, possible genes encoding for other bacteriocins have been identified, and these may be of commercial importance in the future. See the Appendix for BAGEL, a web-based bacteriocin genome mining tool URL.

Genomic analysis of lactic acid bacteria and *Bif. longum* has revealed a number of features of importance to probiotic functions. These include attachment factors (fimbrae, mucus-binding proteins and mannose-specific adhesion proteins), exopolysaccharide synthesis, bacteriocin production, stress and acid tolerance factors (Pridmore *et al.* 2004).

2.9.7 *Listeria* species genome sequence analysis

Phylogenomic analysis (whole genome sequence analysis) of *Listeria* species has shown the same phylogenetic relationship as earlier 16S rDNA sequence comparison studies. The *Listeria* genus has two main branches. One is composed of *L. monocytogenes*, *L. innocua* and *L. welshimeri*. Whereas *L. seeligeri* and *L. ivanovii* form a second branch, with *L. grayii* distant from both branches. The natural habitat of the organism is probably soil and decaying plant material, and it is transferred to humans and animals via ingestion of contaminated food. Those at risk are immunocompromised individuals, pregnant women, neonates and the elderly. Only two species are recognised as pathogenic: *L. monocytogenes* and *L. ivanovii*. They are both facultative intracellular parasites, multiplying in macrophages and other normally non-phagocytic cells (epithelial and endothelial cells and hepatocytes). The organisms cause listeriosis which is an opportunistic infection of humans and animals with severe clinical symptoms such as meningoencephalitis, abortion and septicaemia. *L. monocytogenes* has a host range of mammals and birds, whereas *L. ivanovii* is mostly pathogenic to ruminants. It is questionable whether all strains of *L. monocytogenes* are able to cause human disease as epidemiological studies have revealed that the 13 *L. monocytogenes* serovars vary in their ability to cause infections. About 98% of human cases are due to serovars 1/2a, 1/2b, 1/2c and 4b, with the latter causing >50% of cases worldwide especially feto-maternal cases; see Section 4.3.5. There are three distinct evolutionary lineages of *L. monocytogenes*; lineage I (serovars 1/2b, 3b, 4b, 4d, 4e and 7), lineage II (serovars 1/2a, 1/2c, 3a and 3c) and lineage III (serovar 4a and 4c) (Weidman *et al.* 1997). Strains from foodborne epidemics and sporadic cases are in lineage I. Lineage II contains human and animal cases and no foodborne epidemic strains, and lineage III only contains animal pathogens (Hain *et al.* 2007). Complete genomes sequenced to date include *L. monocytogenes* serotypes 1/2a, 4a and 4b, *L. innocua* (serotype 6a), *L. welshimeri* (serotype 6b), *L. seeligeri* (serotype 1/2b) and *L. ivanovii* (serotype 5), and therefore span the genus divisions. There is a high degree of synteny in gene organisation and content between species. A low occurrence of transposons and insertion sequence elements in the genomes is the reason for a lack of inversions and genome plasticity (Hain *et al.* 2007). Nelson *et al.* (2004) reported that 83 genes were unique to *L. monocytogenes* serotype 1/2a. This included three gene clusters encoding for carbohydrate transport and metabolism, and an operon for the antigenic rhamnose substituents that are located in the serotype 1/2a cell wall associated teichoic acid polymer. There were 51 genes unique to serotype 4b, and lacked insertion sequence elements. These differences probably contribute to the differences in pathogenicity, environmental survival. In general, the *L. monocytogenes* genomes encode for intact glycolysis and pentose phosphate pathways, transportation and utilisation of a number of simple and complex sugars. These traits are associated with the ability to grow in the natural environment, and utilise a wide variety of carbohydrates.

All virulence determinants identified so far are chromosomally located and not plasmid borne. The virulence genes are localised on genetic islands. A 9 kb gene cassette associated with intracellular survival and cell-to-cell spread is known as the 'Listeria pathogenicity island 1' (LIPI-1) (Vazquez-Boland *et al.* 2001). This was previously known as the *hly-* and *prfA*-virulence gene cluster. This region is found in all evolutionary branches of *Listeria*, and is conserved in the core *Listeria* genome due to strong selective pressure. LIPI-1 encodes for four genes. Hly is a pore-forming toxin, and is also known as listeriolysin O in *L. monocytogenes*. This toxin is in the cholesterol-binding 'thiol-activated' family. The PAI also encodes for ActA, a surface protein causing the actin polymerisation and protrusion into the adjacent cell; PlcB, a phospholipase C; and Mpl, a zinc-metalloenzyme. Mpl and PlcB are involved in the bacterium escaping from the vacuole following engulfment into a phagosome. Other virulence determinants are Clp stress tolerance mediators (intraphagosomal survival), the Ami protein (cell attachment), a hexose phosphate transporter (intracellular proliferation), and internalin proteins (internalisation of epithelial cells). The latter contains a domain with a variable number of leucine-rich repeats (LRRs) which form a right-handed helix called a parallel β-helix. These are mostly found in eucaryotes and mediate protein–protein interactions. In prokaryotes, they are associated with virulence proteins; IpaH of *Sh. flexneri* and SspH1 and H2 of *S.* Typhimurium. Thirteen of the 25 *L. monocytogenes*-specific surface proteins (including all but two internalin proteins) are absent in lineage III serovars 4a and 4c, which are animal and not human pathogens. Therefore, surface protein diversity may be linked to host range. A second large (22kb) virulence gene cluster, LIPI-2, has been found in *Listeria ivanovii* which is not found in *L. monocytogenes*. It contains a sphingomyelinase gene and a large number of *inl* genes. Since internalin proteins may be involved in cell and host tropism, their encoding in this gene cluster may account for the more restricted host range of *L. ivanovii* to ruminants compared with *L. monocytogenes*. The LIPI-2 gene is flanked by the *tRNA*^arg and *ydeI* genes. In contrast, *L. welshimeri* has a prophage inserted between these same two genes. Therefore, this region may be a hot spot for genome alterations. *L. welshimeri* has a smaller genome than other *Listeria* species, and has 482 genes less than *L. monocytogenes*. This is probably the result of the loss of genes involved in virulence, intracellular survival and carbohydrate metabolism.

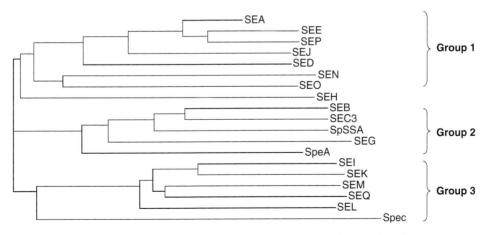

Figure 2.21 Phylogenetic analysis of *St. aureus* enterotoxin and streptococcal pyrogenic toxins.

2.9.8 *Staphylococcus aureus* enterotoxin phylogenetic analysis

St. aureus causes a variety of diseases including boils and carbuncles, but in food-related illness, it is principally the production of heat-stable toxins, termed enterotoxins, that causes vomiting (Section 4.3.7). The enterotoxins are divided serologically into five major types, SEA to SEE, of which SEC is further divided into three variants. Recently, a number of other SEs have been identified: SEG to SEQ. In addition, SEs are members of the pyrogenic toxin (PT) superantigens. Staphylococcal enterotoxins are heat stable and are able to withstand boiling for many hours, and therefore present a health risk as they are not denatured during cooking. Bioinformatic analysis of their protein sequences reveals the close phylogenetic relationship between the enterotoxins. The sequences can also be used to search databases (BLAST searching) for similar regions in other organisms. As shown in Figure 2.21, there are similarities between the enterotoxins of *St. aureus* and the streptococci pyrogenic exotoxins (PT). Phylogenetic analysis reveals three main groups; one group is composed of SEA, SEE, SEP and more distantly related SEJ with SED, and SEN and SEO, with SEH as an outliner to the group. The second group contains SEB, SEC and SEG, as well as streptococcoal PT A and *Strept. pyogenes* superantigen SSA. The third group includes SEI, SEK, SEL, SEM, SEQ with streptococcal pyrogenic exotoxin C as an outliner to the group.

3 The microbial flora of food and its preservation

3.1 Spoilage micro-organisms

Spoilt food is food that tastes and smells 'off'. It is due to the undesirable growth of micro-organisms that produce volatile compounds during their metabolism, and which the human nose and mouth detects. Spoilt food is not poisonous. The food is not of the quality expected by the consumer and therefore it is a quality, not a safety, issue. The terms 'spoilt' and 'unspoilt' are subjective since acceptance is dependent upon consumer expectation and are not related to food safety. Soured milk is unacceptable as a drink, but can be used to make scones. The overgrowth of *Pseudomonas* spp. is undesirable in red meat, yet desirable in hung game birds. Acetic acid production during wine storage is unacceptable, yet acetic acid production from wine (and beer) is necessary for the production of vinegar.

Food spoilage involves any change that renders the food unacceptable for human consumption. It may be due to a number of causes:

1 Insect damage.
2 Physical injury due to bruising, pressure, freezing, drying and radiation.
3 The activity of indigenous enzymes in animal and plant tissues.
4 Chemical changes not induced by microbial or naturally occurring enzymes.
5 The activity of bacteria, yeasts and moulds.

(Adapted from Forsythe & Hayes 1998)

During harvesting, processing and handling operations, food may become contaminated with a wide range of micro-organisms. Subsequently, during distribution and storage, the conditions will be favourable for certain organisms to multiply and cause spoilage. Which micro-organisms will develop or what biochemical and chemical reactions occur is dependent upon the food, intrinsic and extrinsic parameters (Section 2.6.1). Spoilage can be delayed by lowering the storage temperature (Figure 3.1).

For fresh foods, the primary quality changes may be categorised as follows:
- Bacterial growth and metabolism resulting in possible pH changes and formation of toxic compounds, off-odours, gas and slime formation.
- Oxidation of lipids and pigments in fat-containing foods resulting in undesirable flavours, formation of compounds with adverse biological effects or discolouration.

Spoilage is the visible growth of not only micro-organisms but also end-metabolites which are produced resulting in off-odours, gas and slime. The spoilage of milk (Section 3.1.2) illustrates these two aspects with the visible growth of thermoduric micro-organisms (*B. cereus*) and the production of undesirable taste (bitterness due to thermostable proteases).

Although there is much progress in the characterisation of the total microbial flora and metabolites developing during spoilage, not much is known about the identification of specific

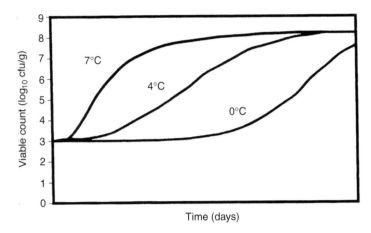

Figure 3.1 Temperature effect on food.

micro-organisms in relation to food composition. Despite the fact that food spoilage is a huge economical problem worldwide, the mechanisms and interaction leading to food spoilage are very poorly understood. Although during storage the total microbial flora may increase, it is specific spoilage organisms which cause the chemical changes and the production of off-odours (Figure 3.2). The shelf-life is dependent upon the growth of the spoilage flora and can be reduced through cold storage to retard microbial growth and packaging (vacuum packaging and modified atmosphere packaging; see Section 3.6.8). The higher the initial microbial load, the shorter the shelf-life due to the increased microbial activities (Figure 3.3).

A wide variety of micro-organisms may initially be present on food and grow if favourable conditions are present. On the basis of susceptibility to spoilage, foods may be classed as non-perishable (or stable), semi-perishable and perishable. The classification depends on the intrinsic factors of water activity, pH, presence of natural antimicrobial agents, and so on.

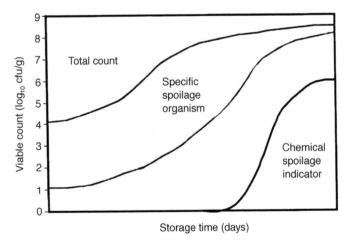

Figure 3.2 Food spoilage indicators.

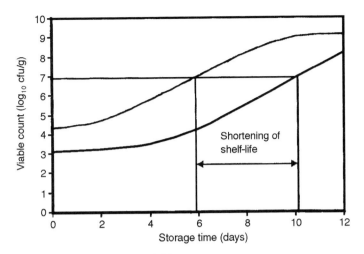

Figure 3.3 Effect of initial microbial load on shelf-life.

Flour is a stable product because of the low water activity. Apples are semi-perishable since poor handling and improper storage can result in slow fungal rot. Raw meat is perishable since the intrinsic factors of pH and water activity favour microbial growth.

3.1.1 Spoilage micro-organisms

Gram-negative spoilage micro-organisms

Pseudomonas, Alteromonas, Shewanella putrefaciens and *Aeromonas* spp. spoil dairy products, red meat, fish, poultry and eggs during cold storage. These foods have high water activity and neutral pH and are stored without a modified atmosphere (i.e. normal levels of oxygen). There are a variety of spoilage mechanisms including the production of heat-stable proteases and lipases which produce off-flavours in milk after the organism has been killed by pasteurisation. They also produce coloured pigments in egg spoilage. *Erwinia carotovora* and various *Pseudomonas* spp. are responsible for ~35% of vegetable spoilage.

Gram-positive non-spore forming spoilage micro-organisms

The lactic acid bacteria (Section 3.7.1) and *Brocothrix thermosphacta* are Gram-positive rods which typically cause spoilage of meats stored under modified atmosphere packaging or vacuum packaging. *Acetobacters* and *Pediococcus* spp. can produce a thick polysaccharide slime in beer, termed 'rope'. The production of diacetyl (see Section 3.7.3) by lactic acid bacteria causes beer spoilage. The organisms can also produce lactic acid in wine giving it a sour taste. The *Acetobacter* produce acetic acid in beer giving it a sour taste.

Gram-positive spore-forming spoilage micro-organisms

The spore-forming *Bacillus* spp. and *Clostridium* spp. can be important spoilage organisms of heat-treated foods, since the spores may survive the process. *B. cereus* can grow in pasteurised milk at 5°C and cause 'sweet curdling' (renin coagulation without acidification) and 'bitty cream'. *B. stearothermophilus* causes 'flat sour spoilage' of canned foods. It grows in the can, producing acids (hence the term 'sour') but without any gas production and subsequently the can does not swell. *Cl. thermosaccharolyticum* spoils canned foods, and due to gas production the can swells. *Desulphotomaculum nigrificans* produces hydrogen sulphide which causes canned foods to swell and smell. This is known as 'sulphur stinker' spoilage. *B. stearother-mophilus, D. nigrificans* and *Cl. thermosaccharolyticum* are thermophilic organisms. Therefore, they are only able to grow at temperatures above ambient. Hence, canned foods are processed to be 'commercially sterile' since under normal storage conditions no microbial growth is expected to occur.

Yeast and mould spoilage micro-organisms

Yeast and moulds are more tolerant of low water activity and low pH than bacteria. Subsequently, they typically spoil foods such as fruits and vegetables and bakery products (Table 3.1). They produce pectinolytic enzymes which soften the plant tissues causing 'rot'. As much as 30% of all fruit spoilage may be due to *Penicillium*. Fungi also produce large quantities of coloured sporangia which visibly colour the food. Bread is spoiled by *Rhizopus nigricans* ('bread mould', black spots), *Penicillium* (green mould), *Aspergillus* (green mould) and *Neurospora sitophila* (red bread). Osmophilic yeasts (*Saccharomyces* and *Torulopsis* spp.) are able to grow in high sugar concentrations (65–70%) and spoil jams, syrups and honey. Fungi also cause various types of meat spoilage such as 'whiskers' due to *Mucor, Rhizopus* and *Thamnidium*.

Standard preservation methods to control fungal growth are high temperatures to denature spores, weak acid preservatives, such as sorbic acid, and antibiotics such as natamycin. However, some fungi have become resistant to sorbic acid and some fungi degrade sorbic acid to the malodourous compound pentadiene.

Table 3.1 Food spoilage fungi

Organisms	Products typically affected
Aspergillus versicolor	Bread and dairy products
A. flavus	Cereals and nuts
A. niger	Spices
Byssochlamys fulva	Cereals in airtight packs
Fusarium oxysporum	Fruit
Mucor spp.	Meat
Neosartorya fischeri	Pasteurised foods
Neurospora sitophila	Bread
Penicillium roqueforti	Meat, eggs and cheese
P. expansum	Fruits and vegetables
P. commune	Margarines
P. discolor	Cheese
Penicillium spp.	Bread
Rhizopus nigricans	Bread
Rhizopus spp.	Meat
Saccharomyces spp.	Soft drinks, jams, syrups and honey
Thamnidium spp.	Meat
Torulopsis spp.	Jams, syrups and honey
Trichoderma harzianum	Margarines
Zygosaccharomyces bailii	Dressings

Reprinted from Brul & Klis 1999, with permission from Elsevier.

3.1.2 Spoilage of diary products

Milk is an ideal growth medium for bacteria and therefore needs to be kept refrigerated. The intrinsic flora ($\sim 10^2$–10^4 cfu/mL) is derived from the cow's milk ducts in the udder and the milking equipment, and so on, during production. The flora includes *Pseudomonas* spp., *Alcaligenes* spp., *Aeromonas* spp., *Acinetobacter-Moraxella* spp., *Flavobacterium* spp., *Micrococcus* spp., *Streptococcus* spp., *Corynebacterium* spp., *Lactobacillus* spp. and lactose-fermenting *Enterobacteriaceae*. Milk spoilage is primarily due to the growth of the psychrophilic micro-organisms which produce heat-stable lipases and proteases which are not denatured during pasteurisation. The pseudomonads, flavobacteria and *Alcaligenes* spp. produce lipases which produce medium- and short-chain fatty acids from milk triglycerides. These fatty acids spoil the milk due to their rancid off-flavour. Proteases are produced by pseudomonads, aeromonads, *Serratia* and *Bacillus* spp. The enzyme hydrolyses milk proteins to produce bitter peptides. Therefore, it is undesirable to have a high microbial count before pasteurisation as the action of the residual enzymes during storage will result in a shortening of the milk's shelf-life.

Pasteurisation at 72°C for 15 seconds is designed to kill all pathogenic bacteria such as *Mycobacterium tuberculosis*, *Salmonella* spp. and *Brucella* spp. at levels expected in fresh milk. Thermoduric organisms are those that survive pasteurisation. These include *Streptococcus thermophilus*, *Enterococcus faecalis*, *Micrococcus luteus* and *Microbacterium lacticum*. The spores of *Bacillus cereus* and *B. subtilis* will survive the heat treatment. *B. cereus* growth causes pasteurised milk spoilage known as 'bitty cream'.

3.1.3 Spoilage of meat and poultry products

Meat (usually muscle) is a highly perishable food product with water activity (Section 2.6.2) suitable for the growth of most micro-organisms. Since meat is highly proteinaceous, it is relatively strongly buffered and the growth of micro-organisms does not significantly lower the pH. Because it is nutritious, it can quickly become spoilt through the growth of micro-organism and even poisonous if contaminated with foodborne pathogens. The meat itself is sterile within the animal's body. However, it can easily become contaminated during slaughter, abattoir practice, handling during processing and improper storage. If micro-organisms such as *Pseudomonas* spp., *Brochothrix thermosphacta* and lactic acid bacteria grow on the meat, then it will become spoilt and unacceptable for eating. It is the growth of foodborne pathogens such as *Salmonella*, toxin producing strains of *E. coli*, *L. monocytogenes*, *Cl. perfringens* and toxin production by *St. aureus* that are of concern with meat and poultry products.

Poultry skin can carry a range of spoilage organisms: *Pseudomonas* spp., *Acinetobacter/Moraxella* spp., *Enterobacter* spp., *Sh. putrefaciens*, *Br. thermosphacta* and *Lactobacillus* spp. The foodborne pathogen *C. jejuni* may also be present on the skin and subsequently transferred to work surfaces. Pathogens such as *S. Entertitidis* can infect the ovaries and the oviducts of hens and subsequently the egg prior to shell formation. Additionally, eggshells become contaminated with intestinal bacteria during passage through the cloaca and the incubator surface. *Cl. perfringens* is isolated in small numbers from raw poultry. It is unable to grow due to the cold temperature of storage and the presence of competitive psychrotrophic organisms. Bacteria found on poultry include *St. aureus*, *L. monocytogenes* and *Cl. botulinum* type C (not toxic to healthy adults). The *St. aureus* strains found on poultry are not pathogenic to man, and staphylococcal food poisoning associated with poultry is normally due to contamination of cooked meat by an infected food handler.

Carcass meat at temperatures above 20°C will readily be spoilt by bacteria originating from the animal's intestines which have contaminated the meat during slaughtering. The spoilage flora is dominated by mesophilic organisms such as *E. coli*, *Aeromonas* spp., *Proteus* spp. and *Micrococcus* spp. At temperatures below 20°C, the spoilage flora will be predominantly psychrotrophs such as *Pseudomonas* spp. (fluorescent and non-fluorescent types) and *Brochothrix thermosphacta*. Poultry meat will also contain small number of *Acinetobacter* spp. and *Sh. putrefaciens*. Refrigeration temperature of 5°C and below the spoilage flora will be dominated by pseudomonads. The pseudomonads are aerobes and hence only grow on the food surface to a depth of 3–4 mm in the underlying tissues. The spoilage is due to the degradation of proteins producing volatile off-flavours such as indole, dimethyl disulphide and ammonia. Chemical oxidation of unsaturated lipids results in a rancid off-flavour.

Fungal growth occurs under prolonged storage periods (Forsythe & Hayes 1998):
- 'Whiskers' are due to *Mucor*, *Rhizopus* and *Thamnidium*.
- Black spot is due to *Cladosporium herbarum* and *Cl. cladosporoides*.
- *Penicillium* spp. and *Cladosporium* spp. cause coloured spoilage due to their yellow- and green-coloured spores.
- 'White spot' is caused by *Sporotrichum carnis* growth.

3.1.4 Fish spoilage

Bacteria can be detected from fresh fish from the slime coat on the skin (10^3–10^5 cfu/cm^2), gills (10^3–10^4 cfu/g) and the intestines (10^2–10^9 cfu/g). Fish from the marine environments of the North Seas will be dominated by psychrotrophs whereas the fish from warmer waters will be colonised from a higher proportion of mesotrophs. The psychrophilic flora will be composed of *Pseudomonas* spp., *Alteromonas* spp., *Sh. putrefaciens*, *Acinetobacter* spp. and *Moraxella* spp. The mesophilic flora will contain micrococci and coryneforms as well as the acinetobacters. The flora of fish landed at ports will also include organisms from the ice used to preserve the fish and the flora of the boat holds. Due to the high psychrotrophic flora, spoilage occurs very rapidly since the low temperature storage favours psychrotrophs. At 7°C for 5 days, the microbial load can be as high as 10^8/g. This is faster than meat spoilage which would take about 10 days to obtain this microbial load. The off-odours of spoiled fish are probably produced by certain strains of pseudomonas producing volatiles esters (ethyl acetate, etc.) and volatile sulphide compounds (methyl mercaptan, dimethyl sulphide, etc.). Trimethylamine oxide is reduced by certain fish spoilage organisms, such as *Sh. putrefaciens*, to produce trimethylamine which gives a characteristic smell of spoiled fish. *Photobacterium phosphoreum* (a luminescent bacterium) causes the spoilage of vacuum-packed cod (Dalgaard *et al.* 1993).

3.1.5 Egg spoilage

Eggs contain a barrage of antimicrobial factors including iron chelating agents (conalbumin) and lysozyme in the albumen (egg white). The shell is covered with a water-repellent cuticle and two inner membranes. In contrast, the yolk does not contain any antimicrobial factors. Egg spoilage is principally due to *Pseudomonas* spp. as well as *Proteus vulgaris*, *Alteromonas* spp. and *Serratia marcescens*. These produce a variety of coloured rots (Table 3.2).

3.2 Shelf-life indicators

The shelf-life is an important attribute of all foods. It can be defined as the time between the production and packaging of the product and the point at which it becomes unacceptable to the consumer. It therefore is related to the total quality of the food and linked to production design, ingredient specifications, manufacturing process, transportation and storage (at retail and in the home). The shelf-life depends upon the food itself (Table 3.3) and it is essential

Table 3.2 Bacterial causes of egg spoilage

Spoilage organism	Type of rot
Pseudomonas spp.	Green
Pseudomonas, Proteus, Aeromonas, Alcaligenes and *Enterobacter* spp.	Black
Pseudomonas spp.	Pink
Pseudomonas and *Serratia* spp.	Red
Acinetobacter-Moraxella spp.	Colourless

for the food manufacturer to identify the intrinsic and extrinsic parameters which limit the shelf-life.

Shelf-life can be determined by combined microbiological and chemical analysis of food samples taken over the products' anticipated shelf-life. There are two approaches to shelf-life determination:

1 *Direct determination and monitoring*: This requires batches of samples to be taken at specified stages in the development of the product. Typically, these samples are taken at intervals equal to 20% of the expected shelf-life to give samples of six different ages. The samples are stored under controlled conditions until their quality becomes unacceptable. Attributes tested range from smell, texture, flavour, colour and viscosity. This method is not ideal for products with shelf lives in the order of 1 year.

2 *Accelerated estimation*: The need to meet product launch dates may require the use of accelerated shelf-life estimations by raising the storage temperature to increase any ageing processes. This method has to be used with care since different microbial floras may develop, which differ in off-flavour formation from the un-accelerated spoilage flora.

Suggested microbiological shelf-life limits (index of spoilage) are given in Table 3.4.

There is a wide range of compounds, principally end products of microbial growth, which have potential for use in the determination and prediction of shelf-life (Dainty 1996).

Chemical indicators of food spoilage are as follows:

1 *Glucose*: Glucose is the main substrate for microbial growth of red meats, including modified atmosphere and vacuum packaged. Spoilage is associated with post-glucose utilisation of amino acids by pseudomonads and hence monitoring glucose depletion can indicate onset of spoilage. The technique is of limited value however due to an ill-defined lag phase before spoilage.

2 *Gluconic and 2-oxogluconic acid*: *Pseudomonas* metabolism of glucose results in the accumulation of gluconic and 2-oxogluconic acids in beef.

Table 3.3 Food products and associated shelf lives

Food product	Typical shelf-life
Bread	Up to 1 week at ambient temperature
Sauces, dressings	1–2 years at ambient
Pickles	2–3 years at ambient
Chilled foods	Up to 4 months at 0–8°C
Frozen foods	12–18 months in freezer cabinets
Canned foods	Unlacquered cans, 12–18 months Lacquered cans, 2–4 years

Table 3.4 Suggested microbiological limits for end of shelf-life

Product	Microbial count	Comments
Raw meat	1×10^6 cfu/g aerobic plate count	Visible deterioration and/or slime at 1×10^7 cfu/g
Ground beef	1×10^7 cfu/g	End of shelf-life, colour begins to fade and slime forms
Vacuum-packaged cooked products	1×10^6 cfu/g aerobic plate count	Represents the point where a pH shift > 0.25 occurs
Fully cooked products	1×10^6 cfu/g aerobic plate count 1×10^3 cfu/g *Enterobacteriaceae* 1×10^3 cfu/g lactic acid bacteria 500 cfu/g yeast and moulds	

3 L- *and* D-*lactic acids, acetic acid and ethanol*: There is an ill-defined lag between maximum microbial cell number and sensory detection of spoilage on vacuum-packed and low oxygen modified atmosphere packed red meats. Therefore, the production of L- and D-lactic acids, acetic acid and ethanol from glucose may be good indicators of the onset of spoilage in certain foods such as pork and beef. Dainty (1996) reported that acetate levels greater than 8 mg/ 100 g meat indicated a microbial flora $>10^6$ cfu/g.

4 *Biologically active amines*: Tyramine is produced by certain lactic acid bacteria and has no sensory property of relevance to spoilage but has vasoactive properties. It is detectable (0.1–1.0 mg/100 g meat) in vacuum-packed beef with high microbial loads ($>10^6$ cfu/g).

5 *Volatile compounds*: The major advantages of volatile compound determination is that no method of extraction from food is required and it enables the simultaneous determination of microbial and chemical activities. Vallejo-Cordoba and Nakai (1994) demonstrated the presence of nine volatile compounds during milk spoilage which could be attributed to spoilage bacteria; 2- and 3-methylbutanals, 2-propanol, ethyl hexanoate, ethyl butanoate, 1-propanol, 2-methylpropanol and 1-butanol. Other volatile compounds could be attributed to chemical oxidation of lipids. Stutz *et al.* (1991) proposed acetone, methyl ethyl ketone, dimethyl sulphide and dimethyl disulphide as indicators of ground beef spoilage. Other volatiles include acetoin and diacetyl for pork spoilage and trimethylamine for fish (Dalgaard *et al.* 1993).

Shelf-life of foods can be determined microbiologically using the following:

1 *Storage trials*: As described above, samples are taken at timed intervals and analysed for total microbial load and specific spoilage organisms such as pseudomonads, *Brochothrix thermosphacta* and lactic acid bacteria. The viable counts are compared with chemical and sensory evaluation of the product and correlations between the variable determined to identify key indicators of early food spoilage.

2 *Challenge tests*: Samples of the food are incubated under conditions which reproduce the large-scale food production and storage period. The food may be inoculated with specified target organisms of interest, such as spore formers, for example *Clostridium sporogenes* as a model for *Clostridium botulinum* survival.

3 *Predictive modelling*: This technique is described in more detail in Section 2.8. Its advantages are that the method can simultaneously predict the growth of micro-organisms over a range of conditions broader than feasible in a microbiology laboratory (Walker 1994). The essential aspect is the need to validate the model using published and in-house laboratory data. A

major limitation at present is that the majority of predictive models are concerned with food pathogens, whereas it is the food spoilage organisms which primarily limit a product's shelf-life.

3.3 Methods of preservation and shelf-life extension

Preservation methods either:
1 prevent pathogen access to the food,
2 inactivate any pathogens which have gained entry, or
3 prevent or slow down the growth of pathogens, if previous methods have failed.
All foods can be spoilt between harvesting, processing and storage before consumption (Gould 1996). Spoilage can be due to physical, chemical and microbial factors. Most preservation methods are designed to inhibit the growth of micro-organisms (Table 3.5).

Methods which prevent or inhibit microbial growth are chilling, freezing, drying, curing, conserving, vacuum packaging, modified atmosphere packaging, acidifying, fermenting and adding preservatives (Tables 2.5). Other methods inactivate the micro-organisms, that is pasteurisation, sterilisation and irradiation. Novel methods include the use of high pressure. The major preservation methods are based upon the reduction of microbial growth due to unfavourable environmental conditions (cf. intrinsic and extrinsic parameters; Section 2.6.1) such as temperature reduction, lowering of pH and water activity and denaturation due to heat treatment. Due to consumer pressure, the trend in recent years has been to use less severe, milder preservation methods including combined methods such as cooked-chill foods (prolonged shelf-life) and modified atmosphere packaging (prolonged quality). The effect on microbial growth with regard to adaptive stress response was considered in Section 2.7.

Table 3.5 Preservation methods and effect on micro-organisms

Effect on micro-organisms	Preservation factor	Method of achievement
Reduction or inhibition of growth	Low temperature	Chill and frozen storage
	Low water activity	Drying, curing and conserving
	Restriction of nutrient availability	Compartmentalisation in water-in-oil emulsions
	Lowered oxygen level	Vacuum and nitrogen packaging
	Raised carbon dioxide	Modified atmosphere packaging
	Acidification	Addition of acids: fermentation
	Alcoholic fermentation	Brewing: vinification: fortification
	Use of preservatives	Addition of preservatives: inorganic (sulphite, nitrite); organic (propionate, sorbate, benzoate, parabens); antibiotic (nisin, natamycin)
Inactivation of micro-organisms	Heating	Pasteurisation and sterilisation
	Irradiation	Ionising irradiation
	Hydrostatic pressure	Application of high hydrostatic pressure

Reprinted from Gould 1996, with permission from Elsevier.

Conventional food processing has relied on heat to kill microbial contaminants, which simultaneously results in physical and chemical changes in the food. Increasingly, the public have requested 'fresh' produce which has helped drive the development of alternative technologies that result in minimal changes in sensory and nutritional characteristics. Methods of predicting food safety have largely been based on thermal treatments, and these must be compared with alternative technologies to evaluate their contribution in food safety.

To validate the effectiveness of a preservation treatment, microbial survival must be evaluated. The term 'surrogate' can be used for micro-organisms added to the food as a 'spike'. Surrogates can be laboratory cultures, which have well-defined characteristics and are non-pathogenic. For examples, surrogates for *Cl. botulinum* include *Cl. sporogenes* and *B. stearothermophilus*. They should also have inactivation characteristics which can be used to predict the survival of the target organism. The validity of a preservation method is usually confirmed using an inoculated test pack which is subject to representative conditions of processing, handling and packaging.

Sublethal cellular damage of bacteria is of particular with preservation methods (cf. Section 5.2.2). Processes aimed to inactivate pathogenic bacteria should also consider the susceptibility of other micro-organisms such as viruses, fungi and parasites to the treatment. Two bacterial genera of importance to food safety produce spores, the aerobic *Bacillus* and anaerobic *Clostridium* species. The spores are resistant to heat, chemicals, irradiation, and other environmental stresses. Though the pasteurisation process inactivates vegetative bacterial cells, it is ineffective for bacterial spores. Therefore, commercial sterilisation processes are designed to inactivate bacterial spores of *B. stearothermophilus* (a thermophile) and *Cl. botulinum*.

3.4 The hurdle concept

Inhibiting microbial growth, as opposed to killing microbes by heat treatment, can be used to preserve foods. Processing with inhibit microbial growth includes low temperature (chilling, freezing), reduced water activity (dried products) and acidification. These can be combined as a series of barriers, or hurdles, to the growth of food spoilage and poisoning organisms. Hence, the hurdle concept combines both physical and chemical parameters to preserve food. The interaction of these factors on microbial growth can be seen in Figure 2.16.

3.5 Preservatives

Preservative agents are required to ensure that manufactured foods remain safe and unspoiled during its whole shelf-life. A range of preservatives are used in food manufacture including traditional foods (Table 3.6). Many preservatives are effective under low-pH conditions: benzoic acid (<pH 4.0), propionic acid (<pH 5.0), sorbic acid (<pH 6.5), sulphites (<pH 4.5). The parabens (benzoic acid esters) are more effective at neutral pH conditions.

3.5.1 Organic acids

Weak organic acids are the most common classical preservative agents. Examples are acetic, lactic, benzoic and sorbic acids which inhibit the growth of both bacterial and fungal cells. Sorbic acid also inhibits the germination and outgrowth of bacterial spores. The addition of 0.2% calcium propionate to bread dough delays *B. cereus* germination sufficiently to reduce the risk of food poisoning to negligible. Benzoic acid at 500 ppm is used to preserve fruit juice-based beverages. Sulphur dioxide concentrations are controlled in European regulation to a limit of 10 ppm.

Table 3.6 Antimicrobial food preservatives

Preservative (typical concentration range, mg/kg)	Examples of use
Weak organic and ester preservatives	
Propionate (1–5000)	Bread, bakery and cheese products
Sorbate (1–2000)	Fresh and processed cheese, dairy products, bakery products, syrups, jams, jellies, soft drinks, margarines, cakes, dressings
Benzoate (1–3000)	Pickles, soft drinks, dressings, semi-preserved fish, jams, margarines
Benzoate esters (parabens, 10)	Marinated fish products
Organic acid acidulants	
Lactic, citric, malic, acetic acids (no limit)	Low-pH sauces, mayonnaises, dressings, salad creams, drinks, fruit juices and concentrates, meat and vegetable products
Inorganic acid preservatives	
Sulphite (1–450)	Fruit pieces, dried fruit, wine, meat sausages
Nitrate and nitrite (50)	Cured meat products
Mineral acid acidulants	
Phosphoric acid, hydrochloric acid	Drinks
Antibiotics	
Nisin	Cheese, canned foods
Natamycin (pimaricin)	Soft fruit
Smoke	Meat and fish

Reprinted from Gould 1996, with permission from Elsevier.

In solution, weak acid preservatives exist in a pH-dependent equilibrium between the undissociated and dissociated state (measured as the pK value). Weak acid preservatives have an optimal inhibitory activity at low pH because of the higher concentration of the uncharged, undissociated state of the acid. The plasma membrane is freely permeable to this molecule (Figure 3.4). Inside the cell, it will dissociate due to the near neutral pH, resulting in the generation of charged anions and protons which accumulate. The inhibition of bacterial growth can therefore be due to several factors: disruption of membrane function and the inhibition of essential metabolic reactions (Bracey *et al.* 1998, Eklund 1985). However in yeast, the inhibitory action may be due to the induction of a stress response to restore the cell homeostasis, and which is energetically demanding (Holyoak *et al.* 1996). The concentration of undissociated acids that inhibit micro-organisms is given in Table 3.7.

3.5.2 Hydrogen peroxide and lactoperoxidase system
The lactoperoxidase system, found in fresh milk, has profound antimicrobial effects against both bacteria and fungi (de Wit & van Hooydonk 1996). The system requires hydrogen peroxide and thiocyanate for optimal activity and is primarily active against micro-organisms producing H_2O_2. Alternatively, hydrogen peroxide can be added to the foods that are to be preserved. Under suitable experimental conditions, the reaction generates short-lived singlet oxygen molecules which is extremely biocidal (Figure 3.5; Tatsozawa *et al.* 1998). Furthermore,

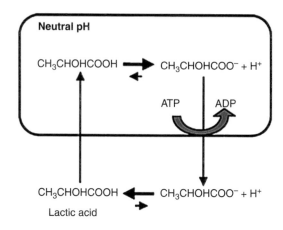

Figure 3.4 Mode of action of organic acids.

Table 3.7 Inhibitory concentration (%) of dissociated organic acids

Acid	*Enterobacteriaceae*	*Bacillaceae*	Yeasts	Moulds
Acetic	0.05	0.1	0.5	0.1
Benzoic	0.01	0.02	0.05	0.1
Sorbic	0.01	0.02	0.02	0.04
Propionic	0.05	0.1	0.2	0.05

Various sources including Borch *et al*. 1996, ICMSF 1996a, Mortimore & Wallace 1994.

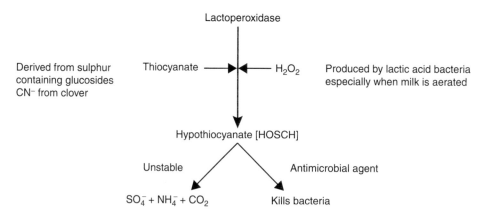

Figure 3.5 The lactoperoxidase system.

during incomplete reduction of molecular oxygen, the superoxide radical is generated. Due to the Fenton reaction, H_2O_2 reacts with the superoxide radical and trace amounts of transition metal ions (e.g. Fe (II)) form the extremely biocidal hydroxyl radical (Luo *et al.* 1994). A wide range of both Gram-negative and Gram-positive bacteria are inhibited by the lactoperoxidase system. However, studies have shown that Gram-negative bacteria were generally found to be more sensitive to lactoperoxidase-mediated (food) preservation than Gram-positive species.

Hydrogen peroxide on its own is also known to be bactericidal, depending on the concentrations applied and on environmental factors such as pH and temperature (Juven & Pierson 1996). Temperature is an extremely important parameter in determining the sporicidal efficacy of hydrogen peroxide. H_2O_2 is weakly sporicidal at room temperatures but very potent at higher temperatures. While the mechanism by which hydrogen peroxide kills spores is not known, the killing of vegetative bacteria and fungi is known to involve DNA damage. In the United States, the regulatory authorities allow the direct addition of hydrogen peroxide to food products such as raw milk for the preparation of certain cheese variants, whey intended for use in modified whey preparation, corn starch and dried eggs, and for the decontamination of packaging material (Juven & Pierson 1996). Several other procedures where H_2O_2 is used as a preservative have been reported, such as fruit and vegetable disinfection and raisin decontamination (Falik *et al.* 1994).

3.5.3 Chelators

Chelators that can be used as food additives include the naturally occurring acid, citric acid and the disodium and calcium salts of ethylenediaminetetraacetic acid (EDTA). EDTA is known to potentiate the effect of weak acid preservatives against Gram-negative bacteria. Citric acid inhibits the growth of proteolytic *Cl. botulinum* due to its Ca^{2+} chelating activity (Graham & Lund 1986).

3.5.4 Natural antimicrobials

In recent years, there has been a change towards more physiologically based approaches to food preservation. In addition to the weak organic acids, H_2O_2 and certain chelators, a few other antimicrobial compounds are permitted by the regulatory authorities for inclusion in foods. Some of these compounds are naturally present in spices (Table 3.8). The compounds are eugenol in cloves, allicin in garlic, thymol in rosemary, cinnamic aldehyde and eugenol in cinnamon, and allyl isothiocyanate in mustard. Essential oils from plants such as basil, cumin, caraway and coriander have inhibitory effects on organisms such as *A. hydrophila*, *Ps. fluorescens* and *St. aureus* (Wan *et al.* 1998). Since many of the essential oils are hydrophobic, they disrupt the cell membrane resulting in the loss of function (Brul & Coote 1999). However, the concentrations at which cell death is achieved are above tolerable taste thresholds for them to be used as major preservatives. Also, it is debatable whether the levels in fresh garlic are significant enough to exert any antibacterial affect; there is no allicin in cooked garlic. In practice, most of the above-mentioned compounds are used at concentrations that may have only limited growth inhibitory effects on micro-organisms. An extensive overview of the antimicrobial (and antioxidant) properties of spices can be found in Hirasa and Takemasa (1998).

3.5.5 Non-acidic preservatives

Nitrate and nitrite are traditional curing salts used to prevent microbial growth, especially *Cl. botulinum.* Many organisms in the microbial flora on the food are able to reduce nitrate

Table 3.8 Concentration of essential oils in some spices and antimicrobial activity of active components

Spice	Essential oil in whole spice (%)	Antimicrobial compounds in distillate or extract	Antimicrobial concentration (ppm)	Organisms
Allspice (*Piementa dioica*)	3.0–5.0	Eugenol Methyl eugenol	1000 150	Yeast *Acetobacter* *Cl. botulinum* 67B
Cassis (*Cinnamomum cassis*)	1.2	Cinnamic aldehyde Cinnamyl acetate	10–100	Yeast *Acetobacter*
Clove (*Syzgium aromaticum*)	16.0–19.0	Eugenol Eugenol acetate	1000 150	Yeast *Cl. botulinum* *V. parahaemolyticus*
Cinnamon Bark (*Cinnamomum zeylanicum*)	0.5–1.0	Cinnamic aldehyde Eugenol	10–1000 100	Yeast, *Acetobacter* *Cl. botulinum* 67B *L. monocytogenes*
Garlic (*Allium sativum*)	0.3–0.5	Allyl sulphonyl Allyl sulphide	10–100	*Cl. botulinum* 67B *L. monocytogenes* Yeast, bacteria
Mustard (*Sinapis nigra*)	0.5–1.0	Allyl isothionate	22–100	Yeast, *Acetobacter* *L. monocytogenes*
Oregano (*Origanum vulgare*)	0.2–0.8	Thymol Carvacrol	100 100–200	*V. parahaemolyticus* *Cl. botulinum* A, B, E
Paprika (*Capsicum annuum*)		Capsicidin	100	*Bacillus*
Thyme (*Thymus vulgaris*)	2.5	Thymol Carvacrol	100 100	*V. parahaemolyticus* *Cl. botulinum* 67B Gram-positive bacteria *Asp. parasiticus* *Asp. flavus* Aflatoxin B$_1$ and G$_1$

Adapted from International Commission on Microbiological Specification for Foods (ICMSF) 1998a, with kind permission of Springer Science and Business Media.

to nitrite, which is subsequently reduced to nitric oxide and ammonia. Nitric oxide in *Cl. botulinum* binds to the pyruvate phosphotransferase enzyme preventing ATP generation. Nitric oxide also binds to the haem ring of myoglobin in meat tissue forming nitrosomyoglobin which produces nitrosylhaemochrome which is the characteristic pink colouration of cooked cured meats.

Nisin and pediocin PA-1 are examples of Class IIa bacteriocins from lactic acid bacteria that are used as food preservatives, primarily in dairy products, and has an established safety record. It has a broad inhibitory spectrum against Gram-positive organisms and even prevents the outgrowth of bacterial spore. Nisin is an antibacterial peptide composed of 34 amino acids antibacterial peptide. It is produced on an industrial scale from certain strains of *Lactococcus lactis* belonging to serological group N. It is a lantibiotic which is a novel class of antimicrobial peptide that contains atypical amino acids and single sulphur lanthionine rings (Figure 3.6). It has two natural variants 'nisin A' and 'nisin Z' which differ in a single amino acid residue at position 27 (histidine in nisin A and asparagine in nisin Z). Their mode of action is primarily

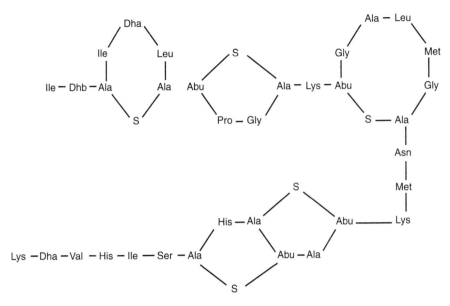

Figure 3.6 The structure of nisin.

at the cytoplasmic membrane of susceptible micro-organisms. The bacteriocins dissipate the proton motive force which stops the generation of ATP via oxidative phosphorylation. Models for nisin interaction with membranes propose that the peptide forms poration complexes in the membrane through a multistep process of binding, insertion and pore formation which requires a trans-negative membrane potential of 50–100 mV.

Intermediate moisture foods have water activity values between 0.65 and 0.85 (Section 2.6.2). This is equivalent to moisture content of ~20–40%, but this conversion must be used with care as free water is not synonymous with available water.

Heat treatment can be used synergistically with water activity. Food with pH values greater than pH 4.5 would normally require a botulinum cook (Section 2.5.2) to ensure safety. However, in the presence of curing salt (nitrate and nitrite), the water activity is reduced and a shelf-stable product with a long, safe ambient stability can be achieved that has had a milder (pasteurisation) heat treatment.

Note, due to the rise in antibiotic resistance amongst bacterial pathogens, antimicrobial agents based on the nisin structure and mode of action are under evaluation for use as therapeutic antibiotics. Current food usage of nisin has not been associated with any acquired resistance, but whether this will change in the future due to the use of nisin-based antibiotics remains to be seen.

3.5.6 Preservation due to weak acids and low pH

Foods can be divided into two groups: low-acid food with pH values above pH 4.5 and high-acid foods with pH values less than pH 4.5. The importance of food's pH value is that *Cl. botulinum* is unable to grow at pH values less pH 4.5. Therefore, processing high-acid foods do not need to take into account the possible growth of *Cl. botulinum*.

Most food preservatives are effective at acid pH values. Organic acids include acetic, propionic, sorbic and benzoic acids. These compounds in their undissociated form are membrane soluble. In the cytoplasm, the acid dissociates with the resultant release of a proton. The expulsion of the proton is energy demanding and therefore restricts cell growth. Carbon dioxide forms a weak acid, carbonic acid, in solution. It prevents the growth of pseudomonads at levels as low as 5%, whereas the lactic acid bacteria are unaffected. Therefore, modified atmosphere packaging (Section 3.6.8) principally changes the spoilage flora.

3.6 Physical methods of preservation

This section covers a wide range of non-chemical methods of preservation, ranging from traditional heat treatment to developments in modified atmosphere packaging.

3.6.1 Preservation by heat treatment

Heat treatment is a ubiquitous means of preparing food and has been in use for thousands of years. This can be in the form of cooking, boiling, roasting, or baking, and can be at the domestic as well as industrial level of food preparation. These will cause physico-chemical changes which may improve the food's taste, smell, appearance and digestibility. In addition, it can reduce the level of microbial contamination. All micro-organisms have a temperature growth range (see Section 2.6.4, Table 2.8). Since most pathogenic bacteria are mesophilic, they will die at cooking temperatures, and heat-labile toxins will be inactivated. In addition, spoilage organisms will be killed and so the heat treatment of food can prolong its shelf-life as well as reduce bacterial pathogens. Heat treatment of food is necessary to kill bacterial pathogens which may be present from environmental and intestinal contamination. Fruits and vegetables similarly need to be heat treated due to possible contamination from water and handling.

Pasteurisation

Milk is often pasteurised using the high-temperature short time (HTST) 72°C, 15 seconds process. This is designed to kill all pathogenic bacteria at levels expected in fresh milk such as follows:

Mycobacterium tuberculosis
Salmonella spp.
Brucella spp.
Campylobacter spp.
Cryptosporidium parvum
Diphtheria, *Corynebacterium diphtheriae*
L. monocytogenes
Poliomyelitis
Q fever, *Coxiella burnetti*
Pathogens, *Streptococcus* spp.

Thermoduric organisms are those that survive pasteurisation. These include *Streptococcus thermophilus*, *Enterococcus faecalis*, *Micrococcus luteus* and *Microbacterium lacticum*. The spores of *Bacillus cereus* and *B. subtilis* will survive the heat treatment. *B. cereus* growth causes pasteurised milk spoilage known as 'bitty cream'. Pasteurisation time and temperature regimes for other food products are given in Table 3.9.

Sterilisation

Milk sterilisation (130°C, at least 1 second) is a well-established method for prolonged milk storage. The heat treatment is severe to kill all micro-organisms present, both spoilage and foodborne pathogens to an acceptable level. There is a statistical chance of an organism

Table 3.9 Pasteurisation time and temperature regimes

Food	Pasteurisation process	Main objective	Secondary effects
Milk	63°C, 30 minutes 71.5°C, 15 seconds	Kill pathogens: *Br. abortis, My. tuberculosis, C. burnetti*	Kill spoilage micro-organisms
Liquid egg	64.4°C, 2.5 minutes 60°C, 3.5 minutes	Kill pathogens	Kill spoilage micro-organisms
Ice cream	65°C, 30 minutes 71°C, 10 minutes 80°C, 15 seconds	Kill pathogens	Kill spoilage micro-organisms
Fruit juice	65°C, 30 minutes 77°C, 1 minute 88°C, 15 seconds	Enzyme inactivation: pectin esterase and polygalacturonase	Kill spoilage yeast and fungi
Beer	65–68°C, 20 minutes (in bottle) 72–75°C, 1–4 minutes (900–1000 kPa)	Kill spoilage micro-organisms: yeast and lactic acid bacteria	

surviving the process, but this is acceptable in the normal sense of safe food production. The 12-D botulinum cook of canned foods is considered in Section 2.5.2.

Sous-vide

Sous-vide products are vacuum packed, undergo a mild heat treatment and have very carefully controlled chilled storage to prevent the outgrowth of spore-forming pathogens, notably *Cl. botulinum*. The heat treatment should be equivalent to 90°C for 10 minutes to kill the spores of psychrotrophic *Cl. botulinum*. Spoilage organisms, especially psychrotrophs, are also reduced during the heat treatment and this prolongs the shelf-life. (i.e. 3 weeks, 3°C). Mesophiles and thermophiles tend to survive mild heat treatments to a greater extent than psychrophiles, but are unable to multiply at the chill storage temperatures.

3.6.2 High-pressure treatment

The use of pressure technology is a novel means of food preservation which is not yet in large-scale usage (Hendrickx *et al.* 1998). High-pressure treatment can be used to preserve food as it kills micro-organisms, including spores (Kalchayanand *et al.* 1998). The advantage of high-pressure treatment is that the nutritional and sensory qualities of the food are only slightly affected, unlike thermal (heat) processes.

The key objective of pressure treatment is the reduction in foodborne pathogens viable count by c. 8 log cycles (8D). Inhibition of microbial growth is found at pressures of 20–130 MPa. Higher pressures of 130–800 MPa (depending on the organism and food matrix) are necessary for microbial cell death. Figure 3.7 shows the recovery of *St. aureus* following high-pressure treatments. The number of survivors depends upon the pressure treatment, and also the recovery media, with greater recovery being achieved using non-selective media. The site of cell damage is possibly the cytoplasmic membrane and ribosomes. Exposure of *E. coli* to high pressure results in the synthesis of HSPs, CSPs and proteins only associated with high-pressure exposure. Intracellular membrane damage is the most likely initial target in high-pressure treatment of yeast cells.

The use of pressure technology has been proposed to inactivate bacterial spores. Spore germination is induced at low pressure (100–250 MPa) followed by inactivation at high pressure

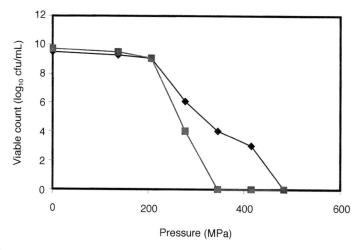

Figure 3.7 *St. aureus* survival following high-pressure treatment. Reprinted from Kalchayanand *et al.* 1998, with permission from Elsevier.

500–600 MPa). A combination of pressure cycling and other preservation methods may be required for the control of *Clostridium* species as the spores of these organisms have greater pressure tolerance than the *Bacillus* species spores. Moderate hydrostatic pressure (270 MPa) at 25°C only reduces the viable count by ~1.3 log cycles. However, 5 log cycle kills can be achieved by combined treatments conditions of 35°C and the inclusion of bactericidal compounds such as pediocin and nisin (Figure 3.6, Kalchayanand *et al.* 1998). The process can be combined (synergistically) with heat treatment such that lower pressures of 100–200 MPa are required.

As high pressure as little effect on chemical constituents, the natural flavour, texture, appearance and nutrient content of the food are similar to fresh or raw food. Hence, this technology has considerable potential in the future for the production of safe food which is acceptable to the consumer.

3.6.3 Ohmic heating and radio frequency

In pasteurising meat, the aim is to eliminate pathogens and reduce the number of spoilage organisms. Consequently, under refrigerated conditions, the meat will have a reasonable shelf-life. However, with large meat products, it is necessary to extend the cooking period as the heat transfer within meat is relatively slow. This can result in overheating of the surface. Alternative heating methods include ohmic heating and radio frequency. With both methods, heat is generated within the product due to ionic friction. In ohmic heating, electrical energy is passed directly into the meat, whereas in radio frequency, the electrical energy is initially converted to electromagnetic radiation which is then transferred to the food. Therefore, radio frequency radiation will pass through plastic packaging materials, without any requirement for direct contact with electrodes, as per ohmic heating.

The heating is due to the generation of an electric field with positive and negative regions. The ions in the food move to the corresponding region of the field. In ohmic heating, the polarity of the field is changing at 50–60 Hz, whereas in radio frequency, the polarity changes at a high frequency of 27.12 MHz. The oscillation of the ions causes the internal heating.

3.6.4 Pulsed electric fields

Pulsed electric fields (PEF) is one of the most promising emergent technologies for achieving a gentle, non-thermal pasteurisation process. The first applications were proposed in the 1960s, but it took until 2005 for the first commercial use. The process can be applied continuously, with very little residence time.

For PEF treatment, a pulse modulator and a treatment chamber are required. The pulse modulator transforms the sinusoidal alternating current to pulses with sufficient peak voltage and pulse energy. Disintegration of plant or animal cells requires ~1 kV/cm electric field, whereas microbial disintegration requires electric strengths above 15 kV/cm. In PEF, 20–100 electric pulses of field strengths from 25 to 70 kV/cm are applied in microsecond periods to the food (Barsotti *et al.* 1999, Jeyamkondan 1997). The PEF generates pores in the membranes of micro-organisms due to dielectric breakdown, resulting in the loss of membrane integrity and function.

3.6.5 Ultrasound

Ultrasound waves inactivate micro-organisms by introducing alternating compression and expansion cycles in a liquid medium. The bactericidal effect of high-intensity ultrasound waves is caused during the expansion phase of ultrasound, when small bubbles grow until they implode. The temperature and pressures reached inside these bubbles can become extremely high. Ultrasound has not been adapted for food preservation, probably because of its adverse effect on the quality of the treated food when applied at the intensity required to kill bacteria. However, by using ultrasound in conjunction with pressure and heat, it may be possible to enhance the lethality of this technology for food applications (Raso *et al.* 1998).

3.6.6 Intense light pulse

Intense light pulse (ILP) is another non-thermal preservation technology (Rowan *et al.* 1999). This uses intense and short-duration pulses of broad spectrum of light 20 000 times more intense than sunlight to inactivate micro-organisms due to photochemical and/or photothermal effect (Wuytack *et al.* 2003). The degree of inactivation is dependent upon the light intensity and the number of pulses delivered.

The method works on vegetative bacterial cells and spores. The process can decontaminate the surface of the food, without penetrating and therefore retains the organoleptic qualities.

3.6.7 Food irradiation

Food irradiation is the exposure of food, either packaged or in bulk, to controlled amounts of ionising radiation (usually gamma rays from ^{60}Co) for a specific period of time to achieve certain desirable objectives. The process does not increase the background radioactivity of the food. It does prevent the growth of bacteria by damaging their DNA. It can also delay ripening and maturation by causing biochemical reactions in the plant tissues. Food irradiation reduces the microbial load at the packaging stage. However, just like other preservation processes, irradiation does not protect against future contamination through handling, storage and preparation of food.

Irradiation technology

The type of radiation used in processing materials is limited to radiations from high-energy gamma rays, X-rays and accelerated electrons. These radiations are also known as ionising radiations because their energy is high enough to dislodge electrons from atoms and molecules and to convert them to electrically charged particles called ions. Gamma and X-rays, like radio waves, microwaves, ultraviolet and visible light rays, form part of the electromagnetic spectrum, occurring in the short wavelength, high-energy region of the spectrum. Only certain radiation sources can be used in food irradiation. These are the radionuclides ^{60}Co or ^{137}Cs, X-ray machines having a maximum energy of 5 million electron volts (MeV) or electron machines having a maximum energy of 10 MeV. Energies from these radiation sources are too low to induce radioactivity in food. ^{60}Co is used considerably more than ^{137}Cs. The latter can only be obtained from nuclear waste reprocessing plants whereas ^{60}Co is produced by neutron bombardment of ^{59}Co in a nuclear reactor. High-energy electron beams can be produced from machines capable of accelerating electrons, but electrons cannot penetrate very far into food, compared with gamma radiation or X-rays. X-rays of various energies are produced when a beam of accelerated electrons bombards a metallic target.

Radiation dose is the quantity of radiation energy absorbed by the food as it passes through the radiation field during processing. It is measured in the unit called the 'gray' (Gy). One gray equals 1 joule of energy absorbed per kilogram of food being irradiated. International health and safety authorities have endorsed the safety of irradiation for all foods up to a dose level of 10 kGy.

Reasons for the application of food irradiation

The reasons for using food irradiation are as follows:
• Decrease the considerable food losses due to infestation, contamination and spoilage
• Concerns about foodborne diseases
• Increase in international trade in food products that must meet strict import standards of quality

The Food and Agriculture Organisation (FAO) of the United Nations has estimated that world-wide about 25% of all food products is lost after harvesting to insects, bacteria and rodents. Irradiation could reduce losses and dependence upon chemical pesticides. Many countries experience considerable losses of grain due to insect infestation, moulds and premature germination. Sprouting is the major cause of losses for roots and tubers. Countries, including Belgium, France, Hungary, Japan, the Netherlands and Russia, irradiate grains, potatoes and onions on an industrial scale.

As described in Chapter 1, even countries such as the United States experience considerable numbers of food poisoning cases, about 76 million cases per year and 5000 deaths. The economical aspect is a loss of $5–$17 billion by the US Food and Drug Administration (FDA). Belgium and the Netherlands irradiate frozen seafoods and dry food ingredients to control foodborne diseases. France uses electron beam irradiation on blocks of mechanically de-boned, frozen poultry products. Spices are irradiated in Argentina, Brazil, Denmark, Finland, France, Hungary, Israel, Norway and United States. In December 1997, the FDA approved irradiation for beef, pork and lamb. The irradiation of poultry was approved by the FDA in 1990 and poultry processing guidelines were approved by the US Department of Agriculture (USDA) in 1992. Annually, about 500 000 tonnes of food products and ingredients are irradiated worldwide. In most countries, irradiated foods must be labelled with the international symbol for irradiation (the radura), simple green petals in a broken circle. This symbol must be accompanied by the words, 'Treated by Irradiation' or 'Treated with Radiation'. Processors may add information explaining why irradiation is used, for example 'treated with irradiation to inhibit spoilage' or 'treated with irradiation instead of chemicals to control insect infestation'. When used as ingredients in other foods, however, the label of the other food does not need to describe these ingredients as irradiated. Irradiation labelling also does not apply to restaurant foods.

The world trade in food has increased and subsequently the health and safety requirements of importing countries must be met. However, there are differences in requirements, for example not all countries allow the importation of chemically treated fruit, some countries (United States and Japan) have banned the use of certain fumigants. Ethylene oxide is extremely toxic. Methyl bromide which is the primary fumigant that allows fruits, vegetables and grain to be exported was banned in 2001. Food irradiation technology may be of particular usefulness to developing countries whose economies rely on food and agricultural production as it is an alternative to fumigation with ethylene bromide and ethylene oxide.

Irradiation technology can be used as an alternative to chemical additives to reduce the number of spoilage organisms, and will also delay sprouting or maturation.

Food irradiation standard

The FDA approved the first use of irradiation on a food product (wheat and wheat flour) in 1963. Currently, the technology of food irradiation has been accepted in about 37 countries for 40 different foods. Twenty-four countries are using food irradiation at a commercial level. The standard covering irradiated food has been adopted by the Codex Alimentarius Commission (Chapter 11.4). It was based on the findings of a Joint Expert Committee on Food Irradiation (JECFI) which concluded that the irradiation of any food commodity up to an overall average dose of 10 kGy presented no toxicological hazard and required no further testing. It stated that irradiation up to 10 kGy introduced no special nutritional or microbiological problems in foods. The accepted dosage (10 kGy) is the equivalent of pasteurisation and cannot sterilise food. The method does not inactivate foodborne viruses such as Norovirus or Hepatitis A, and irradiation does not inactivate enzymes. Therefore, despite the lack of microbial growth, residual enzymes could limit a product's shelf-life. It is very important that foods intended for processing (as for any other preservation method) are of good quality and handled and prepared according to good manufacturing practice established by national or international authorities.

At low doses up to 0.5 kGy, irradiation may be used to kill *Trichinella spiralis* and *Taenia saginata* in meat, or to inactivate metacercariae of *Clonorchis* and *Opisthorchis* in fish. Higher doses of 3–10 kGy will kill non-spore forming bacteria such as *Salmonella, Campylobacter* and *Vibrio*. Irradiation can also be used to reduce the microbial load of spices and dried vegetables and thus prevent the contamination of foods to which these are added. Irradiation is a very useful pathogen control method for foodstuffs which are eaten raw or undercooked. Pork meat may cause half to three-quarters of *Toxoplasma gondii* infections in the United States. Infections due to protozoa and helminths are very common in certain tropical countries, and therefore the irradiation of meat and meat products can significantly reduce these risks. Similarly, the risk of foodborne infections from poultry and seafoods can be reduced by irradiation.

Table 3.10 gives the levels of approved doses for different products. The microbial survival following irradiation for different organisms is given in Figure 3.8.

Testing for irradiated food

Irradiated food can be detected to a limited extent (Haire *et al.* 1997). Tests include thermoluminescence measurement for detection of irradiated spices and electron spin resonance spectroscopy for determining irradiation of meats, poultry and seafoods containing any bone or shells and some specific chemical tests. No single method has however been developed that reliably detects irradiation of all types of foods or the radiation dose level that was used. This

Table 3.10 Approved food irradiation levels

Food	Purpose	Average dose (kGy)
Chicken	Prolong storage life	7
	Reduce the number of certain pathogenic micro-organisms such as *Salmonella* from eviscerated chicken	7
Cocoa beans	Control insect infestation in storage	1
	Reduce microbial load of fermented beans with or without heat treatment	1
Dates	Control insect infestation during storage	1
Mangoes	Control insect infestation	1
	Improve keeping quality by delaying ripening	1
	Reduce microbial load by combining irradiation and heat treatment	1
Onions	Inhibit sprouting during storage	0.15
Papaya	Control insect infestation and to improve its keeping quality by delaying ripening	1
Potatoes	Inhibit sprouting during storage	0.15
Pulses	Control insect infestation in storage	1
Rice	Control insect infestation in storage	1
Spices and condiments, dehydrated onions, onion powder	Control insect infestation	1
	Reduce microbial load	10
	Reduce the number of pathogenic micro-organisms	10
Strawberry	Prolong shelf-life by partial elimination of spoilage organisms	3
Teleost fish and fish products	Control insect infestation of dried fish during storage and marketing	1
	Reduce microbial load of the packed or unpacked fish and fish products[a]	2.2
	Reduce the number of certain pathogenic micro-organisms in packaged or unpackaged fish and fish products[a]	2.2
Wheat and ground wheat products	Control insect infestation in the stored product	1

[a]During irradiation and storage, the fish and fish products should be kept at the temperature of melting ice.

is partly because the irradiation process does not physically change the appearance, shape or temperature of products and causes negligible chemical changes in foods.

After irradiation treatment, the bacterial viability will have decreased by $>10^4$ cfu/g. The dead bacteria can be detected using Direct Epifluorescence Filter Technique (DEFT) or by the Limulus Amoebocyte Lysate (LAL) tests, and the viable organisms enumerated using conventional colony count methods (Anon. 2000, 2004). If the difference is >3 log units, then the sample may be presumptively identified as having been irradiated. Since other processes will cause bacterial cell death, the result should be confirmed using electron spin resonance, thermoluminescence or lipid radiolytic compound. There is still considerably more research required in order to determine whether ingredients have been irradiated, which will have been added at low quantities.

Irradiation should be regarded as complementing good hygienic practice and not a replacement for it.

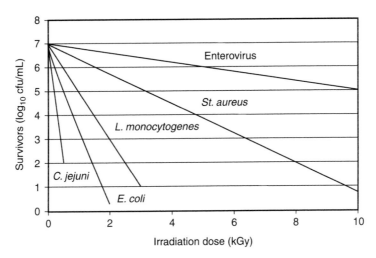

Figure 3.8 Survival of micro-organisms following irradiation.

Acceptability of food irradiation by the consumer

One major factor influencing the large-scale application of food irradiation is public under-standing and acceptance of the technology. Consumer reservation is based upon a number of issues:

1 Fears linked to other nuclear-related technologies. However, irradiated food cannot become radioactive itself.
2 Ingestion of radiolytic product. The maximum dose (10 kGy) produces 3–30 ppb radiolytic products. This would account for 5% of the total diet. The probability of harm occurring from radiolytic product formation from food additives is extremely low.
3 Food irradiation would mask poor quality (i.e. spoilt) food. However, although spoilage bacteria may not be detectable, the off-odours and poor appearance would still be evident.
4 The lack of a suitable detection method is also of concern. However, organically grown produce cannot be identified analytically, nor can unacceptable temperature fluctuations of chilled and frozen foods be determined.
5 Vitamin losses, especially B_1 in pork and vitamin C in fruit. The vitamin B_1 loss is estimated at 2.3% of vitamin B_1 in an American diet and vitamin C is converted into an equally usable form.
6 Induction of chromosomal abnormalities. The publicised polypoidy from the consumption of irradiated wheat in India has never been reproduced by subsequent investigations and the validity of the original data has been disputed by international scientific committees.

Current consumer pressure has hindered the large-scale expansion of irradiation technology despite its promise of safe food production.

3.6.8 Reduced oxygen packaging, modified atmosphere packaging and active packaging

Reduced oxygen packaging (ROP) results in the food being surrounded by an atmosphere which contains little or no oxygen. The term ROP is defined as any packing procedure that results in a reduced oxygen level in a sealed package. It covers a range of packaging processes:

sous-vide, vacuum packaging, cook-chill, controlled atmosphere packaging (CAP), modified atmosphere packaging (MAP).

In cook-chill, a plastic bag is filled with hot cooked food from which the air has been expelled and which is closed with a plastic or metal crimp.

Sous-vide is a specialised process for partially cooked ingredients alone or combined with raw foods that require refrigeration (<3°C) or frozen storage until the package is thoroughly heated immediately before service. The sous-vide process is a pasteurisation step that reduces bacterial load but is not sufficient to make the food shelf-stable.

Vacuum packaging reduces the amount of air from a package and hermetically seals it so that a near-perfect vacuum remains inside.

CAP is an active system in which a desired atmosphere is maintained for the duration of the shelf-life by the use of agents to bind or scavenge oxygen or a gas generating kit. CAP is defined as packaging of a product in a modified atmosphere followed by maintaining subsequent control of that atmosphere.

MAP is defined as the enclosing of food products in an atmosphere inside gas-barrier materials, in which the gaseous environment has been modified to slow down respiration rate, microbial growth and reduce enzymic degradation, with the intention of extending the shelf-life (Parry 1993). Controlled atmosphere storage (CAS) and/or active or passive MAP can be used. In active MAP, a slight vacuum inside the package is generated that is replaced by a chosen mixture of gases. Passive MAP is when the product is packaged with a specific film, and the desired atmosphere develops naturally due to the product's metabolism activity and the diffusion of gases through the film. The choice of the film and storage temperature are crucial, due to their effects on gas diffusion rates. Carbon dioxide is the most important headspace gas. It is a non-combustible, colourless gas that is favoured as a preservative since it leaves no toxic residues in the food. Hence, it can also be used in MAP. A concentration of 20–60% retards the spoilage due to pseudomonads. Recent trends have been the use of argon instead of nitrogen to reduce food spoilage. Also, the application of MAP has been extended to fresh produce, fresh meat, poultry and fin fish, ready meals and bakery products.

Solid CO_2 (dry ice) controls microbial growth by acting as a refrigerant during transport and storage. Then, as the dry ice sublimes, the CO_2 gas further inhibits bacterial growth by displacing the oxygen required by aerobic organisms, as well as by forming carbonic acid and thus possibly lowering the pH of the food to bacteriostatic levels (Foegeding & Busta 1991).

Oxygen is a major cause of food deterioration due to the oxidation of fats and oils, enzymic discolourisation and enabling the growth of many micro-organisms. To scavenge oxygen, a sachet of iron powder may be placed under the label. This can reduce the oxygen levels to <0.01% (MAP is 0.3–3%). In the United States, there are between 1500 and 200 million active packaging units with oxygen scavengers per annum. Carbon dioxide can extend a product's life by retarding microbial growth. Carbon dioxide emitters are frequently sachets containing ascorbic acid, sodium hydrogen carbonate and ferrous carbonate. These can emit an equal amount of gas to the oxygen absorbed, to prevent pack collapse, or can replace the carbon dioxide which has permeated through the packaging. Ethylene scavenging systems may be used to reduce the ripening process of fruits and vegetables. Potassium permanganate is very efficient, but is toxic and therefore must not come in contact with the food. Alternatives are activated carbon and titanium dioxide. Ethanol is antimicrobial and may be incorporated into packaging. The level is very important as too low will be ineffective, but too high will lead to an unwanted taint.

ROP can prevent the growth of aerobic spoilage organisms and hence extend the shelf-life for foods in the distribution chain. However, foods may be subject to temperature abuse during distribution and therefore at least one more barrier (or hurdle, Section 3.4) needs to be incorporated into the ROP process. ROP products commonly do not contain preservatives and do not have the intrinsic factors (pH, a_w, etc.) to prevent microbial growth. Subsequently, to prevent the growth of *Cl. botulinum* and *L. monocytogenes*, the storage temperature must be maintained below 3.3°C (cf. Table 2.8) for an extended shelf-life, otherwise 5°C for the duration of the shelf-life. The temperature treatment for ready-to-eat, including cook-chill, products should achieve a 4-D kill of *L. monocytogenes* (US recommendations; Brown 1991, Rhodehamel 1992). In Europe, sous-vide products should be subjected to a 12–13D *Ent. faecalis* heat treatment. This organism is used on the assumption that thermal inactivation of *Ent. faecalis* would ensure the destruction of all other vegetative pathogens.

In the future, nanomaterials could be used as part of the packaging. Smart sensors with a 'traffic light' systems responding to the food condition can be envisaged. See Section 3.9 for more detail.

3.7 Fermented foods

Fermented foods are one of the oldest means of food processing. It is considered that the civilisations in the fertile crescent of the Middle East ate fermented milk, meats and vegetables. This established traditions of production methods and the incidental selection of microbial strains. In 1910, Metchnikoff (Stanley 1998) proposed that lactic acid bacteria could be used to aid human health, a topic which has resurfaced recently in the area of 'probiotics'; see Section 3.8 for details. Nowadays, the production of fermented foods is practised across the world with many diverse national products (Table 3.11). The important microbial strains have been identified through conventional techniques and their metabolic pathways studied (Caplice & Fitzgerald 1999). Such fermented foods are consequently heavily colonised by bacteria and yeast, depending upon the specific product, yet there is no associated health risk. The reason is that the organisms are non-pathogenic, having been selected during the passage of time through trial and error. In fact, the organisms inhibit the growth of pathogens, and relatively recently, a new range of food products termed 'functional foods', in particular probiotics (i.e. Yakult), have arisen from the ancient practice of fermented food production. The lactic acid bacteria used in the production of most fermented foods produce a range of antimicrobial factors including organic acids, hydrogen peroxide, nisin and bacteriocins. Because lactic acid bacteria have such a proven history of being non-pathogenic, they are termed 'Generally Regarded As Safe' (GRAS) organisms. This title however may come under closer scrutiny with the advent of genetically engineered strains and the area of probiotics. Only the essential aspects of fermented foods as pertaining to their safety will be covered here. For a detailed study of fermented foods, the reader is directed to Wood's (1998) double-volume book entitled *Microbiology of Fermented Foods*.

The main factors resulting in the safety of fermented foods are as follows:
- Their acidity generated through lactic acid production
- The presence of bacteriocins
- High salt concentration
- The anaerobic environment

There are many sources quoting the pH growth range of bacterial pathogens (see Table 2.8). However, it is very important to appreciate that the acid tolerance of bacterial pathogens

Table 3.11 Fermented foods from around the world

Product	Substrate	Micro-organism(s)
Beer		
Ale	Grain	*Sac. cerevisiae*
Lager	Grain	*Sac. carlsbergensis*
	Millet	*Sac. fibulger*
Bread	Grain	*Sac. cerevisiae*, other yeasts, lactic acid bacteria
Bongkrek	Coconut	*Rh. oligosporus*
Cheese		
Cheddar	Milk	*Strep. cremoris, Strep. lactis, Strep. diacetylactis*, lactobacilli
Cottage	Milk	*Step. diacetylactis*
Mould ripened	Milk	*Strep. cremoris, Strep. lactis, Pen. caseicolum*
Swiss	Milk	As per Cheddar, plus *Prop. shermanii*
Coffee	Bean	*Leuconostoc, Lactobacillus, Bacillus, Erwinia, Aspergillus* and *Fusarium* spp.
Cocoa	Bean	*Lb. plantarum, Lb. mali, Lb. fermentum, Lb. collinoides, Ac. rancens, Ac. aceti, Ac. oxydans, Sac. cerevisiae* var. *ellipsoideus, Sac. apiculata*
Fish, fermented	Fish	*B. pumilus, B. licheniformis*
Gari	Cassava	*Corynebacterium manihot, Geotrichum* spp., *Lb. plantarum*, streptococci
Idii	Rice	*Leu. mesenteroides, Ent. faecalis, Torulopsis, Candida, Trichosporon pullulans*
Kimchi	Cabbage, nuts, vegetables, seafoods	Lactic acid bacteria

can be enhanced during processing. The stress response (heat-shock proteins, etc.) of bacterial pathogens is under current investigation and is summarised in Section 2.6. Also, any preformed bacterial or fungal toxins and viruses will persist during storage before ingestion. Examples of foodborne illness related to fermented products are dealt with under individual pathogens in Chapter 4.

3.7.1 Lactic acid bacteria and their metabolism

Lactic acid bacteria are mainly Gram positive, anaerobic, catalase-negative, non-sporing bacteria. They have numerous associations with us, both as part of the indigenous intestinal flora acquired soon after birth, attaching to mucosal surfaces, as well as being part of the microflora of a very wide range of foods, vegetables, wine, milk and meats. They are integral to the production of many well-known foods, not only acidified dairy products and vegetables, but also wine, coffee and cocoa. Some members are pathogenic, and others produce antimicrobial agents called 'bacteriocins'.

The lactic acid bacteria are primarily mesophiles (with a few thermophilic strains) and are able to grow in the range of 5–45°C. They are acid tolerant, and are able to grow at pH values as low as 3.8. Many species are proteolytic with fastidious amino acid growth requirements. The organisms are so called due to their ability to produce lactic acid. The lactic acid may be either or both the L(+) and D(−) isomers. However, the grouping of bacteria as 'lactic acid bacteria' is purely based on biochemical characteristics, and is not taxonomical. They do not comprise a single monophyletic group of bacteria. The majority are in the Gram-positive low

Table 3.12 Lactic acid bacteria groups

Fermentation type	Main products	Lactate isomer	Organism
Homofermentative	Lactate	L(+)	*Lactobacillus bavaricus, Lactobacillus, Enterococcus faecalis*
Homofermentative	Lactate	DL, L(+)	*Pediococcus pentosaceus*
Homofermentative	Lactate	D(−), L(+), DL	*Lactobacillus plantarum*
Heterofermentative	Lactate, ethanol, CO_2	DL	*Lactobacillus brevis*
Heterofermentative	Lactate, ethanol, CO_2	D(−)	*Leuconostoc mesenteroides*

GC% group called *Firmicutes*, although some are in the Gram-positive high GC% group called *Actinobacteria*. Their genetic diversity and genomic aspects are considered in Section 2.9.

The lactic acid bacteria include both homofermenters and heterofermenters (Table 3.12). The homofermentative lactic acid bacteria produce two molecules of lactic acid for each molecule of glucose fermented, whereas the heterofermentative lactic acid produce a variety of end products, including lactic acid, ethanol, acetic acid, carbon dioxide and formic acid (Figure 3.9). The homofermentative lactic acid bacteria are *Pediococcus, Streptococcus, Lactococcus* and some lactobacilli. The heterofermentative lactic acid bacteria are *Weisella, Leuconostoc* and some lactobacilli. The uptake of lactose (a disaccharide) is facilitated either by a carrier, permease or via the phosphoenolpyruvate-dependent phosphotransferase (PTS) system (Figure 3.10). Lactose is cleaved to yield galactose (or galactose-6-phosphate) and glucose. Since lactose is the major sugar in milk, its metabolism has been well studied. Subsequently, it has been demonstrated that the dairy starter culture *Lc. lactis* has the PTS system encoded on a plasmid which may partially explain the instability of certain starter cultures. Although their primary contribution to food is the rapid acidification, they also contribute to the flavour, texture and nutritional content. The lactic acid bacteria are ubiquitous, being found in dairy, meat, vegetable, cereal and plants environments. Historically, they have been used for thousands of years (albeit unknowingly) as starter cultures for the preservation through acidification of dairy, vegetable and meat products.

Citrate metabolism is of particular importance in *Lc. lactis* subsp. *lactis* (biovar *diacetylactis*) and *Lc. mesenteroides* subsp. *cremoris*. These organisms metabolise excess pyruvate via an unstable intermediate, α-acetolactate, to produce acetoin via the enzyme α-acetolactate decarboxylase. However, in the presence of oxygen, α-acetolactate is chemically converted into diacetyl (Figure 3.11). It is the diacetyl which gives the characteristic aroma of butter and certain yogurts.

The proteolytic activity of lactic acid bacteria, especially *Lactococcus*, has been studied in detail (Figure 3.12). Proteolysis is a prerequisite for the growth of lactic acid bacteria in milk, and the subsequent degradation of milk proteins (casein) is of central importance in the production of cheese. Hence, a fundamental understanding of the process is necessary. The growth of *Lc. lactis* is enhanced by the overproduction of the membrane-anchored serine proteinase (PrtP) which can degrade casein. The resultant oligopeptides enter the cell via an oligopeptide transport system (OPP) where they are further degraded by intracellular peptidases. The complete degradation of peptides is achieved by a wide range of intracellular peptidases with overlapping specificities (Mierau *et al.* 1996). The metabolism of proteins and amino acids by the lactic acid bacteria is part of the ripening process of cheese and flavour development.

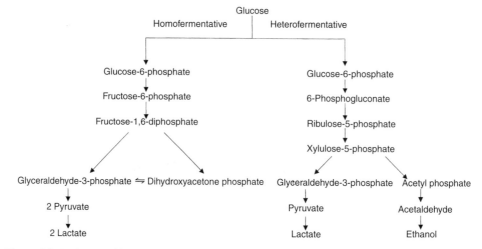

Figure 3.9 Production of lactic acid by homo- and heterofermentative lactic acid bacteria.

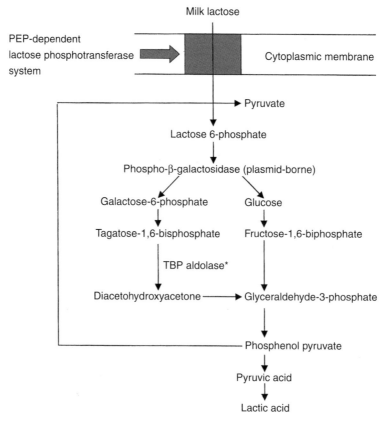

Figure 3.10 Lactose metabolism.

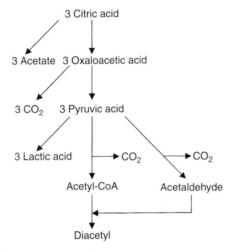

Figure 3.11 Production of diacetyl by lactic acid bacteria.

Lactic acid bacteria produce a range of antimicrobial factors including organic acids, hydrogen peroxide, nisin and bacteriocins. Organic acids such as lactic acid, acetic acid and propionic acid interfere with the proton motive force and active transport mechanisms of the bacterial cytoplasmic membrane (Davidson 1997; Section 3.5.6). The production of hydrogen peroxide is due to the lactic acid bacteria lacking the enzyme catalase. H_2O_2 can consequently cause membrane oxidation of other bacteria (Lindgren & Dobrogosz 1990). Additionally, H_2O_2 activates the lactoperoxidase system of fresh milk causing the formation of antimicrobial hypothiocyanate (Section 3.5.2). Lactic acid bacteria produce four groups of bacteriocins (Klaenhammer 1993, Nes *et al.* 1996). The most well-documented and exploited bacteriocin is nisin (Class 1 bacteriocin), which is a post-translationally modified amino acid (also known

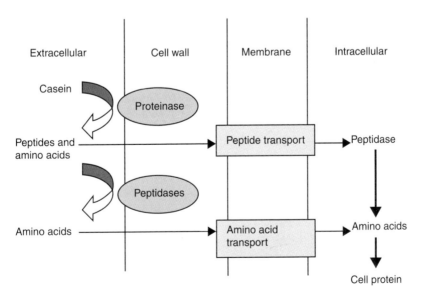

Figure 3.12 Protein metabolism in lactic acid bacteria.

Table 3.13 Lactic acid bacteria and bifidobacteria in the human intestinal tract

Location	Species
Oral cavity	*Streptococcus mutans, Bifidobacterium longum*
Large intestines	*Lactobacillus acidophilus, Lb. gasseri, Lb. johnsonii, Lb. plantarum, Strep. agalactiae, Enterococci faecalis*
Vagina	*Bif. longum, Strep. agalactiae, Lb. crispatus*

as a lantibiotic, Figure 3.6). Nisin is produced by *Lactococcus lactis* subsp. *lactis* and has a broad inhibitory spectrum against Gram-positive bacteria (Section 3.5.5). It is formulated (concentration range 2.5–100 ppm) into a range of food products including cheeses, canned foods and baby foods and is particularly stable in high-acid foods.

Over 100 years ago, Elie Metchnikoff proposed that lactic acid bacteria in fermented milk could promote the development of a healthy intestinal microflora, and prolong the individual's life by preventing putrefaction. Although these ideas are not entirely accurate, they do reflect the current trend of probiotics covered in Section 3.8. Additionally, it is now known that the human intestinal tract is colonised by the lactic acid bacteria, and *Bifidobacterium* species which may also have a role in health; Table 3.13.

3.7.2 Fermented milk products

There are many regional fermented milk products produced around the world (Table 3.13). The production of most fermented milk products (such as cheese and yogurt) requires the addition of starter cultures. These are laboratory-maintained cultures of lactic acid bacteria. The cultures may be well defined or composed of different strains depending upon the scale of production and historical experience (Table 3.14). The cultures will be rotated at regular intervals to avoid the possible accumulation of lactic acid bacteria specific bacteriophage in the production plant. Bacteriophages would slow the fermentation activity of the start cultures and even enable pathogenic organisms to multiply in the fermented food. Typical starter cultures include the mesophiles: *Lc. lactis* subsp. *lactis* and *Lc. lactis* subsp. *cremoris* or thermophilic *Strep. thermophilus, Lb. helveticus* and *Lb. delbrueckii* subsp. *bulgaricus*.

Table 3.14 Fermented milk products around the world

Product	Type	Micro-organisms
Yogurt	Moderate acid	*Strep. thermophilus, Lb. bulgaricus*
Cheddar cheese (United Kingdom)	Moderate acid	*Strep. cremoris, Strep. lactis, Strep. diacetylactis*
Cultured buttermilk (United States)	Moderate acid	*Lc. cremoris, Lc. lactis. citrovorum*
Acidophlus milk (United States)	High acid	*Lb. acidophilus*
Bulgarican milk (Europe)	High acid	*Lb. bulgaricus*
Dahi (India)	Moderate acid	*Lc. lactis, Lc. cremoris, Lc. diacetylactis, Leuconostoc* spp.
Leben (Egypt)	High acid	*Streptococci, lactobacilli, yeast*
Surati cheese (India)	Mild acid	*Streptococci*
Kefir (Russia)	High alcohol	*Streptococci, Leuconostoc* spp., *yeast*
Koumiss (Russia)	High alcohol	*Lb. acidophilus, Lb. bulgaricus, Sacharomyces lactis*

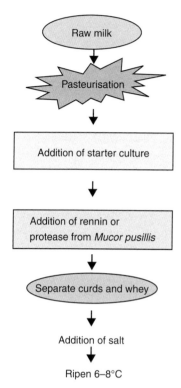

Figure 3.13 Generalised production of cheese.

In cheese manufacture (Figure 3.13), the addition of the protease rennin either from calf stomachs or from the fungus *Mucor pusillis* and acidification from microbial lactic acid production results in the precipitation of casein as 'curds' which are separated from the remaining liquid fraction ('whey'). The curds are salted and pressed into shape and left to ripen at 6–8°C. The characteristics of many regional products are dependent upon secondary bacterial or fungal colonisation (Table 3.15). For example, *Propionibacterium shermanii* causes the production of carbon dioxide gas and the characteristic eyes in Swiss cheese. *Penicillium roqueforti* (a fungus) produces the blue veining in certain cheeses. *Candida utilis* and *C. kefir* are involved in the production of blue cheese and kefir, respectively. Recent application of DGGE analysis (Section 2.9.2) has revealed differences in the microbial ecosystem of the cheese surface and the interior. The aerobic surface flora is dominated by a very diverse range of G+C rich species, whereas the interior is composed of fewer species which are lower in G+C content (Ogier *et al.* 2004).

Pathogenic bacteria are killed by pasteurisation and therefore cheese made from pasteurised milk should be safe. However, inadequate pasteurisation and post-pasteurisation contamination can occur. In addition, some cheeses are made from non-pasteurised milk. Nevertheless, the rapid acidification of the milk due to lactic acid production should inhibit the multiplication of any pathogens present. However, if the starter culture is not fully active, possibly due to bacteriophage infection, then pathogens can multiply to infectious levels. A number of food poisoning incidences linked to cheese have been reported.

Yogurt manufacturing uses the complementary metabolic activities of *Strep. thermophilus* and *Lb. bulgaricus* (1:1 inoculation ratio). The lactobacillus requires the initial growth of the

Table 3.15 Categories of lactic starter strains

Type	Species	Application method
Single strain	*Strep. cremoris* *Strep. lactis* *Strep. lactis* subsp. *diacetylactis*	Single or paired
Multiple strains	As above plus *Leuconostoc* spp.	Defined mixture of two or more strains
Mixed strain starters	*Strep. cremoris*	Unknown proportions of different strains, which may vary on subculturing

streptococci to produce folic acid which is essential for the lactobacillus growth. Subsequently, *Lb. bulgaricus* produces diacetyl and acetaldehyde which add flavour and aroma to the product (Figures 3.11 and 3.14). A slow fermentation by the starter cultures can enable *St. aureus* to multiply to numbers sufficient to produce enterotoxins at levels high enough to induce vomiting.

There are many health benefits attributed to the ingestion of fermented milk products:
- Increased digestibility and the nutritive value of milk itself
- Reduced lactose content, important to a lactose intolerant population
- Increased absorption of calcium and iron

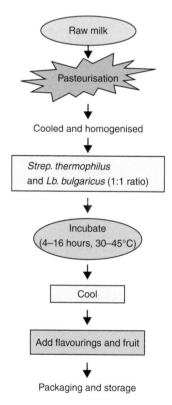

Figure 3.14 Production of yogurt.

Table 3.16 Secondary flora of fermented dairy products

Type	Micro-organism	Cheese types
Bacteria	Cornebacteria	Red smear-ripened, washed rind
	Micrococcus spp.	
	Lactobacillus spp.	Hard and semi-hard
	Pediococcus spp.	
	Propionibacteria	Swiss types
Yeasts	Kluyveromyces lactis	
	Saccharomyces cerevisiae	
	Candida utilis	Blue, soft mould-ripened, red smear-ripened, washed rind
	Debaryomyces hansensii	
	Rhodosporidium infirmominiatum	
	Candida kefir	Kefir
Mould	Penicillium camemberti	Soft mould-ripened (Brie, Camembert, Coulommiers)
	Geotrichum candidum	
	Penicillium roqueforti	Blue cheese (Danish Blue, Roquefort, Stilton)
	Penicillium nalgiovensis	Tomme
	Verticillium lecanii	Tomme

Adapted from Stanley 1998, with kind permission of Springer Science and Business Media.

- Increased content of some B-type vitamins
- Control of the intestinal microbial flora composition
- Inhibition of pathogenic micro-organisms' multiplication in the intestinal tract
- Decreased cholesterol level in blood

(After Oberman & Libudzisz 1998)

Subsequently, there are a number of commercially available fermented milk products which claim to promote health (Table 3.16).

3.7.3 Fermented meat products

Fermented sausages are produced from the bacterial metabolism of meat. The meat is first comminuted and mixed with fat, salt, curing agents (nitrate and nitrite), sugar and spices before being fermented. The sausages have a low water activity (Section 2.6.2) and are generally classified as either dry ($a_w < 0.9$) or semi-dry ($a_w = 0.9–0.95$). The dry sausage is normally eaten without any cooking or other form of processing. In contrast, the semi-dry sausages are smoked and hence receive a heat treatment of between 60°C and 68°C. Sausages produced without the addition of a starter culture have a pH of 4.6–5.0, whereas sausages with starter cultures usually have a lower final pH of 4.0–4.5. These low pH values are inhibitory to the growth of most food pathogens (Table 2.8). The starter cultures are composed of lactobacilli (such as Lb. sakei and Lb. curvatus), pediococci, Strep. carnosus and Micrococcus varians. The fungi Debaryomyces hansenii, Candida famata, Penicillium nalgiovense and P. chrysogenum may also form part of the starter culture (Jessen 1995).

The bacterial pathogens which are potential hazards in fermented sausages are salmonellas, St. aureus, L. monocytogenes and the spore-forming bacteria Bacillus and Clostridium. The high level of salt (~2.5%) inhibits pathogen growth and during the drying process Salmonella (and other Enterobacteriaceae) die off. Controlling salmonellas contamination also controls the growth of *L. monocytogenes*. Although *St. aureus* is resistant to nitrite and salt, it is a poor competitor of the remnant microflora under the acidic anaerobic conditions.

3.7.4 Fermented vegetables

According to Bückenhuskes (1997), there are 21 different commercial vegetable fermentations in Europe. Additionally, there are a large variety of fermented vegetable juices, olives, cucumbers and cabbage. Vegetables have a high microbial load. However if they were pasteurised this would affect their final properties and would be detrimental to the product's quality. Therefore, the majority of vegetable fermentations use lactic acid production and no heat treatment (smoking, etc.). Lactic acid bacteria are present only in low numbers in vegetables and therefore starter cultures are usually used: *Lb. plantarium*, *Lb. casei*, *Lb. acidophilus*, *Lc. lactis* and *Leuc. mesenteroides*.

3.7.5 Fermented protein foods; shoyu and miso

In the Orient, the fungi *Aspergillus* (*A. oryzae*, *A. sojae* and *A. niger*), *Mucor* and *Rhizopus* have been used to ferment grain, soya beans and rice (Table 3.17). This is called the koji process (Figure 3.15). The main microbial activity is amylolytic degradation. The presence of mycotoxins has not been reported in koji products despite extensive investigations. Mycotoxins can potentially be produced after prolonged incubation by commercial strains of *Aspergillus* under specific environmental stress conditions (Table 3.18, Trucksess *et al.* 1987). In contrast, aspergillus fermentations in the koji process seldom exceed 48–72 hours.

3.7.6 Future use of the lactic acid bacteria

The advent of DNA microarrays, application of genomic and proteomic studies will extend the application of lactic acid bacteria in food microbiology (Blackstock & Weir 1999, Schena *et al.* 1998; Section 5.4.6). Genomic science has recently emerged as a powerful means to compare

Table 3.17 Yeast and moulds from koji starter cakes

Starter cake	Moulds	Yeasts
Chinese yeast	*Rhizopus javanicus* R. chinensis, Amylomyces rouxii	*Endomycopsis* spp.
Ragi	*Amy. rouxii, Mucor dubius, M. javanicus, R. oryzae, Aspergillus niger*	*Torula indica, Hansenula anomala, Saccharomyces cerevisiae*
	R. stolonifer, M. rouxii	*Endomycopsis chodati, E. fibuligera*
	R. cohnii, Zygorrhynchus moelleri	*H. subpelliculosa, H. malanga, Candida guilliermondii, C. humicola, C. intermedia, C. japonica, C. pelliculo*
	A. oryzae, A. flavus, R. oligosporus, R. arrhizus Fusarium spp.	
Loog-pang	*Amy. rouxii, Mucor* spp., *Rhizopus* spp., *A. oryzae*	*Endomycopsis fibuligera Saccharomyces cerevisiae*
Murcha	*M. fragilis, M. rouxii, R. arrhizus*	*H. anomala*
Bubod	*Rhizopus* spp., *Mucor* spp.	*Endomycopsis* spp. *Saccharomyces* spp.

Adapted from Lotong 1998, with kind permission of Springer Science and Business Media.

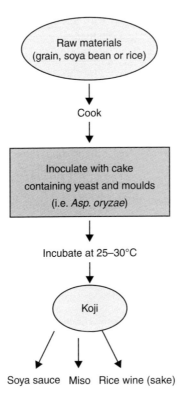

Figure 3.15 Production of Koji.

genomes and examine differentially expressed genes (Kuipers 1999). See Section 2.9.6 for more information regarding genomic analysis of the lactic acid bacteria. Microarray analysis enables differential gene expression of the total genome to be examined (Klaenhammer & Kullen 1999). Initial genetic studies of lactic acid bacteria demonstrated that many important traits were plasmid borne. These included lactose metabolism, bacteriophage resistance and proteolytic activity. Subsequently, stable starter cultures can be developed for large-scale production and have already been applied in the dairy industry. The genomic sequences of over 20 lactic acid bacteria have been published or are in progress (Table 2.15). This increase in genetic knowledge will also improve the use of lactic acid bacteria as cell factories for the production of food-grade

Table 3.18 Mycotoxins produced by koji moulds

Organism	Mycotoxin
A. oryzae	Cyclopiazonic acid, kojic acid, maltoryzine, β-nitropropionic acid
A. sojae	Aspergillic acid, kojic acid
A. tamarii	Kojic acid

Adapted from Rowan et al. 1998, with kind permission of Springer Science and Business Media.

polysaccharides to add texture to food (de Vos 1999, Kuipers *et al.* 1997). It is anticipated that genome sequencing of probiotic cultures will identify gene systems responsible for cell survival and activity and secondly responsive gene systems that enable the organism to adapt to the changing environment within the intestinal tract (Klaenhammer & Kullen 1999).

In the future, molecular engineering of starter cultures will be possible in order to optimise production of end products such as diacetyl, acetate, acetaldehyde and flavour compounds from proteolytic activity (Hugenholtz & Kleerebezem 1999). Other health promoting products produced from lactic acid bacteria will be anti-oxidants and vitamins. The use of lactic acid bacteria to deliver vaccines is currently under investigation (Wells *et al.* 1996). The microbial response to stress (such as heat and pH shock) is an increasing area of genetic and physiological research (Section 2.7). The ability to survive in a low-pH environment (such as the stomach) may be increased due to previous exposure to a heat treatment due to the production of heat-shock proteins.

There is obviously a considerable potential for the scientific exploitation of genetically engineered lactic acid bacteria. However, it must be ensured that the general public regard these organisms, previously 'GRAS', as beneficial to their health and not an example of 'Frankenstein Food' which has recently caused such public reluctance in certain European countries. These factors are of considerable importance in the controversial area of pre- and probiotics.

3.8 Functional foods; prebiotics, probiotics and synbiotics

3.8.1 Functional foods

Functional foods can be defined as foods in which concentrations of one or more ingredients have been modified to enhance their contributions to a healthful diet. The concept is a logical extension of the more conventional viewpoint that a balanced diet is important for health. For example, eating products low in fat, low in cholesterol or sugar, and high in vegetable and fruit intake are encouraged. Currently, food scientists are identifying an increasing list of food components that promote health, and this has led to the general acceptance that individual foods can promote health.

Probiotics, prebiotics and synbiotics are controversial topics in the general area of 'functional foods' (Atlas 1999, Berg 1998, Rowland 1999). Prebiotics are 'non-digestible food components that beneficially affect the host by selectively stimulating the growth and/or activity of one or a limited number of bacteria in the colon, that have the potential to improve host health' (Zoppi 1998). A probiotic can be defined as 'a live microbial food supplement which beneficially affects the host animal by improving its intestinal microbial balance' (Fuller 1989). This definition was broadened by Havenaar and Huis in't Veld (1992) to a 'mono- or mixed-culture of live micro-organisms which benefits man or animals by improving the properties of the indigenous microflora'. The Lactic Acid Bacteria Industrial Platform workshop proposes a different definition for probiotics as 'oral probiotics are living micro-organisms, which upon ingestion in certain numbers, exert health benefits beyond inherent basic nutrition'. This definition means that probiotics may be consumed either as a food component or as a non-food preparation (Guarner & Schaafsma 1998). Synbiotics can be described as 'a mixture of probiotics and prebiotics that beneficially affects the host by improving the survival and implantation of live microbial dietary supplements in the gastrointestinal tract'.

A variety of probiotic foods have been developed and marketed. They can be divided into the following categories:

1 Conventional fermented foods which are consumed primarily for nutritional purpose
2 Food supplements and fermented milk with food formulations which are used to deliver probiotic bacteria
3 Dietary supplements which are in the form of capsules and other formulations

These products usually contained members either of the *Lactobacillus* and/or *Bifidobacterium* genera. As such, these cultures are accepted to be safe for consumption due to their long history of use in conventional fermented foods. The idea of using lactic acid bacteria as an aid to improved health is not new; see Section 3.7 on fermented foods for related stories. However, the current trend is largely driven by consumer demands rather than through scientific investigation. The probiotic preparation needs to be defined as viable, micro-organisms in sufficient numbers, to alter the intestinal microflora. A daily intake of 10^9–10^{10} viable cells is considered the minimum dose. Not all products contain correctly identified species, nor can it be assumed that the organisms are viable. For example, Green *et al.* (1999) demonstrated that two *B. subtilis* oral probitics did not contain *B. subtitlis*. Some products may contain undeclared organisms, non-viable organisms, or even pathogens (Hamilton-Miller *et al.* 1999). Similarly, Fasoli *et al.* (2003) used DGGE (Section 5.5.1) to investigate commercial probiotic yogurts and lyophilised products. There were discrepancies in the identification of *Bifidobacterium* and *Bacillus* spp., as well as the presence of non-declared organisms. The reader is referred to Tannock (1999a) for a fuller coverage of the probiotic topic.

The gastrointestinal tract is the most densely colonised region of the human body (Townsend & Forsythe 2008; Section 4.1). There are approximately 10^{12} bacteria per gram (dry weight) of contents of the large intestine, which is estimated to contain several hundred bacterial species. It is widely accepted that this collection of microbes has a powerful influence on the host in which it resides. The production of bacterial fermentation products (acetate, butyrate and propionate) commonly known as short-chain fatty acids (SCFAs) for colonic health has been documented. For example, butyrate has a trophic effect on the mucosa and is an energy source for the colonic epithelium and regulates cell growth and differentiation. The human intestinal gut-associated lymphoid tissue represents the largest mass of lymphoid tissue in the body, and about 60% of the total immunoglobulin produced daily is secreted into the gastrointestinal tract. The colonic flora is the major antigenic stimulus for specific immune responses at local and systemic levels. Abnormal intestinal response to foreign antigen, as well as local immunoinflammatory reactions, might, as a secondary event, induce impairment of the intestine's function because of breakdown of the intestinal barrier (Diplock *et al.* 1999).

3.8.2 Claims of probiotics

It is implicit in the definition of probiotics that consumption of probiotic cultures positively affect the composition of this microflora and extends a range of host benefits (German *et al.* 1999, Klaenhammer & Kullen 1999, Sanders 1998, Tannock 1999b), including the following:

1 Pathogen interference, exclusion and antagonism (Mack *et al.* 1999)
2 Immunostimulation and immunomodulation (Marteau *et al.* 1997, Schiffrin *et al.* 1995)
3 Anticarcinogenic and antimutagenic activities (Matsuzaki 1998, Reddy & Riverson 1993, Rowland 1990)
4 Alleviation of symptoms of lactose intolerance (Marteau *et al.* 1990, Sanders 1993)
5 Reduction in serum cholesterol
6 Reduction in blood pressure
7 Decreased incidence and duration of diarrhoea (anti-associated diarrhoea, *Clostridium difficile*, travelers and rotaviral) (Biller *et al.* 1995, Isolauri *et al.* 1991, Kaila *et al.* 1992)

8 Prevention of vaginitis

9 Maintenance of mucosal integrity

It has been proposed that prebiotics help the colonic microbial flora to maintain a composition in which bifidobacteria and lactobacilli become predominant in number. This change in flora composition is regarded by believers in probiotics as optimal for health promotion. Non-digestible carbohydrates (e.g. inulin and fructo-oligosaccharides) increase faecal mass both directly by increasing non-fermented material and indirectly by increasing bacterial biomass; they also improve stool consistency and stool frequency. Promising food components for functional foods will be those, such as specific non-digestible carbohydrates, that can provide optimal amounts and proportions of fermentation products at relevant sites in the colon, particularly the distal colon, if they are to help reduce the risk of colon cancer.

3.8.3 Probiotic studies

Many of the specific effects attributed to the ingestion of probiotics, however, remain controversial and it is rare that specific health claims can be made (Ouwehand *et al.* 1999, Sanders 1993). In order to test the claims of pre- and probiotics, there needs to be advances in the detailed characterisation of the intestinal microflora, especially the organisms currently regarded as unculturable. Molecular approaches are required such as 16S ribosomal RNA analysis, denaturing gradient gel electrophoresis (DGGE), temperature gradient gel electrophoresis (TGGE) and fluorescent *in situ* hybridisation (FISH; Muyzer 1999, Rondon *et al.* 1999; see Section 2.9.2). Molecular biology has been used to establish the phylogenetic relationship among members of the *Lactobacillus acidophilus* complex (Schleifer *et al.* 1995; Section 2.9.6). This may in part explain the variation in effects observed in earlier studies. It is plausible that unrelated organisms with identical names were used by different research groups, as the organisms had been misidentified using biochemical (phenotypic) identification techniques.

The lactobacilli and the bifidobacteria constitute the most important probiotic organisms under investigation (Tables 3.19 and 3.20; Reid 1999, Tannock 1998, 1999a, 1999b, Vaughan *et al.* 1999). The reason for this is that they are recognised as part of the human intestinal tract flora (especially in the newborn infant) and they have been safely ingested for many centuries in fermented foods (Section 3.7; Adams & Marteau 1995). Particular lactic acid bacteria strains which have been investigated as probiotics include *Lb. rhamnosus* GG, *Lb. johnsonii* LJ1, *Lb. reuteri* MM53, *Bifidobacterium lactis* Bb12 and *Bif. longum* NCC2705 (Table 3.19, Reid

Table 3.19 Microbial composition of fermented milk products claiming health benefits

Product	Country of origin	Micro-organism
A-38 fermented milk	Denmark	*Lb. acidophilus* and mesophilic lactic acid bacteria
AB-fermented milk	Denmark	*Lb. acidophilus, Bif. bifidum*
Acidophilus milk	United States	*Lb. acidophilus*
Activia	Italy	*Bif. bifidus*
Kyr	Italy	*Lb. acidophilus* and *Bifidobacterium* spp.
Liquid yogurt	Korea	*Lb. bulgaricus, Lb. casei, Lb. helveticus*
Mio	Switzerland	*Bif. lactis*
Miru-Miru	Japan	*Lb. acidophilus, Lb. casei, Bifidobacterium breve*
Real active	United Kingdom	Yogurt culture, *Bif. bifidum*
Yakult	Japan	*Lactobacillis casei* (Shirota)

Table 3.20 Lactic acid bacteria used as probiotics

Genus	Species
Lactobacillus	Lb. acidophilus, Lb. amylovorus, Lb. bulgaricus, Lb. casei, Lb. crispatus, Lb. gallinarum, Lb. gasseri, Lb. johnsonii (strain Lal), Lb. plantarum, Lb. reuteri (strain MM53), Lb. rhamnosus (strain GG), Lb. salivarius
Bifidobacterium	Bif. adolescentis, Bif. animalis, Bif. bifidum, Bif. breve, Bif. infantis, Bif. longum, Bif. lactis (strain Bb12)
Streptococcus	Strep. thermophilus, Strep. salivarius
Enterococcus	Ent. faecium

Reprinted from Klaenhammer & Kullen 1999, with permission from Elsevier.

1999, Vaughan *et al.* 1999). Important attributes of probiotic strains are colonisation factors and conjugated bile salt hydrolysis which aid persistence in the intestinal tract. Additional characteristics include effects on immunocompetent cells and antimutagenicity.

Lb. rhamnosus GG is probably the most studied probiotic preparation. There is some evidence that *Lb. rhamnosus* GG colonises the intestine and reduces diarrhoea (Kaila *et al.* 1995). The oral dose required is greater than 10^9 cfu/day. There are many fermented milk products commercially available which claim to promote health (Table 3.17). *Lb. casei* (Shirota strain) fermented milk is sold as a product called 'Yakult' and is estimated to be consumed daily by 10% of the Japanese population (Reid 1999). On a worldwide basis, it has been predicted that 30 million people ingest this product each day. It is prepared from skimmed milk with the addition of glucose and a *Chlorella* (algae) extract followed by inoculation with *Lb. casei* (Shirota). The fermentation takes 4 days at 37°C. The lactobacillus strain was initially isolated in 1930 and appears to colonise the intestine and improve infant recovery time from rotaviral gastroenteritis (Sugita & Togawa 1994). The strain appears to have antitumour and antimetastasis effects in mice (Matsuzaki 1998). In Canada, the application of probiotics via the urogenital tract has been investigated. Trials with *Lb. rhamnosus* GR-1 and *Lb. fermentum* B-54 (applied as a milk-based suspension or freeze-dried preparations) have indicated that the vaginal microflora can be re-established and the incidence of urinary tract infections reduced (Reid *et al.* 1994).

Although probiotics are promoted as aids to health, it is plausible that there may be some risks associated with the ingestion of probiotic bacteria. The major risk would be the unrestricted stimulation of the immune system in people suffering from autoimmune diseases (Guarner & Schaafsma 1998).

Lactic acid bacteria are also being studied for vaccine delivery in the intestinal tract (Fischetti *et al.* 1993). A strain of *Lb. lactis* containing the luciferase gene has been used as an experimental model to study gene promoter activities in the intestinal tract as a first step to predicting the expression of heterogenous proteins *in vivo* (Corthier *et al.* 1998).

3.9 Nanotechnology and food preservation

Nanotechnology is concerned with compounds in the 1–100 nm size range. Currently, it is a major focus of research by industry, and is receiving considerable government funding for research. There are more than 200 companies involved in this area worldwide, with the leaders being the

United States, followed by Japan and China. By 2010, the market for nanotechnology in food is predicted to be ~$20.4 billion. Nanocomposites have been made as packaging materials or as coatings on plastic containers to control gas diffusion and to prolong the shelf-life (Sozer & Kokini 2009; see Section 3.6.8). Another application is the production of antimicrobial food contact materials. The technology may even lead to 'smart' surfaces which can detect bacterial contamination and inhibit its growth, as well as remove taints and odours.

As for all new compounds used in food production, the potential health risks of nanoscale materials should be evaluated before their use. While the risk of the same material as a macroscale particle may already have been determined, nanoscale materials can behave differently. It is already known that some nanoparticles can cross the blood–brain barrier. One identified problem is the transfer of nanoparticles from packaging material to food. Similarly, food technologists could use nanoparticles of previously approved macroscale ingredients without considering whether there are possibly new toxicological aspects (Bouwmeester *et al.* 2009).

4 Foodborne pathogens

4.1 Introduction

Food poisoning is caused by food which:
1 looks normal
2 smells normal
3 tastes normal

Because the consumer is unaware that there is a potential problem with the food, a significant amount of food is ingested and hence they become ill. Consequently, it is hard to trace which food was the original cause of food poisoning since probably the consumer will not recall noticing anything untoward in their recent meals. Food poisoning is caused by a variety of organisms and the incubation period and duration of illness varies considerably (Table 4.1). The term 'infectious dose' is not a precise threshold value below which a person does not become infected, but is useful as a relative indicator of the infectivity of a foodborne pathogen. Values for a number of foodborne pathogens are given in Table 4.2. Section 9.5.7 covers 'infectious dose' in greater detail, and how it depends upon the organism, host and food.

Food poisoning organisms are normally divided into two groups:
- infections; for example *Salmonella*, *Campylobacter* and pathogenic *Escherichia coli*.
- intoxications; for example *Bacillus cereus*, *Staphylococcus aureus*, *Clostridium botulinum*.

The first group are organisms which multiply in the human intestinal tract, whereas the second group are organisms that produce toxins either in the food or during passage in the intestinal tract. This division is very useful to help recognise the routes of food poisoning. Vegetative organisms are killed by heat treatment, whereas spores may survive and hence germinate if the food is not kept sufficiently hot or cold.

An alternative grouping would be according to severity of illness. This approach is useful in setting microbiological criteria (sampling plans) and risk analysis. ICMSF (1974, revised 1986, 2002) divided the common foodborne pathogens into such groups in order to aid decision-making of sampling plans (Chapter 6). The ICMSF groupings are given in Table 4.3. Detailed descriptions of foodborne pathogens are given in the following sections of this chapter. Initially, however, it is useful to appreciate that some foodborne pathogens are very difficult to detect and hence 'indicator' organisms, whose presence may indicate the possible presence of a pathogen, may be used which are easier to detect.

4.1.1 The human intestinal tract

The human intestinal tract is divided into a number of regions: oesophagus, stomach, small intestine, large intestine and anus (Figure 4.1). The intestinal tract is approximately 30 feet long. In the mouth, food is chewed which breaks it into smaller pieces and mixes it with

Table 4.1 Sources of foodborne pathogens

Food	Pathogen	Incidence (%)
Meat, poultry and eggs	*C. jejuni*	Raw chicken and turkey (45–64)
	Salmonella spp.	Raw poultry (40–100), pork (3–20), eggs (0.1%) and shellfish (16)
	St. aureus[a]	Raw chicken (73), pork (13–33) and beef (16)
	Cl. perfringens[b]	Raw pork and chicken (39–45)
	Cl. botulinum	
	E. coli O157:H7	Raw beef, pork and poultry
	B. cereus[b]	Raw ground beef (43–63), cooked meat (22)
	L. monocytogenes	Red meat (75), ground beef (95)
	Y. enterocolitica	Raw pork (48–49)
	Hepatitis A virus	
	T. spiralis	
	Tapeworms	
Fruit and vegetable	*C. jejuni*	Mushrooms (2)
	Salmonella spp.	Artichoke (12), cabbage (17), fennel (72), spinach (5)
	St. aureus[a]	Lettuce (14), parsley (8), radish (37)
	L. monocytogenes	Potatoes (27), radishes (37), bean sprouts (85), cabbage (2), cucumber (80)
	Shigella spp.	
	E. coli O157:H7	Celery (18) and coriander (20)
	Y. enterocolitica	Vegetables (46)
	A. hydrophila	Broccoli (31)
	Hepatitis A virus	
	Norovirus	
	G. lamblia	
	Cryptosporidium spp.	
	Cl. botulinum	
	B. cereus[b]	
	Mycotoxins	
Milk and dairy products	*Salmonella* spp.	
	Y. enterocolitica	Milk (48–49)
	L. monocytogenes	Soft cheese and pate (4–5)
	E. coli	
	C. jejuni	
	Shigella spp.	
	Hepatitis A virus	
	Norovirus	
	St. aureus[a]	
	Cl. perfringens[b]	
	B. cereus[b]	Pasteurised milk (2–35), milk powder (15–75) cream (5–11), ice cream (20–35)
	Mycotoxins	
Shellfish and fin fish	*Salmonella* spp.	
	Vibrio spp.	Raw seafood (33–46)
	Shigella spp.	
	Y. enterocolitica	
	B. cereus	Fish products (4–9)
	E. coli	

(Continued)

Table 4.1 (*Continued*)

Food	Pathogen	Incidence (%)
	Cl. botulinum[b]	
	Hepatitis A virus	
	Norovirus	
	G. lamblia	
	Cryptosporidium spp.	
	Metabolic by-products	
	Algal toxins	
Cereals, grains,	*Salmonella* spp.	
legumes and nuts	*L. monocytogenes*	
	Shigella spp.	
	E. coli	
	St. aureus[a]	
	Cl. botulinum[b]	
	B. cereus[b]	Raw barley (62–100), boiled rice (10–93), fried rice (12–86)
	Mycotoxins[a]	
Spices	*Salmonella* spp.	
	St. aureus[a]	
	Cl. perfringens[b]	
	Cl. botulinum[b]	
	B. cereus[b]	Herbs and spices (10–75)
Water	*G. lamblia*	Water (30)

[a]Toxin not destroyed by pasteurisation.
[b]Spore-forming organism. Not killed by pasteurisation.
Various including Snyder (1995) (Hospitality Institute of Technology and Management, http://www.hi-tm.com/) and ICMSF (1998a).

saliva. On swallowing, the food passes through the pharynx and oesophagus into the stomach. The stomach produces gastric enzymes (endopeptidases, gelatinase and lipase) to break down the food, and has a low acidic pH (~pH 2). Most digestion and absorption occur in the small intestine which is about 20 feet long. The structure of the small intestine maximises the area available for absorption. The surface of the mucosa is convoluted and folded. The surface is covered with finger-like projections called villi, which are, in turn, covered with absorptive cells. The effective surface area of the mucosal cells is further increased by the microvilli that occur on the luminal membrane of the enterocyte (Figure 4.2). The brush border of the enterocytes contains various enzymes including many disaccharides such as maltase, isomaltase, sucrase and lactase. These are involved in both the digestion and absorption of carbohydrates. Pancreatic enzymes which include trypsin, chymotrypsin, carboxypeptidase, amylase, lipases, ribonuclease, deoxyribonuclease, collagenase and elastase assist in digestion. Also, bile salts secreted from the liver aid the absorption of fats. The tissues underneath the epithelial cells contain blood capillaries which absorb monosaccharides and amino acids, and lymph capillaries to absorb fatty acids and glycerol. The small intestine is where more than 80% of absorption occurs. Mucosal enzyme levels are affected by bacterial activity, and lactose deficiency is a sensitive indicator of the colonisation of the small intestine by pathogenic

Table 4.2 Infectious dose of enteric pathogens

Organism	Estimated minimum infectious dose
Non-spore forming bacteria	
C. jejuni	1000
Salmonella spp.	10^4–10^{10}
Sh. flexneri	10^2–>10^9
Sh. dysenteriae	10–10^4
E. coli	10^6–> 10^7
E. coli O157:H7	10–100
St. aureus	10^5–>10^6/g[a]
V. cholerae	1000
V. parahaemolyticus	10^6–10^9
Y. enterocolitica	10^7
Spore-forming bacteria	
B. cereus	10^4–10^8
Cl. botulinum	10^{3a}
Cl. perfringens	10^6–10^7
Viruses	
Hepatitis A	<10 particles
Norovirus	<10 particles

[a]Viable count able to produce sufficient toxin to elicit a physiological response.

bacteria. Interactions between the cell membranes and the luminal contents are facilitated by the glycocalyx, a complex mucus layer overlaying the enterocytes. Undigested food enters the colon, which has a neutral pH and a slow transit time (up to 60 hours). Finally, food residues and intestinal microbes exit the intestinal tract via the anus.

The intestinal tract is particularly prone to microbial penetration and toxin uptake, due to the large surface area, the level of nutrients and high absorptive capacity. However, it is also equipped with an array of defence mechanisms (O'Hara & Shanahan 2006). The small intestine is the site of major absorption, but is also perturbed by many intestinal infections, leading to acute diarrhoea and dehydration. For example, the cholera toxin (Figure 4.3) inhibits sodium uptake, and stimulates chloride secretion. This results in fluid and electrolyte loss.

4.1.2 Host resistance to foodborne infections

The human body has a variety of non-specific (innate) and specific immune system mechanisms to protect us from foodborne infections. However, many factors can weaken the defences and increase the risk of illness. Additionally, some foodborne pathogens are able to evade the body's defence mechanisms.

The innate immunity is the initial host defence mechanism against micro-organisms in the intestinal tract. The high acidity and pepsin of gastric juices kill many microbial pathogens. The enzymes in pancreatic secretions, and bile acids are also antimicrobial. The epithelial lining is constantly being replaced, and the constant peristaltic movement keeps the intestinal contents and microbial flora in the small intestines low compared with the colon, which is a more static environment. The gastric and intestinal epithelia are lined with a layer of mucus that protects against proteolytic attack and microbial attachment. It also contains antibacterial agents, such as secretory immunoglobulin A and enzymes. Mucus also acts as a barrier preventing large molecular weight compounds from passing into the enterocytes in the epithelial layer.

Table 4.3 Microbiological hazards categorisation according to ICMSF

Effects of hazards	Pathogen
Categorisation of common foodborne pathogens (ICMSF 1986)	
(1) Moderate, direct, limited spread, death rarely occurs	*B. cereus, C. jejuni, Cl. perfringens, St. aureus, Y. enterocolitica, T. saginata, T. gondii*
(2) Moderate, direct, potentially extensive spread, death or serious sequelae can occur. Considered severe	Pathogenic *E. coli,* S. Enteritidis and other salmonellas other than S. Typhi and S. Paratyphi, shigellae other than *Sh. dysenteriae, L. monocytogenes*
(3) Severe, direct	*Cl. botulinum* types A, B, E and F, hepatitis A virus, *Sh. dysenteriae,* S. Typhi and S. Paratyphi A, B and C, *T. spiralis*
Updated categorisation (ICMSF 2002)	
(1) Food poisoning organisms causing moderate, not life-threatening, no sequelae, normally short duration, self-limiting	*B. cereus* (including emetic toxin), *Cl. perfringens* type A, Norovirus, *E. coli* (EPEC, ETEC), *St. aureus, V. cholerae* non-O1 and non-O139, *V. parahaemolyticus*
(2) Serious hazard, incapacitating but not life-threatening, sequelae rare, moderate duration	*C. jejuni, C. coli,* S. Enteritidis, S. Typhimurium, shigellae, hepatitis A, *L. monocytogenes, Cryptosporidium parvum,* pathogenic *Y. enterocolitica, Cyclospora cayetanensis*
(3) Severe hazard for general population, life-threatening, chronic sequelae, long duration	Brucellosis, botulism, EHEC (HUS), S. Typhi, S. Paratyphi, tuberculosis, *Sh. dysenteriae,* aflatoxins, *V. cholerae* O1 and O139
(4) Severe hazard for restricted populations, life-threatening, chronic sequelae, long duration	*C. jejuni* O:19 (GBS), *Cl. perfringens* type C, hepatitis A, *Cryptosporidium parvum, V. vulnificus, L. monocytogenes,* EPEC (infant mortality), infant botulism, *Cronobacter spp.* (*Ent. sakazakii*)

As a stable ecosystem, the normal intestinal flora reduces the opportunities for microbial infection. The normal flora will occupy any binding sites on the enterocytes, and decrease the lumen pH due to the production of volatile fatty acids. Microbes that do penetrate the epithelial barrier are likely to encounter mononuclear phagocytes (blood monocytes or tissue macrophages) and polymorphonuclear phagocytes.

The innate immune system prevents infection by enteric microbes and the entry of large microbial toxins. However, a variety of factors may depress it, such as the use of antacids which suppress gastric acid production, reduction in normal gut flora due to antibiotic usage and damage to the epithelial barrier. This is when the specific immune system becomes important.

The specific immune system recognises antigens which are of high molecular weight (>10 kD) proteins or polysaccharides. These include the bacterial cell wall, capsule, lipopolysaccharide, flagella, fimbriae and toxins. Many specialised cells are involved in the humoral (antibody-mediated) and cell-mediated immune reactions to infection.

Many different micro-organisms cause gastroenteritis or systemic infection by overcoming the intestinal defence mechanisms. There are essentially five pathogenic mechanisms:

1 Bacteria which produce toxin(s), but do not adhere to or multiply in the intestinal tract, that is *B. cereus, St. aureus, Clostridium perfringens* and *Cl. botulinum.*

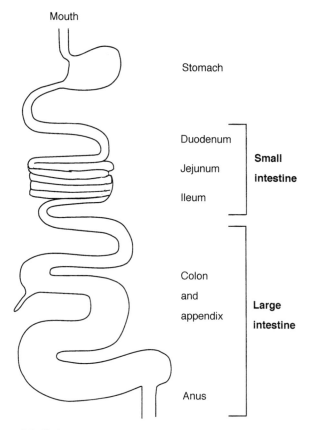

Figure 4.1 The human intestinal tract.

2 Bacteria which adhere to the intestinal tract epithelium and excrete a toxin, that is entero-toxigenic *E. coli* and *Vibrio cholerae*.

3 Bacteria which adhere and damage the brush border microvilli, that is enteropathogenic *E. coli* and enterohaemorrhagic *E. coli*.

4 Bacteria which invade the mucosal layer and initiate intracellular multiplication, that is *Shigella* spp.

5 Bacteria which penetrate the mucosal layer and spread to the underlying lamina propria and lymph nodes, that is *Yersinia* spp.

The antigen sampling process of M cells, in the Peyer's patches, may be an entry route. Motile phagocytes may ingest a bacterium from the lumen, which survives intracellular killing by the lysosomes and spreads systemically.

The immune system clears high molecular weight toxins. However, low molecular weight, non-polar compounds, such as mycotoxins, are not cleared and hence these are rapidly absorbed through the intestinal tract. They are metabolised in the liver by a process called 'biotransformation'. However, biotransformation can make a compound more toxic. For example, aflatoxin B_1 is converted to a reactive epoxide which can react with nuclear DNA causing DNA damage that may lead to liver cancer; see Section 4.7.1.

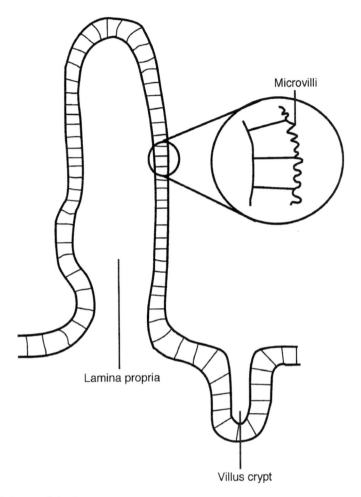

Microvilli

Lamina propria

Villus crypt

Figure 4.2 Gut mucosal structure.

4.1.3 The normal human intestinal flora

In the womb, the intestinal tract of the human foetus is microbiologically sterile. However, during birth, the neonate acquires a microbial flora from the vagina, contact with the environment, and through feeding sources. Consequently, a dense, complex bacterial community becomes established in the intestinal tract. The human intestinal flora is a complex ecosystem and may have a number of effects on the health of the host. The flora changes in relationship to diet, age and disease. The maturation of the immune system requires continual stimulation from the developing gut flora. The lack of development has been linked with an increase in prevalence of atopic diseases; the 'hygiene hypothesis' (Section 1.6).

The use of 16S rRNA gene sequencing and DGGE (Section 2.9.2) has expanded the study of the human intestinal flora because it is culture independent and is therefore not dependent upon the use of numerous media agars and incubation conditions. In general, bacterial colonisation starts soon after birth, and facultative anaerobic bacteria can be detected in faecal samples. These organisms remove oxygen and subsequently lower the redox potential, which enables

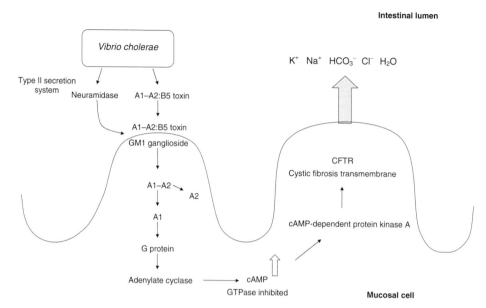

Figure 4.3 Osmotic balance; NaCl flux across gut mucosa.

strict anaerobes to grow. In vaginally delivered neonates, the first colonising bacteria are of maternal origin. In Caesarean section delivered neonates, the environment and hospital staff are a major source of the colonising bacteria. After this initial colonisation, the diversity of the subsequent bacterial flora community is influenced by the diet.

The intestinal microbial flora of formula-fed infants is initially more diverse than that of breastfed infants. Human breast milk is not necessarily sterile, and can have low numbers of streptococci, micrococci, lactobacilli, staphylococci, diphteroids and bifidobacteria. Breastfed infants are often colonisation by staphylococci due to increased contact with the mother's skin during feeding. In breastfed, full-term infants, Gram-positive bifidobacteria, lactobacilli and streptococci dominate. In contrast, in formula-fed infants, the predominant flora is a mixture of Enterobacteriaceae (*E. coli* and *Klebsiella* spp.), *Staphylococcus*, *Clostridium*, *Bifidobacterium*, *Enterococcus* and *Bacteroides* species. *Cronobacter* spp. (*Enterobacter sakazakii*) is associated with neonatal infections through contaminated powdered infant formula. This organism is covered in detail in Section 4.8.2. The intestinal flora diversity changes at weaning with the introduction of solid food, and the two groups become indistinguishable and resembles the adult flora.

Schwiertz *et al.* (2003) studied the intestinal flora of preterm and full-term infants using PCR-DGGE profiles and cultivation from faeces. The DGGE profiles of preterm infants in the first days of life only contained a few DGGE bands, but after 2 weeks, the number of major bands ranged from 5 to 20. *E. coli*, *Enterococcus* spp. and *Klebsiella pneumoniae* were the most frequently identified organisms. Bifidobacteria were detected in breastfed, full-term infants, but not in the preterm infants. The microbial flora was more diverse in the preterm infants than the breastfed, full-term infants.

The adult intestinal flora is a complex ecosystem, which is ~55% of the solids and is dominated by strict anaerobes (Table 4.4). The indigenous flora is composed of the normal flora, micro-organisms which persist in the large intestines and the transient autochthonous flora.

Table 4.4 Microbial flora of the human intestinal tract

Organism	Cell density (cfu/g)[a]
Protozoan parasites	10^6–10^7
Ascaris	10^4–10^5
Enteric viruses	
Enteroviruses	10^3–10^7
Rotaviruses	10^{10}
Adenovirus	10^{12}
Enteric bacteria	
Acidaminococcus fermentans	10^7–10^8
Bacteroides ovatus	
B. uniformis	
B. coagulans	10^9–10^{10}
B. eggerthii	
B. merdae	
B. stercoris	
Bif. bifidum	10^8–10^9
Bif. breve	
Clostridium cadaveris	
Cl. clostridioforme	
Cl. innocuum	10^8–10^9
Cl. paraputrificum	
Cl. perfringens	
Cl. ramosum	
Cl. pertium	
Coprococcus cutactus	10^7–10^8
Enterobacter aerogenes	10^5–10^6
Ent. faecalis	10^5–10^6
E. coli	10^6–10^7
Eubacterium limosum	10^8–10^9
E. tenue	
Fusobacterium mortiferum	
F. naviforme	
F. necrogenes	10^6–10^7
F. nucleatum	
F. prousnitzil	
F. varium	
K. pneumoniae	10^5–10^6
K. oxytoca	
Lactobacillus acidophilus	
Lb. brevis	
Lb. casei	
Lb. fermentum	
L. leichmannii	10^7–10^8
Lb. minutus	
Lb. plantarum	
Lb. rogosa	
Lb. ruminis	
Lb. salovarius	
Megamonas hypermegas	10^7–10^8
M. elsdenii	10^7–10^8
Methanobrevibacter smithii	Undetectable–10^9

(Continued)

Table 4.4 (Continued)

Organism	Cell density (cfu/g)[a]
Methanosphaeraa stadtmaniae	Undetectable–10^9
Morganella morgannii	
Peptostreptococcus asscharolyticus	
P. magnus	10^8–10^9
P. productus	
Proteus mirabilis	10^5–10^6
Salmonella spp.	10^4–10^{11}
Shigella spp.	10^5–10^9
Veillonella parvula	10^5–10^6
Indicator bacteria	
Coliforms	10^7–10^9
Faecal coliform	10^6–10^9

[a]Values are given as concentration per gram of faeces as representative of the large intestinal tract.
From Haas *et al.* (1999) and Tannock (1995).

There are ~1800 genera representing ~16 000 species in the human intestine microbiome. Of these, only ~20% have been cultured (Hattoir & Taylor 2009). As stated above, the adult intestinal flora is primarily composed of a number of strictly anaerobic bacteria. The dominant bacterial species belong to the Gram-negative *Cytophaga–Flavobacterium–Bacteroides* (23%) and the Gram-positive low GC% *Firmicutes* (64%) phyla. The *Proteobacteria* (including *Enterobacteriaceae*) only form a minor component (~8%). Commonly isolated genera are as follows:

1 *Bacteroides* spp., such as *Bacteroides fragilis*: These are Gram-negative, non-spore forming rods. They produce volatile and non-volatile fatty acids (VFA and n-VFA), acetic, succinic, lactic, formic, propionic, *N*-butyric, isobutyric and isovaleric acids.

2 *Bifidobacterium* spp., such as *Bifidobacterium bifidum*: These are Gram-positive, non-spore-forming rods with characteristic club-shaped ends. They produce acetic and lactic acids (3:2 ratio).

3 *Clostridium* spp., such as *Clostridium innocuum*: These are Gram-positive spore-former rods.

4 *Enterococcus* spp., such as *Enterococcus faecalis*: These are Gram-positive cocci which are aerotolerant. They are in Lancefield group D and can grow in 6.5% NaCl and under alkaline conditions <pH 9.6.

5 *Eubacterium* spp.: These are Gram-positive non-spore forming rods. They produce butyric, acetic and formic acids.

6 *Fusobacterium* spp.: These are Gram-negative non-spore forming rods which produce *N*-butyric acid.

7 *Peptostreptococcus* spp.: These are Gram-positive cocci that can degrade peptone and amino acids.

8 *Ruminococcus*: These are Gram-positive cocci that produce acetic, succinic and lactic acids, ethanol, carbon dioxide and hydrogen from carbohydrates.

As given above, the dominant genus of intestinal bacteria in animals and humans are the *Bacteroides* (10^{11} cfu/g), which are Gram-negative, strict anaerobes. The intestinal microflora also contains Gram-positive non-sporing strictly anaerobic rods. There are many other genera of organisms present, including *Bifidobacterium*, *Lactobacillus* and *Clostridium* spp. Gram-positive cocci are numerically important in the intestinal tract, including *Peptostreptococcus* and *Enterococcus* spp. The facultative Gram-negative rods such as *Proteus*, *Klebsiella* and *E. coli* are not numerically important, being outnumbered by *Bacteroides* spp. by ~1000:1.

E. coli are Gram-negative, catalase-positive, oxidase-negative, facultatively anaerobic short rods. They are members of the family *Enterobacteriaceae* and most isolates ferment lactose. The majority of *E. coli* serotypes are not pathogenic and are part of the normal intestinal flora (~10^6 organisms/g). Subsequently, they are used as indicator organisms, as one of the 'coliforms', to indicate faecal pollution of water raw ingredients and foods. Non-pathogenic strains of *E. coli* typically colonise the infant gastrointestinal tract within a few hours after birth. The presence of this bacterial population in the intestine suppresses the growth of harmful bacteria and is important for synthesising appreciable amounts of B vitamins. *E. coli* usually remains harmless when confined to the intestinal lumen. However, in debilitated or immuno-suppressed humans, or when gastrointestinal barriers are violated, even normal, 'non-pathogenic' strains of *E. coli* can cause infection.

As given earlier in Section 2.9.1, the human intestinal flora is the focus of many studies such as the 'human microbiome project' based on metagenomics to reveal the diversity of organisms and their interaction with intestinal function and illness. Over the next few years, our knowledge will considerably increase, and in turn, we will better understand their respective roles in foodborne illness.

4.2 Indicator organisms

The term 'indicator organisms' can be applied to any taxonomic, physiological or ecological group of organisms whose presence or absence provides indirect evidence concerning a particular feature in the past history of the sample. It is often associated with organisms of intestinal origin but other groups may act as indicators of other situations. For example, the presence of members of 'all Gram-negative bacteria' in heat-treated foodstuffs is indicative of inadequate heat treatment (relative to the initial numbers of these organisms) or of contamination subsequent to heating. Coliform counts, since coliforms represent only a subset of 'all Gram-negative bacteria', provide a much less sensitive indicator of problems associated with heat treatment, but are still frequently used in the examination of heat-treated foodstuffs. The term 'index organism' was suggested by Ingram in 1977 for a marker whose presence indicated the possible presence of an ecologically similar pathogen.

Microbial indicators are more often employed to assess food safety and hygiene than quality. Ideally, a food safety indicator should meet certain important criteria. It should:
- be easily and rapidly detectable,
- be easily distinguishable from other members of the food flora,
- have a history of constant association with the pathogen whose presence it is to indicate,
- always be present when the pathogen of concern is present,
- be an organism whose numbers ideally should correlate with those of the pathogen of concern,
- possess growth requirements and a growth rate equalling those of the pathogen,
- have a die-off rate that at least parallels that of the pathogen and ideally persists slightly longer than the pathogen of concern,
- be absent from foods that are free of the pathogen except perhaps at certain minimum numbers.

Common indicator organisms are as follows:
- *E. coli*
- *Enterobacteriaceae*
- Enterococci (formerly known as faecal streptococci)
- Bacteriophage

4.2.1 Coliforms

Coliforms is a general term for Gram-negative, rod-shaped facultatively anaerobic bacteria. They are also known as the 'coli-aerogenes' group. Identification criteria used are the production of gas from glucose (and other sugars) and fermentation of lactose to acid and gas within 48 hours at 35°C (Hitchins *et al.* 1998). The coliform group includes species from the genera *Escherichia*, *Klebsiella*, *Enterobacter* and *Citrobacter*, and includes *E. coli*. Coliforms were historically used as indicator micro-organisms to serve as a measure of faecal contamination, and therefore potentially the presence of enteric pathogens in freshwater. However, since most coliforms are found throughout the environment, they have little direct hygiene significance. In Europe, due to Sanco food criteria, the wider *Enterobacteriaceae* grouping is becoming favoured over 'coliforms', but the later term has been retained here due to its frequent usage. Since coliforms are easily killed by heat, coliform counts can be useful when testing for post-processing contamination.

In order to differentiate between faecal and non-faecal coliforms, the faecal coliform test was developed. Faecal coliforms are defined as coliforms that ferment lactose in EC medium with gas production within 48 hours at 45.5°C (except shellfish isolates, 44.5°C). *E. coli* is the major species in the faecal coliform group. Of the groups of bacteria that comprise the total coliforms, only *E. coli* is generally not found growing and reproducing in the environment. Consequently, *E. coli* is considered to be the species of coliform bacteria that is the best indicator of faecal pollution and the possible presence of enteric pathogens. The use of coliform tests is declining as improved specific detection methods are developed, and the wider range of *Enterobactericaeae* organisms have been used as an indicator group.

4.2.2 *Enterobacteriaceae*

The *Enterobacteriaceae* includes many genera, including those that ferment lactose (i.e. *E. coli*) and those that do not (i.e. *Salmonella*). Therefore, the use of violet red bile glucose agar (VRBGA) is applicable for the *Enterobacteriaceae*, whereas violet red bile lactose agar is more applicable for the coliforms. There has been a general trend away from using coliforms (4.2.1) as indicators of the possible presence of enteric pathogens to *Enterobacteriaceae*. This is in part due to their ease of detection. Water that is contaminated with faecal matter may contain a range of pathogens including *V. cholerae*, *S.* Typhi, other *Salmonella* serovars, *Shigella dysenteriae*, *Campylobacter jejuni*, pathogenic strains of *E. coli* and the protozoan *Giardia lamblia*.

4.2.3 Enterococci

The enterococci include two species found in the human and animal intestines: *Ent. faecalis* and *Enterococcus faecium*. The former is mainly associated with the human intestinal tract, whereas the later is found in both humans and animals. The enterococci are sometimes used as indicators of water contaminated with faeces. The advantage of testing for enterococci is that they die more slowly than *E. coli*, and therefore reduces the risk of false-negative results. Unfortunately, they are found more frequently than *E. coli* in faecal environments and therefore their presence can be inconclusive proof of faecal contamination. It is plausible that they are a better indication of hygienic quality in food as they are more resistant to drying than coliforms. This is especially true for frozen and dried foods, and foods receiving moderate heat treatment. However, their resilience further undermines their value as indicator organisms as their presence in foods may be of little consequence if the pathogens have been killed during processing.

Enumeration of enterococci usually involves sodium azide, thallous acetate of antibiotics as selective agents and raised incubation temperature (45°C). Some media include tetrazolium chloride which leads to red colony formation, and therefore enhances visualisation of enterococci colonies.

4.2.4 Bacteriophage

It is impractical to routinely screen water for the presence of pathogenic viruses. However, bacteriophages (coliphage, F'-specific phage and *Bacteroides* phages) have been proposed as more appropriate indicators of waterborne viruses than bacterial indicators (Lees 2000). Bacteriophages are of similar size and surface characteristics to viruses, are present in human faeces and are easier to detect using the agar overlay technique with a tester bacterial strain, and the enumeration of plaques. This has been a topic of considerable debate, largely due to the lack of carefully validated comparative methodologies.

4.3 Foodborne pathogens, bacteria

4.3.1 C. *jejuni*, *Campylobacter coli* and *Campylobacter lari*

Campylobacters are Gram-negative thin rods (0.2–0.9 μm × 0.2–5.0 μm). They are microaerophiles (requiring 3–5% oxygen and 2–10% carbon dioxide for growth); however, their tolerance to oxygen is strain and species dependent. Their maximum growth is at 42–46°C and no growth occurs below 30°C. Therefore, *Campylobacter* do not multiply at ambient or refrigeration temperatures. Genomic analysis of campylobacters is considered in Section 2.9.3 (Box 4.1).

Box 4.1

Organisms

C. *jejuni* and C. *coli*

Onset of symptoms

2–5 days

Food source

Raw chicken, beef, milk, mushrooms, clams, hamburger, water, cheese, pork, shellfish, eggs, cake icing

Acute symptoms and chronic complications

Abdominal pain, watery diarrhoea (sometimes bloody), fever, malaise, vomiting
Chronic symptoms: Colitis, Guillain–Barré syndrome, Reiter's syndrome

C. jejuni and *C. coli* are commensal in cattle, swine and birds. *C. jejuni* is often the predominant species in poultry, whereas *C. coli* and *C. lari* predominate in swine and birds, respectively. *Campylobacter upsaliensis* has been frequently isolated from domestic cats and dogs.

Globally, campylobacters are the major cause of human bacterial gastroenteritis, totalling as many as 400–500 million cases worldwide each year. Campylobacters were initially recognised as animal pathogens and have only in the past 15–20 years been recognised as human pathogens (Butzler *et al.* 1973, Humphrey *et al.* 2007, Skirrow 1977). Nowadays, the recorded incidence of *Campylobacter* is greater than that of any other food poisoning organisms, and has been linked to serious neurological disorders such as Guillain–Barré syndrome (GBS) and Miller–Fischer syndrome. Worldwide, they infect an estimated 1% of the population of Western Europe. The annual cost of campylobacteriosis in the United States is estimated at $1.3–$6.2 billion (Section 1.7). Reservoirs include several wild animals as well as poultry, cows, pigs and domestic pets. Chickens are a primary risk due to the high level of consumption or improperly handled or undercooked poultry.

The characteristics of campylobacter enteritis are as follows:
- Flu-like illness
- Abdominal pain
- Fever
- Diarrhoea may be profuse, watery and often bloody in children.

The incubation period is 2–10 days, with most people showing symptoms by 4 days (Table 1.18). The disease lasts for about 1 week and is usually self-limiting after a period of bed rest. The organism is excreted in the faeces for several weeks after the symptoms have ceased. Relapses occur in about 25% of cases.

There are 16 described species, six sub-species and two biovars. Reported food poisoning cases are mainly due to *C. jejuni* which causes the majority of outbreaks (89–93%) of cases and *C. coli* (7–10%) (Tam *et al.* 2003). However, *C. upsaliensis* and *C. lari* have only occasionally been implicated. This may in part be due to inefficient isolation methods, since the use of filtration methods has revealed that *C. upsaliensis* is associated with human disease more than previously recognised. These species are commonly known as 'thermophilic campylobacters' since they grow at higher temperatures than other campylobacters. The organism's morphology can change between vibroid (curved), spiral, doughnut, S-shaped and coccoid. The short coccoid morphology is associated with the 'viable but non-culturable' (VNC) state of stressed cells. By definition, it is not recoverable using conventional methods from the VNC condition (Section 5.2.3).

The organism is very sensitive to drying and is readily destroyed by cooking at 55–60°C for several minutes ($D_{50} = 0.88$–1.63 minutes). Temperature, pH and water activity growth ranges and D values for *Campylobacter* are given in Tables 2.3 and 2.9. The infectious dose is approximately 1000 cells and a fuller description of the dose–response relationship is given in Section 9.5.

As given above, the organism does not multiply at room temperature (minimum growth temperature is 30°C, Table 2.8). Therefore, *C. jejuni* and *C. coli* do not multiply in chilled foods, although they will persist under such conditions. This may be why cases of campylobacter gastroenteritis outnumber *Salmonella* in many countries. There is a notable seasonality to *Campylobacter* enteritis, with peak incidences in the summer months (Figure 1.3). The incidence of *Campylobacter* infections has been decreasing in some countries. Figure 1.4 shows the 20-year data for England and Wales from 1986 to 2005.

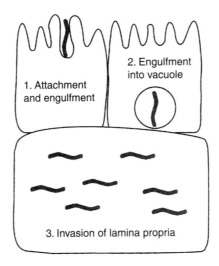

Figure 4.4 *Campylobacter* invasion of the gut wall.

C. jejuni colonises the distal ileum and colon of the human intestinal tract. After colonising the mucus and adhesion to intestinal cell surfaces, the organism perturbs the normal absorptive capacity of the intestine by damaging epithelial cell function. This is due to either direct action, by cell invasion or the production of toxin(s), or indirectly, following the initiation of an inflammatory response (Figure 4.4). Intracellular survival within host cell may be aided by the production of catalase to protect against oxidative stress from lysosomes. The organism may also enter a VNC state which may be of importance in the organism's virulence (Rollins & Colwell 1986).

There is no consensus of opinion concerning the virulence factors of *Campylobacter*. The organism produces at least three toxins (Wassenaar 1997):

1 Heat-labile enterotoxin (60–70 kDa) which increases cyclic AMP levels of intestinal cells and cross-reacts with anti-cholera toxin antibodies. This is also referred to as the 'cholera-like' toxin. It is iron regulated and causes elongation of Chinese hamster ovary cells and binds to GM_1 ganglioside. It has an AB_5 domain structure and there is homology between the B subunit and cholera toxin and *E. coli* heat-labile toxin (LT).

2 Cytoskeletal altering toxin that may also cause diarrhoea. This is also known as the cytolethal distending toxin (CLDT) and has been the most studied toxin from *Campylobacter*. This toxin causes characteristic elongation of cultured mammalian cells followed by distension. It is different from other toxins produced by enteric bacteria in that it is destroyed by heating at 70°C for 15 minutes and by trypsin.

3 Heat-labile protein cytotoxin which is not neutralised by anti-cholera toxin.

DNA microarray analysis of *C. jejuni* strains have correlated the CLDT and haemolysin with the organism's survival under aerobic conditions at room temperature (On *et al.* 2006). The poorly surviving strains represented genotypes that have not been reported in human infections. In contrast, the strains with high surviving potential represented genotypes from human disease cases.

Campylobacter infection is now recognised as the single most identifiable antecedent infection associated with the development of GBS; see Section 1.8 on chronic sequelae. It is the most common cause of acute flaccid paralysis (Allos 1998). Campylobacters are associated with several pathologic forms of GBS, including the demyelinating (acute inflammatory demyelinating polyneuropathy) and axonal (acute motel axonal neuropathy) forms. GBS is an autoimmune disorder of the peripheral nervous system. It is characterised by weakness usually symmetrical, which develops over several days. It occurs worldwide and is the most common cause of neuromuscular paralysis. Of an estimated annual number of 2628–9575 GBS cases in the United States, 526–3830 were triggered by *Campylobacter* infection. Typically, the gastrointestinal symptoms occur 1–3 weeks before the neurological symptoms.

GBS is probably due to an autoimmune response to the ganglioside GM1 on peripheral nerves following infection by *C. jejuni* O19 (although other serotypes may be involved). The peripheral nerves may share epitopes in common with surface antigens of *C. jejuni* LPS (molecular mimicry). Cytokines may induce the inflammatory process and result in nerve demyelination. The complement system may also lead to nerve damage, and an increase in the permeability of the blood–nerve barrier which causes inflammation.

Estimated total costs (in US$) of *Campylobacter*-associated GBS are $0.2–$1.8 billion. A risk assessment example is given in Section 10.2.

Direct enumeration of *Campylobacter* species is rarely possible. Usually, an enrichment step is used to recover low numbers of the organism from processed foods; see Section 5.7.3 for details. On blood agar plates, *Campylobacter* colonies are non-haemolytic, flat, about 1–2 mm in diameter and either spreading with an irregular edge or discrete, circular-convex. Most *Campylobacter* isolates are not speciated in routine analytical laboratories. In contrast, public health laboratories investigating a food poisoning outbreak will 'fingerprint' isolates using a range of procedures; biotyping, phage typing and serotyping; see Section 5.7.3. An increase in ciprofloxacin resistance (a medically important antibiotic) has been reported possibly due to the veterinary use of the structurally related (fluoroquinolone) antibiotic enrofloxacin used in poultry husbandary; see Section 10.2.4.

Source:
- *Livestock*: pigs, cattle and sheep
- *Domestic animals*: cats and dogs
- Poultry
- Raw milk
- Polluted water

The routes of infection are via contaminated water, milk and meat. Poultry is the largest potential source of infectious *Campylobacter*. Consequently, most sporadic infections are associated with improper preparations or consumption of mishandled poultry products. Most *C. jejuni* outbreaks, which are far less common than sporadic illnesses, are associated with the consumption of raw milk or unchlorinated water. *Campylobacter* does not grow at temperatures below 30°C and is sensitive to acidic pH. However, it can survive on cut fruits and vegetables for sufficient time to be a risk to the consumer, and *Campylobacter* enteritis has been associated with the consumption of contaminated raw fruits and vegetables.

Control measures:
- Heat treatment (pasteurisation/sterilisation)
- Hygienic slaughter and processing procedures
- Prevention of cross-contamination
- Good personal hygiene
- Water treatment

Campylobacter can easily cross-contaminate processed food. A contaminated piece of raw meat can leave 10 000 *Campylobacter* cells per cm^2 on a work surface. Since the infectious dose is only ~1000 cells, the residual microbial load must be reduced to <2 per cm^2. *C. jejuni* is quickly killed during cooking at 55–60°C for several minutes and is not a spore former. Hence, the main control mechanisms are adequate cooking regimes and prevention of cross-contamination for raw meats and poultry.

Future control measures may include the following:

1 Competitive exclusion, whereby the chickens are inoculated with a cocktail of non-pathogenic bacteria which colonise the intestines and reduce the incidence of *Salmonella* and *Campylobacter* carriage.
2 Vaccination, as has been applied for the control of *Salmonella*.
3 Bacteriophage treatment of chickens which will reduce the carriage of *Campylobacter* (Scott *et al.* 2007).

4.3.2 *Salmonella* species

Salmonella is an important cause of foodborne disease throughout the world and is a significant cause of morbidity, mortality and economic loss. Salmonellosis ranks as one of the most frequently reported foodborne diseases worldwide. *Salmonella* is a genus in the family *Enterobacteriaceae*. They are Gram-negative, facultatively anaerobic, non-spore-forming short (1–2 μm) rods. The majority of species are motile with peritrichous flagella; *S.* Gallinarum and *S.* Pullorum are non-motile. Salmonellas ferment glucose with the production of acid and gas but are unable to metabolise lactose and sucrose. Their optimum growth temperature is about 38°C and their minimum is about 5°C (Table 2.9). Because they do not form spores, they are relatively heat sensitive, being killed at 60°C in 15–20 minutes ($D_{62.8} = 0.06$ minutes, Table 2.3). Plasmids in the range 50–100 kb have been associated with the *Salmonella* virulence (Slauch *et al.* 1997). *Salmonella* pathogenicity islands are covered in Section 2.3.2 (Table 2.2), and comparative genomic analysis is considered in Section 2.9.4 (Figure 2.19) (Box 4.2).

Box 4.2

Organisms

Salmonella (non-typhoid), *S.* Typhi and *S.* Paratyphi

Food source

Raw poultry, meat, eggs, milk and dairy products, vegetables, fruits, chocolate, coconut, peanuts, fish, shellfish

Onset of symptoms

Salmonella (non-typhoid) 6–48 hours
  S. Typhi and *S.* Paratyphi 7–28 days

Acute symptoms and chronic complications

Salmonella (non-typhoid): Abdominal pain, bloody stools, cold chills, dehydration, diarrhoea, exhaustion, fever, headache and sometimes vomiting. Chronic symptoms: reactive arthritis, Reiter's syndrome
S. Typhi and *S.* Paratyphi: Typhoid-like fever, malaise, headache, abdominal pain, body aches, diarrhoea or constipation

Most human *Salmonella* infections are thought to be associated with foodborne transmission from meat and dairy products. Nevertheless, outbreaks of salmonellosis have been linked to a diversity of fruits and vegetables. *Salmonella* have been isolated from many types of raw fruits and vegetables including bean sprouts, melons, unpasteurised orange juice, apple juice and tomatoes (Bidol *et al.* 2007). The bacterium can grow on the surface of alfalfa sprouts, tomatoes and other possible raw fruits and vegetables. Therefore, it is essential that hygienic practices are observed during handling to reduce produce contamination. *Salmonella* can contaminate cocoa pods during harvesting and fermentation. The pods are lightly roasted (60–80°C) to make chocolate. *Salmonella* can survive due to the minimal processing, low water activity and high fat content which help to protect the organism during transit through the stomach.

There are only two species of *Salmonella*, *Salmonella enterica* and *Salmonella bongori*, which are divided into eight groups (Boyd *et al.* 1996). This is useless with regard to epidemiological investigations and therefore further detailed characterisation is required. Fortunately, there are over 2400 serovars and this is now used as the basis for *Salmonella* nomenclature. For example, *S. enterica* serovar Typhimurium which was formerly referred to *Salmonella typhimurium*. The later term incorrectly inferred that 'typhimurium' is a species, whereas in fact it is a serovar of the species *S. enterica*. Although all serovars can be considered pathogenic to humans, only about 200 are associated with human illness.

Some serotypes were initially named after the place where they were first isolated, for example *S.* Dublin and *S.* Heidelberg. Others were named after the disease and affected animal. For example, *S.* Typhi and the paratyphoid bacteria are normally septicaemic and produce typhoid or typhoid-like fever in humans, whereas *S.* Typhimurium causes typhoid in mice.

Despite their close relatedness, *Salmonella* serovars differ in their host range and the diseases they cause. For example, *Salmonella* Enteritidis is a serovar typically linked to eggs, and to a lesser degree to poultry meat. It is a so-called invasive *Salmonella*, that is a serovar able to enter the bloodstream of the bird and infect the egg. *S.* Typhimurium infects many host species, including humans, poultry and mice. It is often linked to eggs, but more commonly to poultry meat, whereas *S.* Typhi only infects humans. In contrast, *S.* Pullorum and *S.* Gallinarum are both specific to poultry, but they cause distinct diseases. It is plausible that the culling of UK poultry in the early 1970s that was seropositive for *S.* Gallinarum and *S.* Pullorum created a niche that was filled by the antigenically similar *S.* Enteritidis, which increased in the 1980s to become the most common serotype of human isolates.

The *Salmonella* serovars are distinguished by their O, H and Vi antigens using the Kaufmann–White scheme (Brenner 1984, Ewing 1986, Le Minor 1988). The serotypes are then put into serogroups according to common antigenic factors (Table 4.5). The *Salmonella* have a complex lipopolysaccharide (LPS) structure (Figures 2.2 and 2.3; Mansfield & Forsythe 2001) which gives rise to the O antigen. The number of repeating units and sugar composition varies considerably in *Salmonella* LPS and is of vital importance with regard to epidemiological studies. The sugars are antigenic and therefore can be used immunologically to identify *Salmonella* isolates. It is the serotype of the *Salmonella* isolate that aids epidemiological studies tracing the vector of *Salmonella* infections. Further characterisation is required for epidemiological studies and this includes biochemical profiles and phage typing.

Because *Salmonella* cannot tolerate the low pH of the stomach, the infectious (ingested) dose is in the order of 10^5 bacteria, whereas if it is administered intravenously, only 10 cells can kill a mouse.

Antibiotic resistance in *Salmonella* is increasing, and in some Asian countries, more than 90% of *Salmonella* isolates are resistant to the most commonly used antibiotics. The combination

Table 4.5 *Salmonella* serotyping

Serotype	Group	O antigen	H antigen Phase 1	Phase 2
S. Paratyphi A	A	(1), 2,12	a	—
S. Typhimurium	B	(1), 4, (5), 12	i	1, 2
S. Paratyphi C	C₁	6, 7, Vi	c	1, 5
S. Newport	C₂	6, 8	e, h	1, 2
S. Typhi	D	9, 12, Vi	d	—
S. Enteritidis	D	(1), 9, 12	g, m	
S. Anatum	E₁	3, 10	e, h	1, 6
S. Newington	E₂	3, 15		
S. Minneapolis	E₃	(3), (15), 34		

Parenthesis indicate antigen determinant which may be difficult to detect.
Dominant antigenic determinants are underlined.

of increased antibiotic resistance and widespread dissemination of the organism has resulted in strains such as *S.* Typhimurium DT104; see Section 1.9, Table 1.11.

The dominant serotype causing food poisoning has changed over recent decades from *S.* Agona, *S.* Hadar, *S.* Typhimurium to the current *S.* Enteritidis (D'Aoust 1994). In fact, one phage type alone (PT4) of *S.* Enteritidis is the predominant cause of salmonellosis in many countries. The change in serotypes probably reflects changes in animal husbandry and the dissemination in the food chain of new serotypes from increased world trade. Current concern is the increase in multiple antibiotic-resistant serovars such as *S.* Typhimurium DT104.

A wide range of accredited methods for detecting *Salmonella* are available. These usually comprise of three stages: pre-enrichment, enrichment and selection (Mansfield & Forsythe 2000b). Hence, the organism is not enumerated in the foodstuff, but a detection limit of 'Less than one *Salmonella* cell in 25 g of food product' is set. For more details, see Section 5.6.2.

Salmonella typically causes one of the following three diseases:
- *Gastroenteritis*: *S.* Enteritidis and *S.* Typhimurium
- *Enteric fever*: *S.* Typhi and *S.* Paratyphi
- *Invasive systemic disease*: *S.* Cholerasuis

Characteristic symptoms of *Salmonella* food poisoning are as follows:
- Diarrhoea
- Nausea
- Abdominal pain
- Mild fever and chills
- Sometimes vomiting, headache and malaise

Incubation period before illness is between 12 and 36 hours. The illness is usually self-limiting, lasting 4–7 days (Table 1.18). The infected person will be shedding large numbers of salmonellas in the faeces during the period of illness (average of 5 weeks). The numbers of salmonellas in the faeces will decrease but may persist for up to 3 months; approximately 1% of cases become chronic carriers. Children excrete up to 10^6–10^7 salmonellas/g in faeces during convalescence. Occasionally, systemic infections will occur, often due to *S.* Dublin and *S.* Cholerasuis, which may require fluid and electrolyte replacement treatment.

Thorns (2000) estimated the incidence of salmonellosis (per 100 000) as 14 (United States), 38 (Australia), 73 (Japan), 16 (Netherlands) and 120 in parts of Germany. The numbers of salmonellosis cases show a marked seasonal trend with peak incidences in the summer months (Figure 1.4).

Chronic consequences include postenteritis reactive arthritis and Reiter's syndrome may follow 3–4 weeks after onset of acute symptoms; see Section 1.8 on chronic sequelae. Reactive arthritis may occur in 1–2% of cases. Reactive arthritis and Reiter's syndrome are rheumatoid diseases caused by a range of bacteria which induce septic arthritis by hematogenous spread to the synovial space, causing inflammation. Causative organisms include S. Enteritidis, S. Typhimurium and other serotypes such as S. Agona, S. Montevideo, and S. Saint Paul. Non-salmonella causing reactive arthritis include C. jejuni, Shigella flexneri, Shigella sonnei, Yersinia enterocolitica (in particular O:3 and O:9), Yersinia pseudotuberculosis, E. coli and K. pneumoniae. These conditions are related to the major histocompatibility complex (MHC) gene for the Class 1 antigen, HLA-B27 and cross-reaction with bacterial antigen leading to an autoimmune anti-B27 response. Those who are human leukocyte antigen HLA-B27 positive have an 18-fold greater risk for reactive arthritis, a 37-fold greater risk for Reiter's syndrome, and up to a 126-fold greater risk for ankylosing spondylitis than persons who are HLA-B27 negative and have the same enteric infections. However, a lack of correlation between reactive arthritis and HLA-B27 has been reported after S. Typhimurium and S. Heidelberg–S. Hadar outbreaks in Canada (Thomson *et al.* 1995). The condition is immunological and hence patients do not benefit from treatment with antibiotics, but are treated with non-steroid anti-inflammatory drugs.

Initially, it was believed that large numbers of *Salmonella* had to be ingested for infection. However, outbreaks involving cheddar cheese and chocolate were found to be caused by apparently <10 cell, and 50–100 cells, respectively. It is now accepted that the infective dose varies from 20 to 10^6 cells according to age and health of the victim, the food and also the *Salmonella* strain. See Section 9.5.5 for further discussion on dose–response relationship, food and vulnerability of host. It should be noted that the first 50 mL of liquid passes straight through the stomach into the small intestines and is therefore protected from the stomach's hostile acidic environment. Likewise, it is believed that chocolate can protect *Salmonella* while being transient in the stomach, hence reducing the infectious dose. Survival of *Salmonella* spp. through the acidic environment of the human stomach can be enhanced by the induction of acid tolerance due to heat treatment and exposure to short-chain fatty acids (cf. stress responses, Section 2.7.1). Infection is caused by the penetration and passage of *Salmonella* organisms from gut lumen into the epithelium of small intestine where they multiply. Subsequently, the bacteria invade the ileum and even occasionally the colon. The infection elicits an inflammatory response.

Salmonella have over 200 virulence factors which are encoded in at least five pathogenicity islands. They also have a virulence plasmid and many pathogenicity islets; see Section 2.3.2 and Table 2.2. Like many other enteric pathogens, *Salmonella* spp. invade mammalian cells of the lower intestinal tract by inducing actin rearrangements that result in the formation of pseudopods that engulf the bacteria. The role of adhesins is uncertain, but invasion of epithelial cells by *Salmonella* spp. most likely occurs following adhesion to microvilli via adhesins, mannose-specific type 1 fimbriae. *Salmonella* invades intestinal epithelial cells. Although these cells are not phagocytic, *Salmonella* use a Type III secretion system to inject several bacterial factors into the host cell which affect a number of host cell processes (Ly & Casanova 2007). Actin rearrangement occurs underneath the adherent *Salmonella* cell causing a change in the appearance of the surface of the host cell that resembles a droplet splash. Next, the membrane ruffling results in macropinocytosis, and the engulfment of the bacterium inside an endocytic vesicle. The effect of the bacteria on host cells is called ruffling because of the appearance of the deformed host cell membrane (Figure 4.5).

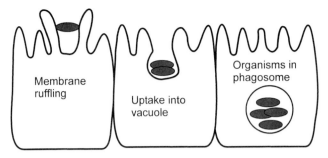

Figure 4.5 *Salmonella* invasion of the gut wall.

Invasion is enhanced under anaerobic conditions, when the cells are in the stationary phase and when osmolarity is high. *Salmonella* spp. have two Type III secretion systems; one for invasion and one for intracellular survival. A large number of gene loci are required for invasion and many are located on SPI-1 (Table 2.2). One of the genes (*inv*) encodes for the formation of surface structures produced when the bacterium adheres to the host cell. After the bacteria are engulfed in a vesicle, the host cell surface and actin filament organisation return to normal. *Salmonella* avoids lysosomal fusion due to a Type III secretion system which causes the vacuole to diverge from the standard pathway that would deliver it to a lysosome. The *Salmonella* remains within the vacuole and multiply. The vesicles may coalesce and this may be followed by further bacterial multiplication which may lead to host cell death. The organism can survive in the phagocytes due to its resistance to oxidative bursts through the production of catalase and superoxide dismutase and resistance to defensins (toxic peptides) due to produces from the *phoP/phoQ* operon. *Salmonella* contain a virulence plasmid which encodes factors needed for prolonged survival within the host.

A fuller treatment of bioinformatics investigation of *Salmonella* can be found in Section 2.9.4, and at the URL supporting this book (http://www.wiley.com/go/forsythe).

The symptoms of typhoid fever are caused by LPS which induces a local inflammatory response during invasion of the mucosa. Unlike *S.* Typhi, it is uncertain if food poisoning strains of *Salmonella* (i.e. *S.* Enteritidis and *S.* Typhimurium) are inside macrophages of the liver and spleen. The symptoms of gastroenteritis probably result from the invasion of mucosal cells. *S.* Typhi causes a systemic infection whereas enteric pathogens such as *S.* Typhimurium rarely penetrate beyond the submucosal tissues.

S. Typhi and *S.* Paratyphi A, B and C produce typhoid and typhoid-like fever in humans. Typhoid fever is a life-threatening illness. The organism multiplies in the submucosal tissue of the ileal epithelium and then spreads throughout the body via macrophages. Subsequently, various internal organs such as the spleen and liver become infected. The bacteria infect the gall bladder from the liver and finally infect the intestines using bile as the transportation medium. If the organism does not progress past the gall bladder, then no typhoid fever develops. Nevertheless, the person may continue to shed the organism in their faeces. *S.* Typhi differs from most *S. enterica* serovars in that it has a capsule and does not carry the virulence plasmid. Presumably, *S.* Typhi has additional virulence factors.

Typical symptoms of typhoid fever are as follows:
- Sustained fever as high as 39–40°C
- Lethargy

- Abdominal cramps
- Headache
- Loss of appetite
- Rash of flat, rose-coloured spots may appear

The fatality rate of typhoid fever is 10% compared to less than 1% for most forms of salmonellosis. A small number of people recover from typhoid fever but continue to shed the bacterium in their faeces. *S.* Typhi and *S.* Paratyphi enter the body through food and drinks that may have been contaminated by a person who is shedding the organism in their faeces.

Salmonella infections in animals differ from the typical gastroenteritis and other sequelae produced in humans. However, although animal models are limited for studying salmonellosis, unlike most other bacterial pathogens, there is a considerable amount of human data on the illness.

Sources:

- Domestic and wild animals: poultry, pigs, cattle, rodents, cats and dogs.
- Infected humans (especially *S.* Typhi and *S.* Paratyphi).

Ninety-six percent of cases are estimated to be caused by a wide range of contaminated foods (see Table 1.2). This includes raw meats, raw or undercooked products of poultry, eggs, products containing raw eggs, milk and dairy products, fish, shrimp, frog legs, yeast, coconut, sauces and salad dressing, cake mixes, cream-filled desserts and toppings, dried gelatin, peanut butter, cocoa and chocolate. In addition, fruits and vegetables are becoming important sources of salmonellosis. Contamination of the foods is through poor temperature control, handling practices or cross-contamination of processed foods from raw ingredients. The organism multiplies on the food to an infectious dose.

Salmonella are zoonotic bacteria with birds as one of the reservoirs. In addition to contaminating eggshells, *S.* Enteritidis can be isolated from the egg yolk due to transovarian infection. The organism travels up the anus from the environment and colonises the ovaries. Subsequently, *S.* Enteritidis infects the egg before the protective shell is formed. An infected unfertilised egg will result in contaminated egg products, whereas an unfertilised egg results in a chronically ill chick with systemic infection and hence a contaminated carcass. A risk assessment conducted by FAO/WHO (2002) noted that the human incidence of salmonellosis transmitted through eggs and poultry meat appeared to have a linear relationship to the observed *Salmonella* prevalence in poultry. This means that, when reducing the prevalence of *Salmonella* in poultry by 50%, the incidence of salmonellosis in humans should also fall by 50%, assuming all other conditions stay constant.

Control:

- Heat treatment (pasteurisation, sterilisation)
- Refrigeration
- Prevention of cross-contamination
- Good personal hygiene
- Effective sewage and water treatment processes

Control of salmonellas is achieved through a number of requirements: the requirement for the absence (<1 *Salmonella* cell in 25 g food) in ready-to-eat foods, temperature control during storage and a processing step (i.e. cooking) to eliminate salmonellas from raw meats.

After rising in the early 1980s, the number of *Salmonella* cases has decreased in some countries; see Figure 1.3. In the United Kingdom, in the past 5 years, there has been a notable decline in *S.* Enteritidis cases which coincide with the introduction of vaccination of egg-laying hens against this serovar. It will be important to ensure other serovars with similar traits do not replace *S*. Enteritidis in the poultry population. *S*. Enteritidis replaced *S*. Typhimurium in the 1980s.

Examples of *Salmonella* risk assessments are given in Section 10.1.

Example of a multistate *Salmonella* outbreak associated with ground meat, US, 2004
(Details taken from Cronquist *et al*. 2006.):

Background: In September 2004, the New Mexico Department of Health received reports from the New Mexico Scientific Laboratory Division that eight *S*. Typhimurium isolates had indistinguishable PFGE patterns (Section 5.5.1) using both *Xba*I and *Bln*I restriction enzymes. The patients were linked by place (three New Mexico counties) and time as the onsets of illness was between 18 and 19 August.

Case definition: A case was defined as a person with a *S*. Typhimurium infection with a PFGE pattern that was indistinguishable from the outbreak profile.

Outbreak description: The *Salmonella* PulseNet database collated a further 31 indistinguishable patient isolates of *S*. Typhimurium. These were from nine states, with an illness onset during 11 August and 2 October 2004. The states were Colorado, Kansas, Minnesota, New Jersey, New Mexico, New York, Ohio, Tennessee, Wisconsin and the District of Columbia.

Epidemiology: Questionnaires were used to collect detailed information on patient's history of food consumption before becoming ill. Several patients reported eating ground beef from the same national chain of supermarkets and therefore this was suspected as the source of the outbreak.

A case–control study was undertaken with 26 of the 31 case-patients and 46 controls. The controls were identified by sequential telephone digit dialling. They had no reported gastrointestinal illness within 7 days before onset of illness of the cases and were matched with cases by age group (2–10, 11–17, 18–60 and >60 years). Details were collected for ground beef consumption, brand, location and date of purchase.

Symptoms reported by the 26 cases included diarrhoea (100%), abdominal cramps (92%), fever (92%), vomiting (65%) and bloody diarrhoea (46%). The median duration of illness was 7.5 days with a range of 2–30 days. A third (35%) of the patients were hospitalised, and there were no deaths. Of the 21 cases who ate ground beef, 15 (71%) purchased the beef from a national chain of supermarkets. This compares with nine (24%) of the controls.

Laboratory analysis: One case-patient had leftover frozen ground beef. It was analysed for *Salmonella*, and *S*. Typhimurium with a PFGE pattern indistinguishable from the outbreak profile was recovered.

Control measures: The source of ground beef and its production date were determined by using their purchase receipts, and so on, from the cases. The traceback showed that the ground beef had been packaged at three processing plants, and that one supplier was common to all three plants. USDA's Food Safety and Inspection Service (FSIS; Section 11.7.1) reviewed the processing plants and concluded that they conformed to current FSIS production guidelines and so no products were recalled.

4.3.3 Pathogenic *E. coli*

E. coli is a Gram-negative, non-spore forming, facultative anaerobe and is a genus in the *Enterobacteriaceae*. Pathogenic strains of *E. coli* are divided according to clinical symptoms and mechanisms of pathogenesis into several groups which vary in their incubation periods and duration of illness (Table 1.18 and Table 4.6). There is also a considerable range of virulence. For example, small doses (10 cells or less) of *E. coli* O157:H7 can cause severe illness, whereas

Table 4.6 Pathogenic strains of *E. coli*. Information kindly supplied by Oxoid, Thermo Fisher Scientific, Basingstoke, UK

Group	Serotypes	Adhesion and invasion characters	Toxins	Disease symptoms
Enterotoxigenic (ETEC)	O6, O8, O15, O20, O27, O63, O78, O80, O85, O115, O128, O139, O148, O153, O159, O167	Adhere uniformly but do not invade	Heat-labile (LT). Heat-stable (ST). LT is similar to cholera toxin and acts on mucosal cells	Cholera-like diarrhoea but generally less severe
Enteropathogenic (EPEC)	O18, O44, O55, O86, O111, O112, O114, O119, O125, O126, O127, O128, O142	Adhere in clumps. Invade host cells, attach and efface	Not apparent	Infantile diarrhoea, vomiting
Enteroinvasive (EIEC)	O124, O143, O152	Invade cells of colon. Spread laterally to adjacent cells	No Shiga-like toxin has been detected	Cell-to-cell spread and disease is similar to dysentery
Enterohaemorrhagic (EHEC)	O6, O26, O46, O48, O91, O98, O111, O112, O146, O157, O165	Adhere tightly. Attach and efface host cells. Invade	Verocytotoxic Shiga-like toxin	Bloody diarrhoea. Haemorrhagic colitis. May progress to haemolytic uraemic syndrome (HUS) and thrombotic thrombocytopenic purpura (TTP)
Enteroaggregative (EAggEC)	Wide range of serotypes, recent outbreaks O62, O73, O134	Adhere in clumps but do not invade	ST-like toxin. Haemolysin. Verocytotoxin reported in some strains	Diarrhoea. Some strains have been reported to cause HUS

enterotoxigenic *E. coli* requires an estimated 10^8–10^{10} cells to cause a mild illness (FDA Bad Bug book, URL in Web Resources section). The six recognised pathogenic groups are as follows:

- Enterohaemorrhagic *E. coli* (EHEC) causes bloody diarrhoea, haemorrhagic colitis, haemolytic uraemic syndrome (HUS) and thrombic thrombocytopenia purpura. This group includes the Shiga-toxin-producing *E. coli* or STEC (formerly known as verotoxigenic *E. coli*, VTEC) serotypes O157:H7, O26:H11 and O111:NM and will be the main focus of this section.
- Enterotoxigenic *E. coli* (ETEC) is commonly known as traveler's diarrhoea. ETEC causes watery diarrhoea, rice water-like and a low-grade fever. The organism colonises the proximal small intestine.
- Enteropathogenic *E. coli* (EPEC) causes watery diarrhoea of infants. EPEC causes diarrhoea, vomiting, fever and diarrhoea which is watery with mucus but no blood. The organism

colonises the microvilli over the entire intestine to produce the characteristic 'attaching and effacing' (A/E) lesion in the brush border microvillus membrane.

- Enteroaggregative *E. coli* (EAggEC) causes persistent watery diarrhoea especially in children, lasting more than 14 days. EAggEC align themselves in parallel rows on either tissue cells or glass. This aggregation has been described as 'stacked brick-like'. They produce a heat-labile toxin antigenically related to haemolysin but not haemolytic and plasmid encoded heat-stable toxin (EAST1) unrelated to the heat-stable enterotoxin of ETEC. It is thought that EAggEC adhere to the intestinal mucosa and elaborate the enterotoxins and cytotoxins, which results in secretory diarrhoea and mucosal damage. EAggEC is associated with malnutrition and growth retardation in the absence of diarrhoea.
- Enteroinvasive *E. coli* (EIEC) causes a fever and profuse diarrhoea containing mucus and streaks of blood. The organism colonises the colon and carries a 120–140 mD plasmid known as the invasiveness plasmid which carries all the genes necessary for virulence.
- Diffusely adherent *E. coli* (DAEC) has been associated with diarrhoea in some studies but not consistently (Boxes 4.3 and 4.4).

The genomic analysis and lineage of pathogenic *E. coli* is considered in Section 2.9.5.

Source:

- Ruminants, and in particular cattle, are considered the primary reservoir of EHEC
- Various other domestic animals and wildlife (sheep, swine, goats and deer)
- Undercooked ground beef products
- Dry fermented meats, cooked and fermented sausages
- Other foodborne sources include milk and dairy products (e.g. unpasteurised milk, cheese from raw milk), fresh produce (e.g. sprouts, salads), drinks (e.g. apple cider/juice) and water
- Raw or minimally processed fruits and vegetables

EHEC was first described in 1977 and recognised as a disease of humans in 1982. They harbour plasmids of various sizes, the most common being 75–100 kb.

Box 4.3

Organism

E. coli O157:H7

Onset of symptoms

1–2 days

Food source

Beef, particularly ground beef, poultry, apple cider, raw milk, vegetables, cantaloupe, hot dogs, mayonnaise, salad bar items

Acute symptoms and chronic complications

Abdominal pain, diarrhoea, fever, malaise
Chronic symptoms: Haemolytic uraemic syndrome, chronic kidney disease, thrombotic thrombocytopenic purpura

Box 4.4

Organisms

Enterotoxigenic *E. coli* (ETEC), enteropathogenic *E. coli* (EPEC), enteroinvasive *E. coli* (EIEC)

Onset of symptoms

ETEC: 24 hours
EPEC: Not known
EIEC: 12–72 hours

Food source

ETEC: Foods contaminated by human sewage or infected food handlers
EPEC: Raw chicken and beef, food contaminated by faeces or contaminated water
EIEC: Food contaminated by human faeces or contaminated water, hamburger meat, unpasteurised milk

Acute symptoms and chronic complications

ETEC: Watery diarrhoea, abdominal cramps, fever, nausea, malaise
EPEC: Watery or bloody diarrhoea
EIEC: Abdominal cramps, vomiting, fever, chills, generalised malise, haemolytic uraemic syndrome

   E. coli O157:H7 was first described in 1977. However, it was not until 1993, following a large outbreak with more than 700 cases infected by eating contaminated fast food hamburgers, that it was recognised as a major food safety issue. Further outbreaks in Japan, Washington and Scotland (Table 1.8) confirmed the severity of infection and the need for additional food safety controls. In 1996, in Japan, the *E. coli* O157:H7 outbreak resulted in ~9523 illnesses in the city of Sakai, three deaths, and ~43 suffered from the aftereffects of the infection. The source was contaminated school lunches. Following the outbreak, the central government compiled a sanitation management manual for cooking facilities that supply meals to large numbers of people. The USDA has established a 'zero tolerance' policy for *E. coli* O157:H7 on ground beef. This policy has not resolved the problem, as an estimated 73 500 cases of *E. coli* O157:H7 (food and non-food related) occur annually in the United States.
   The EHEC belong to many serogroups. It can cause very severe forms of food poisoning resulting in death. It has been postulated that *E. coli* O157:H7 evolved from EPEC and acquired the toxin genes from *Sh. dysenteriae* via a bacteriophage, and that the newly emerging pathogen arrived in Europe from South America (Coghlan 1998; cf. Section 2.9.5). The reported incidence and serotype varies from country to country (Table 4.7).
   In 2001, the incidence of EHEC in New Zealand and Australia was 2 and 0.2 cases per 100 000, respectively. In 2004, the number of laboratory-confirmed cases in the European Union and Norway was 1.3 cases per 100 000 population. This compares with 0.9 cases per 100 000 people in the United States for the same year. EHEC infection and associated diseases can occur in any age group, but in particular, young children are the most vulnerable. The frequency of EHEC,

Table 4.7 Incidence of EHEC in Europe (1996)

Country	EHEC infections	Millions of inhabitants	Per million inhabitants
Spain	4	39.6	0.1
Italy	9	57.1	0.2
Netherlands	10	15.4	0.6
Finland	5	5.1	1.0
Denmark	6	5.2	1.2
Austria	11	8.0	1.4
Germany	314	81.5	3.9
Belgium	52	10.0	5.2
Sweden	118	8.7	13.6
United Kingdom	1180	58.1	20.3
Northern Ireland	14	1.6	8.8
Wales	36	2.9	9.2
England	624	48.5	12.4
Scotland	506	5.1	99.2

Adapted from Anon. 1997b.

and more specifically HUS, appears to be the highest in Argentina. The incidence of HUS in this country is ~22 cases of HUS per 100 000 children aged 6–48 months.

The most well-studied EHEC strain is *E. coli* O157:H7 which predominates in North America, Japan and the United Kingdom, whereas in Central Europe and Australia, serotypes O26:H11 and O111:NM dominate. The 'O157' and 'H7' refer to the serotyping of the strain's O and H antigens, respectively (Section 2.2). These are two distinct genetic lineages; see Section 2.9.5, Figure 2.8. Although EHEC strains cause dysentery similar to that caused by *Shigella* species, they probably do not invade mucosal cells as readily as *Shigella* strains. The low infectious dose is probably due to several reasons, among which are that *E. coli* O157:H7 is more acid tolerant than other *E. coli* strains, and therefore survives the stomach acidity, and that its virulence is influenced by the rest of the intestinal flora by quorum sensing.

EHEC has two major virulence pathways contributing to disease: the Shiga-toxin gene and genes within the Locus of Enterocyte Effacement (LEE) pathogenicity island. This is a 38 kb region of DNA which is inserted into the bacterial chromosome near a tRNA gene. The pathogenicity island encodes for a Type III secretion system and other virulence factors essential for disease; see Section 2.3.2. The LEE pathogenicity island enables the bacterium to adhere to epithelial cells and form a pedestal on the epithelial surface. Effacement of the microvilli occurs, producing the same type of attachment–effacement (A/E) phenomenon as EPEC strains (Figure 2.4). These lesions are characterised by effacement of the brush border microvilli near to the adhering bacteria and intimin-mediated bacterial attachment to the plasma membrane of the host cell. Pedestal formation is the end result of a complex process involving the Type III secretion system, several Type III *E. coli* secreted proteins (Esps) and Tir (Translocated intimin receptor). The latter is transferred from *E. coli* to the host cell membrane where it acts as an anchor for intimin, a 94 kDa bacterial outer membrane protein. This results in bacterial adherence to the host cell. Additionally, Tir recruits host cytoskeletal proteins to cause actin accumulation and pedestal formation.

Because cattle lack the receptor for the Shiga toxin, *E. coli* O157:H7 and related strains can persist in cattle without causing disease. As many as half of all cattle carry the bacterium at

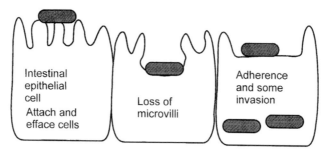

Figure 4.6 Route of infection of pathogenic *E. coli*: EHEC.

some time in their lives, and some are 'super spreaders' of the organism. The organism enters the environment from faeces, manure used as fertiliser, and rainwater runoff can spread the organism to water reservoirs and wells. In addition, faecal contamination of meat during the slaughtering process can occur. The bacterium can be carried by goats and sheep, as well as wild ruminants, such as deer. Following ingestion, the organisms' tolerance to acidic pH may explain the low infectious dose (10–100 cells). This acid tolerance also allows it to survive in low-acid food, and explains the outbreaks traced to unpasteurised apple juice (pH 3.5).

E. coli O157:H7 differs from the majority of *E. coli* strains in that it does not grow or grows poorly at 44°C and does not ferment sorbitol or produce β-glucuronidase. It should be noted that there are other *E. coli* serotypes producing verocytotoxins (Agbodaze 1999). The growth of *E. coli* O157:H7 in the human intestine produces a large quantity of toxin(s) that cause severe damage to the lining of the intestine and other organs of the body (Figure 4.6). These Shiga-like toxins (SLTs) are very similar, if not identical, to the toxins produced by *Sh. dysenteriae*. There are four subgroups: SLT1, SLT2, SLT IIc and SLT IIe. They were formerly referred to as verotoxin (VT), and the equivalent designations were VT1, VT2, VT2C and VT2e, respectively. Strains producing Shiga toxins were initially recognised by the cytotoxicity towards Vero cells (green monkey kidney cell line), and subsequently, the term 'verotoxigenic *E. coli*' or VTEC arose. However, since the purified and sequenced VT has been found to be nearly identical to Shiga toxin, the organisms are referred to as 'Shiga-toxin producing *E. coli*' or STEC. The phage-encoded toxin consists of one A subunit and five identical B subunits (cf. Figure 2.7). The 32 kDa A subunit is an RNA N-glycosidase that removes a specific adenine residue from 28S rRNA. The subunit is non-covalently associated with a pentamer of 7.7 kDa B subunit. The pentamer binds the toxin to plasma-membrane globotriaosyceramide (Gb3) present on susceptible mammalian cells. The toxins destroy the intestinal cells of the human colon and may cause additional damage to the kidneys, pancreas and brain. The dead cells accumulate and block the kidneys, causing HUS. The incubation period for EHEC diarrhoea is usually 3–4 days, although incubation times can be as long as 5–8 days or as short as 1–2 days.

EHEC infections in healthy adults cause thrombic thrombocytopenic purpura (TTP) where blood platelets surround internal organs leading to kidney damage and central nervous system damage. The most vulnerable of the population, children and the elderly, however develop haemorrhagic colitis (HC) which may lead to HUS. HC is a less severe form of *E. coli* O157:H7 infection than HUS. The first symptom of HC is the sudden onset of severe crampy abdominal pains. About 24 hours later, non-bloody (watery) diarrhoea starts. Some victims have a fever of short duration. Vomiting occurs in about half of the patients during the period of non-bloody

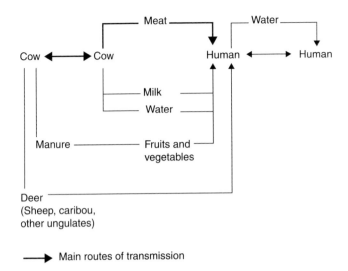

Figure 4.7 *E. coli* O157:H7 transmission.

diarrhoea and/or other times in the illness. After 1 or 2 days, the diarrhoea becomes bloody and the patient experiences increased abdominal pain. This usually lasts between 4 and 10 days. In severe cases, faecal specimens are described as 'all blood and no stool'. In most patients, the bloody diarrhoea resolves with no long-term impairment. Unfortunately, 2–7% patients (up to 30% in certain outbreaks) will progress to HUS and subsequent complications.

In HUS, the patient suffers from bloody diarrhoea, haemolytic anaemia, kidney disorder and renal failure, and require dialysis and blood transfusions. Central nervous disease may develop which can lead to seizures, coma and death. The mortality rate is 3–17%. Acute renal failure is the leading cause of death in children, whereas thrombocytopenia is the leading cause of death in adults.

The SLTs are specific for the glycosphingolipid globotriaosylceramide (Gb3) which is present on renal endothelial cells. Since Gb3 is found in the glomeruli of infants under 2 years of age but not in the glomeruli of adults, the presence of Gb3 in the pediatric renal glomerulus may be a risk factor for development of HUS.

Cattle appear to be the main reservoir of *E. coli* O157:H7. Transmission to humans is principally through the consumption of contaminated foods, such as raw or undercooked meat products and raw milk. Fresh-pressed apple juice or cider, yogurt, cheese, salad vegetable and cooked corn have also been implicated (Figure 4.7). Faecal contamination of water and other foods as well as cross-contamination during food preparation can occur. There is evidence of transmission of this pathogen through direct contact between people. *E. coli* O157:H7 can be shed in faeces for a median period of 21 days with a range of 5–124 days.

According to the Food and Drug Administration of the United States, the infectious dose for *E. coli* O157:H7 is unknown. However, a compilation of outbreak data indicates that it may be as low as ten organisms. The data show that it takes a very low number of micro-organisms to cause illness in young children, the elderly and immune-compromised people.

Most reported outbreaks of EHEC infection have been caused by O157:H7 strains. This suggests that this serotype is more virulent or more transmissible than other serotypes. However, other serotypes of EHEC have been implicated in outbreaks, and the incidence of disease due

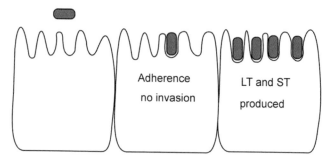

Figure 4.8 ETEC infection of gut mucosal cells.

to non-O157:H7 serotypes seems to be rising. More than 50 of these serotypes have been associated with bloody diarrhoea or HUS in humans. The most common non-O157:H7 serotypes associated with human disease include O26:H11, O103:H2, O111:NM and O113:H21. At least ten of these outbreaks due to these organisms have been reported in Japan, Germany, Italy, Australia, the Czech Republic and the United States. These outbreaks have involved 5–234 people, and for most of them, the source of infection could not be determined. In many countries such as Chile, Argentina and Australia, non-O157:H7 serotypes have been found to be responsible for the majority of HUS cases. Non-bloody diarrhoea has also been associated with some of these non-O157:H7 serotypes.

Infants are particularly prone to EPEC infections. The EPEC cause diarrhoea, vomiting, fever and diarrhoea which is watery with mucus but no blood. The organism colonises the microvilli over the entire intestine to produce the characteristic A/E lesion in the brush border microvillus membrane.

The reason for the low *E. coli* O157:H7 infectious dose (ten cells or less) could be due to several factors; greater acid tolerance resulting in greater survival through the stomach, quorum sensing and its virulence factors are very potent.

ETEC strains resemble *V. cholerae* in that they adhere to the small intestinal mucosa and produce symptoms not by invading the mucosa but by producing toxins that act on mucosal cells to cause diarrhoea (Figure 4.8). Unlike *E. coli* O157:H7 which can cause severe illness even if low numbers of cells (ten or less) are ingested, ETEC requires an estimated 10^8–10^9 cells to cause relatively mild symptoms. ETEC strains produce two types of enterotoxin: a cholera-like toxin called heat-labile toxin (LT) and a second diarrhoeal toxin called heat-stable toxin (ST). Heat-stable being defined as retaining activity after heat treatment at 100°C for 30 minutes. There are two main types of LT: LT-I and LT-II. LT-I shares about 75% amino acid sequence identity with cholera toxin and has an AB_5 structure. The A subunit catalyses the ADP-ribosylation of G_s, which raises host cell cAMP levels, identical to cholera toxin mode of action. The B subunits of LT-I interact with the same receptor as cholera toxin (G_{M1}). However, LT-I is not excreted (as per cholera toxin) but is localised in the periplasm. The toxin leaks out of the organism on exposure to bile acids and low iron concentrations (as would occur in the small intestines).

ST is a family of small ($\sim$2 kDa) toxins which can be divided into two groups: methanol soluble (STa) and methanol insoluble (STb). STa causes an increase in cGMP (not cAMP) level in the host cell cytoplasm which leads to fluid loss. cGMP, like cAMP, is an important signalling molecule in eucaryotic cells, and changes in cGMP affect a number of cellular processes, including activities of ion pumps. LT-I and STa are encoded on plasmids.

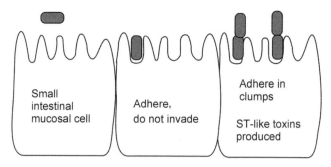

Figure 4.9 EAggEC infection of gut mucosal cells.

The EAggEC resembles ETEC strains in that they bind to small intestinal cells, are not invasive, and cause no obvious histological changes in intestinal cells to which they adhere (Figure 4.9). They differ from ETEC strains primarily in that they do not adhere uniformly over the surface of the intestinal mucosa but tend to clump in small aggregates. EAggEC strains produce a ST-like toxin (EAST) and a haemolysin-like toxin (120 kDa in size). Some strains are reported to produce a Shiga-like (verocytotoxin) toxin.

EPEC do not produce any enterotoxins or cytotoxins. They invade, attach and efface epithelial cells. This induces a characteristic A/E lesion (Figure 4.10) in epithelial cells in which microvilli are lost and the underlying cell membrane is raised to form a pedestal which can extend outward for up to 10 μm. This phenomenon is the result of extensive rearrangement of host cell actin in the vicinity of the adherent bacteria that results in formation of a cuplike pedestal structure under the bacteria. The genes encoding for this virulence are encoded on the LEE, a pathogenicity island (Table 2.2). This island encodes for a Type III secretion system, intimin intestinal colonisation factor and the translocation intimin receptor protein. Contact with epithelial cells results in the secretion of a number of proteins including EspA (25 kDa) and EspB (37 kDa), which subsequently trigger a series of responses in the host cell. These responses include activation of signal transduction pathways, cell depolarisation and the binding of the outer membrane protein intimin (94 kDa). Intimin is encoded by the gene *eae*. Intimin probably binds to Tir, a bacterial protein (78 kDa), which is secreted by EPEC and binds to the host cell. The characteristic diarrhoea of EPEC infections is possibly due to an induced efflux of chloride ions due to activation of protein kinase C. The loss of microvilli would contribute

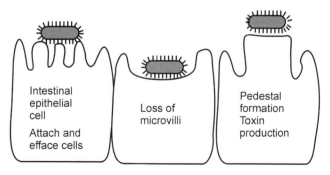

Figure 4.10 EPEC infection of gut mucosal cells.

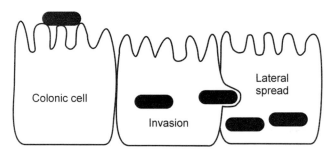

Figure 4.11 EIEC infection of gut mucosal cells.

due to the decreased absorption ability (Figures 4.2 and 4.3). The Type III secretion system is used to avoid being engulfed by the host cell and blocks phagocytosis, enabling the bacterium to remain attached to the host cell surface.

EIEC strains cause a disease that is indistinguishable at the symptomatic level from the dysentery caused by *Shigella* spp. EIEC strains actively invade colonic cells and spread laterally to adjacent cells (Figure 4.11). The steps in invasion and cell-to-cell spread appear to be virtually identical to *Shigella* spp. However, EIEC do not produce Shiga toxin which may account for the lack of haemolytic uraemic syndrome as a complication of EIEC dysentery.

Control measures:
- Effective sewage and water treatment
- Prevention of cross-contamination from raw foods and contaminated water
- Heat treatment: cooking, pasteurisation
- Good personal hygiene

E. coli O157 outbreak in South Wales, UK 2005 (Salmon 2005)

Background: On 16 September 2005, Prince Charles Hospital, Merthyr Tydfil, Wales (United Kingdom) reported nine cases of bloody diarrhoea to the National Public Health Service for Wales and local authorities. The normal annual total for Wales being about 30 cases. The number of cases increased over the following days. These were subsequently confirmed microbiologically as VT-producing *E. coli* O157.

Case definition: A case was defined as any person living in South Wales, who presented with bloody diarrhoea or had a faecal isolate of presumptive STEC O157 in September.

Outbreak description: The date of onset of symptoms ranged from 10 to 20 September. Except for one primary case (adult), they were all school age children attending 26 different schools. A review of the first 15 cases revealed that all had eaten school meals and the adult case was a school meals supervisor. The small numbers of cases across a large number of schools suggested the source of the infection was a centrally distributed product with low levels of contamination rather than a problem in the individual schools. In addition, there was secondary person-to-person spread. Parents were advised to keep children out of school if they develop symptoms of gastroenteritis.

Control measures: The initial early epidemiological investigation focused on a single main supplier of cooked meats to the school meals service. Inspection of the supplier's premises revealed practices that could result in the contamination of cooked meat. Subsequently, local authorities took action on 19 September, and the Food Standards Agency (Wales) issued a food alert on 21 September. Other control measures were to remove ready-to-eat foods (not foods

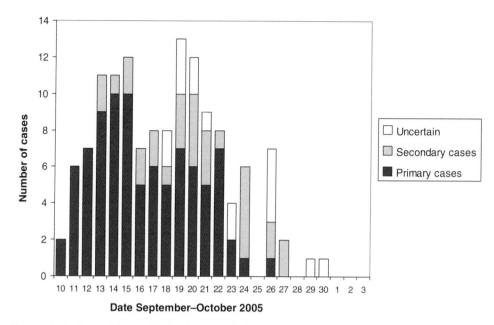

Figure 4.12 Cases of VTEC O157 infection with known date of onset, outbreak in South Wales, UK (*n* = 133). Source of data: Eurosurveillance.

cooked on the school premises) and to reduce any educational activities which might facilitate person-to-person spread.

Epidemiology: Between 16 and 20 September, 10 of the first 18 primary school children cases with early symptom onset dates before 17 September were contacted. Controls were chosen at random from the school register. It was found that all cases had eaten lunch in the school canteen. This compared with 8 out of 13 controls. By the end of the outbreak, 67 males and 90 females had been affected, and 65% of cases (102/157) were children of school age from 42 schools. Dates of onset of symptoms were from 10 to 30 September (Figure 4.12). One boy aged 5 died.

Laboratory analysis: Ninety-seven of the cases were microbiologically confirmed as due to *E. coli* O157 infection. Nearly all of these isolates were phage type (PT) 21/28, and produced Shiga toxin (ST) 2. There was one exception which was PT32 ST2. In addition, four other microbiologically confirmed cases of *E. coli* O157 infection were excluded from the outbreak case list. This was because three were PT1 ST-negative, and the remainder was PT8 ST1+2. These phage types were not associated with the outbreak and there was alternative reason for their infection.

E. coli O157 was isolated from three samples of sliced cooked meat from the supplier. These isolates were typed and confirmed as PT 21/28 and ST2, the same as the isolates from the patients. The PFGE profiles (Section 5.5.1) of food and patient isolates were indistinguishable.

Comment: Please note that at the time of writing (early 2008), legal proceedings concerning these cases were still ongoing. Two issues being investigated were the hygienic practices at the processor and the thoroughness of the local environmental health office inspections. There are parallels with the *E. coli* O157:H7 outbreak in Lanarkshire, Scotland, in 1996. In general, the control of *E. coli* O157 is considered in Section 4.3.3.

E. coli O157 outbreak from spinach, US 2006

Introduction: In September 2006, an *E. coli* O157:H7 outbreak in the United States totalled 205 cases. There were 104 hospitalisations, with 31 cases of kidney failure and 3 deaths. The cause was traced back to contaminated fresh, bagged spinach from one company in California. The product was distributed directly to three further countries. An International Food Safety Authorities Network (INFOSAN; see Section 1.12.1) Emergency Alert was issued to all INFOSAN members because another country received the product through secondary distribution. This outbreak significantly differs to the example in South Wales, UK, due to the wider distribution of the food product, and serves as an example of the need for international surveillance and co-operation.

Outbreak details: On 8 September 2006, the Centers of Disease Control and Prevention (CDC) were notified of three clusters of *E. coli* O157:H7 infections in separate States, which were not all in close proximity to each other. On 12 September, CDC's PulseNet (Section 1.12.3) confirmed that matching *E. coli* O157:H7 strains had been recovered from all the infected patients. On 13 September, the state public health officials notified the CDC that their epidemiological investigations suggested the infections were associated with the consumption of fresh spinach. The following day, Food and Drug Administration (FDA) issued an alert about the multistate outbreak of *E. coli* O157:H7, and advised consumers to avoid this product. Canada and Mexico were also notified due to the large number of people who cross those borders daily and because many of the initial cases came from states on the border with Canada and Mexico. The advice to avoid spinach was revised on 22 September since the FDA had concluded that the contaminated spinach was only from counties in California. By 18 September, it was apparent that the contaminated spinach had been distributed to Canada, Mexico and Taiwan. The relevant public health authorities were notified. Initially, INFOSAN (Section 1.12.1) sent emergency alerts to the INFOSAN emergency contact points in the affected countries. However, on 22 September, the information was sent to the entire INFOSAN network, as it became apparent that it was not possible to trace the full distribution chain.

Based on epidemiological and laboratory evidence, the FDA on 29 September determined that all spinach implicated in the outbreak could be traced back to one company. Four fields on four different ranches were identified as the source of contamination. Subsequently, the company recalled all spinach products, and farmers in the affected area stopped growing spinach and stopped producing ready-to-eat spinach.

Microbiological analysis: PFGE profiles of *E. coli* O157:H7 isolates were analysed around the country as part of PulseNet (Section 1.12.3) and were shown to have the same DNA pattern. In total, there were 204 cases from 26 states. This comprised 104 hospitalisations, of which 31 cases had HUS (Section 4.3.3), and 3 deaths with 2 elderly women and a 2-year-old child. There was one confirmed case in Canada.

Comment: Control of *E. coli* in fresh produce is considered in Section 1.11.2.

4.3.4 *Sh. dysenteriae* and *Sh. sonnei*

Shigella is a highly contagious bacterium that infects the intestinal tract. It is closely related to *E. coli* but can be differentiated from *E. coli* by the lack of gas production from carbohydrates (anaerogenic) and negative lactose fermentation. The genus *Shigella* consists of four species: *Sh. dysenteriae* (serotype A), *Sh. flexneri* (serotype B), *Shigella boydii* (serotype C) and *Sh. sonnei* (serotype D). In general, *Sh. dysenteriae*, *Sh. flexneri* and *Sh. boydii* predominate in developing countries, whereas *Sh. sonnei* is most common and *Sh. dysenteriae* is less common in developed countries. *Shigella* is spread by direct and indirect contact with infected individuals. Most cases

Box 4.5

Organism

Shigella spp.

Food source

Salads, raw vegetables, bakery products, sandwich fillings, milk and dairy products, poultry

Onset of symptoms

12–50 hours

Acute symptoms and chronic complications

Abdominal pain and cramps, diarrhoea, fever, vomiting, blood, pus or mucus in stools

of shigellosis result from the ingestion of food or water contaminated with human faeces. Several large outbreaks have been due to the consumption of contaminated raw fruits and vegetables. Food or water may be contaminated by direct contact with faecal material from infected people. *Shigella* often causes outbreaks in day care centres (Box 4.5).

The main symptoms of shigellosis are as follows:
- Diarrhoea, mild or severe, may be watery or bloody
- Fever and nausea
- Vomiting and abdominal cramping may also occur

The symptoms appear within 12–96 hours after exposure to *Shigella*; the incubation period is typically 1 week for *Sh. dysenteriae*; see Table 1.18. The symptoms from *Sh. sonnei* are generally less severe than the other *Shigella* species. *Sh. dysenteriae* may be associated with serious disease, including toxic megacolon and the HUS. *Shigella* cells are shed in the faeces for 1–2 weeks of infection.

Probably because the organism is not readily killed by the acidity of the stomach, *Sh. dysenteriae* has a low infectious dose (Table 4.2). The organism then colonises the colon and enters the mucosal cells where it multiplies rapidly. Once inside the mucosal cells, the bacterium moves laterally to infect adjacent mucosal cells (Figure 4.13). It provokes an intense inflammatory response in the lamina propria and the mucosal layer, and the inflammatory response causes blood and mucus to be found in the victim's faeces.

The organism produces Shiga toxin, an exotoxin, with AB_5 structure (Figure 2.7; A subunit is 32 kDa, B subunit is 7.7 kDa) and a poorly characterised CLDT. The AB_5 toxin is structurally similar to cholera toxin although there is little similarity in amino acid sequences. The B subunits of Shiga toxin, like those of cholera toxin, recognise a host cell surface glycolipid (Gb3), but the carbohydrate moiety on this lipid is Gal α-1,4-Gal, not sialic acid as for cholera toxin receptor G_{M1}. The Shiga toxin first binds to the surface of mammalian cells and is then internalised by endocytosis. Nicking to activate and translocate the A subunit takes place inside the host cell. In contrast to other AB toxins, the A subunit of Shiga toxin does not ADP-ribosylate a host cell protein, but stops host cell protein synthesis. It inactivates the 60S subunit of host cell ribosomes by cleaving the N-glycosidic bond of a specific adenosine residue

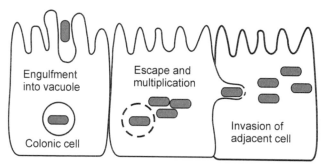

Figure 4.13 *Shigella* spp. infection of gut mucosal cells.

in 28S rRNA, a component of the 60S ribosomal subunit. Cleavage of 23S rRNA at this site prevents binding of aminoacyl-tRNAs to the ribosome and stops elongation of proteins.

Outbreak linked to fresh parsley

In August 1998, multiple reports of *Shigella* infection were received from people who had eaten from two restaurants which were in different cities of Minnesota. These had separate water supplies, and there were no employees in common (CDC 1999). Initial expectations were that the ill food handlers contaminated ice and fresh produce. However, molecular characterisation using PFGE (Section 5.5.1) revealed that both outbreaks were caused by the same strain of *Sh. sonnei*. This strain had not previously been isolated in the state. Since the two outbreaks had a common source, the food ingredients were considered as the likely cause. Further analysis showed an association of illness with those who had eaten chopped parsley. Further matching isolates were identified using PFGE. In total, there were six other outbreaks from July to August: two in California, and one each in the states of Massachusetts, Florida and the provinces of Alberta and Ontario. *Sh. sonnei* isolates were recovered from five of the six outbreaks. Their PFGE patterns matched the Minnesota outbreak pattern. In each outbreak, chopped parsley was sprinkled on the food implicated by a high proportion of cases.

Subsequent traceback identified a farm in Baja California, Mexico, as the most likely source of the parsley in six out of seven outbreaks. Further investigation found that the municipal water used to make the ice that was packed with the parsley was both unchlorinated and susceptible to contamination. The preparation of the parsley in the restaurants further complicated the transmission of *Sh. sonnei*. The parsley was usually chopped in the morning, and then left at room temperature until required. Under these conditions, the bacterium could increase by 1000-fold in 24 hours. Additionally, food handlers became infected and contributed to the ongoing transmission of the bacterium. Therefore, the contamination of a food ingredient was amplified through handling practices, contaminating other ready-to-eat foods and ice.

4.3.5 *Listeria monocytogenes*

Listeria are Gram-positive, non-spore forming bacteria. They are motile by means of flagella and grow between 0°C and 42°C. Therefore, *L. monocytogenes* can multiply slowly at refrigeration temperatures, in contrast to most other recognised foodborne pathogens; see Table 2.8. They are less sensitive to heat compared with *Salmonella*, and hence pasteurisation is sufficient to kill the organism. The genus is divided into six species, of which *L. monocytogenes* is the species of primary concern with regard to food poisoning. *L. monocytogenes* is further subdivided

Box 4.6

Organism

L. monocytogenes

Food source

Soft-ripened cheese, pâté, ground meat, poultry, dairy products, hot dogs, potato salad, chicken, seafood, vegetables

Onset of symptoms

Few days to 3 weeks

Acute symptoms and chronic complications

Influenza-like symptoms, fever, severe headache, vomiting, nausea, sometimes delirium or coma. Chronic symptoms: septicaemia in pregnant women, foetuses, or neonates, internal or external abscesses, meningitis, sepsis, septicaemia

into 13 serotypes. The epidemiologically important serotypes are 1/2a (15–25% of cases), 1/2b (10–35% of cases) and 4b (37–64% of cases). Outbreaks due to serotype 4b are significantly greater in pregnancy cases, whereas serovar 1/2b is more common with non-pregnant cases (Farber & Peterkin 1991, McLauchlin 1990a, 1990b). Those at greatest risk of listeriosis after consuming foods contaminated with *L. monocytogenes* are the foetuses and neonates through an infected mother during pregnancy, the elderly and persons with weakened immune systems. Invasive listeriosis is characterised by a high case-fatality rate between 20% and 30%. Genomic analysis and pathogenomics of *Listeria* are considered in Section 2.9.1 (Box 4.6).

L. monocytogenes has been found in at least 37 mammalian species, both domestic and feral, as well as at least 17 species of birds and possibly some species of fish and shellfish. It is plausible that 1–10% of humans may be intestinal carriers of *L. monocytogenes*. The majority of those who succumb to severe listeriosis are individuals with underlying conditions that suppress their T-cell-mediated immunity. Another factor related to susceptibility is reduced gastric acidity that occurs on ageing, especially in those more than 50 years old. This may explain the age distribution of listeriosis (Figure 4.14).

This ubiquitous bacterium has been isolated from various environments, including decaying vegetation, soil, animal feed, sewage and water. It is present in the intestinal tract of many animals, including humans, and therefore can be found in faeces, sewage, soil and on plants which grow in these soils. The organism also grows on decaying plant materials, and can be isolated from raw fruits and vegetables including cabbages, cucumbers, potatoes and radishes, ready-to-eat salads, tomatoes and cucumbers, and bean sprouts, sliced cucumbers and leafy vegetables.

L. monocytogenes is resistant to diverse environmental conditions and can grow at temperatures as low as 3°C. It is found in a wide variety of foods, both raw and processed, where it can survive and multiply rapidly during storage. These foods include supposedly pasteurised milk and cheese (particularly soft-ripened varieties), meat (including poultry) and meat products, raw vegetables, fermented raw-meat sausages as well as seafood and fish products.

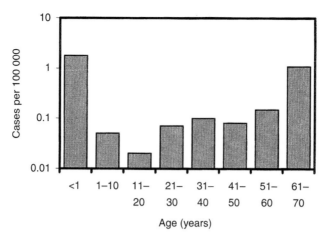

Figure 4.14 Distribution of listeriosis with age. From Buchanan & Linqvist 2000, Microbiological Risk Assessment Series 5, with permission from the Food and Agriculture Organisation of the United States.

L. monocytogenes is quite hardy and resists the deleterious effects of freezing, drying and heat remarkably well for a bacterium that does not form spores. Its ability to grow at temperatures as low as 3°C permits multiplication in refrigerated foods. The foods that pose the highest risk are those which support the growth of *L. monocytogenes*. In contrast, foods that pose the lowest risk of listeriosis are processed or have intrinsic or extrinsic factors that prevent the growth of *L. monocytogenes*. For example, the pH of the food is ≤4.4, the water activity ≤0.92, or it is frozen. *L. monocytogenes* survives adverse environmental conditions longer than many other non-spore forming foodborne pathogens. It is more resistant to nitrite and acidity, tolerates high salt concentrations and survives frozen storage for extended periods.

L. monocytogenes is responsible for opportunistic infections, preferentially affecting individuals whose immune system is perturbed, including pregnant women, newborns and the elderly. Listeriosis is clinically defined when the organism is isolated from blood, cerebrospinal fluid, or an otherwise sterile site, that is placenta and foetus.

Symptoms of listeriosis are:
- meningitis, encephalitis or septicaemia;
- when pregnant women are infected in second and third trimester, it can lead to abortion, stillbirth or premature birth.

The infective dose of *L. monocytogenes* is contentious. *L. monocytogenes* only rarely infects healthy individuals. The majority of infections are of immuno-compromised individuals, and has a high mortality rate (~20%). Whether there is an ingested dose that can be tolerated by the healthy majority of the population is controversial. Studies on the virulence, DNA sequence analysis and ribotyping have separated *L. monocytogenes* into two divisions which differ in their pathogenic potential to humans; see Section 2.9.7.

The CCFH (1999b; Section 10.3) has stated that a concentration of *L. monocytogenes* not exceeding 100 organisms per gram of food at point of consumption is of low risk to the consumers. From cases contracted through raw or supposedly pasteurised milk, it is evident that fewer than 1000 total organisms may cause disease. The incubation period is extremely wide at 1–90 days.

The stomach acidity will reduce the viable number of cells, but it requires exposure times of between 15 and 30 minutes to reduce *L. monocytogenes* numbers by 5-log orders. Additionally,

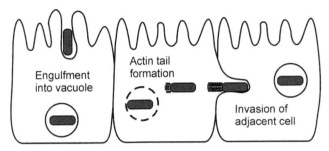

Figure 4.15 *L. monocytogenes* infection of gut mucosal cells.

ingested small liquid volumes of less than 50 mL may pass through the stomach because the pyloric sphincter will not be stimulated to contract. Another factor affecting survival in the stomach is the food matrix, especially the presence of fatty material.

L. monocytogenes attaches to the intestinal mucosa; possibly α-D-galactose residues on the bacterial surface bind to α-D-galactose receptors on the intestinal cells and invades the mucosal cells. In particular, the small intestine, either through the M cells overlaying the Peyer's patches or through non-specialised enterocytes. The bacteria are taken up by induced phagocytosis, whereby the organism is engulfed by pseudopodia on the epithelial cells resulting in vesicle formation. *L. monocytogenes* is encased in a vesicle membrane which it disrupts using listeriolysin O (responsible for the β-haemolysis zone surrounding isolates on blood agar plates). The organism also produces catalase and superoxide dismutase which may protect it from oxidative burst in the phagosome. *L. monocytogenes* also produces two phospholipase C enzymes which disrupt host cell membranes by hydrolysing membrane lipids such as phosphatidylinositol and phosphatidylcholine. In the host cell cytoplasm, the bacterium multiplies rapidly (possibly doubling every 50 minutes). The organism moves through the cytoplasm to invade adjacent cells by polymerising actin to form long tails (Figure 4.15). This form of motility enables it to move at about 1.5 μm/s. The bacterium produces protrusions into adjacent cells, enters its cytoplasm and repeats the cycle of escaping the vacuole and multiplying.

Once the bacterium enters the host's monocytes, macrophages or polymorphonuclear leukocytes, it is bloodborne (septicaemic) and can grow. Its presence intracellularly in phagocytic cells also permits access to the brain and probably transplacental migration to the foetus in pregnant women. The pathogenesis of *L. monocytogenes* centres on its ability to survive and multiply in phagocytic host cells. Listeriosis has a very high mortality rate. When listeric meningitis occurs, the overall mortality may be as high as 70%. Cases of septicaemia have a 50% fatality rate, whereas for perinatal–neonatal infections, the rate is greater than 80%. In infections during pregnancy, the mother usually survives. Infection can be symptomless resulting in faecal excretors of infectious listeria. Consequently, ~1% of faecal samples are positive for *L. monocytogenes* and ~94% of sewage samples.

Control measures:
- Heat treatment of milk (pasteurisation, sterilisation)
- Avoidance of cross-contamination
- Refrigeration (limited period) followed by thorough reheating
- Avoidance of high-risk products (e.g. raw milk) by high-risk populations (e.g. pregnant women)

Most cases of human listeriosis occur sporadically, without any apparent pattern. However, there are a number of foods which can support the growth of *L. monocytogenes* and are ready-to-eat and have been associated with outbreaks or sporadic cases of listeriosis. These include soft cheese made from unpasteurised milk, coleslaw and pate. Outbreaks of listeriosis are often associated with a processing or production failure as illustrated below.

A number of risk assessments for *L. monocytogenes* are given in Section 10.3.

L. monocytogenes outbreak linked to cheese
Between January 1983 and March 1984, a cluster of 25 listeriosis cases was reported in the same medical facility in Switzerland. This consisted of 14 adults and 11 maternal–fetal cases. Fifteen other cases were reported in nearby hospitals. The epidemic was unusual as the patients were otherwise healthy and immuno-competent, yet there was high rate of encephalitis and the mortality rate was 45%. *L. monocytogenes* serotype 4b was isolated for 92% of cases, and were of two unique phage types. In the previous 6 years, only 44% were of these phage types. This indicated the probability of a single source of infection. A further 16 cases were reported in the following November to April (1985).

Following the linking of a listeriosis outbreak in California to the consumption of a Mexican-style cheese, the Swiss health authorities surveyed a variety of dairy products for *L. monocytogenes*. The organism was isolated from 5/25 surface samples of a soft smear-ripened cheese called Vacherin Mont d'Or. This cheese was manufactured from October to March and was primarily consumed in the outbreak area. The five isolates were serotype 4b, and two of the phage types were identical to most of the clinical *L. monocytogenes* strains for the 1983–1986 period. Between 1983 and 1987, there were a total of 122 cases of listeriosis including 34 deaths recorded in Switzerland. The bacterium was traced to the cellars and wooden shelves that the cheeses had been ripened in. After the cellars had been cleaned and the wooden shelves replaced with metal ones, the number of *L. monocytogenes* in the cheese was considerably reduced.

4.3.6 *Y. enterocolitica*
There are three pathogenic species in the genus *Yersinia*, but only *Y. enterocolitica* and *Y. pseudotuberculosis* cause gastroenteritis. *Yersinia pestis*, the causative agent of 'the plague', is genetically very similar to *Y. pseudotuberculosis* but infects humans by routes other than food (Box 4.7).

Box 4.7

Organism

Y. enterocolitica

Food source

Pork products, cured or uncured, milk and dairy products

Onset of symptoms

24–36 hours

Acute symptoms and chronic complications

Abdominal pain, diarrhoea, mild fever, vomiting

Y. enterocolitica, a small (1–3.5 μm × 0.5–1.3 μm) rod-shaped, Gram-negative bacterium, is often isolated from clinical specimens such as wounds, faeces, sputum and mesenteric lymph nodes. Young cultures are oval or coccoid cells. The organism produces peritrichous flagella when it is grown at 25°C but not when grown at 35°C. The organism has an optimal growth temperature of 30–37°C; however, it is also able to grow on food at refrigeration temperatures (8°C). *Y. pseudotuberculosis* has been isolated from the diseased appendix of humans. Both organisms have often been isolated from animals such as pigs, birds, beavers, cats and dogs. Only *Y. enterocolitica* has been detected in environmental and food sources such as ponds, lakes, meats, ice cream and milk. Most isolates have been found not to be pathogenic. However, pigs frequently carry serotypes which are capable of causing human disease.

Typical symptoms of foodborne yersiniosis are:
- abdominal pain;
- fever;
- diarrhoea (lasting several weeks);
- other symptoms may include sore throat, bloody stools, rash, nausea, headache, malaise, joint pain and vomiting.

Yersiniosis is frequently characterised by symptoms such as gastroenteritis with diarrhoea and/or vomiting; however, fever and abdominal pain are the characteristic symptoms (Table 1.17). Yersinia infections mimic appendicitis and mesenteric lymphadenitis, but the bacteria may also cause infections of other sites such as wounds, joints and the urinary tract. The minimum infectious dose is unknown. Illness onset is usually between 24 and 48 hours after ingestion, although the maximum incubation period can be as long as 11 days. Yersiniosis has been misdiagnosed as Crohn's disease (regional enteritis) as well as appendicitis.

There are four serotypes of *Y. enterocolitica* associated with pathogenicity: O:3, O:5, O:8 and O:9. The genes (*inv* and *ail*) encoding for invasion of mammalian cells are located on the chromosome while a 40–50 MDal plasmid encodes most of the other virulence-associated phenotypes. The 40–50 MDal plasmid is present in almost all the pathogenic *Yersinia* species and the plasmids appear to be homologous. A heat-stable enterotoxin has been isolated from most clinical isolates; however, it is uncertain if this toxin has a role in the pathogenicity of the organism.

Y. enterocolitica is ubiquitous. It can be found in meats (pork, beef, lamb, etc.), oysters, fish and raw milk. The exact cause of the food contamination is unknown. However, the prevalence of this organism in soil and water and in animals such as beavers, pigs and squirrels offers ample opportunities for it to enter our food supply. The principal host recognised for *Y. enterocolitica* is the pig.

Yersiniosis does not occur frequently. It is rare unless a breakdown occurs in food processing techniques. Yersiniosis is a far more common disease in Northern Europe, Scandinavia and Japan than in the United States. The major complication with the disease is the performance of unnecessary appendectomies, since one of the main symptoms of infections is abdominal pain of the lower right quadrant. Both *Y. enterocolitica* and *Y. pseudotuberculosis* have been associated with reactive arthritis, which may occur even in the absence of obvious symptoms. The frequency of such postenteritis arthritic conditions is about 2–3%. Another complication is bacteraemia (entrance of organisms into the bloodstream), in which case the possibility of a disseminating disease may occur. This is rare, however, and fatalities are also extremely rare.

The organism is resistant to adverse storage conditions (such as freezing for 16 months).

Poor sanitation and improper sterilisation techniques by food handlers, including improper storage, cannot be overlooked as contributing to contamination. Therefore, primary control of the organism would require changes to current slaughtering practice (Büllte *et al.* 1992). Since the organism is able to grow at refrigeration temperatures, this is not an effective means of control unless combined with the addition of preservatives. The low temperature growth and its presence on raw produce does indicate that there is the potential for salad vegetables to act as vehicles of yersiniosis in humans.

4.3.7 *St. aureus*

St. aureus is a Gram-positive spherical bacterium (coccus) which occur in pairs, short chains, or bunched, grape-like clusters. The organism was first described in 1879. The organism is a facultative anaerobe and is divided into a number of biotypes on the basis of biochemical tests and resistance patterns. The biotypes are then further divided according to phage typing, serotyping, plasmid analysis and ribotyping. *St. aureus* produces a wide range of pathogenicity and virulence factors: staphylokinase, hyaluronidases, phosphatases, coagulases and haemolysins. Food poisoning is specifically caused by enterotoxins. The enterotoxins are low molecular weight proteins (26 000–34 000 Da) which can be differentiated by serology into a number of antigenic types: SEA to SEE, and SEG to SEQ, with three variants of SEC. A toxin previously designated enterotoxin F is now recognised as responsible for toxic shock syndrome and not enteritis (Box 4.8).

SEA is the most common toxin associated with staphylococcal food poisoning ($\sim$77% of outbreaks), and is carried by a temperate bacteriophage. The second most common is SED (38%) and SEB (10%). These toxins are highly heat stable ($D_{98.9} = >2$ hours), resistant to cooking and proteolytic enzymes. They are gastrointestinal toxins, as well as superantigenic, and stimulate monocytes and macrophages to produce cytokines. These two functions are localised on separate domains of the toxin. They share a common sequence (and hence phylogenetic relationship) with streptococcal pyrogenic exotoxins; see Section 2.9.8. These toxins have been associated with toxic shock syndromes, food poisoning, as well as allergenic and autoimmune diseases.

Box 4.8

Organism

St. aureus

Food source

Workers handling foods, meat (especially sliced meat) poultry, fish, canned mushrooms, dairy products, prepared salad dressing, ham, salami, bakery items, custards, cheese

Onset of symptoms

1–7 hours

Acute symptoms and chronic complications

Severe nausea, abdominal cramps, vomiting, retching, prostration, often with diarrhoea

Table 4.8 Conditions for *St. aureus* growth and toxin production

Parameter	Growth	Toxin production
Temperature (°C)	7–48	10–48
pH	4–10	4.5–9.6
Water activity	0.83–0.99	0.87–0.99

Note: To convert to °F, use the equation $°F = (9/5)°C + 32$. As a guidance: $0°C = 32°F$, $4.4°C = 40°F$, $60°C = 140°F$.

A toxin dose of less than 1.0 µg/kg (300–500 ng) in contaminated food will produce symptoms of staphylococcal intoxication. This amount of toxin is produced by 10^5 organisms per gram. The resistance to heat and proteolysis in the intestinal tract means that it is important that foods which may have supported the growth of *St. aureus* are tested for toxin after heat processing (and subsequent bacterial cell death). A comparison of characteristics for *St. aureus* growth and toxin production is given in Table 4.8. The organism is quickly killed by heat ($D_{65.5} = 0.2$–2.0 minutes) but are resistant to drying and are salt tolerant; see Table 2.8.

Staphylococci exist in air, dust, sewage, water, milk and food or on food equipment, environmental surfaces, humans and animals. Humans and animals are the primary reservoirs. Staphylococci are present in the nasal passages and throats and on the hair and skin of 50% or more of healthy individuals. This incidence is even higher for those who associate with or who come in contact with sick individuals and hospital environments. Although food handlers are usually the main source of food contamination in food poisoning outbreaks, equipment and environmental surfaces can also be sources of contamination with *St. aureus*. Human intoxication is caused by ingesting enterotoxins produced in food by some strains of *St. aureus*, usually because the food has not been kept hot enough (60°C, 140°F, or above) or cold enough (7.2°C, 45°F, or below). Foods that are frequently incriminated in staphylococcal food poisoning include meat and meat products; poultry and egg products; salads such as egg, tuna, chicken, potato and macaroni; bakery products such as cream-filled pastries, cream pies and chocolate eclairs; sandwich fillings; and milk and dairy products. Foods that require considerable handling during preparation and that are kept at slightly elevated temperatures after preparation are frequently involved in staphylococcal food poisoning. The organism competes poorly with other bacteria and therefore rarely causes food poisoning in a raw product. Similarly, *St. aureus* can be isolated from fresh produce and ready-to-eat vegetable salads, but because it does not compete well with other micro-organisms present, spoilage will normally occur before the development of sufficiently high populations of *St. aureus* needed for production of staphylococcal enterotoxin.

Symptoms of staphylococcal food poisoning are a rapid onset of:

- nausea,
- vomiting,
- abdominal cramps.

The onset of symptoms in staphylococcal food poisoning is usually rapid within hours of ingestion. The symptoms can be very acute, depending on individual susceptibility to the toxin, the amount of contaminated food eaten, the amount of toxin in the food ingested and the general health of the victim. The most common symptoms are nausea, vomiting and abdominal cramping. Some individuals may not always demonstrate all the symptoms associated with the illness. In severe cases, headache, muscle cramping and transient changes in blood pressure and pulse rate may occur. The illness is usually self-limiting and generally takes 2–3 days. Severe cases will take longer.

Since the staphylococcal toxin is very heat stable, it cannot be inactivated by standard cooking regimes. Therefore, avoiding contamination of food by the organism and maintaining low temperatures is used to eliminate the microbial load.

4.3.8 Cl. perfringens

There are four clinically important *Clostridium* species: *Cl. botulinum*, *Cl. perfringens*, *Clostridium difficile* and *Clostridium tetani*. However, only *Cl. botulinum* and *Cl. perfringens* cause food poisoning. *Cl. perfringens* is an anaerobic, Gram-positive spore-forming rod. 'Anaerobic' means the organism is unable to grow in the presence of free oxygen. It was first associated with diarrhoea in 1895, but the first reports of the organism causing food poisoning were in 1943. It is widely distributed in the environment and frequently occurs in the intestines of humans and animals. Spores of the organism persist in soil, sediments and areas subject to human or animal faecal pollution (Box 4.9).

Cl. perfringens causes two very different types of foodborne diseases due to the production of one or more toxins. The organism can produce over 13 different toxins, but each strain only produces a subset. There are five types (A–E) of *Cl. perfringens* which are divided according to the presence of the major lethal toxins of which three are plasmid borne (Table 4.9). *Cl. perfringens* types A, C and D are human pathogens, whereas types B, C, D and E are animal pathogens. The acute diarrhoea caused by *Cl. perfringens* is due to the production of α-toxin, an enterotoxin (Titbull *et al.* 1999). A more serious but rare illness is also caused by ingesting food contaminated with type C strains. The latter illness is known as enteritis necroticans jejunitis or 'pig-bel disease' and is due to the β-exotoxin.

Box 4.9

Organisms

Cl. perfringens and *Cl. botulinum*

Onset of symptoms

Cl. perfringens: 8–22 hours
Cl. botulinum: 18–36 hours

Food source

Cl. perfringens: Meat, meat stews, meat pies, and beef, turkey and chicken gravies, beans, seafood
Cl. botulinum: Improperly canned or fermented goods

Acute symptoms and chronic complications

Intense abdominal cramps, diarrhoea, nausea
Chronic symptoms: Gas gangrene, necrotising enteritis

Table 4.9 Typing of *Cl. perfringens* according to the presence of toxins and enterotoxin

Cl. perfringens type	α-toxin	β-toxin	ε-toxin	I-toxin	Enterotoxin
A	+	−	−	−	+
B	+	+	+	−	+
C	+	+	−	−	+
D	+	−	+	−	+
E	+	−	−	+	+
Location	Chromosome	Plasmid	Plasmid	Plasmid	Chromosome/Plasmid

Characteristics of perfringens food poisoning are:
- abdominal pain,
- nausea,
- acute diarrhoea,
- symptoms appear 8–12 hours after ingestion of organism.

The common form of *Cl. perfringens* poisoning is characterised by intense abdominal cramps and diarrhoea which begin 8–12 hours after consumption of foods containing large numbers of *Cl. perfringens* bacteria capable of producing the food poisoning toxin (Table 1.18). The illness is usually over within 24 hours but less severe symptoms may persist in some individuals for 1 or 2 weeks. A few deaths have been reported as a result of dehydration and other complications. In most instances, the actual cause of poisoning by *Cl. perfringens* is temperature abuse of prepared foods. Meats, meat products and gravy are the foods most frequently implicated. Small numbers of the organisms are often present as spores after cooking. The spores germinate and the clostridia multiply to food poisoning levels during the cooling period and storage. The cooking process drives off oxygen, hence creating an anaerobic environment favourable to clostridial growth. After ingestion, the enterotoxin is produced in the intestine, after the organism has passed through the stomach. The enterotoxin is associated with sporulation, possibly induced due to the acidic environment of the stomach. The enterotoxin is a heat-sensitive protein 36 000 Da in size. It is destroyed by heat ($D_{90} = 4$ minutes), and a few cases of food poisoning due to ingestion of the toxin have been reported.

Necrotic enteritis (pig-bel) caused by *Cl. perfringens* is often fatal. This disease also begins as a result of ingesting large numbers (greater than 10^8) of *Cl. perfringens* type C in contaminated foods. Deaths from necrotic enteritis (pig-bel syndrome) are caused by infection and necrosis of the intestines and from resulting septicaemia.

The ubiquitous nature of *Cl. perfringens* and its spores make it a common problem in food production. The isolation of low numbers of *Cl. perfringens* from food does not necessarily mean that a food poisoning danger exists. Only when large numbers are present is there a definite hazard and therefore enumeration techniques are essential for this organism. Although the *Cl. perfringens* vegetative cell is killed by chilling and freezing, the spore form may survive. Further details on the detection of clostridia and growth characteristics can be found in Section 5.7.9 and Table 2.8. Control of the organism is mainly achieved through the cooking and cooling steps. Rapid cooling through the temperature range of 55–15°C reduces the opportunity for any surviving clostridia spores to germinate. Thorough reheating of food to 70°C immediately before consumption will kill any vegetative cells present.

4.3.9 Cl. botulinum

Cl. botulinum is a Gram-positive rod-shaped strict anaerobe (0.3–0.7 μm × 3.4–7.5 μm) with peritrichous flagella. It causes the foodborne illness called 'botulism'. This is a food intoxication due to the ingestion of preformed neurotoxins. The organism is ubiquitous in nature, including soil, fish, raw fruits and vegetables.

There are seven types of *Cl. botulinum*: A, B, C, D, E, F and G. These are differentiated on the basis of toxin antigenicity (Table 4.10). Types A, B, E and F are the main causes of human botulism (types C and D in animals). Eight serologically distinct neurotoxins (BoNTA to BoNTG) have been identified and are among the most potent toxins known to man. The toxin is made of two proteins, fragment A (light chain, LC, 50 kDa) and fragment B (heavy chain, HC, 100 kDa), which are linked by a disulphide bridge (Figures 2.5 and 2.6). The LC is responsible for the toxin's effect on nerve cells. The HC contains the membrane translocation domain and the toxin receptor binding moiety (Boquet *et al.* 1998). The organism forms spores which may become airborne and contaminate open jars or cans. Once sealed, the anaerobic conditions favour spore outgrowth and the production of toxins.

Symptoms of botulism are as follows:

- Double vision
- Nausea
- Vomiting
- Fatigue
- Dizziness
- Headache
- Dryness in the throat and nose
- Respiratory failure

The onset of the symptoms is from 12 to 36 hours from ingestion of the bacterial toxins. Botulinum toxins block the release of the neurotransmitter acetylcholine, which results in muscle weakness and subsequently paralysis. The illness may last from 2 hours to 14 days depending upon dose and vulnerability of host. The fatality rate is ~10%.

Botulism is associated with canned (especially home canned) low-acid foods, vegetables, fish and meat products. It is also associated with honey and hence honey should not be given to children under 1 year. Infant botulism is milder than the adult version. The spores germinate

Table 4.10 Distinguishing characteristics of *Cl. botulinum*

	Group			
	I	II	III	IV
Toxin produced	A, B, F	B, E, F	C1, C2, D	G
Proteolysis	+	−	+ or −	+
Lipolysis	+	+	+	−
Glucose fermentation	+	+	+	−
Mannose fermentation	−	+	+	−
Minimum growth temperature	10–12°C	3.3°C	15°C	12°C
Salt inhibition (%)	10	5	3	>3
Volatile fatty acids produced[a]	Ac, iB, B, iV, Ph	A, B	A, P, B	A, iB, iV, Pa

[a]Ac, acetic acid; iB, isobutyric acid; B, butyric acid; iV, isovaleric acid; Ph, phenylpropionic acid; Pa, phenylacetic acid.

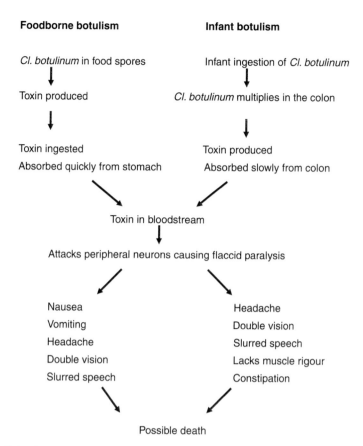

Figure 4.16 *Cl. botulinum* intoxication.

in the intestinal tract and the bacterium produces the toxins causing 'floppy baby syndrome'. A comparison of foodborne botulism and infant botulism is shown in Figure 4.16.

Heat-treating low-acid canned foods to 121°C for 3 minutes or equivalent will eliminate the *Cl. botulinum* spores. *Cl. botulinum* cannot grow in acid or acidified foods with a pH less than 4.6 (Table 2.8). Botulism has been linked to coleslaw prepared from packaged, shredded cabbage and chopped garlic in oil. The continued metabolism of packed salad vegetables can result in an anaerobic environment supporting the growth of *Cl. botulinum* and toxin production. Therefore, the permeability of packaging films must minimise the possible development of anaerobic conditions. An additional control measure is storage at less than 3°C.

4.3.10 B. cereus

B. cereus is a spore-forming food pathogen that was first isolated in 1887. The spores can survive many cooking processes. The organism grows well in cooked food because of the lack of a competing microflora. *B. cereus* is a Gram-positive, facultatively aerobic spore former whose cells are large rods and whose spores do not swell the sporangium. These and other characteristics, including biochemical features, are used to differentiate and confirm the presence of *B. cereus*. However, these characteristics are shared with *B. cereus* var. *mycoides*, *Bacillus thuringiensis* and

Box 4.10

Organism

B. cereus

Onset of symptoms

Diarrhoeal: 6–15 hours
Emetic: 0.5–6 hours

Food source

Diarrhoeal: Meats, milk vegetables and fish
Emetic: Rice products, starchy foods (e.g. potato, pasta, and cheese products)

Acute symptoms and chronic complications

Diarrhoeal: Watery diarrhoea, abdominal cramps and pain
Emetic: Nausea and vomiting

Bacillus anthracis. Differentiation of these organisms depends upon determination of motility (most *B. cereus* are motile), presence of toxin crystals (*B. thuringiensis*), haemolytic activity (*B. cereus* and others are β-haemolytic, whereas *B. anthracis* is usually non-haemolytic) and rhizoid growth which is characteristic of *B. cereus* var. *mycoides*. *Bacillus subtilis* and *Bacillus licheniformis* have been proposed as food poisoning organisms but the evidence to date is not complete. Nevertheless, the UK HPA (formerly PHLS) ready-to-eat guidelines do include them in its sampling plan; see Section 6.11 (Box 4.10).

B. cereus is ubiquitous in nature, being isolated from soil, vegetation, freshwater and animal hair. It is commonly found at low levels in food ($<10^2$ cfu/g) which is considered acceptable. Food poisoning outbreaks usually occur when the food has been subjected to time–temperature abuse which was sufficient for the low level of organisms to multiply to a significant (intoxication) level ($>10^5$ cfu/g).

Sources:
- *Ubiquitous*: soil, vegetation, meat, milk, water, rice, fish

Characteristics of temperature and water activity growth range and D values are given in Table 2.8. Isolation methods are covered in Section 5.6.10.

There are two recognised types of *B. cereus* food poisoning: diarrhoeal and emetic (Table 4.11). Both types of food poisoning are self-limiting and recovery is usually within 24 hours. *B. cereus* produces the diarrhoeal toxins during growth in the human small intestines. In contrast, the emetic toxins are preformed on the food. A wide variety of toxins have been identified (Granum & Lund 1997b, Kramer & Gilbert 1989):
- diarrhoeal enterotoxin
- emetic toxin
- haemolysin I
- haemolysin II
- phosphatase C

Table 4.11 Characteristics of food poisoning caused by *B. cereus*

	Emetic syndrome	Diarrhoeal syndrome
Infective dose	10^5–10^8 cells/g	10^5–10^7 total
Toxin produced	Preformed	Produced in the small intestine
Type of toxin	Cyclic peptide (1.2 kDa)	Three protein subunits (L$_1$, L$_2$, B; 37–105 kDa)
Toxin stability	Very stable (126°C, 90 min; pH 2–11)	Inactivated at 56°C, 30 minutes
Incubation period	0.5–6 hours	8–24 hours
Duration of illness	6–24 hours	12–24 hours
Symptoms	Abdominal pain, watery diarrhoea, nausea	Nausea, vomiting and malaise, watery diarrhoea
Foods most frequently implicated	Fried and cooked rice, pasta, noodles and potatoes	Meat products, soups, vegetables, fish, puddings, sauces and milk and dairy products

Adapted from Granum & Lund 1997a with permission of Blackwell Publishing Ltd.

Symptoms of *B. cereus* diarrhoeal food poisoning are as follows:
- watery diarrhoea
- abdominal cramps and pain
- nausea, rarely vomiting

The diarrhoeal type of illness is caused by a large molecular weight protein composed of three subunits (L1, L2 and B; 37–105 kDa) (Table 4.11). This toxin is inactivated at 56°C for 30 minutes. The symptoms of *B. cereus* diarrhoeal type food poisoning mimic those of *Cl. perfringens* food poisoning. The onset of watery diarrhoea, abdominal cramps and pain occurs 8–24 hours after consumption of contaminated food. Nausea may accompany diarrhoea, but vomiting (emesis) rarely occurs. Symptoms persist for 24 hours in most instances, during which time the organism is excreted in large numbers.

Symptoms of *B. cereus* emetic food poisoning are as follows:
- Nausea
- Vomiting
- Abdominal cramps and diarrhoea may occur

The vomiting (emetic) type of illness is caused by a low molecular weight (1.2 kDa), heat-stable cyclic peptide. It is similar to the potassium ionophore valinomycin in that it is ring shaped with three repeating units of the amino and/or oxy acids '-D-O-leucine-D-alanine-O-valine-L-valine-'. It is very resistant to heat (126°C for 90 minutes) and pH values 2–11. It is produced in the stationary phase of growth, but it is uncertain if its production is linked with spore formation. The emetic type of food poisoning is characterised by nausea and vomiting within 0.5–6 hours after consumption of contaminated foods. Occasionally, abdominal cramps and/or diarrhoea may also occur. Duration of symptoms is generally less than 24 hours. The symptoms of this type of food poisoning resemble those caused by *St. aureus* foodborne intoxication. Some strains of *B. subtilis* and *B. licheniformis* have been isolated from lamb and chicken incriminated in food poisoning episodes. These organisms demonstrate the production of a highly heat-stable toxin which may be similar to the vomiting type toxin produced by *B. cereus*.

A wide variety of foods including meats, milk, vegetables and fish have been associated with the diarrhoeal type food poisoning. The vomiting-type outbreaks have generally been associated with rice products; however, other starchy foods such as potato, pasta and cheese products have also been implicated. Food mixtures such as sauces, puddings, soups, casseroles, pastries and salads have frequently been incriminated in food poisoning outbreaks. The presence of large numbers of *B. cereus* (greater than 10^6 organisms/g) in a food is indicative of active growth and proliferation of the organism and is consistent with a potential hazard to health.

Since the organism is ubiquitous in the environment, low numbers commonly occur in food. Therefore, the main control mechanism is to prevent spore germination and multiplication in cooked, ready-to-eat foods. Storage of foods below 10°C will inhibit *B. cereus* growth.

Control measures:
- Temperature control to prevent spore germination and outgrowth
- Storage temperature either >60°C or <10°C, unless other parameters (i.e. pH, a_w) prevent bacterial growth
- Avoid storage of pre-cooked foods

4.3.11 *Vibrio cholerae*, *V. parahaemolyticus* and *V. vulnificus*

Vibrio species are frequently isolated from estuarine waters and are therefore associated with fish and various seafoods. Although there are 12 human pathogenic *Vibrio* species, only *V. cholerae*, *V. parahaemolyticus* and *V. vulnificus* are of major concern to human infection. *V. cholerae* causes cholera and is the best known *Vibrio* infection. Carriage of *V. cholerae* by infected humans is important in the disease transmission, as water used for drinking or washing food can become contaminated by faecal matter (Box 4.11).

Box 4.11

Organisms

V. cholerae, *V. parahaemolyticus*, *V. vulnificus*

Onset of symptoms

V. cholerae: 1–3 days
V. parahaemolyticus: 9–25 hours
V. vulnificus: 12 hours–3 days

Food source

V. cholerae: Seafood, vegetables, cooked rice, ice
V. parahaemolyticus: Raw or undercooked fish and fishery products
V. vulnificus: Seafood, particularly raw oysters

Acute symptoms and chronic complications

V. cholerae: Profuse watery diarrhoea
V. parahaemolyticus: Profuse watery diarrhoea free from blood or mucus, abdominal pain, vomiting, and fever
V. vulnificus: Profuse diarrhoea with blood in stools

In the small intestine, *V. cholerae* attaches via cell-associated adhesins to the mucosal surface of the small intestine and produces an exotoxin, cholera toxin, that acts on intestinal mucosal cells. The toxin is 84 kDa in size and is composed of A1 subunit (21 kDa) and A2 subunit (7 kDa) covalently linked to five B subunits (10 kDa each). The B subunit binds to a receptor ganglioside on the mucosal cells and the A subunit enters the cell activating adenylate cyclase and subsequently increasing the concentration of cAMP (Figure 2.7).

Mucosal cells have a set of ion transport pumps (for Na^+, Cl^-, HCO_3^- and K^+) that normally maintain a tight control over ion fluxes across the intestinal mucosa; see Section 4.1.1. Because water can pass freely through membranes, the only way to control the flow of water in and out of tissue is to control the concentration of ions in different body compartments. Under normal conditions, the net flow of ions is from lumen to tissue (Figure 4.3), resulting in a net uptake of water from the lumen. The effect of cholera toxin on mucosal cells is to alter this balance. Cholera toxin causes no apparent damage to the mucosa, but the increased level of cAMP decreases the net flow of sodium into tissue and produces a net flow of chloride (and water) out of tissue and into the lumen, resulting in massive diarrhoea and electrolyte imbalance.

V. parahaemolyticus means 'dissolving blood' and the organism was first isolated in 1951. The organism is not isolated in the absence of additional NaCl (2–3%) and was therefore not cultivated in early studies of gastroenteritis. *V. parahaemolyticus* is now recognised as a major cause of foodborne gastroenteritis in Japan. This is because the organism is associated with the consumption of seafoods which is a significant part of the average diet in Japan.

Typical symptoms of *V. parahaemolyticus* food poisoning:

- Diarrhoea
- Abdominal cramps
- Nausea
- Vomiting
- Headaches
- Fever and chills

The incubation period is 4–96 hours after the ingestion of the organism, with a mean of 15 hours (Table 1.18). The illness is usually mild or moderate, although some cases may require hospitalisation. On average, the illness lasts about 3 days. The infective dose is possibly greater than one million organisms. Virulence is associated with the production of a thermostable direct haemolysin (TDH), the ability to invade enterocytes and possibly the production of an enterotoxin (possibly Shiga-like). The production of the *tdh* gene is the only reliable trait that currently distinguishes between pathogenic and non-pathogenic strains. Although the demonstration of the Kanagawa haemolysin was initially regarded as indicative of pathogenicity, this is now uncertain.

There are usually fewer than 10^3 cfu/g of *V. parahaemolyticus* on fish and shellfish, except in warm waters where the count may increase to 10^6 cfu/g. Infections with this organism have been associated with the consumption of raw, improperly cooked, or cooked, recontaminated fish and shellfish. A correlation exists between the probability of infection and warmer months of the year. Improper refrigeration of seafoods contaminated with this organism will allow its proliferation, which increases the possibility of infection. The organism is very sensitive to heat, and outbreaks are therefore frequently due to improper handling procedures and temperature abuse. Control of the organism is therefore through prevention of the organism's multiplication after harvesting by chilling ($<5°C$) and cooking to an internal temperature of $>65°C$. Isolation of any *Vibrio* species from cooked food indicates poor hygienic practice since the organism is rapidly killed by heat.

V. vulnificus was first reported in 1976 as 'lactose-positive vibrios'. 'Vulnificus' means 'wound inflicting' which reflects the organism's ability to invade and destroy tissue. The organism is therefore associated with wound infections and fatal septicaemia (Linkous & Oliver 1999). It has the highest death rate of any foodborne disease agent (Todd 1989b) and causes 95% of all seafood-related deaths in the United States.

Typical symptoms of *V. vulnificus* food poisoning are as follows:

* Fever
* Chills
* Nausea
* Skin lesions

The onset of symptoms takes about 24 hours (range: 12 hours to several days) after ingestion of contaminated raw shellfish (especially oysters) by vulnerable people. Individuals most susceptible to infection include the elderly, the immuno-compromised, those suffering with chronic liver disease and chronic alcoholism. The organism differs from other pathogenic vibrios in that it invades and multiplies in the bloodstream. The mortality is between 40% and 60% of cases. The organism is highly invasive and produces various factors which protect it from the host immune system, including a serum resistance factor, capsular polysaccharide and the ability to acquire iron from iron-saturated transferrin. It produces a range of exoenzymes including a heat-labile haemolysin/cytolysin and an elastolytic protease which probably causes the cellular damage. The LPS is endotoxic. Since a large amount of shellfish is consumed each year and only relatively small numbers *V. vulnificus* food poisoning is reported, it seems reasonable to predict that not all strains are pathogenic.

V. vulnificus is isolated from shellfish and coastal water. The organism is rarely isolated from seawater at <10–15°C, but numbers rise when the water temperature is >21°C. The principal routes of infection are through wounds and septicaemia after ingestion. It is not a significant cause of food poisoning among healthy adults. Therefore, the main means of prevention is for immuno-compromised individuals to avoid eating raw shellfish, in particular oysters. Isolation of any *Vibrio* species from cooked food indicates poor hygienic practice since the organism is rapidly killed by heat and has not been reported in processed foods. Control of the organism is primarily by ceasing oyster harvesting if the water temperatures exceed 25°C, and by chilling and holding oysters to <15°C post-harvesting. See Section 10.6 for risk assessment of *V. parahaemolyticus* in molluscan shellfish.

4.3.12 *Brucella melitensis, Br. abortus* and *Br. suis*

Brucella is a strictly aerobic, Gram-negative coccobacillus which causes brucellosis. This organism is carried by animals and causes incidental infections in humans. The four species that infect humans are named after the animal they are commonly isolated from: *Br. abortus* (cattle), *Br. suis* (swine), *Br. melitensis* (goats) and *Br. canis* (dogs). The cattle and dairy industries are the prime source of infection. Brucella can enter the body via the skin, respiratory tract or the digestive tract. Once there, the intracellular organism can enter the blood and the lymphatics where it multiplies inside phagocytes and eventually cause bacteraemia (bacterial infection of blood). Symptoms vary from patient to patient but can include high fever, chills and sweating (Box 4.12).

Worldwide brucellosis remains a major source of disease in humans and domesticated animals. The incidence of brucellosis has decreased in North America and Western Europe. However, it remains an important health problem in a number of other countries: the Middle

Box 4.12

Organism

Brucella spp.

Food source

Raw or unheated processed foods of animal origin, for example milk, milk products, cream, cheese, butter

Onset of symptoms

Days to weeks

Acute symptoms and chronic complications

Sweating, headache, lack of appetite, fatigue, fever

East, the Mediterranean, Mexico, Peru, some regions of China, the former USSR and India. Some countries in the Middle East are experiencing an increasing incidence.

In humans, *Br. melitensis* is the most important clinically apparent disease. Pathogenicity is related to the production of LPSs containing poly *N*-formyl perosamine O chain, Cu–Zn superoxide dismutase, erythrulose phosphate dehydrogenase, stress-induced proteins related to intracellular survival, and adenine and guanine monophosphate inhibitors of phagocyte functions.

Prevention of brucellosis depends on the eradication or control of the disease in animal hosts, hygienic precautions to limit exposure to infection through occupational activities and the heat treatment of dairy and other potentially contaminated foods.

4.3.13 *Aeromonas hydrophila*, *A. caviae* and *A. sobria*

The genus *Aeromonas* was proposed in 1936 for rod-shaped *Enterobacteriaceae* but possessing a polar flagellum. *A. hydrophila* is present in all freshwater environments and in brackish water. It has frequently been found in fish and shellfish and in market samples of red meats (beef, pork, lamb) and poultry. The organism is able to grow slowly at 0°C. It is presumed that given the ubiquity of the organisms that not all strains are pathogenic. Some strains of *A. hydrophila* are capable of causing illness in fish and amphibians as well as in humans who may acquire infections through open wounds or by ingestion of a sufficient number of the organisms in food or water. *A. hydrophila* may cause gastroenteritis in healthy individuals or septicaemia in individuals with impaired immune systems or various malignancies. *A. caviae* and *A. sobria* also may cause enteritis in anyone or septicaemia in immuno-compromised persons or those with malignancies (Box 4.13).

There is controversy as to whether *A. hydrophila* is a cause of human gastroenteritis. The uncertainty is because volunteer human feeding studies (10^{11} cell dose) have failed to demonstrate any associated human illness. However, its presence in the stools of individuals with diarrhoea, in the absence of other known enteric pathogens, suggests that it has some role in disease and has been included in this survey of foodborne pathogens. Similarly, *A. caviae*

Box 4.13

Organism

A. hydrophila

Food source

Seafood (fish, shrimp, oysters), snails, drinking water

Onset of symptoms

24–48 hours

Acute symptoms and chronic complications

Watery stools, stomach cramps, mild fever and vomiting

and *A. sobria* are putative pathogens associated with diarrhoeal disease, but as of yet, they are unproven causative agents.

General symptoms of *A. hydrophila* gastroenteritis are as follows:

- Diarrhoea
- Abdominal pain
- Nausea
- Chills and headache
- Dysentery-like illness
- Colitis
- Additional symptoms include septicaemia, meningitis, endocarditis and corneal ulcers

Two distinguishable types of gastroenteritis have been associated with *A. hydrophila*: a cholera-like illness with a watery (rice and water) diarrhoea and a dysenteric illness characterised by loose stools containing blood and mucus. The infectious dose of this organism is unknown. *A. hydrophila* may spread throughout the body in the bloodstream and cause a general infection in persons with impaired immune systems. Those at risk are individuals suffering from leukaemia, carcinoma and cirrhosis, and those treated with immuno-suppressive drugs or who are undergoing cancer chemotherapy. *A. hydrophila* produces a cytotonic enterotoxin(s), haemolysins, acetyl transferase and a phospholipase. Together, these virulence factors may account for the organism's pathogenicity.

4.3.14 *Plesiomonas shigelloides*

Pl. shigelloides is a facultatively anaerobic, rod-shaped Gram-negative bacterium. In contrast to the taxonomy of *Aeromonas*, the *Plesiomonas* genus comprises a single species *Pl. shigelloides*, which is now classified in the family *Plesiomonadaceae*. *Plesiomonas* shares characteristics with both *Vibrio* and *Aeromonas*. It also contains the *Enterobacteriaceae* common antigen. It has been isolated from freshwater, freshwater fish and shellfish and from many types of animals including cattle, goats, swine, cats, dogs, monkeys, vultures, snakes and toads. Most human *Pl. shigelloides* infections are suspected to be waterborne. The organism may be present in contaminated or untreated water which has been used as drinking water and recreational water. *Pl. shigelloides* does not always cause illness after ingestion but may reside temporarily as a transient, non-infectious member of the intestinal flora. It has not only been isolated from the stools of patients with diarrhoea, but is also sometimes isolated from healthy individuals (0.2–3.2% of population). As with *A. hydrophila*, *Pl. shigelloides* is not a proven foodborne pathogen but nevertheless has been included in this survey as a few studies have linked water and food contamination with *Pl. shigelloides* gastroenteritis outbreaks (Claesson *et al.* 1994).

Typical symptoms of *Pl. shigelloides* gastroenteritis are as follows:
- Diarrhoea
- Abdominal pain
- Nausea
- Chills
- Lesser extent fever, headaches and vomiting

The symptoms may begin 20–24 hours after consumption of contaminated food or water and last from 1 to 9 days; see Table 1.18. Diarrhoea is watery, non-mucoid and non-bloody. In severe cases, however diarrhoea may be greenish-yellow, foamy, and blood tinged.

The infectious dose is presumed to be quite high, at least greater than one million organisms. The pathogenesis of *Pl. shigelloides* infection is not known. The organism produces a heat-stable cytotonic enterotoxin and possibly haemolysins, proteases and endotoxins. Its significance as an enteric (intestinal) pathogen is presumed because of its predominant isolation from stools of patients with diarrhoea.

The organism is ubiquitous in the environment. Most *Pl. shigelloides* infections occur in the summer months and correlate with environmental contamination of freshwater (rivers, streams, ponds, etc.). The usual route of transmission of the organism in sporadic or epidemic cases is by ingestion of contaminated water or raw shellfish. Most *Pl. shigelloides* strains associated with human gastrointestinal disease have been from stools of diarrhoeic patients living in tropical and subtropical areas. Such infections are rarely reported in the United States or Europe because of the self-limiting nature of the disease. Subsequently, it may be included in the group of diarrhoeal diseases 'of unknown etiology' which are treated with and respond to broad-spectrum antibiotics. Since the organism is heat sensitive, the main means of control is the proper cooking of shellfish before ingestion.

4.3.15 *Streptococcus* and *Enterococcus* species

The genus *Streptococcus* comprises of Gram-positive, microaerophilic cocci (round), which are non-motile and occur in chains or pairs. The genus is defined by a combination of antigenic, haemolytic and physiological characteristics into Groups A, B, C, D, F and G. Groups A and D can be transmitted to humans via food. Group A contains one species (*Strep. pyogenes*) with 40 antigenic types. Group D contains five species: *Ent. faecalis*, *Ent. faecium*, *Strep. durans*, *Strep. avium* and *Strep. bovis*. The link between scarlet fever and septic sore throat was linked to the consumption of contaminated milk almost 100 years ago. Outbreaks of septic sore throat and scarlet fever were numerous before the advent of milk pasteurisation (Box 4.14).

Symptoms of Group A streptococcal infection are as follows:
- Sore and red throat
- Pain on swallowing
- Tonsillitis
- High fever and headache
- Nausea and vomiting
- Malaise
- Runny nose
- Rash

Box 4.14

Organism

Group A *Streptococcus* spp. (*S. pyogenes*)

Food source

Temperature-abused milk, ice cream, eggs, steamed lobster, ground ham, potato salad, egg salad, custard, rice pudding, shrimp salad

Onset of symptoms

Group A: 24–72 hours
Group D: 2–33 hours

Acute symptoms and chronic complications

Sore, red throat, pain on swallowing, tonsillitis, high fever, headache, nausea, vomiting, malaise, rhinorhea

Group A streptococci cause septic sore throat and scarlet fever as well as other pyrogenic and septicaemic infections. The organism may also cause toxic shock syndrome (Cone *et al.* 1987). The onset of illness is 1–3 days. The infectious dose is probably quite low at less than 1000 organisms. Food sources include milk, ice cream, eggs, steamed lobster, ground ham, potato salad, egg salad, custard, rice pudding and shrimp salad. In almost all food poisoning cases, the foodstuffs had been subjected to temperature abuse between preparation and consumption. Contamination of the food is often due to poor hygiene, ill food handlers or the use of unpasteurised milk. Streptococcal sore throat is very common, especially in children. Usually, it is successfully treated with antibiotics. Complications are rare and the fatality rate is low. Most current outbreaks have involved complex foods (i.e. salads) which were infected by a food handler with a septic sore throat. One ill food handler may subsequently infect hundreds of individuals.

The pathogenicity of Group A streptococci is partially due to the presence of M proteins in the fibrillae on the cell surface. These adherence factors enable the organism to adhere to epithelial cells and evade phagocytosis. The organism produces haemolysin(s) as evidence by the β-type haemolysis on blood agar.

Group D streptococci may produce a clinical syndrome similar to staphylococcal intoxication.

General symptoms of Group D streptococcal infection are as follows:
- Diarrhoea
- Abdominal cramps
- Nausea and vomiting
- Fever and chills
- Dizziness

Symptoms occur in 2–36 hours after ingestion of suspect food. The infectious dose is probably greater than 10^7 organisms. The diarrhoeal illness is poorly characterised, but is acute and self-limiting. Food sources include sausage, evaporated milk, cheese, meat croquettes, meat pie, pudding, raw milk and pasteurised milk. Entrance into the food chain is due to underprocessing and/or unhygienic practices during food preparation.

Control of foodborne streptococcal infections is by strict control of personal hygiene and excluding staff with sore throats from the production area.

Strep. parasanguinis is an emerging foodborne pathogen. This organism was isolated from two sheep in Spain during a bacteriological survey for determining the prevalence of subclinical mastitis. Since the organism has been associated with the development of experimental endocarditis, its presence at relatively high concentrations in apparently healthy sheep milk may pose a health risk in persons with predisposing heart lesions.

4.4 Foodborne pathogens, viruses

Non-bacterial gastroenteritis due to a number of enteric viruses has been established since the first discoveries of Norovirus (formerly Norwalk virus) in 1972 and rotavirus in 1973. Viruses can be excreted in large numbers by infected individuals. Viruses cannot grow in or on foods; however, they serve as vehicles for infection. Sentinent studies (Section 1.7) have estimated that in the United States, approximately 30.9 million (80%) of the 38.6 million total foodborne illnesses per year can be attributed to viruses. Previously, these infections have been grossly underestimated due to under-reporting of mild gastroenteritis and by a lack of reliable detection methods. Fortunately, molecular biology is greatly helping in the development of analytical procedures to analyse food and aid epidemiological studies. However, these methods have not yet evolved to the same level of application as conventional food bacteriology.

Viral gastroenteritis is usually characterised by the following symptoms:
1 Onset after a 24–36-hour incubation period
2 Vomiting and/or diarrhoea lasting a few days
3 A high attack rate (average 45%)
4 A high number of secondary cases (Kaplan *et al.* 1982)

The infectious dose is not known but is presumed to be low (10^0 to 10^2). Viruses are transmitted by the faecal–oral route via person-to-person contact or ingestion of contaminated foods and water. Ill food handlers may contaminate foods that are not further cooked before consumption. Enteric adenovirus may also be transmitted by the respiratory route.

Foodborne viral pathogens range in size from 15 to 400 nm. As obligate intracellular parasites, they require mammalian (even as specific as human) cells to replicate in and have properties to protect the genome when outside the cell and aid transmission. Most food- and waterborne viruses are non-enveloped and are relatively resistant to heat, disinfection and pH changes. They can be transmitted in different ways, for example inhalation of water droplets (aerosol) from a cough and faecal contamination of a work surface. Therefore, contamination can occur anywhere in the farm to fork sequence. Unfortunately, little work has been done on the thermal inactivation of enteric viruses in foods other than shellfish.

General features of foodborne viral infections are as follows:
- Only a few particles are needed to produce an illness.
- High numbers of viral particles are shed in the faeces from infected persons (up to 10^{11} particles/g faeces reported for rotaviruses). Subsequently, raw sewage can contain 10^3–10^5 infectious particles/L.
- Viruses do not replicate in food or water.
- Foodborne viruses are typically quite stable outside the host and are acid resistant.

Table 4.12 Viral foodborne pathogens

Symptoms		
Gastroenteritis	Hepatitis	Other illnesses
Adenovirus	Hepatitis A	Coronavirus
Aichi virus	Hepatitis E	Coxsackievirus Groups A and B
Astrovirus		Cytomegalovirus
Enteric adenovirus types 40 &41		Echovirus serotypes 68–71
Norovirus		Parvovirus
Rotavirus Groups A–C		Poliovirus
Sapporo viruses		
Small round viruses		

After Koopmans & Duizer (2002) and Koopmans *et al.* (2002), with permission of Blackwell Publishing Ltd.

Current food hygiene procedures have been developed to prevent bacterial foodborne pathogens and are therefore probably inadequate for the control of viral pathogens present before processing. An additional problem with foodborne viral pathogens is that the most common ones (i.e. Noroviruses) grow poorly, or not at all in cell culture.

Viral foodborne pathogens can be divided into three groups:

1 *Viruses that cause gastroenteritis*: Rotavirus (Groups A, B and C), enteric adenovirus, Noroviruses and Sapporo viruses (both caliciviruses), adenovirus (types 40 and 41) and astrovirus (serotypes 1–8).
2 *Faecal–orally transmitted hepatitis viruses*: Hepatitis A virus (HAV) and hepatitis E virus (HEV).
3 *Those that cause other illnesses*: Cytomegalovirus, parvo-like viruses, coronavirus and the enteroviruses (polio 1–3, Coxsackie A and B, echo 68–71).

After Koopmans & Duizer (2002) and Koopmans *et al.* (2002); see Table 4.12.

The most common cause of viral foodborne illness are Noroviruses and HAV. In the United States, an estimated 32–42% of foodborne infections are caused by viruses. These viruses are principally transmitted via human-to-human, and no zoonotic (animal-to-human) transfer is recognised. Noroviruses have been found in a large proportion of cattle herds and some pigs but are genetically distinct from those infecting humans (Van Der Poel *et al.* 2000).

Routes by which foods can be contaminated with pathogenic viruses are as follows:

1 Shellfish contaminated by faecally polluted marine water
2 Human sewage pollution of drinking and irrigation water
3 Ready-to-eat and prepared foods contaminated as a result of poor personal hygiene by infected food handlers
4 Aerosolisation of vomit
5 Contact with contaminated surfaces (fomite)

Viruses have a strict intracellular replication requirement and cannot replicate in food or water. Hence, their numbers do not increase during processing, transport or storage. Additionally, they are more resistant to heat, disinfection and pH changes than bacterial pathogens.

Although there is no correlation between the presence of viruses and indicator bacteria (i.e. coliforms, *E. coli*), no specific microbiological criteria for enteric viruses are included in statutory sanitary control measures for shellfish. This is of concern since depuration does not reduce the levels of viral contaminants as effectively as bacterial contaminants.

Box 4.15

Organism

Viral gastroenteritis; Norovirus, adenovirus, rotavirus
Viral hepatitis A

Food source

Gastroenteritis: Food and drinking water contamination, filter-feeding shellfish
Hepatitis A: Shellfish, raw fruits and vegetables, bakery products

Onset of symptoms

Gastroenteritis: 15–50 hours
Hepatitis A: 2–6 weeks

Acute symptoms and chronic complications

Gastroenteritis: Diarrhoea and vomiting, often severe and projectile
Hepatitis A: Loss of appetite, fever, malaise, nausea, vomiting, dark urine, pale stools and jaundice

---

Detection methods are either based on ELISA-based kits or molecular probes (PCR of specific gene sequence) and are used in public health laboratories rather than food microbiology laboratories (Atmar & Estes 2001). There is a foodborne virus network as a rapid alert system for the European Union (Section 1.12.5; see the internet directory for URL) (Box 4.15).

4.4.1 Norovirus (formerly known as Norwalk-like viruses and small round structured viruses)

The Noroviruses were formerly known as 'Norwalk-like virus' (NLV) and 'small round structured viruses' (SRSV). These viruses cannot be grown in cell culture. The taxonomy of the Noroviruses has changed frequently, but it is generally agreed that they belong within the *Caliciviridae* family. The human enteric caliciviruses used to be divided into three genogroups:
1 *Genogroup I*: NLV
2 *Genogroup II*: Snow Mountain agent
3 *Genogroup III*: Sapporo-like virus (SLV)
   However, this was revised into the four genera (Van Regenmortel *et al.* 2000):
1 *Vesiviruses*: Feline calicivirus
2 *Lagoviruses*: Rabbit haemorrhagic disease virus
3 *NLV*: Norwalk virus
4 *SLV*: Sapporo virus
   The family *Caliciviridae* should now be described as follows (Green *et al.* 2000):
• *Genus*: Lagovirus; type species: Rabbit haemorrhagic disease virus
• *Genus*: Vesivirus; type species: Feline calicivirus
• *Genus*: Norovirus; type species: Norwalk virus
• *Genus*: Sapovirus; type species: Sapporo virus

The first two genera are not associated with human infections, whereas Norovirus and Sapporo virus are responsible for epidemic gastroenteritis. Noroviruses account for 67% of foodborne-associated gastroenteritis cases, 33% of hospitalisations and 7% of deaths (Mead *et al.* 1999). Noroviruses cause diarrhoea and vomiting and is the single most common cause of gastroenteritis in all age groups. Sapporo virus predominantly causes illness in children. There are 15 distinct Noroviruses genotypes due to genetic differences and differences in protein composition. Immunity is developed to one genotype and is short lived. Hence, multiple Norovirus infections can occur.

Norovirus is as common as rotavirus in patients who visit general practitioners (Koopmans *et al.* 2002, Wheeler *et al.* 1999; see Section 1.7 for sentinel studies), and is second to the common cold in reported cases in the United States. The incidence is highest in young children (Figure 1.1), though illness does also occur in adults, and of relevance to hygienic food production in the home and factory as asymptomatic infections are common. Outbreaks occur associated with schools, hospitals, restaurants and nursing homes. The proportion of infections that are due to the ingestion of contaminated food is unknown though it is expected to be common. The virus is typically spread by direct person-to-person contact, the secondary (indirect) route is via contaminated food, water or environments. It should be noted that the virus does not multiply on food or in water, and hence cold storage cannot be used to control its presence.

Early descriptions of epidemic non-bacterial gastroenteritis or 'winter vomiting disease' can be traced back to 1929. But the causative agent was not identified until an outbreak in 1968 at an elementary school in Norwalk (Ohio, US) when the immune electron microscopy was used on faecal samples. The non-enveloped virus was the first virus to be associated with acute gastroenteritis outbreaks and was named after the location of the outbreak (Kapikan *et al.* 1972). The practice of naming the virus after the outbreak area continued until recently, and subsequently, the serologically distinct groups of viruses had location names such as Hawaii virus, Snow Mountain virus and Toronto virus. This practice has now ceased since the geographic names might be disputed because of national and/or local sensitivities.

The Norovirus has a positive-strand RNA genome of 7.3–7.6 kb encoding for a set of non-structural proteins at the 5'-end and the major structural protein at the 3'-end. The 28–35 nm viral particles have a buoyant density of 1.39–1.40 g/mL in CsCl, and the capsid is composed of 180 copies of a single structural protein. Unfortunately, the Norovirus cannot be cultivated in the laboratory which until the development of molecular studies severely hampered our understanding and detection of this virus.

For genotyping, the common sequences used for multiple alignments are region A, which is part of the polymerase gene, and region C, which is part of the capsid gene. See Section 1.12.5 for reference to variants in circulation in 2007.

Norovirus gastroenteritis is self-limiting and mild. The characteristics of Norovirus gastroenteritis are as follows:
- Low-grade fever
- Projectile vomiting
- Diarrhoea
- Headache
- Chills, muscle aches and weakness may also occur

The illness is considered mild and self-limiting. The infectious dose is unknown but presumed to be low. A mild and brief illness usually develops 24–48 hours after contaminated food or water is consumed and lasts for 24–60 hours; see Table 1.18. The virus invades and damages the gastrointestinal tract, resulting is mucosal lesions of the small intestinal tract. The inflammatory response in the lamina propria (Figure 4.2) is similar to that caused by rotavirus. Severe illness or hospitalisation is very rare. The virus is shed in faeces (as high as 10^8 virus particles/g faeces) and vomit, starting during the incubation period and lasting up to 10 days or longer. Thirty percent of cases shed viruses for up to 3 weeks after infection. Norovirus infections are highly contagious with over 45% attack rate. Many different Norovirus genotypes circulate in the general population causing sporadic cases and outbreaks. The strain sequence of different isolates within an outbreak being nearly identical, unless there has been a common sewage-contaminated source. Immunity to the specific infecting genotype only lasts for a short period, and hence the development of long-lasting vaccine may not be possible.

Routes of transmission include the following:

- *Water (most common source of outbreaks)*: Municipal supplies, wells, recreational lakes, swimming pools, and water stored aboard cruise ships.
- *Contaminated food*: Directly by food handler or by contaminated washing and irrigation water.
- Projectile vomiting (common feature of Norovirus infection) and aerosolised vomit is a vehicle of transmission via environment contamination and airborne transmission (Marks *et al.* 2000).

Noroviruses can survive outside the host, and are more resistant to common disinfectants than bacteria. Norovirus is reported to be resistant to pH as low as 2.7, heating (60°C, 30 minutes) and free chlorine (up to 1 mg free chlorine/L, 30 minutes exposure time). The virus is inactivated at higher concentrations (>2 mg/L free chlorine). Only limited studies on the effect of other disinfectants on Norovirus have been undertaken due to the lack of *in vitro* cultivation methods. Drinking water processing can reduce the amount of Norovirus by 4 log orders.

Filter-feeding shellfish and salad ingredients are the foods most often implicated in Norovirus outbreaks. Shellfish can filter up to 10 gallons (38 L) of water/hour and hence concentrate viruses from the surrounding water. Unfortunately, depuration is about 100 times less effective at removing viral pathogens as bacterial pathogens (Power & Collins 1989). Due to being heat resistant, even steaming may not prevent Norovirus gastroenteritis. Although feline caliciviruses (a model for Norovirus, though from a different genera) are completely inactivated in shellfish after a heat treatment of 70°C for 5 minutes or boiling for 1 minute, ingestion of raw or insufficiently steamed clams and oysters poses a high risk for infection with Norovirus.

The inability to grow the virus in the laboratory severely hinders studies of inactivation and detection. Nevertheless, control of the Norovirus in food is the same as for HAV. It is achieved through avoidance of food contamination from contaminated water and infectious personnel.

Berg *et al.* (2000) reported a multistate outbreak associated with oysters that was traced to a sick oyster harvester whose vomit had been disposed overboard in the oyster bed area. Foods other than shellfish are contaminated by ill food handlers. Unfortunately, little work has been done on the thermal inactivation of enteric viruses in foods other than shellfish.

Frozen raspberries have been the source of several other Norovirus outbreaks in recent years. Hjertqvist *et al.* (2006) reported on four outbreaks in Sweden where contaminated raspberries were the source of infection (Table 4.13). All of the suspected raspberries were from the same brand, from the same distributor, being imported from China. A high percentage of the outbreaks with 'unknown etiology' are probably outbreaks of viral gastroenteritis. One retrospective study confirmed the presence of Noroviruses in 90% of non-bacterial gastroenteritis outbreaks (Fankhauser *et al.* 1998). In another retrospective review of 712 foodborne outbreaks from 1982 to 1989 which had been reported to CDC, 48% met the epidemiologic criteria for outbreaks of Noroviruses. Recently, the 'Food-borne viruses in Europe' network was established

Table 4.13 Norovirus outbreaks in Sweden associated with raspberry desserts

Outbreak (2006)	Description	Incidence and symptoms	Laboratory analysis
23 June	15 people at a private party	1–2 days after the party, 12 people were ill with gastroenteritis	Stool samples from two patients were positive for Norovirus by PCR
	Ate homemade cake containing cream and raspberries		
2 August	11 people at a family gathering	Following day, 10 people were ill with signs of Norovirus infection	No faecal samples due to significant time before authorities notified
	Cheese cake and raspberries eaten by everyone	There were also two secondary cases	Raspberries were of the same brand as the first outbreak, and were imported from China
24 August	School class, 30 pupils aged 13	26 or 27 August, parent reported ill child	Faecal samples from two children were positive for Norovirus. Raspberries were the same brand as the previous outbreaks
	Drinks prepared with raspberries	Cohort study showed 12 children had been ill	
		Incubation period 24–36 hours, symptoms of vomiting, fever, diarrhoea and headache. Duration of illness 1–3 days	
25 August	Meeting with nine participants. Guests ate a homemade raspberry parfait	Eight people became ill	One stool sample was tested, and found positive for Norovirus. The berries were the same brand as in previous outbreaks

(Section 1.12.5) which monitors Norovirus activity in ten countries, and reports on strain variations; see Web Resources section for URL.

4.4.2 Hepatitis A

HAV is classified with the genus *Hepatovirus* and is in the *Piconaviridae* family (Cuthbert 2001). Many other picornaviruses cause human disease, including polioviruses, coxsackieviruses, echoviruses and rhinoviruses (cold viruses). HAV is a non-enveloped spherical virus composed of a single molecule of RNA (7.5 kb) surrounded by a small (27 nm diameter) protein capsid and a buoyant density in CsCl of 1.33 g/mL. A large polyprotein is encoded by the RNA and is processed into four structural and seven non-structural proteins by proteases. The genome encodes the structural proteins at the 5'-end, and non-structural proteins at the 3'-end. Seven genotypes have been recognised, four of which occur in humans. Genetic diversity has been used to study food- and waterborne outbreaks.

The infectious dose is unknown but is presumed to be 10–100 virus particles. The incubation period is dependent upon the number of infectious particles consumed and varies from 10 to 50 days (mean 30 days) with virus shedding starting 10–14 days before the onset of symptoms. Typical symptoms of hepatitis A are as follows:

- Fever
- Headache
- Chills
- Malaise
- Loss of appetite
- Nausea
- Jaundice
- Dark urine
- Light-coloured stools
- Abdominal pain in liver area

The signs of hepatitis start 1–2 weeks after the general symptoms of fever, headache and malaise; see Table 1.18. The virus can be spread from early in the incubation period to about a week after jaundice develops, especially in the middle of the incubation period which is before the symptoms develop. The virus enters the body through the intestinal tract and is transported to the liver where it replicates in the hepatocytes and may be shed in the bile, though it has also been isolated from spleen, kidneys, tonsils and saliva. Diagnosis of acute hepatitis A is by the detection of IgM antibody to the capsid proteins. Only one serotype has been observed among HAV isolates collected from various parts of the world, though the incidence varies considerably among and within countries (Mast & Alter 1993).

HAV has a worldwide distribution occurring in both epidemic and sporadic fashions. HAV is primarily transmitted by person-to-person contact through faecal contamination, but common-source epidemics from contaminated food and water also occur. Poor sanitation and crowding facilitate transmission. Outbreaks of HAV are common in institutions, crowded house projects, and prisons and in military forces in adverse situations. In developing countries, the incidence of disease in adults is relatively low because of exposure to the virus in childhood. Outbreaks under these conditions are rare since young children generally remain asymptomatic. Most individuals of 18 and older demonstrate an immunity (IgG anti-HAV antibodies) that provides lifelong protection against reinfection.

Many infections with HAV do not result in clinical disease, especially in children. When disease does occur, it is usually mild and self-limiting, and recovery is complete in 1–2 weeks. Occasionally, the symptoms are severe and convalescence can take up to 6 months. Patients suffer from feeling chronically tired during convalescence, and their inability to work can cause financial loss. Approximately 15% of patients require hospitalisation and up to 20% may have a relapse and be impaired for as long as 15 months. The fatality rate is 0.5–3% in adults 15–40 years old, and >1.8% in >59 years old people. There is an unusually high fatality rate in pregnant women (15–20%). Whilst HAV infection is regarded as one of the most severe of foodborne illnesses, in most cases recovery is complete and lifelong immunity is acquired.

HAV is excreted in faeces (up to 10^9 particles/g) of infected people and can produce clinical disease when susceptible individuals consume contaminated water or foods. Cold cuts and sandwiches, fruits and fruit juices, milk and milk products, vegetables, salads, shellfish and iced drinks are commonly implicated in outbreaks. Water, shellfish and salads are the most frequent sources. Since shellfish can filter up to 10 gallons (38 L) of water/hour, they can concentrate the virus 100-fold from contaminated water and depuration is about 100-fold less effective at removing viral pathogens as it is for bacterial pathogens (Power & Collins 1989). Subsequently, the use of indicator bacteria for faecal bacteria in food and water is unreliable as an indicator for viral contamination. Heating shellfish to >85°C for 1 minute will result in a 4 log decrease in infectious HAV (Lees 2000). Although heat inactivation of viruses has been studied in few foods other than shellfish, it is known that a heat treatment for milk of <0.5 minute at 85°C is sufficient to cause a 5 log drop in infectious HAV particles.

HAV is resistant to pH as low as 1, survives 60°C for 1 hour and is relatively resistant to free chlorine when organic mater is present. It is more resistant to heat and drying than most other human enteric viruses. It can survive for long periods (weeks) in seawater, water and on-work surfaces. For example, only a 2 log decrease was detected after 50 days at 4°C in water. Similarly, the virus infectivity at 4°C on an aluminium surface only decreased by 2 log orders after 60 days. Subsequently, contamination of foods by infected workers in food processing plants and restaurants is common. More than 1000 virus particles can easily be transferred from faecally contaminated fingers to foods and surfaces (Bidawid et al. 2000). Viruses do not multiply on food; it is simply the vehicle of transmission. HAV outbreaks can be large. The largest reported outbreak was in Shanghai when 250 000 people were infected with HAV after consuming contaminated clams (Halliday et al. 1991). A smaller outbreak in the United States was due to contaminated frozen strawberries distributed in school lunches. These caused 213 people to developed hepatitis A (Hutin et al. 1999). Hepatitis A infection has also been linked to the consumption of lettuce, diced tomatoes and raspberries. The efficacy of disinfectants to kill HAV and other viruses on the surface of fruits and vegetables is poorly studied due to the greater technical difficulties in laboratory procedures compared with bacteria.

A vaccine is available for HAV and contacts can be treated within 2 weeks of exposure. However, there is no consensus as to whether all food handlers should be vaccinated; instead personal hygiene is emphasised.

A fundamental problem with investigating HAV and other viral cases of foodborne infection is because of the long incubation period, any suspected food is unlikely to be still available for analysis. In addition, there are no methods available for the routine analysis of food for viruses. Molecular methods involving PCR are being developed for the detection of HAV in water and clinical specimens. These could also be applicable to food samples.

4.4.3 Hepatitis E
HEV is the major etiological agent of enterically transmitted non-A, non-B hepatitis worldwide. It is spherical, non-enveloped, single-stranded RNA that is approximately 32–34 nm in diameter. HEV has provisionally been classified in the *Caliciviridae* family. This virus enters the body through water or food, especially raw shellfish that has been contaminated by sewage. Anti-HEV activity has been determined in the serum of a number of domestic animals such as pigs in areas with a high endemicity of human infection, indicating that this may be an emerging zoonosis (Meng et al. 1997).

4.4.4 Rotaviruses
Rotaviruses are classified with the *Reoviridae* family. They have a genome consisting of 11 double-stranded RNA segments surrounded by a distinctive two-layered protein capsid. Particles are 70 nm in diameter and have a buoyant density of 1.36 g/mL in CsCl. Six serological groups have been identified, three of which (Groups A, B and C) infect humans (Desselberger 1998).

Rotavirus gastroenteritis in the developed world is usually a self-limiting, mild to severe disease characterised by vomiting, watery diarrhoea and low-grade fever (Ciarlet & Estes 2001, Hart & Cunliffe 1999). However, worldwide rotavirus is one of the worst infectious organisms and kills up to 2000 children a day; 20–40 deaths/year in the United States compared with 15 000–30 000/year in Bangladesh. Rotavirus kills 1 in 200–250 Indian children before they are 5 years old.

The infective dose is presumed to be 10–100 infectious viral particles (Lundgren & Svensson 2001, Shaw 2000). Because a person with rotavirus diarrhoea often excretes large numbers of virus (10^8–10^{10} infectious particles/mL of faeces), infectious doses can be readily acquired through contaminated hands, objects or utensils (Ponka *et al.* 1999). Asymptomatic rotavirus excretion has been well documented and may play a role in perpetuating endemic disease.

Rotaviruses are transmitted by the faecal–oral route. Person-to-person spread through contaminated hands is probably the most important means by which rotaviruses are transmitted in close communities such as pediatric and geriatric wards, day care centres and family homes. Infected food handlers may contaminate foods that require handling and no further cooking, such as salads, fruits and hors d'oeuvres (Richards 2001). Rotaviruses are quite stable in the environment and have been found in estuary samples at levels as high as 1–5 infectious particles/gal. Sanitary measures adequate for bacteria and parasites seem to be ineffective in endemic control of rotavirus, as similar incidence of rotavirus infection is observed in countries with both high and low health standards (Fleet *et al.* 2000, Inouye *et al.* 2000, Mead *et al.* 1999, Sethi *et al.* 2001). Rotaviruses survive in aerosols for up to 9 days at 20°C, persist on surfaces (aluminium, china, polystyrene) for 1–60 days with 100-fold reduction in infectivity and over 1 year in mineral water at 4°C (Beuret *et al.* 2000, Biziagos *et al.* 1988).

Group A rotavirus is endemic worldwide. It is the leading cause of severe diarrhoea among infants and children, and accounts for about half of the cases requiring hospitalisation. Almost all children acquire serum antibodies by 5 years of age. Over 3 million cases of rotavirus gastroenteritis occur annually in the United States. In temperate areas, it occurs primarily in the winter, but in the tropics, it occurs throughout the year. The number attributable to food contamination is unknown. Group B rotavirus, also called adult diarrhoea rotavirus or ADRV, has caused major epidemics of severe diarrhoea affecting thousands of persons of all ages in China. Group C rotavirus has been associated with rare and sporadic cases of diarrhoea in children in many countries. However, the first outbreaks were reported from Japan and England. The incubation period ranges from 1 to 3 days. Symptoms often start with vomiting followed by 4–8 days of diarrhoea. Temporary lactose intolerance may occur. Recovery is usually complete. However, severe diarrhoea without fluid and electrolyte replacement may result in severe diarrhoea and death. Childhood mortality caused by rotavirus is relatively low in the United States, with an estimated 100 cases/year, but reaches almost 1 million cases/year worldwide. Association with other enteric pathogens may play a role in the severity of the disease.

The virus has not been isolated from any food associated with an outbreak, and no satisfactory method is available for routine analysis of food. Control of the organism is the same as for hepatitis A and Norovirus, that is the prevention of food contamination by polluted water and an infected food handler.

4.4.5 Small round viruses, astroviruses, SLVs, adenoviruses and parvoviruses

Although the rotaviruses and Noroviruses are the major causes of viral gastroenteritis, a number of other viruses have been implicated in outbreaks, including astroviruses, enteric adenoviruses and parvovirus. Viruses with smooth edge and no discernible surface structure are designated 'featureless viruses' or 'small round viruses' (SRVs). These agents resemble enterovirus or parvovirus and may be related to them.

Astroviruses cause sporadic gastroenteritis in children under 4 years of age and account for about 4% of the cases hospitalised for diarrhoea. Most American and British children over 10 years of age have antibodies to the virus. Astroviruses are unclassified viruses which contain a single positive strand of RNA of about 7.5 kb surrounded by a protein capsid of 28–30 nm diameter. A five or six pointed star shape can be observed on the particles under the electron microscope. Mature virions contain two major coat proteins of about 33 kDa each and have a buoyant density in CsCl of 1.38–1.40 g/mL. At least five human serotypes have been identified in England. The Marin County agent found in the United States is serologically related to astrovirus type 5.

Sapporo viruses infect children between 6 and 24 months of age and account for about 3% of hospital admissions for diarrhoea. By 6 years of age, more than 90% of all children have developed immunity to the illness. Sapporo viruses are classified in the family *Caliciviridae*. They are sometimes termed 'typical caliciviruses' to separate them from the other foodborne infectious caliciviruses (Noroviruses). They contain a single strand of RNA surrounded by a protein capsid of 31–40 nm diameter. Mature virions have cup-shaped indentations which give them a 'Star of David' appearance in the electron microscope. The particle contains a single major coat protein of 60 kDa and has a buoyant density in CsCl of 1.36–1.39 g/mL. Four serotypes have been identified in England.

The enteric adenovirus causes 5–20% of the gastroenteritis in young children, and is the second most common cause of gastroenteritis in this age group. The virus may be transmitted by the respiratory route as well as oral–faecal, person–person and contaminated food. By 4 years of age, 85% of all children have developed immunity to the disease. Enteric adenoviruses represent serotypes 40 and 41 of the family *Adenoviridae*. These viruses contain a double-stranded DNA surrounded by a distinctive protein capsid of about 70 nm diameter. Mature virions have a buoyant density in CsCl of about 1.345 g/mL.

Parvoviruses belong to the family *Parvoviridae*, the only group of animal viruses to contain linear single-stranded DNA. The DNA genome is surrounded by a protein capsid of about 22 nm diameter. The buoyant density of the particle in CsCl is 1.39–1.42 g/mL. The Ditchling, Wollan, Paramatta and cockle agents are candidate parvoviruses associated with human gastroenteritis. Shellfish have been implicated in illness caused by a parvo-like virus. Parvo-like viruses have been implicated in a number of shellfish-associated outbreaks, but the frequency of disease is unknown. A mild, self-limiting illness usually develops 10–70 hours after contaminated food or water is consumed and lasts for 2–9 days. The clinical features are milder but otherwise indistinguishable from rotavirus gastroenteritis. Co-infections with other enteric agents may result in more severe illness lasting a longer period of time. Only a parvovirus-like agent (cockle) has been isolated from seafood associated with an outbreak.

4.4.6 Human enteroviruses

The human enteroviruses are classified within the *Picornaviridae* and can be found in human faeces and hence sewage. They include the polioviruses, Groups A and B coxsackieviruses, echoviruses and enteroviruses serotypes 68–71. The virus is a smooth, round, non-enveloped particle ~27 nm diameter with a positive sense, single-stranded non-segmented RNA genome. Poliovirus can be transmitted by contaminated water and unpasteurised milk. Human enteroviruses are the most common viruses detected in shellfish. A small number of outbreaks have been associated with coxsackie and echoviruses. Unfortunately, as previously stated, there is no correlation between the presence of coliforms (indicator organisms) and human enteroviruses. Poliovirus has been proposed as a suitable viral pathogen indicator since it is more easily detected than other human enteroviruses, because of techniques developed alongside vaccine production.

4.5 Seafood and shellfish poisoning

There are a number of causes of food poisoning originating from seafoods and shellfish (Table 4.14). Seafood poisoning can be caused by ciguatera poisoning, a toxin from microalgae which has been accumulated in fish flesh. It is believed that scombroid poisoning was due to the consumption of fish flesh containing high levels of histamine from bacterial histidine dehydrogenase activity on mackerel and similar fish. Incriminated bacteria are *Morganella morganii*, *Proteus* spp., *Hafnia alvei* and *Klebsiella pneumonia*. However, this has not been conclusively proven since human volunteers ingesting histamine do not always produce the characteristic scombroid poisoning symptoms. Possibly, there are other biogenic amines present. Shellfish poisoning is caused by a group of toxins produced by planktonic algae (dinoflagellates, in most cases) upon which the shellfish feed. The toxins are accumulated and sometimes metabolised

Table 4.14 Micro-organisms and toxins associated with seafood and shellfish poisoning

Disease	Micro-organism	Toxin	Incriminated seafood
Paralytic shellfish poisoning	*Alexandrium catenella* *Alexandrium tamarensis* Other *Alexandrium* spp. *Pyrodinium bahamense* *Gymnodinium catenatum*	Saxitoxin Neosaxitoxin Gonyautoxins Other saxitoxin derivatives	Mussels, oysters, clams, planktonivorous fish
Diarrhoeic shellfish poisoning	*Dinophysis fortii* *Dinophysis acuminata* *Dinophysis acuta* *Dinophysis mitra* *Dinophysis norvegica* *Dinophysis sacculus* *Prorocentrum lima* Other *Prorocentrum* spp.	Okadaic acid Dinophysis toxin Pectenotoxin Yessotoxin	Mussels, scallops, clams, oysters
Neurotoxic shellfish poisoning	*Gymnodinium breve*	Brevetoxins	Oysters, mussels, clams, scallops
Amnesic shellfish poisoning	*Pseudonitzschia pungens*	Domoic acid	Mussels
Ciguatera	*Gambierdiscus toxicus* *Ostroepsis lenticularis*	Ciguatoxin Maitotoxin Scaritoxin	Reef-associated fish
Scombroid poisoning	*Morganella morganii*, *Proteus* spp., *Hafnia alvei*, *K. pneumoniae*, and other bacteria capable of decarboxylating amino acids to biogenic amines	Histamine and other biogenic amines	Scombroid fish species, Mahi mahi, bluefish, tuna, sardines

Adapted from ICMSF 1988.

by the shellfish. Ingestion of contaminated shellfish results in a wide variety of symptoms, depending upon the toxins(s) present, their concentrations in the shellfish and the amount of contaminated shellfish consumed. Paralytic shellfish poisoning is better characterised than the symptoms associated with diarrhoeic shellfish poisoning, neurotoxic shellfish poisoning and amnesic shellfish poisoning. All shellfish (filter-feeding molluscs) are potentially toxic. Paralytic shellfish poisoning is generally associated with a range of shellfish: mussels, clams, cockles and scallops. Neurotic shellfish poisoning is associated with shellfish harvested along the Florida coast and the Gulf of Mexico. Diarrhoeic shellfish poisoning is associated with mussels, clams, oysters and scallops, whereas amnesic shellfish poisoning is only associated with mussels.

4.5.1 Ciguatera poisoning
Ciguatera poisoning is characterised by the following symptoms:
- Prickling of the lips, tongue and throat
- Headache
- Severe pain in arms, legs and eyes
- Impaired vision
- Skin disorders: blisters, stinging sensation and erythema

The majority of cases recover within days to weeks and mortality is low. A lipid-soluble ciguatera toxin has been identified (Figure 4.17, Murata *et al.* 1990). This polyether is similar in structure to the brevetoxins and is known to affect thermoregulation and sensory, motor, autonomic and muscular activities. The scaritoxin which is less potent than ciguatera toxin may actually be a derivative of ciguatera toxin. A third toxin has been implicated in scombroid poisoning called maitotoxin. It activates Ca^{2+} channel, releases neurotransmitters and increases the contraction of smooth, cardiac and skeletal muscles. Since there are a variety of symptoms, it is possible that there is a range of toxins involved in the illness.

4.5.2 Scombroid poisoning
Scombroid poisoning symptoms are as follows:
- Metallic, sharp or peppery taste in the mouth
- Intense headache
- Dizziness
- Nausea and vomiting
- Facial swelling and flushing
- Epigastric pain
- Rapid and weak pulse

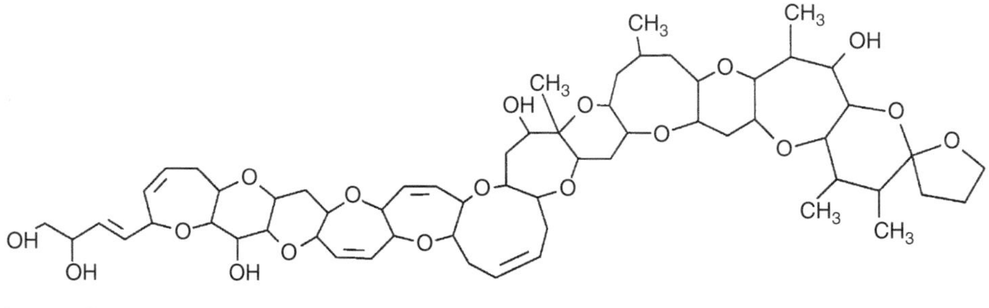

Figure 4.17 Ciguatoxin structure.

- Itching skin
- Burning throat and difficulty in swallowing

Usually, recovery is within 12 hours. Fatalities are rare, and are usually due to other predisposing factors. Initially, it was believed that the symptoms were caused by histamine intoxication. The histamine is being bacterially produced during storage. However, studies on human volunteers have failed to show a correlation between amounts of histamine in fish flesh and scombrotoxicosis. Therefore, the causative agent is still unknown.

4.5.3 Paralytic shellfish poisoning
Symptoms of paralytic shellfish poisoning are primarily neurological and include tingling, burning, numbness, drowsiness, incoherent speech and respiratory paralysis. Paralytic shellfish poisoning is due to 20 toxins which are all derived from saxitoxin produced by dinoflagellates (Figure 4.18).

4.5.4 Diarrhoeic shellfish poisoning
Diarrhoeic shellfish poisoning is normally a mild gastrointestinal disorder, that is nausea, vomiting, diarrhoea and abdominal pain accompanied by chills, headache and fever. Onset of diarrhoeic shellfish poisoning may be as little as 30 minutes to 3 hours depending on the dose of toxin ingested. The symptoms may last as long as 2–3 days. Recovery is complete with no aftereffects and the disease is generally not life-threatening. It is probably caused by high molecular weight polyethers, including okadaic acid, the dinophysis toxins, the pectenotoxins and yessotoxin produced by dinoflagellates.

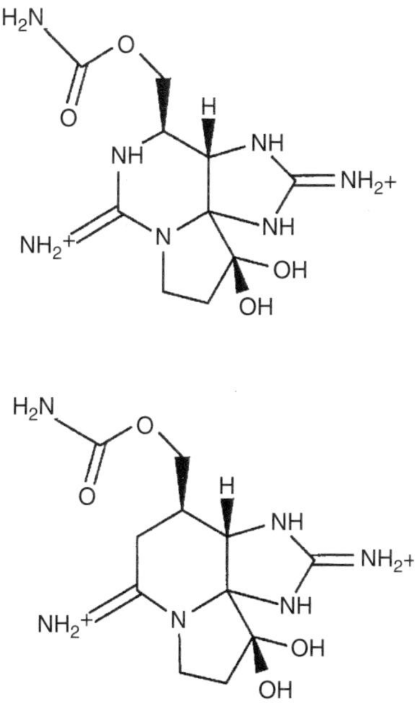

Figure 4.18 The structure of saxitoxin (upper) and neosaxitoxin (lower).

4.5.5 Neurotoxic shellfish poisoning

Neurotoxic shellfish poisoning causes both gastrointestinal and neurological symptoms, including tingling and numbness of lips, tongue and throat, muscular aches, dizziness, reversal of the sensations of hot and cold, diarrhoea and vomiting. Onset of neurotoxic shellfish poisoning occurs within a few minutes to a few hours. The illness duration is fairly short, from a few hours to several days. Recovery is complete with few aftereffects; no fatalities have been reported. The poisoning is due to exposure to a group of polyethers called brevetoxins produced by dinoflagellates.

4.5.6 Amnesic shellfish poisoning

Amnesic shellfish poisoning is characterised by gastrointestinal disorders (vomiting, diarrhoea, abdominal pain) and neurological problems (confusion, memory loss, disorientation, seizure, coma). The gastroenteritis symptoms occur within 24 hours, whereas neurological symptoms occur within 48 hours. The toxicosis is particularly serious in elderly patients, and includes symptoms reminiscent of Alzheimer's disease. All fatalities to date have involved elderly patients. It is caused by the presence of an unusual amino acid, domoic acid, as the contaminant of shellfish from diatoms (Figure 4.19). Hence, it is also known as domoic acid poisoning.

4.6 Foodborne pathogens: eucaryotes

There are a number of protozoans and other eukaryotic organisms which are of major significance to human health. Their principal modes of transmission include ingestion of contaminated water, and food as well as person-to-person contact.

The protozoans *Giardia lamblia*, *Cryptosporidium spp.* and *Entamoeba histolytica* cause persistent diarrhoea. As usual, those most at risk are young child and those with immune deficiency. *Toxoplasma gondii* infection is a particular hazard to pregnant women and the foetus, and is a significant cause of morbidity or mortality. Raw or undercooked meat and raw vegetables contaminated with cat faeces are recognised routes of infection.

The epidemiology of infections by *Cyclospora*, *Toxoplasma* and *Cryptosporidium* is not fully understood. However, although their life cycles are known to differ, a common feature is that they all require passage through an animal or human intestinal tract. Cysts or spores are dispersed via faeces, which can contaminate water. Subsequently, raw fruits and vegetables can become contaminated during irrigation, washing or handling.

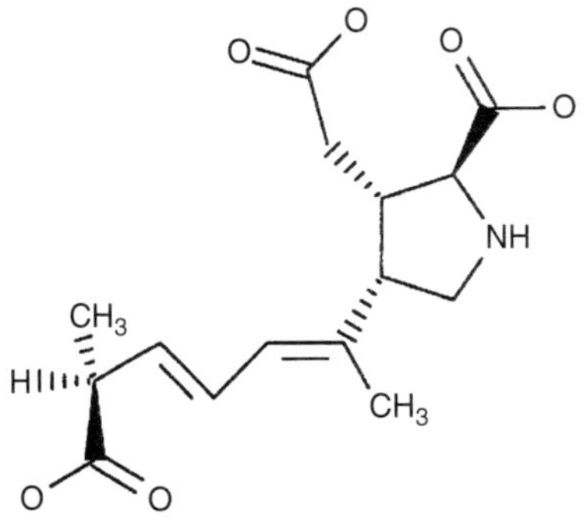

Figure 4.19 The structure of domoic acid.

4.6.1 *Cyclospora cayetanensis*

This coccidian parasite occurs in tropical waters worldwide and causes a watery and sometimes explosive diarrhoea in humans. The first known human cases were reported in 1979. It was initially associated with waterborne transmission but has also been linked to the consumption of raspberries, lettuce and basil or basil-containing products. *Cyclospora* infects the small intestine. The incubation period is 1 week after the ingestion of the contaminated food and the agent is shed in the faeces for more than 3 weeks. *Cyclospora* is spread by people who ingest contaminated water or food. Direct person-to-person transmission is unlikely because excreted oocysts require time (days to weeks) under favourable environmental conditions before they become infectious (i.e. sporulate). The natural host for this parasite is unknown. Nevertheless, contaminated water which is used for irrigation and pesticide application as well as poor worker hygiene are the most likely routes of contamination. The illness lasts from a few days to a month or longer. Relapses may occur one or more times.

Typical symptoms of cyclosporiasis are as follows:

- Watery diarrhoea
- Frequent, sometimes exposive bowel movements
- Loss of appetite
- Substantial weight loss
- Abdominal bloating
- Increased gas and abdominal cramps
- Nausea
- Vomiting
- Muscle aches
- Low-grade fever
- Fatigue

4.6.2 *Cryptosporidium parvum*

Cryptosporidium was first described in the early 1900s. However, it was not considered to be medically important until the 1970s when it was linked to diarrhoea in calves. The first human case was in 1976, who had been receiving immuno-suppressive chemotherapy. Further human cases were also immuno-suppressed or immuno-deficient. In the 1980s, the first cases of cryptosporidiosis in people with normal functioning immune systems and exposure to calves were reported. Also, outbreaks in child day centres and waterborne cryptosporidiosis were recorded. In 1993, a massive waterborne outbreak in Milwaukee (Wisconsin, US) due to contaminated water caused more than 400 000 illnesses (Box 4.16).

Symptoms of crytosporosis include fever, diarrhoea, abdominal pain and anorexia. The disease usually lasts less than 30 days but may be prolonged in immuno-deficient individuals and can cause death. Genotyping can distinguish between human and bovine strains, although humans are susceptible to bovine strains. The mode of transmission is faecal to oral via water and foods. Reservoirs include man and domestic animals, including cattle. Oocysts can survive and remain infective in the environment for long periods of time. Although they are resistant to chemicals used to purify drinking water, they are removed by filtration.

4.6.3 *Anisakis simplex*

Anisakiasis is an infection of the human intestinal tract caused by the ingestion of raw or undercooked fish containing larval stages of the nematodes *Anasakis simplex* or *Pseudoterranova decipiens*. Infections caused by the latter roundworm are not a serious threat to human health,

Box 4.16

Organism

Cryptosporidium parvum

Food source

Raw milk, drinking water, apple cider

Onset of symptoms

2–14 days

Acute symptoms and chronic complications

Diarrhoea, nausea, vomiting, abdominal pain, sometimes influenza-like illness and fever

but those caused by *A. simplex* are more serious in that this agent penetrates the gastrointestinal tissue and causes disease that is difficult to diagnose. The primary hosts are warm-blooded marine mammals such as seals, walruses and porpoises. Their larvae pass via krill to fish such as cod, pollack, halibut, rockfish, flat fish, mackerel, salmon and herring (Box 4.17).

4.6.4 *Taenia saginata* and *Taenia solium*

Tapeworm (cestodes) infections in man are caused by the beef tapeworm (*T. saginata*) and pork tapeworm (*T. solium*) through the consumption of raw or undercooked pork and beef,

Box 4.17

Organisms

Anisakis spp., *T. solium, T. saginata*

Food source

Anisakis spp.: Raw fish dishes, sushi, sashimi, herring, cebiche
T. solium: Raw or undercooked beef
T. saginata: Raw or undercooked pork

Onset of symptoms

Anisakis spp.: Days to weeks
T. solium, T. saginata: Days to years

Acute symptoms and chronic complications

Anisakis spp.: Ulceration of stomach wall, nausea, vomiting
T. solium, T. saginata: Nervousness, insomnia, anorexia, weight loss, abdominal pain

respectively, which is infected with the parasite larvae. Both tapeworms are obligate parasites of the human intestine. The organisms have a complex life cycle. The larval form is ingested in infected beef or pork meat and develops into the adult form (several metres in length) which attaches to the intestinal wall and produces hundreds of proglottids which are shed in the faeces. The proglottids produce eggs in the environment and in the intestine which are the main vector in cattle and pig infection. In the otherwise healthy adult, taeniasis is not severe and may not show any symptoms. However, people infected with *T. solium* are also potential transmitters of cysticercosis. Cerebral lesions due to infection can lead to neurological and mental symptoms. Taeniasis is endemic in some countries such as Ethiopia, Kenya, Zaire, former Yugoslavia and central Asia. Breaking the life cycle of the organism is the main control measure, through thorough meat inspection and adequate cooking (>60°C) (Box 4.17).

4.6.5 *Toxoplasma gondii*
T. gondii is the causative agent of toxoplasmosis which can be found in undercooked and raw meats such as pork, lamb, beef and poultry. The primary hosts are cats, and human infection takes place when contact is made with their faeces. This can also take place by the ingestion of raw or undercooked meat from intermediate hosts such as rodents, swine, cattle, goats, chicken and birds. Toxoplasmosis in humans often produces mononucleosis-type symptoms, but transplacental infection can result in foetal death if it occurs early in pregnancy. The organism causes hydrocephalus and blindness in children, the symptoms being less severe in adults. In immuno-compromised individuals, it can cause pneumonitis, myocarditis, meningoencephalitis, hepatitis or chorioretinitis or combinations of these. Cerebral toxoplasmosis is often seen in AIDS patients. Proper cooking of meat will kill the organism. The incidence of the disease worldwide is unknown but it is reported to be the most common parasitic infection in the United Kingdom (Box 4.18).

Box 4.18

Organism

T. gondii, T. spiralis

Food source

T. gondii: Raw or undercooked meat, vegetables, goat's milk, food and water contaminated with cat faeces
T. spiralis: Pork, horse, wild boar, game

Onset of symptoms

T. gondii: 5–23 days
T. spiralis: 8–21 days

Acute symptoms and chronic complications

T. gondii: Abortion or stillbirth, brain damage
T. spiralis: Nausea, vomiting, diarrhoea, fever

4.6.6 Trichinella spiralis

T. spiralis is a nematode worm that is transmitted principally through raw or undercooked pork, as well as game (e.g. bear, wild boar). It causes trichinosis, also known as trichiniasis and trichinelliasis. It is mainly associated with the ingestion of contaminated pork. The organism is a roundworm (nematode) which lives in the upper two-third part of the small intestine. The female is viviparous (giving birth to living larvae) which are deposited into the mucosa. About 1500 larvae are produced before the adult is expelled due to the host's immune system. The larvae are spread around the body via the bloodstream. Those which invade striated muscle (other than cardiac muscle) continue to develop. The larvae are digested by humans and subsequently invade the duodenal mucosa and become adult in 3–4 days, whereupon the life cycle continues.

Main symptoms of trichinosis are as follows:
- *First week*: Enteritis
- *Second week*: Irregular fever (39–41°C), muscle pain, difficulty in breathing, talking or moving
- *Third week*: High fever, swollen eyelids, muscle pain
- *Fourth week*: Fever and muscle pains subside

The larvae can be killed by a number of methods: heating to 65.5°C (150°F); freezing at −15°C (5°F) for 3 weeks; or freezing at −30°C (−22°F) for 1 day.

4.7 Mycotoxins

Mycotoxins are of concern because of their acute toxicity and their potential carcinogenicity. They are the toxic products of certain microscopic fungi which, in some circumstances, develop on or in foodstuffs of plant or animal origin (Table 4.15). They are ubiquitous and widespread at all levels of the food chain. Hundreds of mycotoxins have been identified and are produced by some 200 varieties of fungi. Mycotoxins are secondary metabolites which have been responsible for major epidemics in man and animals. Mycotoxins are produced by the fungal genera *Aspergillus*, *Fusarium* and *Penicillium*. These fungi are ubiquitous and are

Table 4.15 Toxicity of mycotoxins

Mycotoxin	Food	Fungus species	Biological affect	LD_{50} $((mg\ kg)^{-1})$
Aflatoxins	Maize, groundnuts, milk	*Asp flavus*, *Asp. parasiticus*	Hepatotoxin, carcinogen	0.5 (dog), 9.0 (mouse)
Cyclopiazonic acid	Cheese, maize, groundnuts	*Asp. flavus*, *Pen. aurantiogriseum*	Convulsions	36 (rat)
Fumonisin	Maize	*Fus. moniliforme*	Equine encephalomalacia, pulmonary oedema in pigs	Unknown
Ochratoxin	Maize, cereals, coffee beans	*Pen. verrucosum*, *Asp. ochraceus*	Nephrotoxin	20–30 (rat)
Zearalenone	Maize, barley, wheat	*Fus. graminearum*	Oestrogenic	Not acutely toxic

Adapted from Adams & Motarjemi (1999), with kind permission from World Health Organization. Geneva.

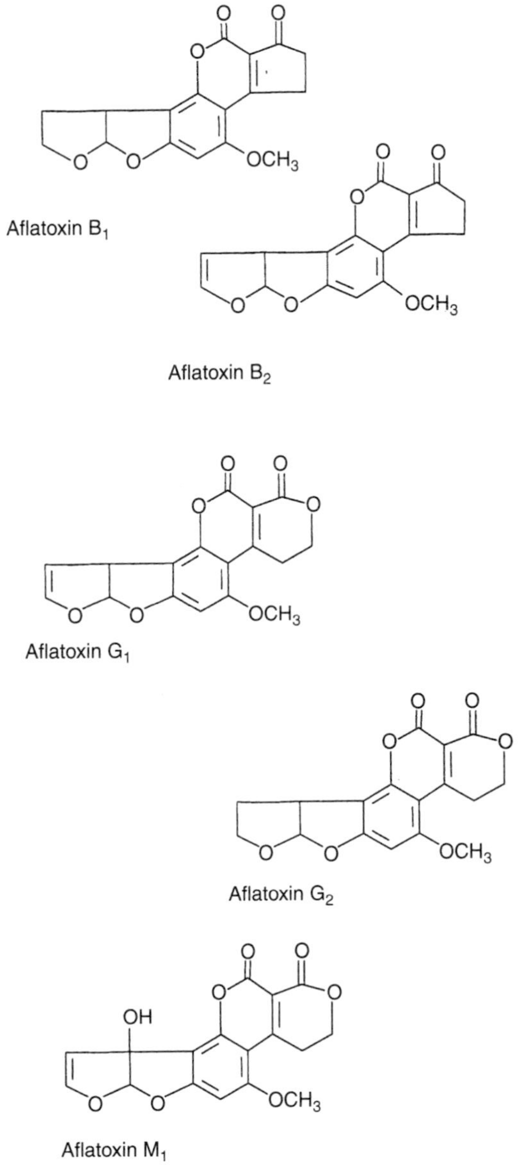

Aflatoxin B₁

Aflatoxin B₂

Aflatoxin G₁

Aflatoxin G₂

Aflatoxin M₁

Figure 4.20 The structure of aflatoxins.

part of the normal flora of plants. Annually, over 1000 million tonnes of cereals are at risk of mycotoxin contamination.

The aflatoxins (produced by *Aspergillus* spp.) range from single heterocyclic rings to six- or eight-membered rings (Figure 4.20). *Penicillium* produces a range of 27 mycotoxins such as patulin (an unsaturated lactone) and penitrem A (nine adjacent rings composed of 4–8 atoms). Ergotism, alimentary toxic aleukia, stachybotryotoxicosis and aflatoxicosis have killed thousands of humans and animals in the past century.

Control of mycotoxins is very difficult as it is due to preharvest invasion by the fungi through seeds, soil or even air. Adequate drying and storage are useful provided there are good farm management practices beforehand. UV screening procedures (for fluorescence of the aflatoxins) are of use in corn, cottonseed and figs, but not peanuts since they autofluoresce.

There are four types of toxicity:
- *Acute*: Resulting in liver or kidney damage
- *Chronic*: Resulting in liver cancer
- *Mutagenic*: Causing DNA damage
- *Teratogenic*: Causing cancer in the unborn child

4.7.1 Aflatoxins

The aflatoxins have been studied in more detail than other mycotoxins (Figure 4.20, Table 4.15). The aflatoxins are a group of structurally related toxic compounds produced by certain strains of *Asp. flavus* and *Asp. parasiticus* under favourable conditions of temperature and humidity. These fungi grow on certain foods and feeds, resulting in the production of aflatoxins. The most pronounced contamination has been encountered in tree nuts, peanuts and other oilseeds, including corn and cottonseed. The major aflatoxins of concern are designated B_1, B_2, G_1 and G_2 by the blue (B) or green (G) fluorescence given when viewed under a UV lamp. These toxins are usually found together in various foods and feeds in various proportions. However, aflatoxin B_1 is usually predominant and is the most toxic. When a commodity is analysed by thin-layer chromatography, the aflatoxins separate into the individual components in the order given above; however, the first two fluoresce blue when viewed under ultraviolet light and the second two fluoresce green. Aflatoxin M is a major metabolic product of aflatoxin B1 in animals and is usually excreted in the milk and urine of dairy cattle and other mammalian species that have consumed aflatoxin-contaminated food or feed. Lifetime exposure to aflatoxins in some parts of the world, commencing *in utero*, has been confirmed by biomonitoring.

Aflatoxins produce acute necrosis, cirrhosis and carcinoma of the liver in a number of animal species. No animal species is resistant to the acute toxic effects of aflatoxins; hence, it is logical to assume that humans may be similarly affected. A wide variation in LD_{50} values has been obtained in animal species tested with single dose of aflatoxins. For most species, the LD_{50} value ranges from 0.5 to 10 mg/kg body weight. Animal species respond differently in their susceptibility to the chronic and acute toxicity of aflatoxins. The toxicity can be influenced by environmental factors, exposure level and duration of exposure, age, health and nutritional status of diet. Aflatoxin B_1 is a very potent carcinogen in many species, including non-human primates, birds, fish and rodents. In each species, the liver is the primary target organ of acute injury. Metabolism plays a major role in determining the toxicity of aflatoxin B_1. This aflatoxin requires metabolic activation to exert its carcinogenic effect and these effects can be modified by induction or inhibition of the mixed function oxidase system.

The discovery of aflatoxins in the 1960s led to extensive surveying of koji moulds for mycotoxin production. Although under laboratory conditions mycotoxins can be produced by *A. oryzae*, *A. sojae* and *A. tamari*, no aflatoxins have been demonstrated in commercial production strains (Table 3.1, Trucksess *et al.* 1987). The moulds used in cheese manufacture have also been tested for toxin production. *P. roqueforti* produces trace amounts of patulin, roquefortine C, whereas *P. camembertii* produces low levels of cyclopiazonic acid. These toxins are only produced under laboratory-induced stress conditions and it is reported that levels in cheese are 'extremely low' (Rowan *et al.* 1998).

In well-developed countries, aflatoxin contamination of food rarely occurs at levels that could cause acute aflatoxicosis. Instead, human toxicity studies have focused on their carcinogenic potential. Studies in Africa and Southeast Asia have shown an association between the incidence of hepatoma and dietary aflatoxin intake. Bowers *et al.* (1993) and Merican *et al.* (2000) studied the possible link between aflatoxin exposure, hepatitis B prevalence and primary liver cancer in China. The best estimates of lifetime cancer potency of aflatoxin in the United States were 9 and 230 $(mg/kg/day)^{-1}$ for hepatitis B negative and positive populations, respectively. This is lower than the 75 $(mg/kg/day)^{-1}$ estimate for Africa and Southeast Asia due to the greater prevalence of hepatitis B.

4.7.2 Ochratoxins

Ochratoxins are produced by *A. ochraceus*, *Penicillium verrucosum* and *P. viridicatum*. Ochratoxin A is the most potent of these toxins (Figure 4.21). The main dietary sources are cereals but significant levels of contamination may be found in grape juice and red wine, coffee, cocoa, nuts, spices and dried fruits. Contamination may also carry over into pork and pig blood products and into beer. Ochratoxin is potentially nephrotoxic and carcinogenic, the potency varying markedly between species and sexes. It is also tetratogenic and immuno-toxic.

4.7.3 Fumonisins

Fumonisins are a group of *Fusarium* mycotoxins occurring worldwide in maize and maize-based products. Their casual role in several animal diseases has been established. Available epidemiological evidence has suggested a link between dietary fumonisin exposure and human oesophageal cancer in some locations with high disease rates. Fumonisins are mostly stable during food processing.

4.7.4 Zearalenone

Zearalenone (Figure 4.21) is a fungal metabolite mainly produced by *Fusarium graminearium* and *F. culmorum*, which are known to colonise maize, barley, wheat, oats and sorghum. These compounds can cause hyperoestrogenism and severe reproductive and infertility problems in animals, especially in swine, but their impact in public health is hard to evaluate.

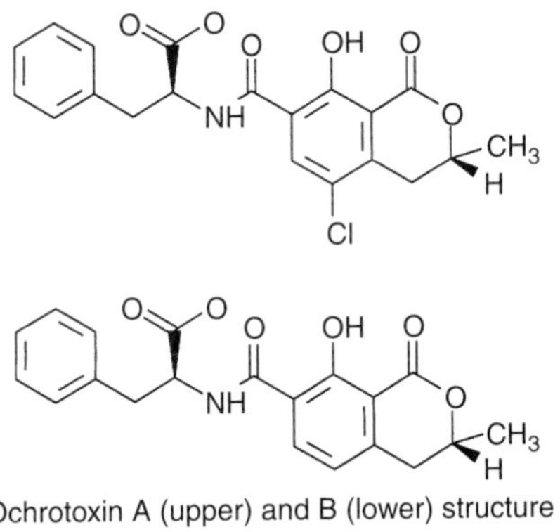

Ochrotoxin A (upper) and B (lower) structure

Figure 4.21 The structure of mycotoxins, other than aflatoxins.

4.7.5 Trichothecenes

Trichothecenes are produced by many species of the genus *Fusarium*. They occur worldwide and infect many different plants, notable of which are the cereal grains, especially wheat, barley and maize. There are over 40 different trichothecenes but the most well known are deoxynivalenol and nivalenol. In animals, they cause vomiting and feed refusal, as well as affecting the immune system. In humans, they cause vomiting, headache, fever and nausea. A large number of deoxynivalenol poisoning cases have occurred in China and India. In food poisoning cases due to deoxynivalenol, severe gastrointestinal disturbance was the primary symptom.

4.8 Emerging and uncommon foodborne pathogens

For various reasons, the number of identified foodborne pathogens has increased (Tauxe 1997, 2002). Emerging (and re-emerging) infections have been defined as 'new, recurring, or drug-resistant infections whose incidence in humans has increased in the last decades or whose incidence threatens to increase in the near future' (NRC 1993). Current surveillance methods only detect pathogens in 40–60% of patients suffering from gastroenteritis (de Wit *et al.* 2001). Hence, there is a considerable scope for new pathogens to be discovered or 'emerge'. Some emerging foodborne disease are well characterised, but are considered 'emerging' because their reporting has recently (10–15 years) become more common. Table 4.16 lists the various emerging food- and waterborne pathogens and toxins.

The (re-)emergence of certain foodborne pathogens and toxins is due to a number of causes:
1 Weakened or collapsed public health infrastructure for epidemic disease control due to economic problems, changing health policies, civil strife and war
2 Poverty, uncontrolled urbanisation and population displacements
3 Environmental degradation and water and food sources contamination
4 Ineffective infectious disease control programmes
5 Newly appeared in the microbial population such as the rise of antimicrobial resistance resulting from inappropriate use of antibiotics including those used in animal production
6 Diseases crossing from animal to human populations with increasing frequency especially as humans exploit new ecological zones and as intensification of animal food production (including fish) and industrialisation of food processing and distribution become global practices
7 Increased potential for spread of disease through globalisation of travel and trade including that of processed and raw foodstuffs of vegetal and animal origin

Table 4.16 Emergent pathogens and toxins

Microbial group	Organisms
Bacteria	Enteroaggregative *E. coli* (EAggEC), *V. cholera*, *V. vulnificus*, *Strep. parasanguinis*, *Mycobacterium paratuberculosis*, *Arcobacter* spp., *Cronobacter* spp. (*Ent. sakazakii*)
Viruses	Hepatitis E
Protozoa	*Cyclospora cayetanensis*, *T. gondii*, *Cryptosporidium parvum*
Hemilinths	*Anisakiasis simplexi* and *Pseudoterranova decipiens*
Prions	Bovine spongiform encephalitis, variant CJD
Mycotoxins	Fumonisins, zearalenone, trichothecenes, ochratoxins

Table 4.17 Factors which can affect the epidemiology of emergent food pathogens

1	Microbial adaptation through natural selection; antibiotics usage can select for antibiotic-resistant strains, that is *S.* Typhimurium DT104
2	New foods and food preparation technologies, that is BSE and nvCJD
3	Changes in host susceptibility, that is increase in population age
4	Changes in lifestyle such as an increased consumption of 'convenience food' and subsequent risk to *L. monocytogenes*
5	Increasing international trade and travel; enabling the rapid spread of pathogens worldwide, that is *E. coli* O157:H7
6	New food vehicles of transmission being recognised, that is *My. paratuberculosis* (plausible)

8 Dispersed to new vehicles of transmission
9 Recently identified due to increased knowledge or methods of identification, though previously widespread

Lindsay 1997

The factors that can affect the epidemiology of emerging food pathogens are given in Table 4.17.

Micro-organisms have evolved many adaptation mechanisms to survive and persist in otherwise 'unfavourable' growth conditions (Lederberg 1997). They can exchange genetic material (e.g. conjugation, transduction and transformation) and hence acquire new gene sequences. Horizontal gene transfer is now recognised as one route by which toxin genes can be distributed to new bacterial strains (i.e. origin of *E. coli* O157:H7). Gene sequences called 'pathogenicity islands' which encode for specific virulence factors can be recognised by their GC% content which differs from the rest of the bacterial genome. Likewise, though less well understood, bacteriophage may interact with host genomes and emerge in new susceptible populations and vehicles, with less than predictable outcomes. Foodborne pathogens often have an animal reservoir from which they can spread to humans, although they frequently do not cause illness in the primary host. Because of the considerable increase in international travel and trade, foodborne pathogens can be spread rapidly worldwide.

There has been considerable concern over the emergence of antibiotic-resistant foodborne pathogens such as *S.* Typhimurium DT104 and *C. jejuni*. One of the most publicised concerns is the use of fluoroquinolone antibiotics in both veterinary and medical practice. The risk assessment for fluoroquinolone resistance in *C. jejuni* is covered in Section 10.2.4.

The emergence of certain food pathogens is due to a number of causes:
• Newly appeared in the microbial population
• Dispersed to new vehicles of transmission
• Rapidly increasing incidence or geographic range; *V. cholerae* in southern US coastal waters in 1991
• Recently identified due to increased knowledge or methods of identification, though previously widespread

Adapted from Van de Venter 1999

4.8.1 Prions

Transmissible spongiform encephalopathies in animals and humans are caused by an unconventional virus or prion. These conditions include scrapie in sheep, bovine spongiform

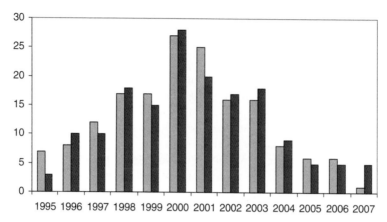

Figure 4.22 Number of vCJD deaths in the United Kingdom. Histogram of number of vCJD cases diagnosed (□) and deaths (■) over the period 1995–2007. Please note that one death in 2003 and two deaths in 2006 were secondary cases linked to blood transfusions. http://www.cjd.ed.ac.uk/figures.htm, accessed on 23 April 2008.

encephalopathy (mad cow disease) in cattle and Creutzveld Jacob Disease in humans. It is commonly accepted that BSE was first caused in Britain when cattle were fed carcass meal from scrapie-infected sheep. It is also accepted that humans contracted the non-classic form of CJD called 'variant CJD' (vCJD) after consuming cattle meat, in particular nerve tissue. The average age at death from vCJD is 28.5 (range 14–74). The median period between onset and diagnosis is 328 days, and death in 413 days. It has been proposed that the BSE-contaminated milk may cause sporadic CJD.

On 20 March 1996, the Secretary of State for Health in the British Government announced that the most likely cause of ten new cases in humans was from the ingestion of meat from cattle suffering from bovine spongiform encephalopathy (BSE; Will *et al.* 1996). This resulted in the Specified Offal Ban which stopped the recycling of potentially infectious material (such as the spinal chord) to cattle through food supplements and greater inspections on abattoirs. These restrictions and the decrease in BSE cases mean that there should now be almost no exposure to the UK public of the infectious agent.

Evidence that the infectious agent was a prion (abbreviation for 'proteinaceous infectious particles') came from the studies of Collinge *et al.* (1996), Bruce *et al.* (1997) and Hill *et al.* (1997). Prions are modified forms of a normal protein called PrP^c which is referred to as PrP^* or PrP^{Sc}. The proteins accumulate in the brain causing holes or plaques and the subsequent clinical symptoms leading to death. The ultimate number of vCJD cases in the United Kingdom has been a matter of controversy. Some groups claimed the number could not be estimated (Ferguson *et al.* 1999), whilst Thomas & Newby (1999) estimate that the value will not exceed 'a few hundred, and is most likely to be a hundred or less', whereas Cousens *et al.* (1997) gave a value of 80 000. There have been 163 vCJD deaths out of 166 cases in the United Kingdom. It appears that the peak was passed in 2000 with 28 deaths (see Figure 4.22), although there could be further peaks in other genetic groups.

4.8.2 *Cronobacter* spp.

The genus *Cronobacter* is associated with infections of immuno-compromised individuals especially neonates, as well as adults with underlying illness. Formerly, *Ent. sakazakii* was used to describe these bacteria which were believed to be closely related to *Enterobacter cloacae*.

However, detailed analysis has led to a recent (2008) taxonomic revision, with the naming of the new genus Cronobacter which is currently split into five species in the genus *Cronobacter* (Iversen *et al.* 2008). It is therefore difficult to refer to the many previous studies with respect to which *Cronobacter* species was being studied. Nevertheless, the majority of isolates are *Cronobacter sakazakii*, and although clinical isolates are found in all described species, neonates cases have been linked to *Cr. sakazakii*, *Cr. malonaticus* and *Cr. turicensis*. In neonates, the infections include necrotising enterocolitis, meningitis and bacteraemia. Bowen & Braden (2006) considered 46 neonatal cases and reported that the symptoms of very low birth weight neonates tend to be bacteraemia, whereas those of birth weight ca. 2000 g tended to suffer from meningitis. The number of neonatal infections from 1958 to date reported globally exceeds 109, but it is likely the number of cases is underestimated. Due to the association of some *Cronobacter* spp. cases with contaminated powdered infant formula, this product has come under considerable attention with regard to its microbial safety. Consequently, the microbiological criteria for these products have been revised, and stricter limits have been adopted. This issue is covered later in Section 10.7 with respect to the FAO/WHO risk assessments.

Ent. sakazakii was designated a unique species in 1980 and had previously been referred to as 'yellow-pigmented *Ent. cloacae*' (Farmer *et al.* 1980). In 2008, it was further divided into *Cr. sakazakii*, *Cr. malonaticus*, *Cr. turicensis*, *Cr. muytjensii* and *Cr. dublinensis*. It is inevitable that when a previously understudied group of bacteria is investigated, taxonomic revisions will occur. The difficulty is for regulatory authorities to ensure that the appropriate control measures and detection methods are implemented to protect human health. The yellow pigment description was used in the early isolation methods of the FDA and ISO, but is now recognised as inaccurate as non-pigmented strains can be isolated using other methods (Fanning & Forsythe 2008). The bacterium has been implicated in a rare yet severe form of neonatal meningitis with a high mortality rate (40–80%, Caubilla-Barron *et al.* 2007). It is also associated with neonatal necrotising enterocolitis and septicaemia (Townsend *et al.* 2007, 2008). Reconstituted powdered infant formula has been implicated in some outbreaks and sporadic cases of *Cronobacter* spp. infection. It can survive in the desiccated state for over 2 years (Caubilla-Barron & Forsythe 2007). The organism is ubiquitous in the environment, and also causes infections in adults who are immuno-compromised. However, it is the infection of neonates which has raised the profile of this bacterium.

Cronobacter spp. has been isolated from 0% to 12% dried infant formula samples (FAO/WHO 2006a), but has never been reported at levels greater than 1 cell per gram. Therefore, an increased risk of neonatal infection from contaminated reconstituted feeds may be attributed to temperature abuse. The organisms' minimum growth temperature is ~5°C, and the doubling time is ~40 minutes at room temperature and 5 hours at 10°C. The FAO/WHO have recommended that these powders are reconstituted at >70°C to reduce the number of any enteric pathogens to an acceptable number. However, this raised temperature can cause vitamin losses and clumping. A temperature of ~55°C or less is frequently practiced in neonatal units and at home, which do not result in any significant loss in bacterial viability. Subsequent incubation at room temperature or prolonged storage in refrigerators can result in significant bacterial multiplication and contribute to infant infections. The FAO/WHO have produced a risk assessment for *Cronobacter* spp. and *Salmonella* in infant formula (FAO/WHO 2004, 2006b), and an online risk model has been developed. See Section 10.7 for more detail. In addition, other *Enterobacteriaceae* and *Acinetobacter* species have been categorised as 'Causality plausible, but not yet demonstrated' with respect to causing neonatal infection through contaminated formula powder, and further monitoring has been advised. The FAO/WHO (2004, 2006b) also produced a third category of 'Organisms causality less plausible, or not yet demonstrated'. This covers *B. cereus, Cl. botulinum, Cl. difficile, Cl. perfringens, L. monocytogenes, St. aureus* and coagulase-negative staphylococci. For a more extensive coverage of *Cronobacter*, the reader should refer to the author's web page (http://www.wiley.com/go/forsythe) and the recent ASM publication (Farber & Forsythe 2008).

4.8.3 Mycobacterium paratuberculosis and pasteurised milk, an emerging pathogen?

Mycobacterium paratuberculosis (commonly referred to as 'MparaTB') is a sub-species of *Mycobacterium avium*. It causes the chronic enteritis in cattle known as Johne's disease. In North America and Europe, the organism is highly prevalent in a subclinical form in dairy herds and domestic livestock such as sheep and goat. It can also infect a range of wild animals such as rabbits and deer. The symptoms are diarrhoea, weight loss, debilitation and, since it is incurable, death. The prevalence of Johne's disease in the United States is 2.6% of the dairy herd; this is comparable to the 2% of cattle being clinically infected in England (Çetinkaya *et al.* 1996). *My. paratuberculosis* is excreted at 10^8 cfu/g and has been isolated from milk of asymptomatic carriers at a level of 2–8 cfu/50 mL milk (Sweeney *et al.* 1992). It has been reported that the organism can survive pasteurisation conditions of 72°C and 15 seconds (Chiodini & Hermon-Taylor 1993, Grant *et al.* 1996, Stabel *et al.* 1997). Hence, it has been proposed that the organism is the causative agent of the human equivalent called 'Crohn's disease' (Acheson 2001, FSA 2001, Hermon-Taylor 2001, 2009, Hermon-Taylor *et al.* 2000). Crohn's disease is a gastrointestinal disease in a highly debilitating chronic inflammation of gastrointestinal tract in humans. Most commonly, the distal ileum and colon are affected. It is a lifelong disease with no cure and usually affects young people. The highest incidence rate is in the age group of 15–24 years. Genetic and immunologic factors may have an important role in the occurrence of the disease. A multicentre European study has reported an incidence rate of 5.6 per 100 000 individuals per year (Shivananda *et al.* 1996).

The link between Crohn's disease and *My. paraTB* proposal is controversial because (i) the organism has not been detected from commercial pasteurised milk sources and (ii) it is not the consensus of opinion that Crohn's disease is caused by *My. paratuberculosis*. If it is proven, by ongoing research (as of the date of this book), that *My. paratuberculosis* is transferred from infected cattle to humans, resulting in Crohn's disease via pasteurised milk, then the standard pasteurisation time and temperature will have to be re-evaluated. Milk and water are considered as potential sources of exposure. Other dairy products, for example cheese as well as meat, are also considered a potential vehicle for this organism. Another feature of this organism, which makes its control difficult, is its ability to survive in the environment as well as methods of culturing partly due to its slow growth. The problem with studying this organism is that it is very difficult to grow in the laboratory. Colony formation is only just visible after 4 weeks' incubation and requires confirmatory tests. Hence, large-scale experiments are extremely time-consuming. Surveys carried out in the United Kingdom have shown the presence of the organism in a small proportion of raw as well as commercially available pasteurised milk samples (FSA 2001).

4.8.4 Arcobacter genus

The *Arcobacter* genus was formerly known as aerotolerant campylobacters and *Campylobacter*-like organisms (CLO). Nowadays, the organism has been formally recognised as a separate

1 (a)

Enter the Calculator

To get started, choose a pathogen below. *Salmonella* and shiga toxin-producing *E. coli* O157 (STEC O157) are currently online. Check back as we build the system to include more pathogens and estimates.

Pathogen	CDC estimate of annual number of cases	ERS cost estimate (2006 dollars)
▸ *Campylobacter* (foodborne sources)	2,000,000	
▸ *Salmonella* (all sources)	1,397,187	$2,467.322,866
▸ Shiga toxin-producing *E. coli* O157 (STEC O157) (all sources)	73,480	$445,857,703
▸ Non-O157 shiga toxin-producing *E. coli* (non-STEC O157) (all sources)	31,229	
▸ *Listeria* (all sources)	2,797	

First Time Users

Help is just a click away, Get detailed instructions on how to proceed.

Related Resources

Food Safety briefing room

1 (b)

Data Sets

Foodborne Illness Cost Calculator: STEC O157

Return to table

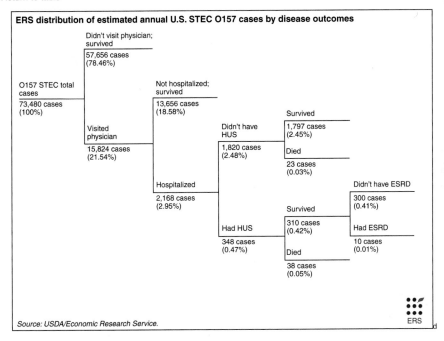

Plate 1 USDA ERS foodborne illness calculator. Website image reproduced with thanks to USDA's Economic Research Service (http://www.ers.usda.gov/Data/FoodbornelIlness).

1 (c)

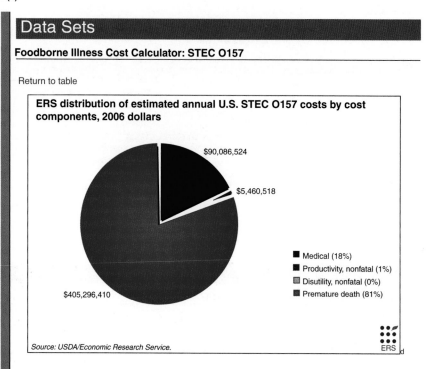

Plate 1 (*Continued*)

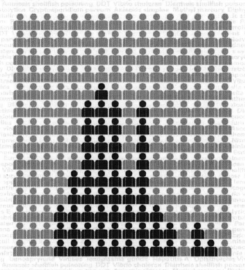

FOODBORNE DISEASE OUTBREAKS

Guidelines for Investigation and Control

World Health Organization

Plate 2 Foodborne disease outbreaks. Guidelines for investigation and control. Reproduced with kind permission from WHO: http://www.who.int/foodsafety/publications/foodborne_disease/fdbmanual/en/index.html.

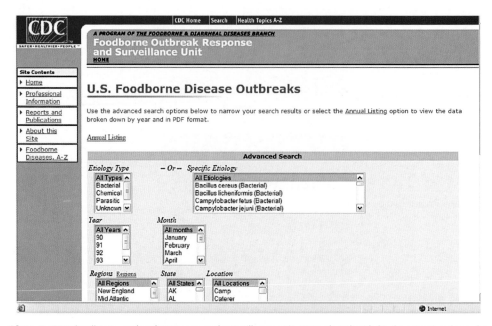

Plate 3 CDC foodborne outbreak response and surveillance unit. Reproduced with kind permission from the CDC: www2.cdc.gov/ncidod/foodborne/fbsearch.asp.

Strain

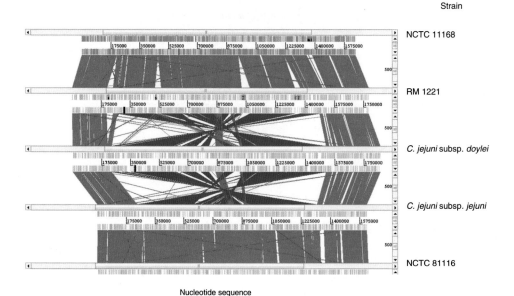

NCTC 11168

RM 1221

C. jejuni subsp. *doylei*

C. jejuni subsp. *jejuni*

NCTC 81116

Nucleotide sequence

Plate 4 Comparison of five *C. jejuni* genome sequences, using WebACT.

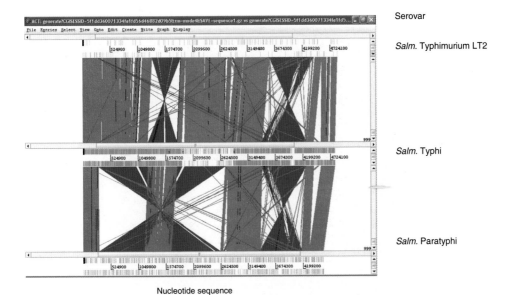

Serovar

Salm. Typhimurium LT2

Salm. Typhi

Salm. Paratyphi

Nucleotide sequence

Plate 5 Comparison of genome sequence for three *Salmonella* serovars, using WebACT.

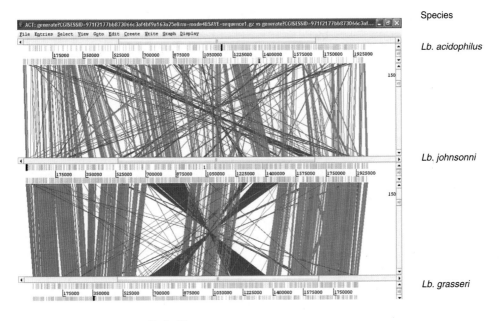

Species

Lb. acidophilus

Lb. johnsonni

Lb. grasseri

Nucleotide sequence

Plate 6 Genome sequence comparison of three species in the *Lactobacillus acidophilus* complex, using WebACT.

Plate 7 *Bacillus cereus* media – mannitol fermentation. © Oxoid Ltd (part of Thermo Fisher Scientific); reproduced with their kind permission.

Structure	Description	Colour
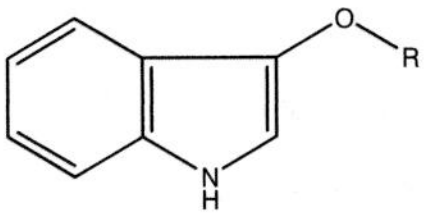	3-Indolyl-R	
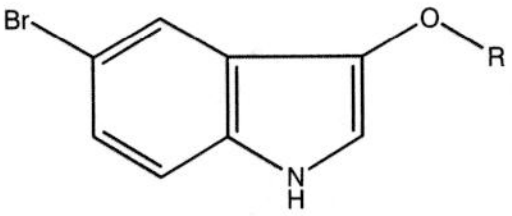	5-Bromo-3-indolyl-R	
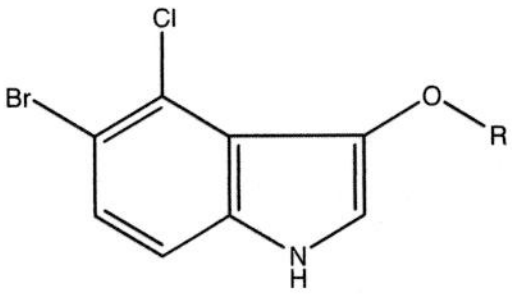	5-Bromo-4-chloro-3-indolyl-R	
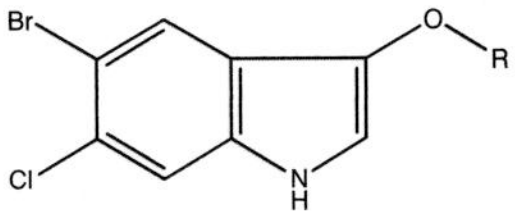	5-Bromo-6-chloro-3-indolyl-R	
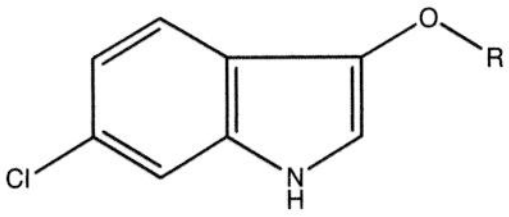	6-Chloro-3-indolyl-R	
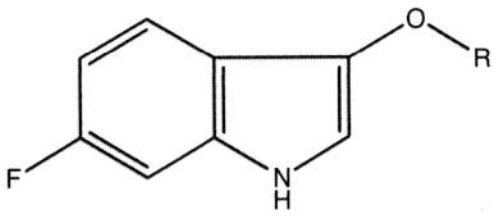	6-Fluoro-3-indolyl-R	

Plate 8 Chromogen diagram. © Oxoid Ltd (part of Thermo Fisher Scientific); reproduced with their kind permission.

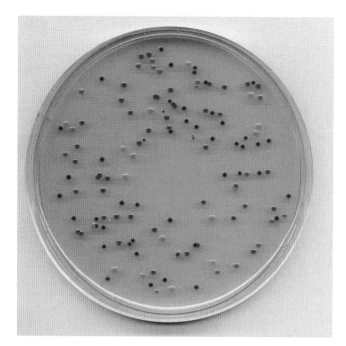

Plate 9 Chromogenic Cronobacter (*E. sakazakii*) agar (DFI). © Oxoid Ltd (part of Thermo Fisher Scientific); reproduced with their kind permission.

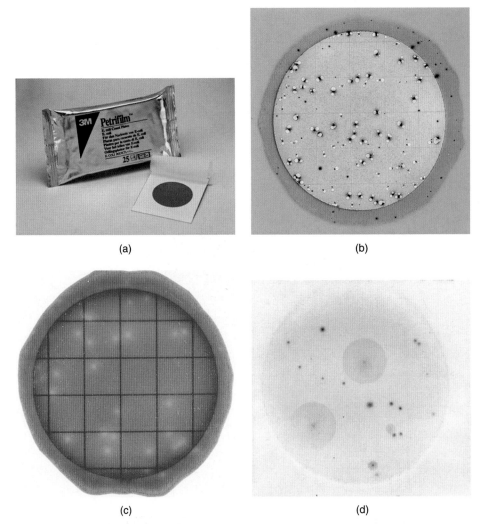

Plate 10 3M™ Petrifilm™. (a) Plate materials. (b) *E. coli*/coliform plate count 9. (c) Rapid coliform 6h plate count. (d) Yeast and mould count. Images copyright: 3M Microbiology, St. Paul, MN, US. Reproduced with permission of 3M™.

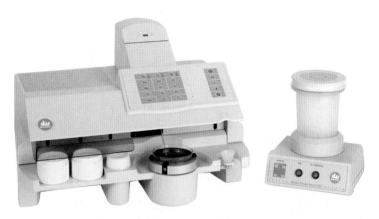

Plate 11 WASP – Spiral Plater. Photograph courtesy of Don Whitley Scientific Ltd.

(a)

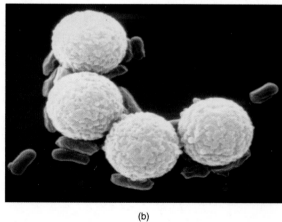

(b)

Plate 12 Immunomagnetic separation. (a) Bead retriever. (b) Bacteria attached to beads. Images courtesy of Invitrogen Ltd.

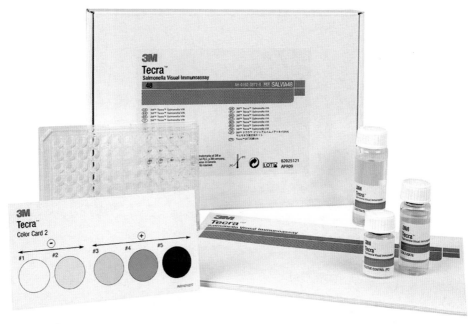

Plate 13 3M™ Tecra™ *Salmonella* VIA kit. Image copyright: 3M Australia Pvt. Ltd. Reproduced with permission.

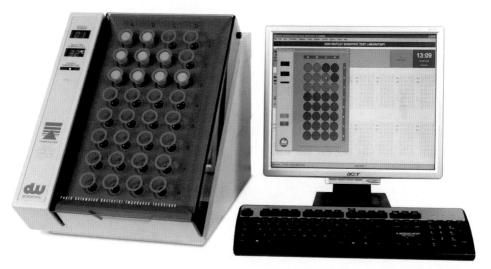

Plate 14 Rapid automated bacterial impedance technique (RABIT). Photography courtesy of Don Whitley Scientific Ltd.

(a)

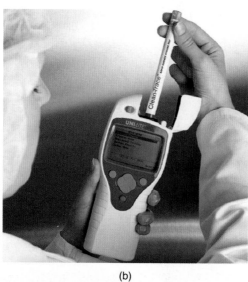

(b)

Plate 15 (a) 3M Clean-Trace ™ Surface ATP Test. Image copyright: 3M Health Care Limited. Reproduced with permission. (b) 3M Clean-Trace ™ NG Luminometer. Image copyright: 3M Health Care Limited. Reproduced with permission.

(a)

(b)

(c)

Plate 16 Microarray analysis. (a) 4 bay hybridisation incubator. (b) 3 plex. (c) 4 plex microarray. © 2008 Roche NimbleGen, Inc. All rights reserved.

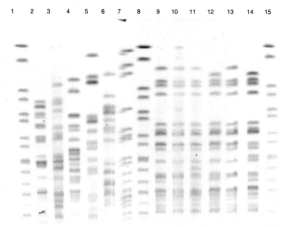

Plate 17 Pulse field gel electrophoresis. Genomic DNA banding pattern of various *Enterobacteriaceae* following SpeI restriction enzyme digestion. Reproduced with permission of Juncal Caubilla-Barron (Nottingham Trent University).

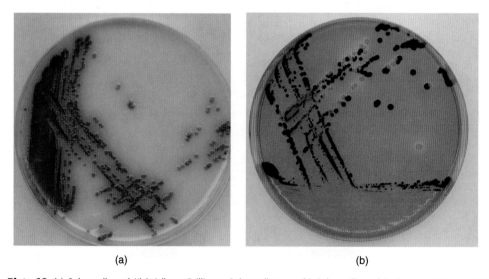

Plate 18 (a) *Salmonella* and *Klebsiella* on Brilliance *Salmonella* agar. (b) *Salmonella* and *Proteus* on XLD agar. © Oxoid Ltd (part of Thermo Fisher Scientific); reproduced with their kind permission.

Plate 19 *L. monocytogenes* on PALCAM agar. © Oxoid Ltd (part of Thermo Fisher Scientific); reproduced with their kind permission.

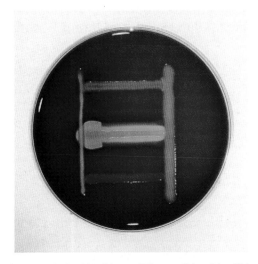

Plate 20 CAMP reaction of *Listeria*. © Oxoid Ltd (part of Thermo Fisher Scientific); reproduced with their kind permission.

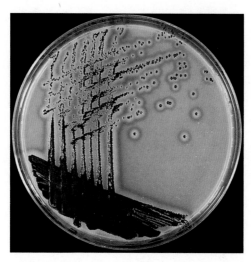

Plate 21 *St. aureus* on Baird Parker and egg yolk agar. © Oxoid Ltd (part of Thermo Fisher Scientific); reproduced with their kind permission.

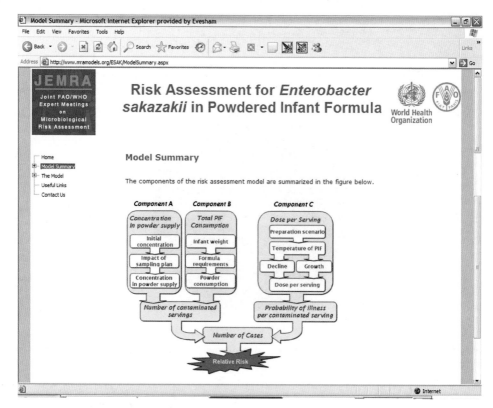

Plate 22 JEMRA risk assessment for Cronobacter spp. (*E. sakazakii*) in powdered infant formula (2004). Reproduced with kind permission of the Food and Agriculture Organisation of the United Nations; sourced from http://www.mramodels.org/ESAK/ModelSummary.aspx.

genus and divided into seven species: *A. butzleri*, *A. cryaerophilus*, *A. skirrowii*, *A. nitrofigilis*, *A. cibarius*, *A. mytili* and *Candidatus* A. sulphidicus.

A. butzleri, *A. cryaerophilus* and *A. skirrowii* are veterinary pathogens causing porcine abortions, whereas *A. nitrofigilis* has been reported only from the roots of *Spartina alterniflora*, a salt marsh plant. *A. butzleri* serotypes 1 and 5 are regarded as the primary human pathogens; however, no epidemiological studies have yet shown the transmission of the organism through the food chain to humans. The situation is, however, reminiscent of *C. jejuni* and *L. monocytogenes*. These organisms which were recognised as veterinary pathogens many years before the medical microbiologists used suitable isolation media to samples from patients suffering from gastroenteritis.

Arcobacters appear resistant to antimicrobial agents typically used in the treatment of diarrhoeal illness caused by *Campylobacter* spp., for example erythromycin, other macrolide antibiotics, tetracycline and chloramphenicol. Isolation of arcobacters requires selective media such as mCCDA and CAT (Forsythe 2006). Identification can subsequently be achieved using 16S rRNA probes (Mansfield & Forsythe 2000a).

The occurrence of arcobacter-related diseases may be underestimated due to the lack of surveillance and optimised detection procedures. This is reminiscent of the situation with *C. jejuni* and *L. monocytogenes* 4b. Both these organisms were initially recognised as veterinary pathogens for several years, prior to the application of specific detection methods by the clinical microbiologists and subsequently the food microbiologists. Examination of human and veterinary clinical specimens for the presence of *Arcobacter* species is rarely performed, and in most cases, suboptimal procedures are used (Houf *et al.* 2001). In addition, little is known about the risk factors for human infection. The most extensive study to date on this matter has been conducted by Vandenberg *et al.* (2004) who studied a total of 67 599 stool samples over an 8-year period. *A. butzleri* was the fourth most common *Campylobacter*-like organism isolated. It was more frequently associated with symptoms of a persistent, watery diarrhoea than *C. jejuni*. For a more detailed consideration of *Arcobacter*, the reader should consult Forsythe (2006).

4.8.5 Nanobacteria
Nanobacteria are the smallest bacteria that have been described. They have been found in marine limestone, freshwater streams, springs, water pipes and caves as well as human and cow's blood (Folk 1999). During their growth, nanobacteria form carbonate apatite on their cell envelope, which resemble the smallest apatite units in the kidney stone. In a study of Finnish patients with kidney stones, 97.25% of cases were positive for nanobacteria (Ciftcioglu *et al.* 1999, Hjelle *et al.* 2000), hence indicating that the organism may be linked with the formation of kidney stones. However, this is lightly controversial with alternative explanations proposed (Cisar *et al.* 2000). Further studies are needed to test this hypothesis and the ecology and route of transmission of these organisms, which could conceivably include food and water transmission.

5 Methods of detection and characterisation

5.1 Prologue

Initially, microbial analysis of food was primarily end product testing, which screened finished products before release. Negative test batches went into distribution, while the positive test batches being re-processed or discarded. Although this approach reduced the release of contaminated products onto the market, it did not help in preventing product contamination or improving production efficiency. These days, the more effective food safety approaches aim to eliminate foodborne pathogens by proactive screening from ingredients to finished product with the Hazard Analysis and Critical Control Point (HACCP) approach; see Chapter 8. By preventing pathogens from entering the production process or reducing them to acceptable levels, and having critical microbial controls in place, then the finished product should be within design specifications. Therefore, current food safety focus is on prevention and process control, and not retrospective control following the detection of processing failure.

Alocilja & Radke (2003) estimate that the pathogen detecting biosensor market was $563 million with annual growth of 4.5%, of which the food industry accounted for $192 million. According to Tom Weschler of Strategic Consultants (http://www.strategic-consult.com), the worldwide food microbiology market in 2005 represented over 629 million tests with a market value in excess of $1.65 billion and a rate of growth of ~7–9%. In 2008, the number of tests was about 738 million, with 80% being general routine tests (i.e. total viable counts), and 138 million for specific pathogens (market value approximately $1 billion). *Salmonella* is the most frequently tested pathogen, followed by *Listeria*, *E. coli* O157 and *Campylobacter*. In 2008, the microbial testing of food represented a market worth over $2 billion. By 2010, this will have grown to 822.8 million tests and a market value of $2.4 billion. Part of the increase is the increasing use of rapid methods (molecular and immunological based) which account for about 35% of the total number of tests. Although they are more expensive per test, they offer a faster throughput than conventional methods. The choice of test varies per organism. For example, conventional methods predominate for *Campylobacter* spp., whereas the opposite is true for *E. coli* O157. Around 68.5 million *Salmonella* tests are carried out annually, but the results may take several days. Hence, there is still a need for even more rapid testing which would benefit both food producers and consumers. This chapter reviews the various methods available to detect the major foodborne pathogens.

Analysing food and environmental samples for the presence of food poisoning and food spoilage bacteria, fungi and toxins is standard practice for ensuring food safety and quality. Despite the considerable collection of detection methods that have been developed, the interpretation of results in food microbiology is far more difficult than that is normally appreciated. Not only is the specificity and sensitivity of the method to be appreciated, but

Table 5.1 Confidence limits associated with numbers of colonies on plates (Cowell & Morisetti 1969)

Colony count	95% Confidence intervals for the count	
	Lower	Upper
3	<1	9
5	2	12
10	5	18
12	6	21
15	8	25
30	19	41
50	36	64
100	80	120
200	172	228
320	285	355

also how representative was the sample which was analysed. This chapter considers the major groups of detection and characterisation methods available to the food microbiologist, and the issues of sampling plans and statistical representation of samples are covered in Chapter 9.

The various reasons for caution in interpreting microbiological results include the following:

- Micro-organisms are in a dynamic environment in which multiplication and death of different species occur at differing rates. This means that the result of a test is only valid for the time of sampling.
- Viable counts by plating out dilutions of food homogenate onto agar media can be misleading if no micro-organisms are cultivated as preformed toxins or viruses may be present. For example, staphylococcal enterotoxin is very heat stable and will persist through the drying process in the manufacture of powdered milk.
- Homogeneity of food, however, is rare, especially with solid foods. Therefore, the results for one sample may not necessarily be representative of the whole batch. However, it is not possible to subject a whole batch of the food to such examination for micro-organisms as there would be no product left to sell.
- Colony counts are only valid within certain ranges and have confidence limits (Table 5.1).

Because of the reasons above, microbiological counts obtained through random sampling can only form a small part of the overall assessment of the product.

There are a number of issues related to the recovery of micro-organisms from food which must be addressed in any isolation procedure:

1 If solid food, then a liquidised homogenate is necessary for dilution purposes.
2 The target organism is normally in the minority of the microbial population.
3 The target organism is present at low levels.
4 The target organism may be physically and metabolically injured.
5 The target organism may not be uniformly distributed in the food.
6 The food may not be of a homogenous composition.

Plate counts are obtained for three purposes and groups of organisms:

1 The basic aerobic plate count (APC) indicates the general microbial load and hence the shelf-life of the product. The APC is very useful in the food industry as the technique is easy to perform and can provide a threshold for acceptance or rejection decisions for samples taken regularly at the same point under the same conditions.

2 The presence of faecal organisms (i.e. *Enterobacteriaceae*) to indicate if the food has been inadequately heat processed or has been mishandled and contaminated post-processing.

3 Specific pathogens associated with the raw ingredients of processed food.

A degree of assurance is only obtained when tests on uniform quantities of representative samples of the food by standard methods prove negative. The methods therefore must be reliable, robust and accredited. These aspects are considered in the following sections. Only representative examples of detection methods can be covered here, and fuller details can be found in various sources such as those listed below. The reader should consult the most recent edition of these for up-to-date protocols.

Useful sources of approved protocols include the following:
- Association of Official Analytical Chemists (AOAC) International
- International Organisation for Standardisation (ISO)
- Food and Drug Administration (FDA) Bacteriological Analytical Manual
- USDA Microbiology Laboratory Guidebook
- Compendium of Methods for the Microbiological Examination of Foods
- Practical Food Microbiology (Roberts & Greenwood 2003).

The URLs for most of these organisations and laboratory manuals are given in the Web Resources section.

The validity of a method can be defined as the ability of the test to do what it is intended to do. This involves determining the 'sensitivity' and 'specificity' of the method. Sensitivity is the probability of a sample testing positive if it is contaminated. Specificity is the probability of a sample testing negative if the organism is truly absent. For many methods available for foodborne pathogens, the sensitivity and specificity are both in the order of 95%.

New detection methods are constantly being announced, but their acceptance by industry depends upon three criteria: speed, accuracy and ease-of-use. Industry requires testing with the shortest time between sample availability and a test result. The accuracy of a method includes its sensitivity, specificity and limit of detection, whereas ease-of-use refers to the level of technical expertise and equipment. Essentially, microbial detection methods can be divided into conventional cell cultivation, and more rapid methods based on immunological and molecular methods (Table 5.2). However, this is a very simplified categorisation as frequently immunological and molecular methods require pre-cultivation to enrich and amplify the target organism to a detectable level.

Conventional cell culture methods normally require several days before a colony of the target organism is visible and therefore an actionable test result is obtained. Despite this limitation, many conventional methods are recognised as approved for international use by ISO or FDA, and are the gold standard procedures by which all others are compared. Conventional methods are relatively easy to use, and require an appropriately trained laboratory staff for sample preparation and interpretation of results. Good laboratory practices (GLPs) are necessary for sample and media preparation, and most of the materials required are disposable consumables. There are a large number of companies which produce dehydrated media, and pre-poured, sample-ready plates.

Immunological methods are mostly based on the antigen/antibody-binding affinity using the enzyme-linked immunosorbent assay (ELISA) technology. Antibodies with high specificity to an antigen of the target organism are anchored to a solid surface, such as the wells of a microtitre tray. These can be stored for a long period of time until required for use. If the target antigen is present in the sample, then it will bind to antibody, and non-target organisms will be removed by subsequent washing steps. A second antibody conjugated to an enzyme (i.e. horseradish peroxidase) is then added which will bind to the pre-existing

Table 5.2 Selection of detection methods for major foodborne pathogens

Organism	Method	Technique	Manufacturer
Campylobacter spp.	Simplate *Campylobacter* CI	Biochemical assay	BioControl
	O.B.I.S. campy	Phenotyping profile	Oxoid Thermo Fisher Scientific
	Dryspot *Campylobacter*	Latex agglutination	Oxoid Thermo Fischer Scientific
	Pathatrix	IMS (sample circulation)	Matrix Microscience
	Campylobacter VIA	ELISA	TECRA/Biotrace
	VIDAS *Campylobacter*	Automated ELFA	bioMérieux
	Accuprobe *Campylobacter*	DNA hybridisation	Gen-Probe
	BAX®	PCR	DuPont Qualicon
	Hybriscan®	Hybridisation	Sigma-Aldrich®
Salmonella serovars	Bactiflow *Salmonella* spp.	Automated immunoassay	AES Chemunex
	FastrAK™	IMS+bioluminescence+phage capture	Alaska
	S.P.R.I.N.T. *Salmonella*	Culture media	Oxoid Thermo Fisher Scientific
	Brilliance™ *Salmonella*	Chromogenic agar	Oxoid Thermo Fisher Scientific
	BBL CHROMagar	Chromogenic agar	BBL CHROMagar
	ISO-GRID *Salmonella* spp.	Hydrophobic grid membrane	Neogen Corp.
	API 20E, ID32 E	Biochemical test kits	bioMerieux
	O.B.I.S. *Salmonella*	Phenotyping profile	Oxoid Thermo Fisher Scientific
	Assurance *Salmonella*	Immunoassay	BioControl
	Anti-*Salmonella* Dynabeads	IMS	In Vitrogen
	Pathatrix	IMS (sample circulation)	Matrix Microscience
	Salmonella Unique	Immunoassay	TECRA/Biotrace
	VIDAS *Salmonella*	Automated ELFA	bioMerieux
	GENE-TRAK *Salmonella*	DNA hybridisation	Neogen Corporation
	MicroSeq	DNA sequence	Applied Biosystems
	PROBELIA Protocol	PCR	Bio-Rad Laboratories
	BAX®	PCR	DuPont Qualicon
	Hybriscan®	Hybridisation	Sigma-Aldrich®
	DuPont™ Lateral Flow System	Lateral flow system	DuPont Qualicon
E. coli	Brilliance™ agar	Chromogenic agar	Oxoid Thermo Fisher Scientific
	TEMPO® EC	MPN culture	bioMérieux
E. coli O157	Harlequin™	Chromogenic agar	LabM
	VIDAS® UP *E. coli* O157		bioMérieux
	Anti-*E. coli* O157 Dynabeads	IMS	In Vitrogen

Table 5.2 (*Continued*)

Organism	Method	Technique	Manufacturer
	Captivate™	IMS	LabM
	FastrAK™	IMS+bioluminescence+ phage capture	Alaska
	Foodproof	PCR	Merck KgaA
	BAX®	PCR	DuPont Qualicon
	DuPont™ Lateral Flow System	Lateral flow system	DuPont Qualicon
St. aureus	TEMPO® STA	MPN culture	bioMérieux
	CHROMagar™ Staph aureus	Chromogenic agar	BBL
	VITEK® Gram-positive	Phenotyping profile	bioMérieux
	RapID™ STAPH PLUS	Phenotyping profile	Oxoid Thermo Fisher Scientific
	BAX®	PCR	DuPont Qualicon
St. aureus toxins	TST-RPLA & SET-RPLA	RPLA	Oxoid Thermo Fisher Scientific
L. monocytogenes	BAX®	PCR	DuPont Qualicon
	DuPont™ Lateral Flow System	Lateral flow system	DuPont Qualicon
	ALOA®	Chromogenic agar	AES Chemunex
	GENE-TRAK Listeria	DNA hybridisation	Neogen Corporation
	VITEK® Gram-positive	Phenotyping profile	bioMérieux
	O.B.I.S. mono	Phenotyping profile	Oxoid Thermo Fisher Scientific
	Hybriscan®	Hybridisation	Sigma-Aldrich®
B. cereus	Duopath® cereus enterotoxins	Lateral flow device	Merck KGaA
	Bacillus-ID	Phenotyping profile	Microgen Bioproducts Ltd
	Phenotype Microarray™	Phenotyping profile	Biolog Inc.
Cl. perfringens	CP ChromoSelect agar	Chromogenic agar	Sigma-Aldrich®
	m-CP	Chromogenic agar	Oxoid Thermo Fisher Scientific
	PET-RPLA	RPLA enterotoxin	Oxoid Thermo Fisher Scientific

cell–antibody–surface complex, forming what is known as an 'antibody–antigen–antibody sandwich'. The presence of target cell is visualised by adding the enzyme's substrate resulting in a colorimetric or fluorescent reaction, the intensity of which determines the presence or absence of the target organism in the original sample. The accuracy of the method depends upon the specificity of the antibodies, and false-positive and false-negative results may occur due to cross-reactivity with related non-target organisms, or the absence of the target antigen in some strains of the target organism. The detection limit of the ELISA method is $\sim 10^4–10^6$

cells and therefore the sample must be enriched in a growth medium (a time-consuming step) to amplify the target organism to a detectable level beforehand. Therefore, the speed of ELISA technology is determined by the target and the duration of the sample enrichment. Immunological kits for various pathogens are commercially available. The cost-per-test for consumables, including enrichments and reagents, is in the \$4–6 range.

Molecular methods are primarily based on the PCR reaction. Being DNA based, the methods can be of highly accurate and are not subject to variation in protein expression as can occur with immunological methods targeting cell surface structures. Nevertheless, the detection is commonly after a time-consuming sample enrichment step in order to increase the target of organisms to $\sim 10^4$. As for the other methods, trained laboratory staff are required for sample analysis. The average test cost is in the \$5–10 range. Although this is more expensive than conventional and immunological tests, since multiple samples can be processed simultaneously using automated procedures, sample analysis productivity is greater.

Other emerging rapid methods are being developed using laser-based technologies such as flow cytometry and Raman spectroscopy in which unique, target-specific spectral signals are developed. In flow cytometry, the laser may excite specific nucleic acid sequences or proteins in the target organism and offers very short detection times. Being able to detect multiple targets simultaneously, giving the desirable speed, accuracy and ease-of-use of reliable real-time results may be achievable due to current developments in biosensors. In the future, multiple organisms and toxins could be analysed in a single test, and the unit could be reusable. Currently, the closest real-time result based analysis is probably ATP testing for hygienic monitoring. Even though this method does not detect any specific pathogens, it is normal practice for >37% of all food processing plants, and over 30 million ATP monitoring tests are performed annually (Weschler, Strategic Consulting). One can only imagine the market for a real-time test which covers at least the major foodborne pathogens.

5.2 Conventional methods

A number of steps are required in order to successfully isolate target organisms from food.
- Choose representative samples to test the batch of ingredients/food.
- Where practical, homogenise the food before sampling. Otherwise, take representative samples from the different phases (liquid/solid). Volumes of food analysed are often 1 g when direct enumeration is used, or 25 g when a large sample size is required for presence/absence testing. Multiple numbers of samples may be required from each batch of food, depending on the appropriate microbiological criteria (Chapter 6).
- Homogenise solid ingredients/food using a Stomacher™ or Pulsifier® machine.
- A pre-enrichment step may be required to allow any injured cells to repair their damage membranes and metabolic pathways. Injury may have occurred during processing (cooking, desiccation, etc.).
- Enrich the target organisms from within the mixed flora using media which encourages the growth of the target organisms and suppresses the growth of other micro-organisms.

Conventional methods are frequently plate counts obtained from homogenising the food sample, diluting and inoculating specific media to detect the target organism (Figure 5.1). The first step is normally preparing a 1:10 dilution of the food. The sample is usually homogenised in order to release attached micro-organisms from the food surface. The methods are very sensitive, relatively inexpensive (compare to rapid methods) but require incubation periods of at least 18–24 hours for visible colony formation.

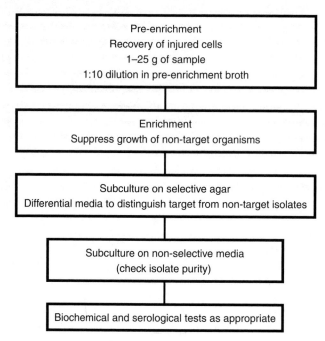

Figure 5.1 General sequence of isolation of foodborne pathogens.

The target organism, however, is often in the minority of the food microbial flora and may be sublethally injured due to processing (cooking, etc.). Therefore, the above procedure is frequently modified to allow a recovery stage for sublethally injured cells, or to enrich for the target organism. For example, the recovery of *Salmonella* spp. from ready-to-eat foods is in the stages of pre-enrichment, enrichment, selection and detection and uses a large sample size (25 g). The procedure is covered in more detail in Section 5.6.2. As referred to above, this approach is 'bacteriological' rather than 'microbiological' in that the presence of toxins, protozoa and viruses will not be revealed.

Specific examples of methods for the detection of key target organisms are given in later sections.

5.2.1 Culture media

Conventional food microbiology requires the use of broths and agar plates to cultivate the target organism(s). These media must meet the organism's nutritional and physiological requirements. Hence, the media must be designed with sufficient protein, carbohydrates, minerals as well as a suitable pH, and incubated under favourable conditions of temperature and oxygen availability for an adequate length of time. In general terms, media whether as a broth or in a solid agar form may be (a) non-selective and able to grow most organisms in a sample, (b) selective to favour the growth of a target organism, or (c) semi-selective and differential where the target organism is presumptively identified based on colony morphology including colour in the presence of other organisms. Such presumptive isolates require further confirmatory tests, which are frequently phenotypic (biochemical profiles). Selective media may be designed to suppress the growth of non-target organisms while enabling the differential growth of the target organism. If the target organism outgrows the non-target organism by >100-fold in broth culture, then there is a good chance that it will be isolated as a pure culture on plating out. Ideally, selective media will be non-inhibitory to the target organism. However, this is not always achieved, but preferably the media will recover more than 50% of the initial population.

Fermentation of carbohydrates is frequently used in differential agar; Plate 7. Violet red bile lactose agar (VRBA) contains the indicator neutral red causing lactose-fermenting organisms (commonly referred to as coliforms) to acquire a pink-red colour. This differentiates them from other bile resistant Gram-negative organisms. Although pH indicators have a long history of use, they suffer in that the surrounding medium may change colour and the observer may be unable to select the target organism when large numbers of colonies have grown on the plate.

The advantage of incorporating fluorogenic and chromogenic substrates into growth media is that they generate brightly coloured or fluorescent compounds after bacterial metabolism (Manafi 2000). Therefore, colonies of the target organism can be spotted on a mixed flora plate, even when considerably outnumbered. The main fluorogenic enzymes substrates in use are based on 4-methylumbelliferone, such as 4-methylumbelliferyl-β-D-glucuronide (MUG) (Figure 5.2). Although these compounds are very specific for the distinguishing enzyme activity, they can diffuse into the medium making the target colony less noticeable in a mixed flora. In addition, they require the medium to be slightly alkaline, and require a UV light source for visualisation. Consequently, the application of MUG substrates has been limited. Chromogenic

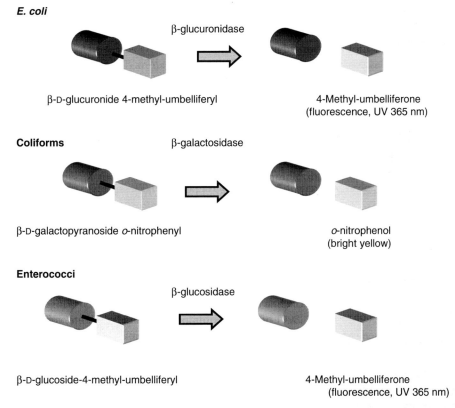

Figure 5.2 Fluorogenic substrates for specific detection of food pathogens.

substrates commonly used are indoxyl substrates, for example 5-bromo-4-chloro-3-indolyl-β-D-glucuronic acid (BCIG), which in the presence of oxygen form coloured aglycones; examples are shown in Plates 8 and 9. In contrast to MUG substrates, these do not diffuse into the agar. TBX agar (Oxoid Thermo Fisher, Merck and LabM) is an example of a chromogenic agar for the detection of *E. coli* which is based on the splitting of BCIG by β-D-glucouronidase enzyme activity to form blue-green colonies. Unlike the majority of *E. coli* strains, those in the serovar group O157:H7 do not ferment sorbitol or rhamnose in the presence of sorbitol. They are β-D-glucouronidase negative and do not grow at 45.5°C. This has enabled the design of specific chromogenic agars such as CHROMagar O157 (CHROMagar) and Fluorocult *E. coli* O157:H7 (Merck) which can differentiate between the two types of *E. coli* based on sugar fermentation by colony colour.

Although very popular by media manufacturers, the indoxyl substrates require the presence of oxygen or other oxidants for colour formation, and may produce toxic intermediates. Recently, new chromogens and fluorogens (ALDOL^{TM}) which do not require oxidation have been developed by Biosynth AG (Switzerland). These undergo intramolecular aldol condensation to form insoluble dyes. Consequently, they can be incorporated into media for use under both aerobic and anaerobic incubation conditions which was not feasible with indoxyl substrates.

Media such as the modified semi-solid Rappaport–Vassiliadis medium and diagnostic semi-solid *Salmonella* (DIASALM) agar have used bacterial motility as a means to enrich the target organism. This principle has been applied to the improved detection of *Salmonella* serovars (as per above examples), *Campylobacter* spp. and *Arcobacter* (de Boer *et al.* 1996, Wesley 1997). The semi-solid Rappaport medium isolates motile *Salmonella* as they migrate through the medium ahead of competing organisms. This medium however will not isolate non-motile *Salmonella* strains.

Compact Dry and the Petrifilm® system (manufactured by 3 M) are alternatives to the conventional agar plate. The Petrifilm® system uses a dehydrated mixture of nutrients and gelling agent on a film. The additional 1 mL of sample rehydrates the gel which enables the colony formation of the target organism. Colony counts are performed as per the standard agar plate method. The throughput of samples is estimated to be double that of conventional agar plates. Petrifilm systems are available for various applications including APCs, yeast, coliforms and *E. coli*; see Plate 10.

5.2.2 Sublethally injured cells

Sublethal injury implies damage to structures within the cells which causes some loss or alteration of cellular functions, the leakage of intracellular material and making them susceptible to selective agents. Changes in cell wall permeability can be demonstrated by the leakage of compounds from the cytoplasm (increase absorbance at 260 nm of culture supernatants) and the influx of compounds such as ethidium bromide and propidium iodide.

Conditions that can generate sublethally injured cells include the following:
- *Moderate heat*: Pasteurisation
- *Low temperature*: Refrigeration
- *Low water activity*: Dehydration
- *Radiation*: Gamma rays
- *Low pH*: Organic and inorganic acids
- *Preservatives*: Sorbate inclusion
- *Sanitisers*: Quaternary ammonium compounds
- *Pressure*: High hydrostatic pressure
- *Nutrient deficiencies*: Clean surfaces

Cells in the exponential phase of growth are generally less resistant than cells in the stationary phase due to the synthesis of stress resistance proteins.

'Metabolic' injury is often taken as the inability to form colonies on minimal salt media while retaining colony forming ability on complex nutrient media. In contrast, 'structural' injury can be taken as the ability to proliferate or survive in media containing selective agents that have no apparent inhibitory action upon non-stressed cells. Injury is reversible by repair but only if the cells are exposed to favourable resuscitation conditions such as a non-selective nutrient-rich medium under optimal growth conditions.

In practical analytical food microbiology, the phenomenon of injury may present considerable problems. Many of the physical treatments including heat, cold, drying, freezing, osmotic activity and chemicals (disinfectants, etc.) may generate injured cells causing variations in plate counts. The injured cells that may remain undetected as selective media usually contains ingredients such as increasing salt concentrations, deoxycholate lauryl sulphate, bile salts, detergents and antibiotics. The injured cells are 'viable' but are not metabolic activity enough to achieve cell division. Subsequently, microbiological examination for quality control can indicate low plate counts, when in fact the sample contains a high number of injured cells. An example of the difference between plate counts on selective and non-selective agar can be seen in Figure 3.7 where food pathogens have been exposed to high pressure.

In food and beverage products, once the stress causing injury is removed, these injured cells are often able to recover. The cells regain all of their normal capabilities which include pathogenic and enterotoxin properties. Therefore, important food poisoning organisms may be undetected by analytical testing, but may cause a major food poisoning outbreak. For these reasons, substantial efforts need to be made to develop improved analytical procedures that will detect both injured and uninjured cells.

In *Salmonella* detection (Section 5.6.2), the sample is incubated overnight in buffered peptone water (BPW) or lactose broth to allow injured *Salmonella* to recover and multiply to detectable levels. However, it is uncertain if BPW is the best recovery medium since other organisms can suppress the growth of low numbers of salmonellas and also there is a problem of 'how do you know if injured *Salmonella* were present in the sample if you do not detect a colony on a plate?'

For other organisms which might be sublethally injured, it has been recommended that food samples should be resuscitated in a non-inhibitory medium for an hour or two, allowing injured cells to resuscitate yet prevent the population size increasing. This generalised approach is far from optimised and leaves plenty of opportunity for oversight in the detection of potentially pathogenic food poisoning organisms. Hence, such techniques need to be appropriately validated.

Extreme environments usually kill the majority of the bacterial population, and can result in the selection of resistant mutants. It is possible that stress induces hypermutability and leads to the greater chance of survivors. The cross-protection effect in which exposure to one stress induces the resistance to another stress is of particular concern for food processing. For example, acid-adapted *E. coli* O157:H7 are more heat tolerant. Similarly, heat shocking *L. monocytogenes* increases its resistance to ethanol and salt.

5.2.3 Viable but non-culturable bacteria (VNC)

It has been proposed that many bacterial pathogens are able to enter a dormant state (Dodd *et al.* 1997). Such cells are not culturable yet remain viable (as demonstrated by substrate uptake) and virulent. Hence, the term 'viable but non-culturable' or VNC was derived. This phenomenon has been shown in *Salmonella* spp., *C. jejuni*, *E. coli* and *V. cholerae*. For example, in the human intestine, previously non-culturable vibrios were shown to regain their ability to multiply (Colwell *et al.* 1996). Therefore, VNC bacterial pathogens pose a potential threat to health and are of considerable concern in food microbiology since a batch of food might be released due to the negative presence of pathogens, yet contain infectious cells.

The VNC state may be induced due to a number of extrinsic factors such as temperature changes, low nutrient level, osmotic pressure, water activity and pH. Of these, the most important factor seems to be temperature changes. Hence, current methods may not be recovering all the pathogens from foods and water. Therefore, alternative end-detection methods need to be further developed, such as those based on immunology (ELISA) and DNA sequences (PCR).

The VNC concept is not accepted by all microbiologists. Some argue that it is a matter of time before we design the most appropriate recovery media and others believe that the cells have self-destructed due to an oxidative burst causing DNA damage (Barer 1997, Barer *et al.* 1998, Bloomfield *et al.* 1998).

5.3 Rapid methods

New pathogen identification technologies are faster than conventional methods and increasingly automated. No single method is appropriate for all circumstances, so the selection of the most appropriate method is necessary. Conventional procedures are by nature labour-intensive and time-consuming. Therefore, a plethora of alternative, rapid methods have been developed to shorten the time between taking a food sample and obtaining a result. These methods aim to either replace the conventional enrichment step with a concentration step (i.e. immuno-magnetic separation) or to replace the end-detection method with one that requires shorter time period (i.e. impedance microbiology and ATP bioluminescence).

Major improvements have been in three areas:
1 Sample preparation
2 Separation and concentration of target cell, toxins or viruses
3 End detection
Sometimes a rapid technique will involve one or more of the above aspects, that is the hydrophobic grid membrane both concentrates and enumerates the organisms on specific detection agar media.

5.3.1 Sample preparation
Agar slides containing selective or non-selective agar can be pressed against the surface to be examined and directly incubated. This obviates the need for sampling and the errors inherent in releasing organisms from cotton wool swabs. Samples can be put onto agar surfaces in a spiral format allowing several effective dilutions to be countable on one plate; see Plate 11.

Another improvement in recent years in sample preparation is the automatic diluter. This enables the operator to take a food sample of approximately 25 g, and then an appropriate volume of diluent is added to give an accurate 1:10 dilution factor.

5.3.2 Separation and concentration of target
Separation and concentration of target organisms, toxin or viruses can shorten the detection time and improve specificity of a test procedure. Common methods include the following:
• Immunomagnetic separation (IMS)
• Direct Epifluorescence Filter Technique (DEFT)
• Hydrophobic grid membrane

IMS

IMS is increasingly being used as it significantly reduces the detection period due to the elimination of a culturing enrichment step. It uses superparamagnetic particles (3–5 μm diameter) which contain γ-Fe_2O_3 and coated with antibodies against the target organism. Hence, the target organism is 'captured' in the presence of a mixed population due to the antigen–antibody specificity. This has removed the need for an overnight enrichment broth incubation period in *Salmonella* isolation, and for *Escherichia coli* O157, the enrichment step is only 6 hours (see Section 5.6.4). A generalised procedure is given in Figure 5.3 and Plate 12. Commercially available IMS kits target key food and water pathogens *Salmonella* spp., *E. coli*

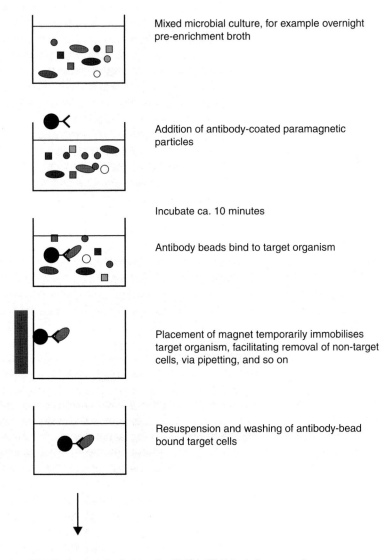

Mixed microbial culture, for example overnight pre-enrichment broth

Addition of antibody-coated paramagnetic particles

Incubate ca. 10 minutes

Antibody beads bind to target organism

Placement of magnet temporarily immobilises target organism, facilitating removal of non-target cells, via pipetting, and so on

Resuspension and washing of antibody-bead bound target cells

Pipetted onto selective media, ELISA, DNA techniques, and so on

Figure 5.3 Immunomagnetic separation technique.

Table 5.3 Applications of immunomagnetic separations

Organism	Application
E. coli O157	Food and water microbiology
Salmonella spp.	
L. monocytogenes	
St. aureus	
Cryptosporidium parvum	
Legionella spp.	
Yersinia pestis	Clinical microbiology
Chlamydia trachomatis	
HIV	
Erwinia chrysanthemi	Plant pathogen detection
Erwinia carotovora	
Saccharomyces cerevisiae	Biotechnology
Mycobacterium spp.	

Adapted from Safarík et al. 1995.

O157:H7, *L. monocytogenes* and *Cryptosporidium* (Table 5.3). IMS can enrich for sublethally injured micro-organisms which would otherwise be missed using the standard enrichment broth and plating procedures. These organisms might be killed in the enrichment broth due to changes in cell wall permeability (Section 5.2.2). Dead cells can be detected using a combined IMS and PCR procedure. For reviews of IMS in medical and applied microbiology, see Olsvik *et al.* (1994) and Safarík *et al.* (1995). IMS can be combined with almost any end-detection method: culture media, ELISA, DNA probe.

The IMS *Salmonella* detection method is as efficient as the enrichment broth selection stage which is the most efficient of the ISO procedures (Mansfield & Forsythe 1996, 2000b; Section 5.6.2). The selective enrichment step (overnight incubation) is replaced with the IMS (10 minutes). Hence, the technique reduces the total time required for sampling and detection by 1 day. In addition, IMS can have a greater recovery of stressed *Salmonella* than ISO protocols.

Commonly, the first step in isolating a pathogen is pre-enrichment to aid the recovery of damaged cells, followed by enrichment to encourage the growth of the target organism and suppress non-target cells. Yet at each stage, only a small aliquot is taken, leaving the bulk of the microbial culture to be discarded. To increase sensitivity, a variation on the IMS technique is to circulate the total volume over the magnet during incubation. This approach has been demonstrated by Pathatrix (Matrix MicroScience) to be very effective and rapid for detecting the presence of *Salmonella* and *E. coli* O157 from a variety of food samples.

DEFT and hydrophobic grid membrane
Membrane filters can be used to shorten the overall detection time because of the following reasons:

1 They can concentrate on the target organism from a large volume to improve detection limits.
2 They can remove growth inhibitors.
3 They can transfer organisms to a different growth media without physical injury through centrifugation and resuspension.

The membranes can be made from nitrocellulose, cellulose acetate esters, nylon, polyvinyl chloride and polyester. Because they are only 10 μm in thickness, the cells can be visualised by directly mounted on a microscope.

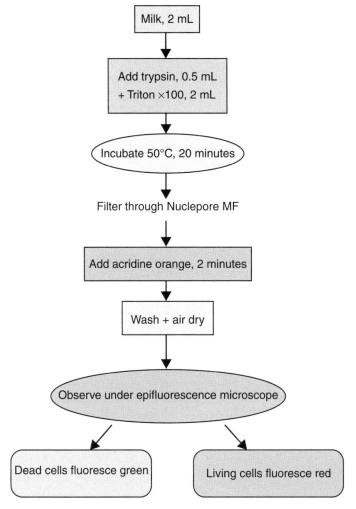

Figure 5.4 Direct Epifluorescence Filter Technique (DEFT) for the detection of bacteria in milk.

The DEFT method concentrates cells on a membrane before staining with acridine orange (Figure 5.4). Acridine orange fluoresces red when interchelated with RNA and green with DNA. Subsequently, viable cells fluoresce orange-red whereas dead cells fluoresce green.

The DEFT count has gained acceptance as a rapid, sensitive method for enumerating viable bacteria in milk and milk products. The count is completed in 25–30 minutes and detects as few as 6×10^3 bacteria/mL in raw milk and other dairy products which is three to four orders of magnitude better than direct microscopy. Because it is a microscopic technique, one is able to distinguish whether the micro-organisms present are yeasts, moulds or bacteria.

The hydrophobic grid membrane filter (HGMF) is a filtration method which is applicable to a wide range of micro-organisms (Entis & Lerner 1996, 2000). The pre-filtered food sample (to remove particulate matter >5 µm) is filtered through a membrane filter which traps micro-organisms on a membrane in a grid of 1600 compartments, due to hydrophobicity effects. The membrane is then placed on an appropriate agar surface and the colony count is determined after a suitable incubation period.

5.4 Rapid end-detection methods

Improvements in end-detection methods include the following:
- Immunoassays, ELISA, latex agglutination
- Impedance microbiology, also known as conductance microbiology
- ATP bioluminescence
- Gene probes linked to the polymerase chain reaction

5.4.1 ELISA and antibody-based detection systems

ELISAs are widely used in food microbiology. ELISA is most commonly performed using mono- and polyclonal antibiodies-coated microtitre trays to capture the target antigen (Figure 5.5); Plate 13. The captured antigen is then detected using a second antibody (usually monoclonal for specificity) which is conjugated to an enzyme. The addition of the enzyme substrate enables the presence of the target antigen to be visualised. ELISA methods offer considerable specificity and can be automated.

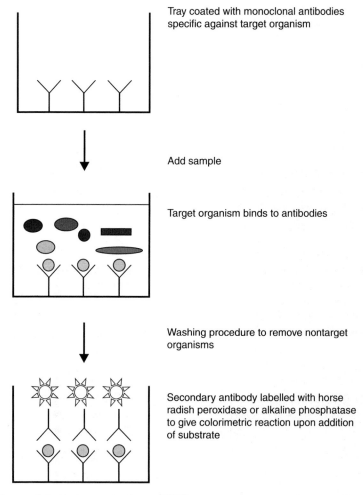

Tray coated with monoclonal antibodies specific against target organism

Add sample

Target organism binds to antibodies

Washing procedure to remove nontarget organisms

Secondary antibody labelled with horse radish peroxidase or alkaline phosphatase to give colorimetric reaction upon addition of substrate

Figure 5.5 Enzyme-linked immunosorbent assay (ELISA).

A wide range of ELISA methods are commercially available, especially for *Campylobacter*, *Salmonella* spp. and *L. monocytogenes*. The technique generally requires the target organism to be 10^6 cfu/mL, although a few tests report a sensitivity limit of 10^4. Hence, the conventional pre-enrichment and even selective enrichment might be required prior to testing. The VIDAS system (bioMerieux) has predispensed disposable reagent strips. The target organism is captured in a solid-phase receptacle coated with primary antibodies and then transferred to the appropriate reagents (wash solution, conjugate and substrate) automatically. The end-detection method is fluorescence which is measured using an optical scanner. The VIDAS system can be used to detect most major food poisoning organisms. The bioMerieux VIDAS® UP *E. coli* O157 uses bacteriophage binding sites for specific capture and detection of the target organism.

5.4.2 Reversed passive latex agglutination (RPLA)
RPLA is used for the detection of microbial toxins such as the Shiga toxins (from *Sh. dysenteriae* and EHEC), *E. coli* heat-labile (LT) and heat-stable (ST) toxins (Figure 5.6). Latex particles are coated with rabbit antiserum which is reactive towards the target antigen. Therefore, the particles will agglutinate in the presence of the antigen forming a lattice structure. This settles to the bottom of a V-bottom microtitre well and has a diffuse appearance. If no antigen is present, then a tight dot will appear.

5.4.3 Impedance (conductance) microbiology
Impedance microbiology is also known as conductance microbiology; impedance is the reciprocal of conductance and capacitance. It can rapidly detect the growth of micro-organisms by two different methods (Silley & Forsythe 1996); see Plate 14.
1 Directly due to the production of charged end products
2 Indirectly from carbon dioxide liberation
In the direct method, the production of ionic end products (organic acids and ammonium ions) in the growth medium causes changes in the conductivity of the medium. These changes are measured at regular intervals (usually every 6 minutes) and the time taken for the impedance value to change is referred to as the 'time to detection'. The greater the number of organisms, the shorter the detection time. Hence, a calibration curve is constructed and thereafter the equipment automatically determines the number of organisms in a sample.

The indirect technique is a more versatile method in which a potassium hydroxide bridge (solidified in agar) is formed across the electrodes. The test sample is separated from the potassium hydroxide bridge by a headspace. During microbial growth, carbon dioxide accumulates in the headspace and subsequently dissolves in the potassium hydroxide. The resultant potassium carbonate is less conductive and it is this decrease in conductance change which is monitored. The indirect technique is applicable to a wide range of organisms including *St. aureus*, *L. monocytogenes*, *Ent. faecalis*, *B. subtilis*, *E. coli*, *P. aeruginosa*, *A. hydrophila* and *Salmonella* serovars. Standard selective media or even an agar slant can be used for fungal cultures.

The time taken for a conductance change to be detectable ('time to detection') is dependent upon the inoculum size. Essentially, the equipment has algorithms which determine when the rate of conductance change is greater than the preset threshold. Initially a reference calibration curve constructed using known numbers of the target organism. Subsequently, the microbial load of subsequent samples will be automatically determined. The limit of detection is a single viable cell since, by definition, the viable cell will multiply and eventually cause a detectable conductance change.

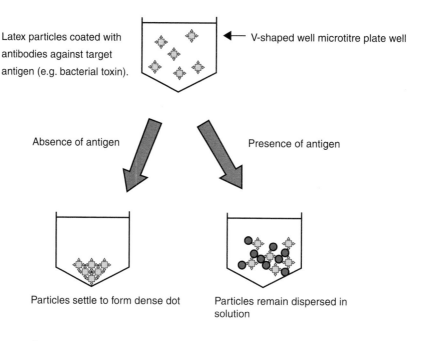

Latex particles coated with antibodies against target antigen (e.g. bacterial toxin).

V-shaped well microtitre plate well

Absence of antigen

Presence of antigen

Particles settle to form dense dot

Particles remain dispersed in solution

Appearance viewed from above microtitre plate:

Figure 5.6 The principle of reversed passive latex agglutination (RPLA).

Microbes frequently colonise an inert surface by forming a biofilm; for more detail, see Section 7.6. Biofilms can be 10- to 100-fold more resistant to disinfectants than suspended cultures and therefore the efficacy of disinfectants for their removal is very important. Impedance microbiology can be used to monitor microbial colonisation and efficacy of biocides (Druggan *et al.* 1993).

5.4.4 ATP bioluminescence techniques and hygiene monitoring

The molecule adenosine triphosphate (ATP) is found in all living cells (eucaryotic and procaryotic). Therefore, the presence of ATP indicates that living cells are present. The limit of detection is ~1 pg ATP, which is equivalent to approximately 1000 bacterial cells based on the assumption of 10^{-15} g ATP per cell. Since a sample is analysed in seconds or minutes, it is considerably faster than conventional colony counts for the detection of bacterial, yeast and fungi. Additionally, food residues which act as the loci of microbial growth will also be detected rapidly (Kyriakides 1992). Hence, ATP bioluminescence is primarily used as a hygiene monitoring method and not for the detection of bacteria *per se*. In fact, in a food factory there

Table 5.4 Applications of ATP
bioluminescence in the food industry

(1) Hygiene monitoring
(2) Dairy industry
    Raw milk assessment
    Pasteurised milk, shelf-life prediction
    Detection of antibiotics in milk
    Detection of bacterial proteases in milk
(3) Assessing microbial load
    Poultry carcasses
    Beef carcasses
    Mince meat
    Fish
    Beer

will not necessarily be a correlation between plate counts and ATP values for identical samples, since the latter will additionally detect food residues (Plate 15).

ATP is detected using the luciferase–luciferin reaction:

$$ATP + luciferin + Mg^{2+} \rightarrow Oxyluciferin + ADP + light\ (562\ nm)$$

The firefly (*Photinus pyralis*) is the source of the luciferase and the reagents are formulated such that a constant yellow-green light (maximum 562 nm) is emitted.

ATP bioluminescence measurement requires a series of steps to sample an area (usually 10 cm^2). Many instruments currently have the extractants and luciferase–luciferin reagents encased with the swab in a 'single-shot' device. This saves the preparation of a series of reagents and the associated pipetting errors.

ATP bioluminescence can be used as a means of monitoring the cleaning regime especially at a critical control point of a HACCP procedure (Section 8.5). See Table 5.4 for a list of examples of ATP bioluminescence applications. There are three food production processes which are not amendable to ATP bioluminescence. These are milk powder production, flour mixes and sugar because the cleaning procedures do not remove all food residues.

It has been noted that the luciferase–luciferin reaction can be affected by residues containing sanitisers (free chlorine), detergents, metal ions, acid and alkali pH, strong colours, many salts and alcohol (Calvert *et al.* 2000). Hence, the commercially available ATP bioluminescence kits may contain detergent neutralisers such as lecithin, Tween 80 and cyclodextrin. Enhancement and inhibition of the luciferase–luciferin reaction can lead to errors of decision and hence an ATP standard should be used to test the activity of the luciferase. An improvement has been the use of caged ATP as an internal ATP standard whereby a known quantity of ATP is released into solution upon exposing the swab to high-intensity blue light (Calvert *et al.* 2000).

5.4.5 Protein detection
An alternative to ATP detection for hygiene monitoring is the detection of protein residues. The method uses the Biuret reaction (Figure 5.7). There are many simple kits available now which are able to detect ~50 µg protein on a work surface within 10 minutes. The surface is sampled either by swabbing or a dipstick and reagents added. The development of a green colour indicates a clean, hygienic surface; grey is caution and purple is 'dirty'. The technique

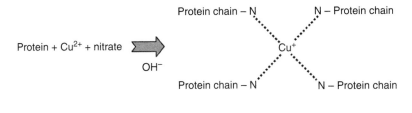

Purple colour

Figure 5.7 The Biuret reaction.

is more rapid than conventional microbiology and less expensive than ATP bioluminescence since no capital equipment is required. It is however less sensitive than ATP bioluminescence.

5.4.6 Flow cytometry

Flow cytometry is based on light scattering by cells and fluorescent labels which discriminate the micro-organisms from background material such as food debris (Figure 5.8). Fluorescence-labelled antibodies have been produced for the major food poisoning organisms such as *Salmonella* serovars, *L. monocytogenes*, *C. jejuni* and *B. cereus*. The level of detection of bacteria is limited to approximately 10^4 cfu/mL due to interference and autofluorescence by food particles. Fluorescent labels include fluorescein isothiocyanate (FITC), rhodamine isothiocyanate and

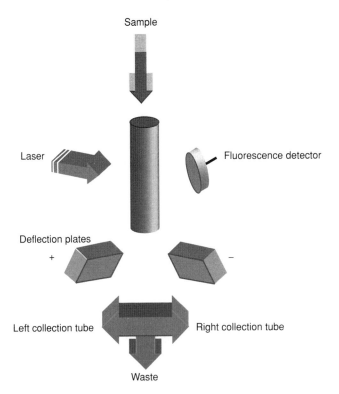

Figure 5.8 Flow cytometry with cell sorting.

phycobiliproteins such as phycoerythrin and phycocyanin. These emit light at 530, 615, 590 and 630 nm, respectively. Viable counts are obtained using carboxyfluorescein diacetate which intracellular enzymes will hydrolyse releasing a fluorochrome. Fluorescent-labelled nucleic acid probes, designed from 16S rRNA gene sequences, enable a mixed population to be identified at genus, species or even strain level. However, as the organism might be non-culturable, it is uncertain whether the organism was viable in the test sample and subsequently questions whether its detection is of any significance. The method has been used for the detection of viruses in seawater (Marie *et al.* 1999).

5.4.7 Nucleic acid probes and the polymerase chain reaction (PCR)
The use of DNA probes for selected target organisms is increasingly being used in the food industry (Scheu *et al.* 1998). The advantage is that food pathogens are detected without such an emphasis on selective media, and is much less time-consuming than culture methods. However, the presence of DNA does not demonstrate the presence of a viable organism which is capable of multiplying to an infectious level. The key method is the use of the PCR to amplify trace amounts of DNA to detectable levels (Figure 5.9). Specificity is obtained by the design of appropriate DNA probes. The PCR technique uses a heat-stable DNA polymerase, *Taq* or *Pfu*, in a repetitive cycle of heating and cooling to amplify the target DNA.

The procedure is essentially as follows:

1 The sample is mixed with the PCR buffer, *Taq* or *Pfu* polymerase, deoxyribonucleoside triphosphates and two primer DNA sequences ($\sim$20–30 nucleotides long).
2 The reaction mixture is heated to 94°C for 5 minutes to separate the double-stranded target DNA.
3 Mixture is cooled to approximately 55°C for 30 seconds. During this time, the primers anneal to the complementary sequence on the target DNA.
4 The reaction temperature is raised to 72°C for 2 minutes and the DNA polymerase extends the primers, using the complementary strand as a template.
5 The double-stranded DNA is separated by reheating to 94°C.
6 The replicated target sites act as new templates for the next cycle of DNA copying.
7 The cycle of heating and cooling is repeated 30–40 times. The PCR will have amplified the target DNA to a theoretical maximum of 10^9 copies, though usually the true amount is less due to enzyme denaturation. The amount of amplified DNA is approximately 100 μg.
8 The DNA is stained with ethidium bromide or preferably for safety SYBR® Safe (In VItrogen™) and visualised by agarose gel electrophoresis with UV transillumination at 312 nm.

Negative control samples omitting DNA must be used in order to check for contamination of the PCR reaction by extraneous DNA.

The gene for the ribosomal RNA (rRNA) molecule, especially the 16S rRNA, can be used as the target for highly specific nucleic acid probes (Amann *et al.* 1995; see Section 2.9.2). The gene encoding for the rRNA molecule contains regions which are highly conserved and other regions which are highly variable. The chromosomal gene encoding for the ribosomal RNA molecule is the target for DNA amplification and not rRNA in the ribosome. The Ribosomal Database Project II is an online database specifically dedicated to this gene and currently has >900,000 sequences of the gene. See http://www.wiley.com/go/forsythe for an online tutorial on selecting sequences of interest. Another target for DNA probes is the 16S–23S rRNA intergenic spacer (ITS) region.

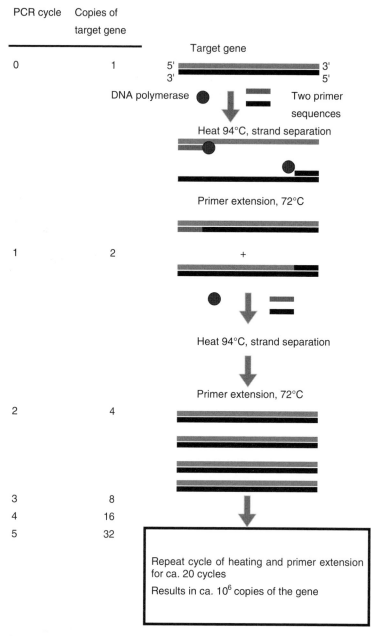

Figure 5.9 The polymerase chain reaction (PCR).

Numerous detection kits have been developed for the detection of food pathogens. However, PCR is not usually directly performed on food samples for several reasons:
- Technique would not distinguish between viable and non-viable cells in the sample.
- The reaction is inhibited by some food components.
- The target cell number may be too low for detection.

Instead, the target organism is usually detected after an enrichment broth step. Apart from the principal method of amplifying a single target site, there are three other main PCR methods. In multiplex PCR, different primers are used to amplify different DNA regions simultaneously. In contrast, the previous methods give presence/absence for the target cell in a sample; real-time PCR is quantitative. It uses the fluorescent tag on the amplicon. Consequently, the increase in fluorescence is proportional to the number of target organisms in the sample; various methods have been developed including the molecular beacon and TaqMan. Reverse transcriptase PCR (RT-PCR) can be used to ensure only viable cells are detected. Certain genes are specifically expressed in the growth phases and these are the targets for the reverse transcriptase enzyme which transcribes the mRNA to single-stranded DNA, which can be amplified by PCR and subsequently detected.

One variation on the PCR technique is 'DIANA' which stands for 'Detection of Immobilised Amplified Nucleic Acids'. The main difference is that DIANA uses two sets of primers for PCR of which only the inner set of primers is labelled. One of the primers is biotinylated on the 5'-end, the second one is labelled with a tail of a partial sequence of the *lac* operator (*lac*Op) gene. The target DNA is first amplified with the outer set of primers (30–40 cycles) to generate a large amount of DNA. Then the inner set of labelled primers is amplified for 10–20 cycles. Streptavidin-coated magnetic beads are used to selectively isolate the amplified biotinylated primary DNA. After washing the magnetic particles, the label is detected appropriately by the addition of a chromogenic substrate for the *lac* gene.

In common with other detection methods, results based on PCR techniques may not be comparable with results from other laboratories due to different protocols. Therefore, inter-laboratory proficiency tests, collaborative trials and standardised protocols are needed; see Section 5.8.

5.4.8 Microarrays

Microarrays can simultaneously measure the transcription level or presence of every gene within a cell. They consist of large arrays of oligonucleotides on a solid support originating from an amplification (using PCR methodology) of genes within the selected genome or DNA dataset (Plate 16; Schena *et al.* 1998, Graves 1999). In general, they are prepared by one of the two following methods:

1 Growing oligonucleotides on the surface, base by base. This is called a Genechip™.
2 Linking presynthesised oligonucelotides or PCR products to a surface.

The DNA sequences are 'spotted' onto the chips by a robot and act as the reference for comparison. The spots are positioned in a grid pattern, where each spot contains many identical copies of an individual gene. The position of the DNA sequence is recorded by spot location, so that the appropriate gene sequence can be identified any time a probe hybridises with, or binds to, its complementary DNA strand on the chip.

The applications of microarrays are as follows: (i) studies of genomic structure, using comparative genomic hybridisation, and (ii) studies of active-gene expression, or 'transcriptomics' (Figure 5.10). In comparative genome hybridisation, the presence of a gene sequence from a test strain is compared with the reference strain on the array. The array is exposed to red and green (Cy3- and Cy5)-labelled reference DNA and comparator sequences allowed to hybridise, which takes 1–10 hours (Ramsey 1998). This has the advantage that large numbers of genes can be compared simultaneously; however, the disadvantage is that one does not identify genes which are present in the test strain and which are absent in the reference strain. The second application is 'transcriptomics', in which the mRNA expression is measured under the given

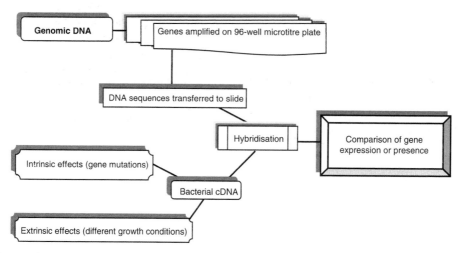

Figure 5.10 Applications of bacterial genomics.

conditions. Cells are grown under two different conditions: the experimental condition and the reference (control) condition. The mRNAs from the bacteria grown under the two conditions are extracted (separately), and the enzyme reverse transcriptase is used to convert the mRNA into complementary DNA (cDNA). One set of cDNA will be labelled with a green fluorescent dye (Cy3) and the other with a red fluorescent dye (Cy5). Therefore, cDNA from the two growth conditions can be distinguished by fluorescence (O'Donnell-Maloney et al. 1996). The two sets of cDNA are then incubated with the DNA array, during which time complementary regions with the array will bind together. The array is then scanned twice: once to detect the spots containing cDNAs labelled with green dye, and then to detect the spots containing red-labelled cDNAs. By merging the scan images, yellow spots correspond with DNA regions which bound both the red- and green-labelled cDNAs. Therefore, genes (transcripts) that are expressed under both sets of growth conditions are identified. Similarly, based on the merged spot colours, genes which are up- or down-regulated are also identified.

DNA array technology can also simultaneously detect different sequences in mixed DNA samples. Consequently, it will be possible in the future to detect and genotype different bacterial species in a single food sample. It is accepted that in the near future, sequencing whole genomes will also be more affordable and hence compete with microarray analysis. However, the bottleneck remains of accurate genome annotation which current databases are unable to achieve automatically. The area of bioinformatics and DNA sequence analysis is covered in Section 2.9.

Multipathogen microarrays have been reported for food and biodefence analysis. Sergeev et al. (2004) announced the FDA-1 microarray for the simultaneous detection of four *Campylobacter* spp. (*C. jejuni*, *C. coli*, *C. lari*, and *C. upsaliensis*) using *glyA* and *fur* genes. Six *Listeria* species were distinguished using the *iap* gene. Sixteen different *St. aureus* set genes encoding SEA-SEE and SEG-SEQ, and six *Cl. perfringens* toxin genes (*cpb1*, *cpb2*, *etxD*, *cpe*, *cpa*, *Iota*) were covered. The reported sensitivity was 30–200 colony forming units.

5.4.9 Biosensors

Biosensors are analytical devices which incorporate a biological material integrated with a physicochemical transducer. They are currently the fastest growing pathogen detection technology and have the promise of multiple sample analysis, in real-time with high specificity and sensitivity (Lazcka *et al.* 2007). The advantages of small-scale devices are reduced cost per unit, small sample volume requirement, shorter analysis times and possible multitarget analysis. Currently, many biosensor systems use either DNA probes or specific antibodies to give the high specificity. A greater variety of end-detection methods are used: optical, electrochemical and piezoelectric. Optical biosensors are very sensitive and specific, with methods based on fluorescence and the more recent technology of surface plasmon resonance (refractive index measurement) being developed. Although optical system may be more sensitive, electrochemical based biosensors detecting current or impedance changes may be preferable for turbid samples. Cell-based biosensors based on mammalian and higher eucaryotic cells or cell components for the detection of bacterial pathogens and toxins are emerging (Banerjee & Bhunia 2009). Mammalian cells generate the initial response which is converted via electrochemical or optical system to a detectable signal. Examples include gangliosides to detect *E. coli* LT-II toxin, E- cadherin for the detection of *L. monocytogenes* and β1 integrins for *Y. enterocolitica*. Cell receptors can be the ligands for a number of pathogens and toxins and are therefore used as the detector.

There is a demand by industry for affordable high-throughput methods (Hyytiä-Trees *et al.* 2007). However so far only a few biosensors have been developed for foodborne pathogens. Muhammad-Tahir & Alocilja (2003) reported on a conductometric biosensor to detect both *E. coli* O157:H7 and *Salmonella* spp. and Lin *et al.* (2008) developed a disposable immunosensing amperometric strip for the quantitative detection for *E. coli* O157:H7. There are a number of different transducer-based immunosensors under development which are promising for real-time detection of pathogens. The three main types are surface plasmon resonance (SPR), quartz crystal microbalance (QCM) and cantilever-based sensors. They are comparable in that target binds to immobilised antibodies on the sensor surface and this causes a direct measurable signal. In SPR, when the target molecule binds to the gold sensor surface, it causes a change in the angle of reflected light. A SPR sensor can simultaneously detect *E. coli* O157:H7, *S.* Typhimurium, *Listeria monocytogenes* and *Campylobacter jejuni* (Taylor *et al.* 2006). QCM sensors are composed of a thin quartz disc with electrodes. The key aspect is that an oscillating electric field is applied to the disc, which induces an acoustic wave with a specific resonance frequency. Consequently, by coating the disc with a capture layer (antibodies, nucleic acids, etc.), it will then respond to the presence of the target by a change in the resonance frequency. These have been developed for *Salm.* Typhimurium, *B. cereus* and *L. monocytogenes*. Cantilever-based sensors using antibodies are also being developed. There are essentially two modes of cantilever. In the static mode, the bending of the cantilever on binding the target is measured, whereas the dynamic mode is similar to QCM sensors in that resonance frequency changes are monitored to determine when target molecule binds. There is considerable interest in using cantilevers for the detection of pathogenic micro-organisms. For example, Campbell & Mutharasan (2007) detected *E. coli* O157:H7 at 1 cell/mL without the conventional requirement of cultivation to increase target cell numbers. Detection times of 10 minutes for 10 *E. coli* O157:H7 cfu/mL have been given. It should be noted that these experiments are under ideal laboratory conditions in the absence of interfering food matrix. Antibody microarrays can analyse many samples in parallel for foodborne pathogenic bacteria and biomolecules (Gehring *et al.* 2008; Karoonuthaisiri *et al.* 2009). However, with a few exceptions, most current methods do not have particularly low detection limits and may only equate to a conventional ELISA method; 10^5–10^7 cfu/mL of target organism. Hence, in general, biosensors are not yet comparable to conventional methods for detection limit. One SRP sensor that has been commercialised is Spreeta™. These small (15×8 cm) sensors are very portable due to their low weight (600 g) and power source (9 V battery). They have been used for *Campylobacter*, *E. coli* and *L. monocytogenes* detection (Nanduri *et al.* 2007, Wei *et al.* 2007).

5.5 Molecular typing methods

Typing micro-organisms is an important tool in outbreak investigations and surveillance both at national and international levels. Typing methods must have high discriminatory power in order to distinguish between related and unrelated isolates. A standardised method should be used which enables results to be comparable between national and international laboratories. Therefore, it can be used for international surveillance purposes. Until recently, determining the relatedness between bacteria relied on phenotyping and chemotyping methods. However, these are limited to well-studied organisms, and growth conditions, and so on, must be carefully standardised to ensure reproducible and reliable results. Despite serotyping and phage typing being well-recognised methods for *Salmonella* and *E. coli*, they have not been developed for many other organisms. Instead, DNA-based methods of genotyping are more appropriate, being applicable across the microbial world. The major advantages of genotyping are as follows:

1 DNA can be extracted from all organisms, including those which cannot be cultivated.
2 It is not dependent upon growth conditions.
3 Closely related strains can be distinguished.
4 Identical methods can be applied to different bacterial species.
5 DNA profiles can be digitised and easily distributed.
6 DNA profiles can be used for comparative analysis.

The molecular typing techniques can be designed to target different areas of the genome. However, the stability of the target areas must be considered in case there is too much variability for typing purposes. The most common genotyping methods include the following:

- Pulsed-field gel electrophoresis (PFGE)
- DNA-probe-based hybridisations such a ribotyping
- PCR-based methods, such as random-amplified polymorphic DNA (RAPD)
- Sequence-based methods, including multilocus sequence typing (MLST)

5.5.1 Pulsed-field gel electrophoresis (PFGE)

PFGE has been applied to a number of bacteria and is one of the most widely used surveillance methods; see Section 1.12.3. It is frequently the genotyping method of choice, or 'gold standard' by which other genotyping methods are compared. The CDC, like other health authorities, have standardised PFGE methods for specific pathogens. It is the method currently used as the basis of PulseNet for the active surveillance of foodborne infections due to *Salmonella*, *E. coli* O157, *Shigella* spp., *L. monocytogenes, C. jejuni* and Norovirus (see Section 1.12.3; Graves & Swaminathan 2001, Ribot *et al.* 2001).

Genomic DNA is digested with a rare cutting restriction enzyme, such as *Xba*I for *Enterobacteriaceae*. The resulting DNA fragments will be too big to be separated by conventional gel electrophoresis. Instead, the polarity of the electric field is briefly reversed periodically throughout the run. The field reversal causes the DNA molecules to be re-orientated in order to move through the pores of the gel in the opposite direction. The longer the molecule is, the slower this process will be. Consequently, the net migration of the DNA fragments will be size dependent. It is important that the starting DNA is intact or at least very large fragments. To achieve this, the bacterial cells are embedded in agarose before lysis and restriction digestion. Then, the agarose plug is inserted into the gel for electrophoresis (Plate 17).

5.5.2 Restriction fragment length polymorphism (RFLP)

If a DNA probe is applied to a restricted total DNA preparation and it hybridises to a DNA sequence that is present only as a single copy in the genome, then only a single band will be detected on a Southern blot. The size of the fragment will depend on the position of the restriction sites flanking the detected sequence. The band will be of the same size between two strains if the structure and location of that gene are the same. However, if there is variation in the distance between the restriction sites, then there will be a difference in the size of the detected band. The variation can be caused by loss of a restriction site due to a point mutation (less frequent), and the insertion or deletion of a DNA region. This is called RFLP. The most useful polymorphisms are due to duplication or transposition of repetitive sequences. These are short sequences which occur twice or more in succession in the genome. The technique is applicable to both bacteria and eucaryotes, and *Campylobacter* (Messens *et al.* 2009) and *Toxoplasma gondii* (Velmurugan *et al.* 2009).

5.5.3 Multiple-locus variable-number tandem-repeats (MLVA)

One of the limitations of RFLP is that it requires a relatively large amount of DNA for a band to be detected. Alternatively, PCR can be used to amplify sample DNA. Instead of using a probe and Southern blotting, primers are used to hybridise either side of the tandem repeat. The size of the resultant DNA band on the gel will depend on the number of copies of the repeated sequence. Each strain can be given a number corresponding to the number of tandem repeats. The MLVA method is based on the amplification of the variable number of tandem repeat (VNTR) areas. By identifying several loci that contain VNTRs, the number of copies at each position is obtained. These numbers generate a profile which is used for genotyping. MLVA has high discriminatory power and is under consideration for adoption by PulseNet to complement PFGE. MLVA schemes have been developed for Enterobacteriaceae, *E. coli* O157, *Salm.* Typhi and *Salm.* Typhimurium. It has been reported to be better than PFGE for both surveillance and outbreak investigations of the latter (Torpdahl *et al.* 2007).

5.5.4 Multilocus sequence typing (MLST)

MLST of bacteria is based on DNA sequence variation within (usually) seven housekeeping genes. The maximum length of the internal fragments is usually between 450 and 500 bp, due to the limitation of DNA sequencer. For each housekeeping gene, the different sequences present within a bacterial species are assigned as distinct alleles and, for each isolate, the alleles at each of the seven loci define the allelic profile or sequence type (ST). There is sufficient sequence variation within bacterial housekeeping genes to generate a number of alleles per locus. Each isolate can therefore be unambiguously characterised by a series of seven integers which correspond to the alleles at the seven housekeeping loci. The procedure can be applied to DNA extracted from non-culturable organisms.

A considerable advantage of MLST is that the sequence data are unambiguous and the allelic profiles can be compared with those in a large online central database via sites such as 'PubMLST' (http://pubmlst.org) and http://www.mlst.net These sites list ~30 bacteria for which MLST profiling is available. As an example, for *E. coli*, the MLST protocol uses PCR primers to specifically amplify the seven genes: asparate aminotransferase, caseinolytic protease, acyl-CoA synthetase, isocitrate dehydrogenase, lysine permease, malate dehydrogenase and β-glucoronidase. Sequence data for each allele generates a profile which is designated as a ST, and can be compared with other pathogenic *E. coli* strains in the EcMLST database (http://www.shigatox.net/cgi-bin/mlst7/index).

An example of MLST application is Sopwith *et al.* (2006). They used MLST for the first continuous population-based survey of campylobacteriosis. They demonstrated that MLST could identify variations in the epidemiology of campylobacteriosis between distinct populations over a 3-year period and described the distribution of major subtypes of interest. It was based on 493 cases, from which 93 distinct MLST sequence types of *C. jejuni* were obtained. The most common type was ST-21 (102 cases) which was isolated three times more frequently than the next complex ST-45. The clonal complex ST-21 has previously been reported from humans, cattle, chicken, milk, sand and water. Hence, ST-21 is associated with environmental and foodborne transmission, whereas ST-45 is principally from humans and chickens.

5.6 Specific detection procedures

Standardised protocols for the isolation of most foodborne pathogens have been defined by various regulatory and accreditation bodies such as the ISO and the FDA. There is however not always a single ideal method for each pathogen and countries may vary with their preferred technique. Subsequently, only an overview of techniques will be given here, and specific instructions must be obtained from the relevant authoritative source before using the methods. The procedures for a number of regulatory authorities, such as the FDA and Health Canada, are available online and are listed in the Food Safety Resources at the end of the book. A consequence of the variation in methodology is the uncertainty of how confidently one can compare food poisoning statistics between countries; see Table 1.12 for food poisoning outbreaks by country.

Because batches of media can vary in their composition, as a means of monitoring personnel proficiency, good laboratory management requires that standardised positive and negative control organisms are used to confirm the selectivity of the media. The control organisms originate from national and international culture collections such as the National Collection of Type Cultures (NCTC) and American Type Culture Collection (ATCC). This is indicated by the culture collection index numbering, that is *St. aureus* ATCC® 25923. These are well-characterised strains available to all quality control laboratories and act as a international standard for referencing.

5.6.1 Aerobic plate count (APC)
The APC is commonly used to determine the general microbial load of the food and not specific organisms. It uses a complex growth medium containing vitamins and hydrolysed proteins which enable the growth of non-fastidious organisms. The agar plates may be inoculated by a variety of techniques (Miles-Misra, spread plate, pour plate) which vary in the volume of sample (20 μL to 1 mL) applied. The plates are generally incubated at 30°C for 48 hours before enumerating the number of colonies. The accuracy of colony numbers is given in Table 5.1.

5.6.2 *Salmonella* serovars
Standardised procedures (i.e. ISO 6579) have been established for the isolation of *Salmonella* serovars from food (Table 5.2). Usually the detection criterion for ready-to-eat foods is the isolation of one *Salmonella* cell from 25 g of food. Subsequently, the protocols require a number of steps to recover *Salmonella* cells which are present at low initial numbers. Additionally, the cells may have been injured during processing and hence the initial step is resuscitation (see Section 5.3.2 on IMS technique). The flow chart for ISO 6579 is given in Figure 5.11. The general outline for both procedures is given as follows:
1 *Pre-enrichment*: To enable injured cells to resuscitate. Resuscitation requires a nutritious non-selective medium such as BPW and lactose broth. These may be modified if large numbers

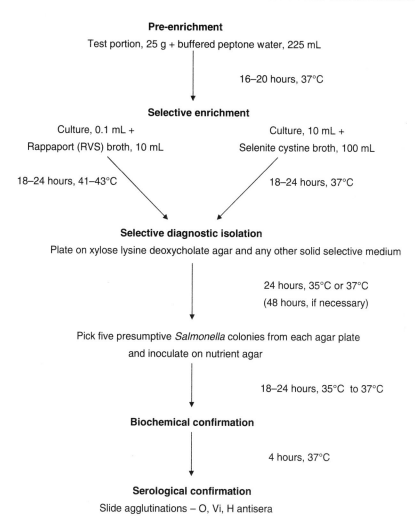

Pre-enrichment

Test portion, 25 g + buffered peptone water, 225 mL

16–20 hours, 37°C

Selective enrichment

Culture, 0.1 mL +
Rappaport (RVS) broth, 10 mL

Culture, 10 mL +
Selenite cystine broth, 100 mL

18–24 hours, 41–43°C

18–24 hours, 37°C

Selective diagnostic isolation

Plate on xylose lysine deoxycholate agar and any other solid selective medium

24 hours, 35°C or 37°C
(48 hours, if necessary)

Pick five presumptive *Salmonella* colonies from each agar plate
and inoculate on nutrient agar

18–24 hours, 35°C to 37°C

Biochemical confirmation

4 hours, 37°C

Serological confirmation

Slide agglutinations – O, Vi, H antisera

Figure 5.11 ISO 6579 *Salmonella* isolation procedure.

of Gram-positive bacteria are present by the addition of 0.002% brilliant green or 0.01% of malachite green. Since milk products are so highly nutritious, the resuscitation broth can be distilled water plus 0.002% brilliant green. Normally, 25 g of food is homogenised and added to 225 mL pre-enrichment broth and incubated overnight. If the food is highly bacteriostatic, this can be overcome by the addition of sodium thiosulphate (in the case of onion) or increased dilution factor, for example 25 g in 2.25 L (Table 5.5).

2 *Selective enrichment*: To suppress the growth of non-*Salmonella* cells and enable *Salmonella* cells to multiply. This is achieved by the addition of inhibitors such as bile, tetrathionate, sodium biselenite (this compound is very toxic) and either brilliant green or malachite green dyes. Selectivity is enhanced by incubation at 41–43°C. Selective broths are selenite cystine broth, tetrathionate broth, lactose broth and Rappaport–Vassiliadis (RVS) broth. More than one selective broth is used because the broths have different selectivities towards the 2500+ *Salmonella* serovars.

Table 5.5 Selection of pre-enrichment media

Medium	Commodity
Buffered peptone water (BPW)	General purpose
BPW + casein	Chocolate, and so on
Lactose broth	Egg and egg products, frog legs, food dyes pH > 6
Lactose broth + Tergitol 7 or Triton X-100	Coconut, meat, animal substances – dried or processed
Lactose broth + 0.5% gelatinase	Gelatin
Non-fat dry milk + brilliant green	Chocolate, candy and candy coatings
Tryptone soya broth	Spices, herbs, dried yeast
Tryptone soya broth + 0.5% potassium sulphate	Onion and garlic powder, and so on
Water + brilliant green	Dried milk

3 *Selective diagnostic isolation*: To isolate *Salmonella* cells on an agar medium to enable single colonies to be isolated and identified. The media contain selective agents similar to the selective broths such as bile salts and brilliant green. *Salmonella* colonies are differentiated from non-*Salmonella* by detection of lactose fermentation and H_2S production. Selective agars include xylose lysine desoxycholate (XLD) agar, Rambach agars, Brilliant Green agar (BGA), Lysine Iron agar (LIA) and mannitol lysine crystal violet brilliant green (MLCB agar); Plate 18. As for the selective enrichment stage, more than one agar medium is used since the media differ in their selectivities.

4 *Biochemical confirmation*: To confirm the identity of presumptive *Salmonella* colonies.

5 *Serological confirmation*: To confirm the identity of presumptive *Salmonella* colonies and to identify the serotype of the *Salmonella* isolate which is useful in epidemiology.

5.6.3 *Campylobacter*

There have been a considerable variety of methods, based on both cultivation and DNA amplification for *Campylobacter* detection. It is likely, however, as is probably true for many other organisms, that no single method is ideal. Some of the major methods and protocols have been reviewed by Corry *et al.* (2002) and are summarised in Table 5.2 and Figure 5.12. Readers should consult the original articles for precise details.

Campylobacter cells can be stressed during processing and hence a pre-enrichment stage to enable injured cells to be resuscitated is commonly used prior to selection for the organism. Frequently selective agents are added as supplements, and lower incubation temperatures are used. In order to aid the growth of the organism ferrous sulphate, sodium metabisulphite and sodium pyruvate (FBP) are added to growth media to quench toxic radicals and increase the organisms' aerotolerance. The organism is microaerophilic and is unable to grow in normal air levels of oxygen. The preferred atmosphere is 6% oxygen and 10% carbon dioxide. This is achieved in gas jars by using gas sachets which generate the required gases.

PCR-based detection methods include the detection of the genus *Campylobacter* using the highly conserved 16S rRNA gene sequence as a target for PCR, and the more specific *C. jejuni* probe targeting the hippurate gene (Wang 2002).

There are a number of typing methods for campylobacters. Traditional typing methods based on differences of key surface antigens such as LPS and flagella have been applied to campylobacters, but are problematic as their surface structures are variable even within an individual strain. The biotyping (biochemical activity profiling) scheme of Preston is more

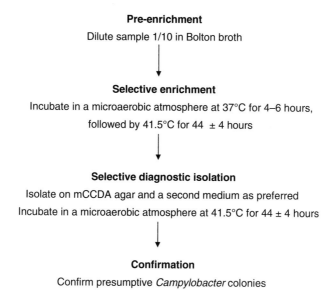

Pre-enrichment

Dilute sample 1/10 in Bolton broth

Selective enrichment

Incubate in a microaerobic atmosphere at 37°C for 4–6 hours,
followed by 41.5°C for 44 ± 4 hours

Selective diagnostic isolation

Isolate on mCCDA agar and a second medium as preferred
Incubate in a microaerobic atmosphere at 41.5°C for 44 ± 4 hours

Confirmation

Confirm presumptive *Campylobacter* colonies

Figure 5.12 Horizonal method for detection and enumeration of *Campylobacter* spp. (ISO 2006).

extensive than that of Lior. Two serotyping procedures are used: the Penner and Hennessy scheme for the heat-stable antigens (lipopolysaccharide) and the Lior scheme for the heat-labile antigens (flagella). Phage typing can differentiate strains within a serotype and is a simple enough technique to apply to a large number of strains simultaneously.

C. jejuni and C. coli genomes have been sequenced (Parkhill *et al.* 2000) and these help to support two DNA-based methods of typing: MLST (Section 5.5.5) and DNA microarrays (On *et al.* 2006, Section 5.4.6). An advantage of MLST is that it is unambiguous and more reliable than phenotyping. However, this is also a disadvantage as common clones will require further differentiation. As a consequence, the MLST method uses a core of seven housekeeping genes. Two genes involved in the synthesis of flagella proteins may also be included as they are more variable (Dingle *et al.* 2002). Different clonal groupings of C. jejuni have been identified in specific animals. However, some are also widespread and can be found in humans. Real-time PCR Taqman assays to detect single-nucleotide polymorphisms specific for the six major MLST clonal complexes have been developed (Best *et al.* 2005). This can be applied to the rapid detection of C. jejuni and clonal allocation. *Campylobacter* typing methods have been reviewed by Wassenar & Newell (2000) and the UK Advisory Committee for the Microbiological Safety of Food (ACMSF 2005).

5.6.4 *Enterobacteriaceae* and *E. coli*

E. coli and other *Enterobacteriaceae* are often initially detected together in liquid media and then differentiated by secondary tests of indole production, lactose metabolism, gas production and growth at 44°C (Table 5.2). E. coli produces acid and gas at 44°C within 48 hours. MacConkey broth is a commonly used medium for the presumptive detection of lactose-fermenting *Enterobacteriaceae* (formerly 'coliforms') from water and milk. It selects for lactose fermenting, bile-tolerant organisms, which in the past have been given the general term 'coliforms'. Acid formation from lactose metabolism is shown by a yellow colouration of the broth (due to a pH indicator dye, neutral red or bromocresol purple), and gas formation is indicated by gas being trapped in an up-turned Durham tube. Lauryl tryptose broth (also known as lauryl sulphate broth) can be used for the detection of lactose-fermenting *Enterobacteriaceae* from food. Initially, the inoculated medium is incubated at 35°C, and afterwards, presumptive positive tubes are used to inoculate duplicate tubes, one for incubation at 35°C and the other at 44°C. Both broths can be supplemented with 4-methylumbelliferyl-β-D-glucuronide (Section 5.2.1) to enhance *E. coli* detection. EE broth, also known as buffered glucose-brilliant green bile broth, is an enrichment medium for *Enterobacteriaceae* from food. The broth is inoculated with samples which have been incubated at 25°C in aerated tryptone soya broth (1:10 dilution) for resuscitation of any injured cells. There are numerous solid media employed for the detection of *E. coli* coliforms and *Enterobacteriaceae*. For example, there are two types of violet red bile agars: (1) VRBA for lactose-fermenting *Enterobacteriaceae* in food and dairy products and (2) violet red glucose agar for the general detection of *Enterobacteriaceae*. These select for bile-tolerant organisms, a predicted trait for intestinal bacteria. Other media include MacConkey agar, china blue lactose agar, desoxycholate agar and eosin methylene blue agar. These have different differentiation efficiencies and regulatory approval. A recent trend has been the inclusion of chromogenic and fluorogenic substrates in particular to detect β-glucuronidase which is produced by ∼97% of *E. coli* strains (Section 5.2.1).

The ISO 21528:2004 standardised methods for the detection and enumeration of *Enterobacteriaceae* include the MPN technique and colony counts. They involve several steps:

- *Pre-enrichment*: A portion of the food is added to BPW in the ratio of 1:9, and incubated at 37°C for 18 hours to resuscitate cells. For the MPN procedure, triplicate tubes are incubated with three different sample volumes equivalent to neat, 10^{-1} and 10^{-2} dilutions.
- *Enrichment*: One millilitre of incubated BPW and sample is pipetted into 10 mL of *Enterobacteriaceae* enrichment (EE) broth which is incubated at 37°C for 24 hours. This suppresses the growth of non-*Enterobacteriaceae*.
- *Selective plating*: By streaking the EE broth on VRBG agar, incubated at 37°C for 24 hours. Presumptive *Enterobacteriaceae* colonies (pink to red or purple, with or without precipitation halo) are streaked on nutrient agar, incubated at 37°C for 24 hours.
- *Confirmation of Enterobacteriaceae*: Five presumptive colonies are selected and streaked for purity on nutrient agar plates before confirmation as *Enterobacteriaceae* according to negative oxidase reaction and glucose fermentation.

For direct enumeration, the pour plate method (ISO 21528-2:2004) is used with 1 mL sample volume (in duplicate) in VRBGA, with an overlay of VRBGA to prevent spreading colonies and encourage anaerobic conditions during colony growth. Presumptive *Enterobacteriaceae* colonies, streaked on nutrient agar for purity checking and standardisation, are confirmed using the oxidase and glucose fermentation tests.

E. coli can be detected using chromogenic agar (TBX) for the β-glucuronidase reaction (i.e. ISO 16649-2:2001). The medium contains the chromophore BCIG. A pour plate method is used, in which 1 mL test sample is pipetted into a sterile Petri dish (in duplicate) and molten TBX agar (at 44–47°C) is added. The plates are incubated at 44°C for 18–24 hours. Typical *E. coli* colonies are blue. In order to resuscitate cells, the sample is initially spread onto a cellulose membrane (pore size 0.45–1.2 μm, 85 mm diameter) on two minerals-modified glutamate agar (MMGA), which are then incubated for 4 hours at 37°C. Afterwards, the membranes are transferred to TBX agar plates for incubation at 44°C, for 18–24 hours. This method will not detect *E. coli* strains which do not grow at 44°C, and those which are β-glucuronidase negative. Hence, *E. coli* O157 will not be detected.

5.6.5 Pathogenic *E. coli*, including *E. coli* O157:H7

Since *E. coli* is a commensal organism in the human large intestine, there is a problem in isolating and differentiating pathogenic strains from the more numerous non-pathogenic varieties. The key differentiation traits have been based on the observation that, unlike most non-pathogenic *E. coli* strains, *E. coli* O157:H7 does not ferment sorbitol, does not possess β-glucuronidase and does not grow above 42°C. Subsequently, MacConkey agar was modified to include sorbitol in place of lactose as the fermentable carbohydrate (SMAC). This medium has been further modified by the inclusion of various other selective agents such as tellurite and cefixime. Pre-enrichment in a modified BPW broth or modified tryptone soya broth is used to resuscitate injured cells before plating onto solid media. Because the cell surface antigenic determinant (O157:H7) is indicative of pathogenicity (though not 100%), the IMS technique (Section 5.3.2) greatly increases the recovery of *E. coli* O157:H7 (Chapman & Siddons 1996). The IMS technique is used worldwide and is recognised as one of the most sensitive methods for *E. coli* O157:H7. It has been approved for use in a number of countries.

The toxins of *E. coli* O157:H7 can be detected using cultured Vero cells and RPLA which are sensitive to 1–2 mg/mL culture filtrate (Figure 5.6). Polymyxin B is added to the culture to facilitate the release of the verocytotoxins/Shiga toxins. ELISA methods specific for pathogenic strains of *E. coli* have been developed. DIANA (Section 5.4.2) has been applied to a number of IMS assays including enterotoxigenic *E. coli*. The assay can detect five ETEC cells in 5 mL without interference from 100-fold excess of SLI negative strains.

The ISO horizontal method for the detection of *E. coli* O157 (ISO 16654:2001) is applicable for a large variety of foods. It has four steps:

- **Enrichment** in modified tryptone soya broth containing novobiocin. This is usually a 1:9 ratio of sample to broth. The mixture is incubated at 41.5°C for 6 hours before IMS, followed by a further 12–18 hours after which, the IMS can be repeated.
- **Separation and concentration** of the target organism are achieved using IMS as previously described (Section 5.3.2).
- **Isolation** by plating the IMS mixture onto cefixime tellurite sorbitol MacConkey agar (CT-SMAC) and a second agar of the user's choice. *E. coli* O157 colonies are about 1 mm in diameter, transparent with a pale yellowish-brown appearance. As stated in the ISO standard method, the enrichment broth may give heavy bacterial growth after overnight incubation, and this can give rise to problems recognising *E. coli* O157 colonies even on selective agars. Therefore, modified techniques such as plating out dilutions of the IMS mixture or plating smaller volumes may be used, while also accepting that this will decrease the sensitivity of the procedure. Five typical colonies are selected from each plate and streaked on nutrient agar, followed by incubation for 18–24 hours at 37°C.
- **Confirmation** is by detection of indole formation and agglutination with *E. coli* O157 antiserum.

5.6.6 *Shigella* spp.

The differentiation between *E. coli* and *Shigella* is troublesome since the two organisms are genetically closely related. *Shigella* and enteroinvasive strains of *E. coli* are closely related phenotypically and they have a close antigenic relationship. Serologically, *Sh. dysenteriae* 3 and *E. coli* O124, *Sh. boydii* 8 and *E. coli* O143, and *Sh. dysenteriae* 12 and *E. coli* O152 appear identical. Distinguishing traits are given in Table 5.6. Other useful differential traits are as follows:

- Cultures that ferment mucate, utilise citrate or produce alkali on acetate agar are likely to be *E. coli*.

Table 5.6 Major differentiation characteristics between *E. coli* and *Shigella* spp.

Property	*E. coli*	*Shigella* spp.
Motility	+	−
Lactose fermentation	+	−
Indole fermentation	+	−
Gas from glucose	+	−

- Cultures that decarboxylate ornithine are most likely to be *Sh. sonnei*.
- Cultures that ferment sucrose are likely to be *E. coli*.

Direct plating onto selective media is unlikely to be successful due to the close relationship of *Shigella* and *E. coli*. Hence, the detection of *Shigella* is generally through use of the distinguishing biochemical differences (Table 5.6). For example, on MacConkey agar, *Shigella* colonies initially appear as non-lactose fermenters.

The ISO method for *Shigella* spp. (ISO 21567:2004) has several stages.

- **Enrichment** of sample in 9 × volume of *Shigella* broth containing novobiocin, followed by anaerobic incubation at 41.5°C for 16–20 hours.
- **Selective plating** on MacConkey agar, XLD agar and HE agar. These are incubated at 37°C for 20–24 hours. If no suspect colonies are visible, then the plates are incubated for further 18–24 hours. Five suspect colonies are picked and streaked on nutrient agar plates followed by incubation for 20–24 hours at 37°C.
- **Identification** is by biochemical tests followed by serological analysis of positive isolates. Biochemical tests include growth on Triple Sugar Iron (TSI) agar for H_2S and gas production, motility and a number of other phenotyping tests which can be achieved using commercial biochemical test kits (i.e. API 20E and Microbact). Polyvalent antisera is used with biochemically identified *Shigella* isolates to identify the following species: *Shigella flexneri*, *Shigella dysenteriae*, *Shigella boydii* and *Shigella sonnei*.

An abbreviated version of the FDA/BAM method is given in Figure 5.13 for illustrative purposes. The FAD/BAM method also uses anaerobic incubation as under these conditions shigella cells can compete against other *Enterobacteriaceae*, and novobiocin is added to the media as a selective agent.

Shiga toxins are virtually identical to the verocytotoxin produced by EHEC and therefore detection methods are applicable to both groups of foodborne pathogens.

5.6.7 *Cronobacter* spp.

As previously described in Section 4.8.2, this organism is an emergent pathogen primarily associated with severe, albeit rare, neonatal infections. It does also cause infections in all other age groups. However, the link of some cases with contaminated reconstituted powdered infant formula has led to a revision in hygienic practices and microbiological criteria for this product. Similar to *Salmonella* isolation, the recovery of *Cronobacter* by conventional microbiology involves the three steps of pre-enrichment, enrichment and plating on selective media. Pre-enrichment is in BPW (10 g + 90 mL broth), followed by overnight enrichment in EE broth (or specific *Cronobacter* selective broth), and plating onto a chromogenic agar such as Druggan–Forsythe–Iversen (DFI) agar (Iversen *et al.* 2004; Iversen & Forsythe 2007); see Plate 9. This medium is manufactured by several companies, but was initially designed by Patrick Druggan

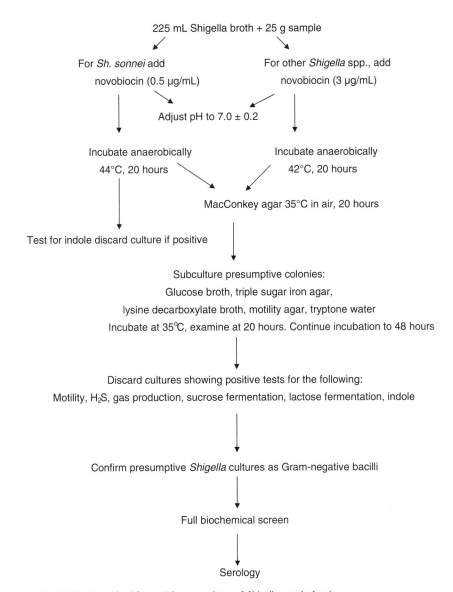

225 mL Shigella broth + 25 g sample

For *Sh. sonnei* add
novobiocin (0.5 µg/mL)

For other *Shigella* spp., add
novobiocin (3 µg/mL)

Adjust pH to 7.0 ± 0.2

Incubate anaerobically
44°C, 20 hours

Incubate anaerobically
42°C, 20 hours

MacConkey agar 35°C in air, 20 hours

Test for indole discard culture if positive

Subculture presumptive colonies:
Glucose broth, triple sugar iron agar,
lysine decarboxylate broth, motility agar, tryptone water
Incubate at 35°C, examine at 20 hours. Continue incubation to 48 hours

Discard cultures showing positive tests for the following:
Motility, H_2S, gas production, sucrose fermentation, lactose fermentation, indole

Confirm presumptive *Shigella* cultures as Gram-negative bacilli

Full biochemical screen

Serology

Figure 5.13 FDA/BAM method for enrichment culture of *Shigella* spp. in foods.

of Oxoid Thermo Fisher Scientific (United Kingdom). The agar contains the chromogen 5-bromo-4-chloro-3-indolyl-α-D-glucopryranoside (X-αGlc) as *Cronobacter* have α-glucosidase activity which results in blue-green colony formation after 18 hours at 37°C. A number of other *Enterobacteriaceae*, including *Salmonella* and *Proteus*, have α-glucosidase activity and are H_2S producers. So, the medium also contains sodium thiosulphate and ferric ammonium citrate which act as a H_2S indicator and distinguishes these organisms from *Cronobacter* by their black colonies. Sodium desoxycholate is also present to suppress the growth of Gram-positive organisms. The vertical ISO method (DTS 22964) uses the similar chromogenic agar ESIA (AES

Enrichment

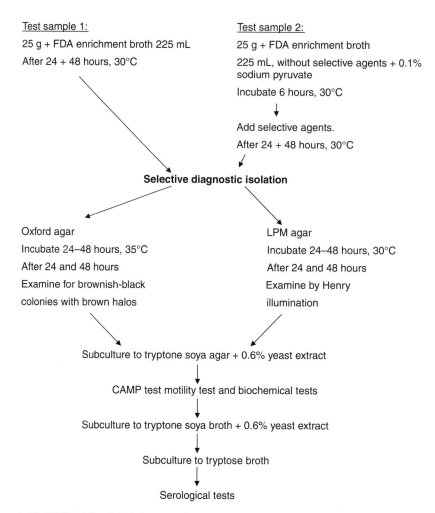

Test sample 1:

25 g + FDA enrichment broth 225 mL

After 24 + 48 hours, 30°C

Test sample 2:

25 g + FDA enrichment broth

225 mL, without selective agents + 0.1% sodium pyruvate

Incubate 6 hours, 30°C

Add selective agents.

After 24 + 48 hours, 30°C

Selective diagnostic isolation

Oxford agar

Incubate 24–48 hours, 35°C

After 24 and 48 hours

Examine for brownish-black

colonies with brown halos

LPM agar

Incubate 24–48 hours, 30°C

After 24 and 48 hours

Examine by Henry

illumination

Subculture to tryptone soya agar + 0.6% yeast extract

CAMP test motility test and biochemical tests

Subculture to tryptone soya broth + 0.6% yeast extract

Subculture to tryptose broth

Serological tests

Figure 5.14 FDA/BAM *Listeria* isolation procedure.

Laboratoire) which is incubated at 44°C for 24 hours. ESIA contains the chromogen X-αGlc as well as crystal violet and *Cronobacter* colonies are green to blue-green in colour. At the time of writing, international trials are underway to standardise *Cronobacter* isolation methodology for the FDA and ISO agencies. Conventional and DNA-based methods of isolation, identification and typing have been recently reviewed by Fanning & Forsythe (2008).

5.6.8 *L. monocytogenes*

Unlike the isolation procedures for salmonellas, *E. coli* and *Cronobacter* spp., pre-enrichment is not commonly used for the isolation of *Listeria* spp. This is because other organisms present will outgrow listeria cells. Instead, various enrichment media have been developed and have regulatory approval. A common enrichment broth is Fraser broth (modified from

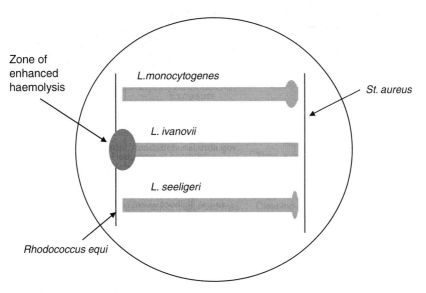

Figure 5.15 The CAMP test for haemolysis testing of *L. monocytogenes*.

UVM broth) which employs aesculin hydrolysis coupled with ferrous iron as an indicator of presumptive *Listeria* spp. Enrichments are streaked onto agars such as ALOA, Oxford and PALCAM. Oxford agar is often incubated at 30°C, whereas PALCAM agar is incubated at 37°C under microaerophilic conditions (Plate 19). A large number of selective agents are used in listeria media such as acriflavin, cycloheximide, colistin and polymixin B as *L. monocytogenes* can be quickly outgrown by competing flora. Typical *L. monocytogenes* colonies are surrounded by black zone due to black iron phenolic compounds. On PALCAM agar, the centre of the colony may have a sunken centre after 48 hours incubation.

Presumptive *L. monocytogenes* colonies are confirmed using biochemical and serological testing. Most non-listeria isolates can be eliminated using the motility test, catalase test and Gram staining. *Listeria* spp. are short Gram-positive rods, catalase positive and are non-motile if incubated above 30°C. Motility of cultures grown at room temperature are characterised by a tumbling action. *L. monocytogenes* is β-haemolytic on horse blood agar. The CAMP test (named after Christie, Atkins, Munch and Peterson) is used for species differentiation (Plate 20). The listeria isolates are streaked on sheep blood agar and *St. aureus* NCTC 1803 and *Rhodococcus equi* NCTC 1621 are streaked in parallel close to the listeria streaks (Figure 5.15). The phenomena of enhanced zones of haemolysis are observed (Table 5.7).

Table 5.7 CAMP reactions for *Listeria* spp.

Listeria spp.	*St. aureus*	*Rh. equi*
L. monocytogenes	+	−
L. seeligeri	+	−
L. ivanovii	−	+

The ISO 11290 method for *L. monocytogenes* is in two parts: detection and enumeration methods. The detection method has four stages:

- **Primary enrichment** is in a selective enrichment broth which has a reduced concentration of the selective agents acriflavine and naladixic acid (i.e. half Fraser broth). *Note*: This medium uses lithium chloride which gives a strong exothermic reaction with water, and irritates the mucous membranes. Normally, the ratio of test sample to broth is 1:9. This is incubated at 30°C for 24 hours before inoculating secondary enrichment broth and also plating onto selective agar.
- **Secondary enrichment** in selective enrichment broth with full strength selective agents (i.e. Fraser broth), for 48 hours at 35°C or 37°C. Typically, 0.1 mL of the incubated primary enrichment mixture is added to 10 mL Fraser broth. After incubation, the mixture is streaked onto selective agars.
- **Plating out and presumptive identification** on two selective agars: ALOA (or equivalent formulation) and a second agar according to the user's choice, such as Oxford agar or PALCAM. ALOA plates are incubated at 37°C for 24 hours, and up to 48 hours, if necessary. As given above, the selective agars are inoculated with primary and secondary enriched cultures. On ALOA typical *Listeria* colonies are green-blue surrounded by an opaque halo.
- **Confirmation** using morphological, physiological and biochemical tests. Five presumptive *Listeria* colonies are picked from each plate, and streaked on tryptone soya yeast extract agar (TSYEA), and incubated at 35°C or 37°C for 18–24 hours. Confirmation tests include Gram stain, catalase production and motility of culture grown at 25°C. *Listeria* are small Gram-positive rods, which are catalase positive and are motile with a tumbling motion. *L. monocytogenes* isolates are confirmed by positive β-haemolytic reaction on sheep blood agar plates, utilisation of rhamnose but not xylose sugars, and enhanced β-haemolysis by *Staphylococcus aureus* and not *Rhodococcus equi* in the CAMP test (Plate 20). When observed by the Henry illumination tests, *Listeria* spp. colonies have a bluish colour and a granular surface. This test involves observing the colonies which are illuminated underneath by reflected white light. Quality assurance strains include *L. monocytogenes* 4b ATCC 13922, *L. monocytogenes* 1/2a ATCC 19111, and for specificity and selectivity *Listeria innocua* ATCC 33090, *E. coli* ATCC 25922 or 8739 and *Enterococcus faecalis* ATCC 29212 or 19433. Equivalent strains can be obtained from other culture collections.

The ISO enumeration method has several stages:

- **Sample preparation** in the ratio 1:9 in BPW or half-Fraser broth (without selective agents).
- **Resuscitation** for 1 hour at 20°C before inoculation of selective agar given below. After which the selective agents (lithium chloride, acriflavine and naladixic acid) can be added and the sample incubated as given above for the detection method.
- **Inoculation** of PALCAM agar; usually 0.1 mL or 1 mL if low numbers of *L. monocytogenes* are expected. Incubate plates at 35°C or 37°C for 24 hours, or incubate a further 24 hours if only slight growth is obtained.
- **Identification and enumeration** by counting colonies with typical *Listeria* morphology.
- **Confirmation** of *Listeria* spp., and *L. monocytogenes* by selecting five presumptive colonies from each plate. Same tests are applied as previously described in the detection method.

5.6.9 *St. aureus*

Large numbers of *St. aureus* cells are required to produce sufficient amounts of heat-stable toxin. Therefore, small numbers of *St. aureus* in food is of little significance, and therefore an enrichment step is not used for the organism's isolation. Consequently, tests for viable cells are applicable for samples before heat treatment, and tests for the enterotoxin and heat-stable thermonuclease for heat-treated samples. The Baird–Parker agar is the most widely accepted selective agar for *St. aureus* (Plate 21). This medium includes sodium pyruvate to aid the resuscitation of injured cells. The selectivity is due to the presence of tellurite, lithium chloride and glycine. *St. aureus* forms black colonies due to tellurite reduction and clearance of egg yolk due to lipase activity. Glycine acts as a growth stimulant and is an essential component of the staphylococcal cell wall. An alternative medium, mannitol salt agar, has better recovery efficiency of *St. aureus* from cheese. The selective agent is salt (7.5%) and mannitol fermentation is indicated by the pH indicator phenol red (reddish-purple zones surrounding *St. aureus* colonies).

The coagulase test (clotting of diluted mammalian blood plasma) is a reliable test for pathogenic *St. aureus*. DNAse production correlates with the coagulase test and is therefore also indicative of pathogenicity. Testing for DNAse activity used to involve growing the organism on agar containing DNA and then flooding with HCl to visualise zones of DNA degradation, clear zone due to the lack of precipitated DNA. However, the acid kills the organism resulting in a non-viable culture. Current alternatives use indicator dyes, toluidine blue and methyl green, which are included in the agar medium. The dyes form coloured complexes with the DNA and so presumptive pathogenic *St. aureus* colonies will show colour changes on DNA hydrolysis: toluidine blue produces pink zones, whereas methyl green goes almost colourless. Staphylococcal enterotoxins can be detected using RPLA (Section 5.4). The limit of sensitivity is about 0.5 ng of enterotoxin per gram food. A number of enzyme immunoassays are available for staphylococcal enterotoxin detection. ELISA kits are also available which have a detection limit of >0.5 μg toxin per 100 g food and require 7 hours to obtain the result.

5.6.10 *Clostridium perfringens*

Cl. perfringens is a strict anaerobe which produces spores that can survive heating processes. Therefore, enrichment is required to detect low numbers of clostridia cells which may be outnumbered by other organisms (Figure 5.16). Numerous media include sulphite and iron which result in a characteristic blackening of *Cl. perfringens* colonies (Table 5.2). However, this blackening reaction is not limited to *Cl. perfringens*, and hence the term 'sulphite reducers' is often used instead of *Cl. perfringens*. The lecithinase (phospholipase C) activity of *Cl. perfringens* is also a common test in diagnostic media resulting is opaque zones surrounding the colonies. Selectivity is by the inclusion of cycloserine or neomycin. All media are incubation under anaerobic conditions which are either generated using an anaerobe jar or an anaerobic cabinet.

Tests to distinguish *Cl. perfringens* from other anaerobic sulphite reducers are microscopy of the Gram stain, metronidazole sensitivity, the Nagler reaction and the reversed CAMP test. The Nagler test uses *Cl. perfringens* type A antitoxin to neutralise lecithinase activity. The reversed CAMP test involves streaking sheep blood agar with *St. agalactiae* and the test isolate at right angles, without touching. After anaerobic incubation at 37°C for 24 hours, a positive result is indicated by arrow-shaped areas of synergistic enhanced haemolysis at the junction of the two streaks (Figure 5.17). Production of 'stormy clot' in litmus milk and the detection of acid phosphatase are useful confirmatory tests. A number of biological methods are available including the rabbit ligated ileal loop tests which though very effective and widely used does require live animal testing. To date, few commercially produced kits are available for the detection of the extracellular toxins produced by *Cl. perfringens*. RPLA (Section 5.4) is available for *Cl. perfringens* enterotoxin.

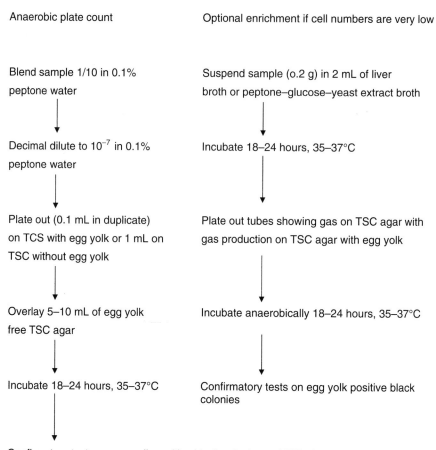

Anaerobic plate count

Blend sample 1/10 in 0.1% peptone water

↓

Decimal dilute to 10^{-7} in 0.1% peptone water

↓

Plate out (0.1 mL in duplicate) on TCS with egg yolk or 1 mL on TSC without egg yolk

↓

Overlay 5–10 mL of egg yolk free TSC agar

↓

Incubate 18–24 hours, 35–37°C

↓

Optional enrichment if cell numbers are very low

Suspend sample (o.2 g) in 2 mL of liver broth or peptone–glucose–yeast extract broth

↓

Incubate 18–24 hours, 35–37°C

↓

Plate out tubes showing gas on TSC agar with gas production on TSC agar with egg yolk

↓

Incubate anaerobically 18–24 hours, 35–37°C

↓

Confirmatory tests on egg yolk positive black colonies

Confirmatory tests on egg yolk positive black colonies on TSC with egg yolk and black colonies on TSC without egg yolk

Figure 5.16 Procedure for the isolation and quantification of *Cl. perfringens*.

5.6.11 *Bacillus cereus, B. subtilis* and *B. licheniformis*

Enrichment is not normally used for the detection of *B. cereus* since low numbers of the organism are not regarded as being of significance (Figure 5.18). Direct plating onto selective media containing the antibiotic polymyxin B is often used. The two key distinguishing features incorporated into media design are the demonstration of phospholipase C activity and the inability to produce acid from the sugar mannitol as shown in Plate 7. If it is necessary to only count spores, then the vegetative cells must be killed by heat treatment (1:10 dilution, 15 minutes, 70°C) or alcohol treatment (1:1 dilution in 95% ethyl alcohol, 30 minutes at room temperature).

B. subtilis and *B. licheniformis* can be isolated easily using routine non-selective media. They have similar appearance on PEMBA medium, but are distinguishable from *B. cereus*. ELISA and RPLA tests are commercially available for *Bacillus* diarrhoeal enterotoxin. They have a sensitivity limit of 1 ng toxin/mL of material and take approximately 4 hours to obtain a result. However, no test has been developed for the emetic toxin due to problems of purification.

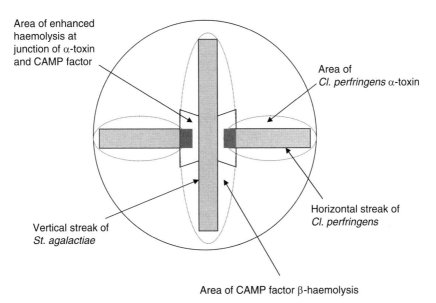

Area of enhanced
haemolysis at
junction of α-toxin
and CAMP factor

Area of
Cl. perfringens α-toxin

Horizontal streak of
Cl. perfringens

Vertical streak of
St. agalactiae

Area of CAMP factor β-haemolysis

Figure 5.17 Reversed CAMP test for *Cl. perfringens* haemolysis.

5.6.12 Mycotoxins

ELISA and latex agglutination assays are commercially available for the detection of aflatoxins (Li *et al.* 2000). The toxins are visualised under UV light since the four naturally produced aflatoxins B_1, B_2, G_1 and G_2 are named after their blue and green fluorescence colours produced. ELISA assays can be used for the quantitative detection of moulds in foods. Compared to ELISA tests, which require 5–10 hours to complete, the latex agglutination method is much faster taking only 10–20 minutes. Trichothecenes, fumonisins and moniliformin (from *Fusarium* spp.) can be detected using high performance liquid chromatography.

5.6.13 Viruses

In most studies of food- and waterborne viruses, samples have been screened for viruses by electron microscopy, DNA probe, ELISA or tissue culture. But not all these techniques are feasible with all viruses, that is Norovirus viruses cannot be grown in tissue culture. Electron microscopy has a limit of sensitivity of 10^5–10^6 particles/mL of faecal suspension. ELISA-based assays have been developed for group A rotavirus and adenovirus in clinical samples only. Although the presence of bacterial indicators (*Enterobacteriaceae*, *E. coli*) does not correlate with the presence of viral pathogens, there are no established viruses detection methods for foods other than shellfish. Consequently, foodborne viral gastroenteritis is not usually diagnosed and detection methods for viruses and viral genomic material have not been adopted by routine food analysis laboratories.

For epidemiological studies, large volumes (25–100 g) of food need to be analysed for enteric viruses since the level of food contamination is assumed to be low. If tissue culture is to be used to detect enteric viruses, then the viruses need to be separated from the bulk of the food. The viruses may be enumerated using plaque assays with dilution end point assays. Cell culture technique, however, is not applicable for Noroviruses (formerly Norwalk-like virus and small round-structured viruses) due to the lack of host cell line, and is only moderately successful

Vegetative cells

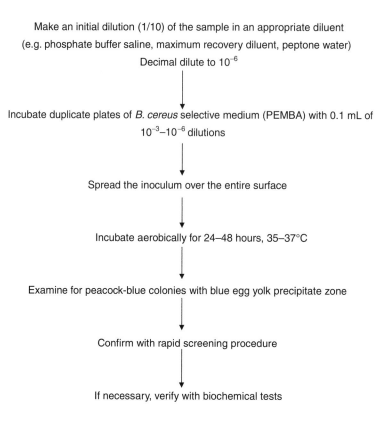

Make an initial dilution (1/10) of the sample in an appropriate diluent
(e.g. phosphate buffer saline, maximum recovery diluent, peptone water)
Decimal dilute to 10^{-6}

Incubate duplicate plates of *B. cereus* selective medium (PEMBA) with 0.1 mL of
10^{-3}–10^{-6} dilutions

Spread the inoculum over the entire surface

Incubate aerobically for 24–48 hours, 35–37°C

Examine for peacock-blue colonies with blue egg yolk precipitate zone

Confirm with rapid screening procedure

If necessary, verify with biochemical tests

Spores

If necessary to count spores, then first treat with heat or alcohol to destroy the vegetative cells

Heat treatment: Heat the initial 1/10 suspension for 15 minutes 70°C. Proceed as for detection of vegetative cells

Alcohol treatment: Dilute the initial suspension 1:1 in 95% ethyl alcohol. Leave for 30 minutes, room temperature. Proceed as for detection of vegetative cells, adjusting the dilutions to account for the 1:1 dilution of the sample suspension in alcohol

Figure 5.18 Typical procedure for detection of *B. cereus* (http://www.fda.gov/privacy.html#linking).

with hepatitis A virus (HAV). The sequencing of Norovirus and HAV genomes has enabled the reverse transcriptase polymerase chain reaction (RT-PCR) to be developed for detection and characterisation of the viruses in faecal, vomit and shellfish samples. The Norovirus can be characterised by sequencing an amplified products of the test (amplicons), and this has enabled outbreaks in different locations to be linked to a single source. Electron microscopic analysis for Norovirus is only of use with samples during the first 2 days of symptoms, whereas RT-PCR is usable within 4 days of symptoms appearing (Atmar & Estes 2001). HAV genome is detectable by RT-PCR in artificial sterile seawater for 232 days, compared with only 35 days for cell culture. Therefore, the RT-PCR is not a reliable indicator of infectious HAV.

5.7 Accreditation schemes

In order to have confidence that the method of choice could have detected the target organism, the methods used should be accredited. Therefore, the method must be validated against standards tests using collaborative studies. Validation of laboratory procedures will form part of a company's quality system. There are various international bodies which validate detection methods. The Association of Official Analytical Chemists (AOAC) International (Anon. 1999b) and ISO are the most widely accepted; others include UKAS (United Kingdom), EMMAS (European), AFNOR (French), DIN (Germany) and the European MICROVAL. Appropriate control organisms should always be run to ensure the media is conforming to its specification.

6 Microbiological criteria

6.1 Background to microbiological criteria and end-product testing

Testing food at the end of production for micro-organisms ('end-product testing') has been standard practice in the food industry for decades. However, a statistical appreciation of its usefulness has been largely overlooked. In 1974, the International Commission on Microbiological Specifications for Foods (ICMSF) wrote an excellent text regarding the setting of microbiological criteria. The book was written at a time of increasing global food transportation and was primarily aimed at being applied to food entering a country (port of entry) with no known history. Nevertheless, these criteria have been applied within industry for their own products despite in-house knowledge of the product.

It is very important nowadays to recognise that microbiological testing of foods should be carried out under the umbrella of Hazard Analysis Critical Control Point (HACCP) as part of the verification principle (Section 8.5). In other words, **end-product testing in itself does not guarantee a safe food product**, but could support the HACCP plan implementation. Nevertheless, microbiological criteria (levels of microbes acceptable in a particular food) are required by regulatory authorities and between companies in a supply chain. The details of these criteria are often historical, and not necessarily the most appropriate. In the case of criteria between companies, they are often confidential. These criteria may eventually be replaced by 'Food Safety Objectives' which are currently being evaluated (Section 9.7).

6.2 International Commission on Microbiological Specifications for Foods (ICMSF)

In the 1960s, the role of micro-organisms in foodborne disease had become well recognised, and in addition, international trade had significantly increased. However, microbiological testing of food was hampered due to the lack of standardised methods and the use of sampling plans which lacked statistical validity. Subsequently, in 1962, the ICMSF was formed by the International Committee on Food Microbiology and Hygiene. The ICMSF aims were to:
1 assess data on the microbiological safety and quality (including spoilage) of foods;
2 determine if the use of microbiological criteria would both improve and assure the microbiological safety of certain foods;
3 recommend methods of sampling and examination.

These objectives have evolved as knowledge of foodborne micro-organisms and their control have increased and is reflected in the series of ICMSF books and papers. The first ICMSF book (initially published in 1968 and revised in 1978, 1982 and 1988) was centred upon encouraging a comparison of worldwide testing methods in order to obtain agreed methods for use in international trade. In 1974, the ICMSF published 'Microorganisms in Foods 2. Sampling for Microbiological Analysis: Principles and Specific Applications'. This recognised the need for scientifically based sampling plans for foods in international trade. It described two- and three-class sampling plans and 'choice of case'. The sampling plans were originally designed for application at port of entry, that is when there is no prior knowledge on the history of the food. This pioneering work set forth the principles of sampling plans for the microbiological evaluation of foods and is also known as attributes and variables sampling depending on the extent of microbiological knowledge of the food. The second edition of the book in 1986 took note of the successful application of the acceptance sampling plans on a worldwide basis, not only at an international level but also at national and local levels by both industry and regulatory agencies. The ICMSF book has subsequently been updated in 2002. The third (two volume) ICMSF book was on the 'Factors affecting the life and death of microorganisms' (Volume 1, 1980a) and 'Food commodities (Volume 2, 1980b). In addition, this detailed the microbial load, spoilage patterns and microbial hazards of 14 food commodity groups. The use of HACCP as the most assured means of safety food production was the subject of the Commission's fourth book (ICMSF 1988). In order to support HACCP and Good Hygienic Practices (GHP), in 1996, the Commission produced their fifth book as a reference source on the growth and death response of foodborne pathogens (ICMSF 1996a). Due to increased knowledge of foodborne pathogens, in 1998 the Commission updated its earlier 1980b publication on 16 commodity foods. This was again with regard to typical microbial loads, spoilage patterns and control measures. More recently (in 2002), the ICMSF has produced book 7 (*Microbiological Testing in Food Safety Management*) which is a revision of the 1986 publication. It has a threefold purpose:

1 Supporting the use of statistically valid sampling plans at port of entry (i.e. where there is no prior knowledge concerning the processing conditions).
2 Demonstrating the application of HACCP and GHP for structured food safety management, along with end-product testing (cf. Chapter 8).
3 Recommending the incorporation of 'food safety objectives' (Section 9.7) as a means of translating 'risk' into the principles of HACCP and GHP (cf. Section 8.9).

6.3 Codex Alimentarius principles for the establishment and application of microbiological criteria

The Codex Alimentarius Commission has become the reference for international food safety requirements (cf. Section 11.2 for more detail). The Codex (1997c) definition of a microbiological criterion is as follows:

A microbiological criterion for food defines the acceptability of a product or a food lot, based on the absence or presence, or number of micro-organisms including parasites, and/or quantity of their toxins/metabolites, per unit(s) of mass, volume, area or lot.

Codex Alimentarius Commission (1997c) requires that a microbiological criterion consists of the following:

A statement of the micro-organisms of concern and/or their toxins/metabolites and the reason for that concern.

The analytical methods for their detection and/or quantification.

A plan defining the number of field samples to be taken and the size of the analytical unit.

Microbiological limits considered appropriate to the food at the specified point(s) of the food chain.

The number of analytical units that should conform to these limits.

A microbiological criterion should also state the following:
- The food to which the criterion applies.
- The point(s) in the food chain where the criterion applies.
- Any actions to be taken when the criterion is not met.

The value of microbiological testing as a control measure varies along the food chain. Therefore, a microbiological criterion should be established and applied only when there is a definite need and when its application is practical. Such need is demonstrated, for example, by epidemiological evidence that the food under consideration may represent a public health risk and that a criterion is meaningful for consumer protection, or as the result of a risk assessment. The criterion should be technically attainable by applying Good Manufacturing Practices (Codes of Practice). Criteria should be reviewed periodically for relevance with respect to emerging pathogens (Section 5.7), changing technologies and new understandings of science.

A sampling plan includes the sampling procedure and the decision criteria to be applied to a lot, based on examination of a prescribed number of sample units and subsequent analytical units of a stated size by defined methods. A well-designed sampling plan defines the probability of detecting micro-organisms in a lot, but it should be borne in mind that no sampling plan can ensure the absence of a particular organism. Sampling plans should be administratively and economically feasible (Codex Alimentarius Commission 1997b, 1997c).

In particular, the choice of sampling plans should take into account the following:
1 Risks to public health associated with the hazard.
2 The susceptibility of the target group of consumers.
3 The heterogeneity of distribution of micro-organisms where variables sampling plans are employed.
4 The acceptable quality level (AQL) and the desired statistical probability of accepting a non-conforming lot.

The AQL is the percentage of non-conforming sample units in the entire lot for which the sampling plan will indicate lot of acceptance for a prescribed probability (usually 95%). For many applications, two- or three-class attribute plans may prove useful.

Microbiological criteria should be based on scientific analysis and advice, and where sufficient data are available, on a risk analysis appropriate to the foodstuff and its use (CAC 1997c). These criteria may be relevant to the examination of foods, including raw materials and ingredients of unknown or uncertain origin, or when no other means of verifying the efficacy of HACCP-based systems and GHP are available. Microbiological criteria may also be used to determine that processes are consistent with the general principles of food hygiene. Microbiological criteria are not normally suitable for monitoring critical limits as defined in the HACCP system.

The purpose of establishing microbiological criteria is to protect the public's health by providing food which is safe, sound and wholesome, and to meet the requirements of fair trade practices. The presence of criteria, however, does not protect the consumer's health since it is possible for a food lot to be accepted which contains defective units. Microbiological criteria may be applied at any point along the food chain and can be used to examine food at the port of entry and at the retail level.

The statistical performance characteristics or operating characteristics curve should be provided in the sampling plan; see Section 6.4. Performance characteristics provide specific information to estimate the probability of accepting a non-conforming lot. The time between taking the field samples and analysis should be as short as reasonably possible, and during transport to the laboratory, the conditions (e.g. temperature) should not allow increase or decrease of the numbers of the target organism, so that the results reflect – within the limitations given by the sampling plan – the microbiological conditions of the lot.

6.4 Sampling plans

In addition to the ICMSF (1986, 2002) on sampling plans, Harrigan & Park (1991) wrote an excellent book on their practical mathematics. Just as it is impractical to test a sample for every possible food pathogen, it is also impractical to destructively test 100% of an ingredient or end product. Whilst it is accepted that no sampling plan can guarantee the absence of a pathogen in a batch of food (food lot) and therefore food safety, there is a need to use sampling plans to appropriately test a batch of material and give a statistical basis for acceptance or rejection of a food lot.

Microbiological sampling plans are frequently used in food production, import control and in contractual agreements with suppliers and customers. Sampling plans are used to check the microbiological status of a commodity, its compliance to safety requirements and adherence to Good Hygiene Practices (GHP; Section 8.11) during or after manufacture. The results from single-sample examinations may provide valuable baseline data which can be used for trend analysis, particularly where samples form part of a specific survey. However, statistical principles should be observed when sampling particular food commodities many of which (usually end products) are heterogeneous, even when they have a similar formulation. In situations where a food inspector might be concerned about a particular food, a sample taken for microbiological analysis may provide evidence that food hygiene regulations have been contravened or may provide the basis for additional inspection and/or examination. The single-sample concept is likely to retain a role in assessing food safety in small-scale food production businesses which will have fewer resources for implementation of HACCP but will nevertheless have to take proper account of the risk to public health posed by each individual operation.

There are two types of sampling plans:

1 Variables plans, when the microbial counts conform to a 'log-normal' distribution; see Section 6.5. This data would be known by a producer and is not applicable to an importer at the port of entry situation.
2 Attribute plans, when no prior knowledge of the distribution of micro-organisms in the food is known, that is at port of entry or the distribution of target organism is not log-normal; see Section 6.6.

Attributes sampling plans can be either according to a two-class plan or a three-class plan. The two-class plan is used almost exclusively for pathogens, whereas a three-class plan is frequently used to examine for hygiene indicators. The main advantage of using sampling plans is that they are statistically based and provide a uniform basis for acceptance against defined criteria.

The type of sampling plan required can be decided using Figure 6.1.

Attributes plans also involve the concept of 'choice of case', based on microbiological risk. 'Case' is a classification of sampling plans ranging from 1 (least stringent) to 15 (most stringent). The choice of case, and therefore the sampling plan, depends on the following:

• The relative severity of the hazard to food quality or consumer health on the basis of the micro-organisms involved. See Chapter 5.
• The expectation of their destruction, survival or multiplication during normal handling of the food. See Chapter 2.

Table 6.1 and the decision trees of Figures 6.1 and 6.2 should be referred to aid deciding on the appropriate sampling plan. For example, cases 1–3 refer to utility applications, such as

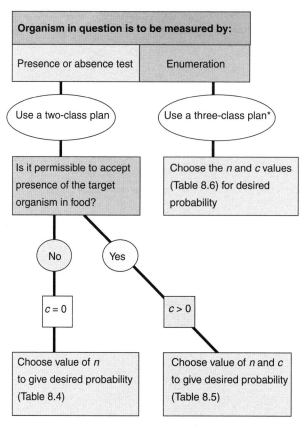

Figure 6.1 Decision tree for choosing a sampling plan. Reprinted with permission from ICMSF and Health Canada. In the figure, the symbol '∗' denotes that a variable plan may be applicable if the organism is distributed in a log-normal fashion.

shelf-life, whereas cases 13, 14 and 15 refer to severely hazardous foodborne pathogens. The severity of the microbiological hazard has been covered in Chapter 4 and the foodborne pathogens grouped (1986 version) to assist in referring to Table 4.3.

Sampling plans and recommended microbiological limits were published by ICMSF (1986) for the following foods:

1 Raw meats, processed meats, poultry and poultry products
2 Pet foods
3 Dried milk and cheese
4 Pasteurised liquid, frozen and dried egg products
5 Seafoods
6 Vegetables, fruit, nuts and yeast
7 Cereals and cereal products
8 Peanut butter and other nut butters
9 Cocoa, chocolate and confectionery
10 Infant and certain categories of dietetic foods
11 Bottled water

Table 6.1 Sampling plans in relation to degree of health hazard and conditions of use

Type of hazard	Conditions in which food is expected to be handled and consumed after sampling		
	Reduce degree of hazard	Cause no change in hazard	May increase hazard
No direct health hazard	Case 1	Case 2	Case 3
Utility, for example reduced shelf-life and spoilage	Three-class, $n = 5, c = 3$	Three-class, $n = 5, c = 2$	Three-class, $n = 5, c = 1$
Health hazard	Case 4	Case 5	Case 6
Low, indirect (indicator)	Three-class, $n = 5, c = 3$	Three-class, $n = 5, c = 2$	Three-class, $n = 5, c = 1$
Moderate, direct, limited spread	Case 7	Case 8	Case 9
	Three-class, $n = 5, c = 2$	Three-class, $n = 5, c = 1$	Three-class, $n = 10, c = 1$
Moderate, direct, potentially extensive spread	Case 10	Case 11	Case 12
	Two-class, $n = 5, c = 0$	Two-class, $n = 10, c = 0$	Two-class, $n = 20, c = 0$
Severe, direct	Case 13	Case 14	Case 15
	Two-class, $n = 15, c = 0$	Two-class, $n = 30, c = 0$	Two-class, $n = 60, c = 0$

Adapted from ICMSF 1986 and reprinted with permission from the University of Toronto Press.

6.5 Variables plans

Variables plans can be applied when the number of micro-organisms in the food are distributed 'log-normally'; that is, the logarithms of the viable counts conform to a normal distribution (Figure 6.3; Kilsby *et al.* 1979). This applies to certain foods which have been analysed over a period of time by the producer and therefore does not apply at port of entry.

If the micro-organisms' distribution within a lot is log-normal, then sampling plans can be used to develop acceptance sampling plans. The sample mean (x) and standard deviation (s) are determined from previous studies. The sample mean and standard deviation are used to decide whether a 'lot' of food (Section 6.3) should be accepted or rejected.

In addition:

- The proportion (p_d) of units in a lot which can have a concentration above the limit value, V, must be decided.
- The desired probability P can be chosen where P is the probability of rejecting a lot which contains at least a proportion p_d above V.

The lot of food is rejected if $x + k_1 s > V$.

where:

k_1 is obtained from reference tables (Table 6.2) according to the p_d and P values. It is therefore dependent upon the stringency of the sampling plan and number of sample units, n, analysed.

V is the microbial count as a log-concentration which has been set as a safety limit.

Deciding that a lot would be rejected if 10% ($p_d = 0.1$) of samples exceeded the value V, with a probability of 0.95 and taking five sample units (n) gives $k_1 s$ as 3.4. The more samples (n) are taken, the lower the chance of rejecting an acceptable lot of food.

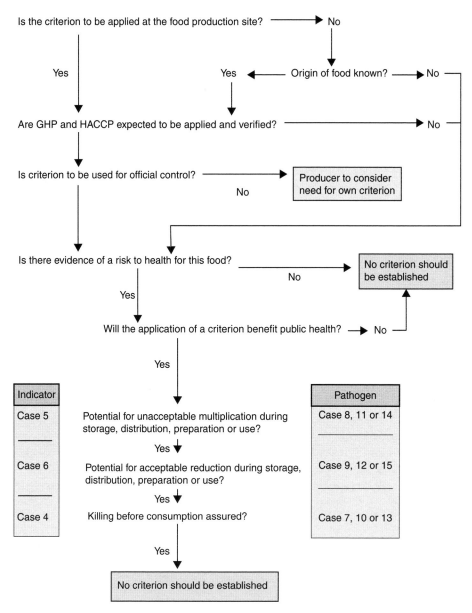

Figure 6.2 Decision tree for choice of criteria for microbiological pathogens and indicator organisms.

The value of V is set by the microbiologist from previous experience. It can be similar to M in the three-class plan; see Section 6.6.2.

For example,

the aerobic plate count for ice cream from the old milk products directive 92/46/EEC gives $M = 500\,000$ cfu/g.

$500\,000 = \log 5$.

Therefore, $V = 5$.

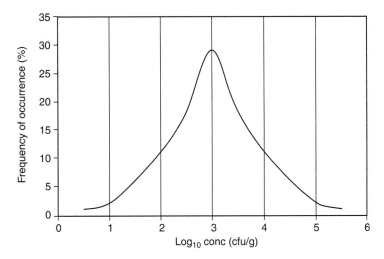

Figure 6.3 Normal log distribution of a micro-organism.

Previous analysis gave mean log value of 5.111 and standard deviation of 0.201. Therefore, deciding:

1 $p_d = 0.1$, probability that a lot would be rejected if 10% samples exceeded V.
2 Probability of rejection $= 0.95$.
3 Number of sample units $= 3$.

Gives $k_1 = 6.2$

Hence:

$$x + k_1 s = 5.111 + (6.2 \times 0.201) = 6.3572$$

Since $V = 5$, the lot would be rejected.

Variables plan can be applied to Good Manufacturing Practice (GMP) standards using k_2 values in Table 6.3 (ICMSF 1986, Kilsby 1982).

A similar formula is applied where the lot is accepted if $x + k_2 s < v$.

The value of P and p_d are decided as before, v is similar to 'm' in the three-class plan (Section 6.6.2) and the GMP values of IFST (1999) can be used.

Table 6.2 Safety and quality specification (reject if $x + k_1 s > V$)

		Number of sample units			
Probability (P) of rejection	Proportion (p_d) exceeding V	3	5	7	10
0.95	0.05	7.7	4.2	3.4	2.9
	0.1	6.2	3.4	2.8	2.4
	0.3	3.3	1.9	1.5	1.3
0.90	0.1	4.3	2.7	2.3	2.1
	0.25	2.6	1.7	1.4	1.3

Adapted from ICMSF 1986 and reprinted with permission from the University of Toronto Press.

Table 6.3 Determining the GMP limit (accept if $x + k_2 s < v$)

Probability (P) of rejection	Proportion (p_d) exceeding v	Number of sample units			
		3	5	7	10
0.90	0.05	0.84	0.98	1.07	1.15
	0.1	0.53	0.68	0.75	0.83
	0.3	−0.26	−0.05	0.04	0.12
0.75	0.01	1.87	1.92	1.96	2.01
	0.1	0.91	0.97	1.01	1.04
	0.25	0.31	0.38	0.42	0.46
	0.5	−0.47	−0.33	−0.27	−0.22

Adapted from ICMSF 1986 and reprinted with permission from the University of Toronto Press.

For example:
GMP value for the aerobic plate count for raw poultry (IFST 1999) = $<10^5$ = 5.0.
Previous analysis had given $x = 4.3$ with a standard deviation (s) = 0.475.
Therefore, deciding:
1 Proportion of acceptance, $P = 0.9$
2 Proportion exceeding v, $p_d = 0.1$
3 Number of sample units = 7
Gives $k_2 = 0.75$.
Hence:

$$x + k_2 s = 4.3 + (0.75 \times 0.475) = 4.65625$$

Therefore, the lot of raw chicken is below the GMP value.

6.6 Attributes sampling plan

The attributes sampling plan(s) is applied when there is no previous microbiological knowledge of the distribution of micro-organisms in the food or the micro-organisms are not distributed 'log-normally'. There are two types of attributes sampling plans as defined by ICMSF (1986):
• Two-class plan; $n = 5$, $c = 0$ or $n = 10$, $c = 0$
• Three-class plan; $n = 5$, $c = 1$, $m = 10^2$, $M = 10^3$
The two-class plan is used almost exclusively for pathogens, whereas a three-class plan is often applied for indicator organisms. The main advantage of using sampling plans is that they are statistically based and provide a uniform basis for acceptance against defined criteria.

6.6.1 Two-class plan
A two-class plan consists of the specifications n, c and m.
Where:
n = The number of sample units from a lot that must be examined.
c = The maximum acceptable number of sample units that may exceed the value of m. The lot is rejected if this number is exceeded.
m = The maximum number of relevant bacteria/g. Values greater than this are either marginally acceptable or unacceptable.
For example:

$$n = 5, \ c = 0$$

This means that five sample units are analysed for a specific pathogen (e.g. *Salmonella*). If one unit contains *Salmonella*, then the complete batch is unacceptable. Each sample unit analysed is normally 25 g for *Salmonella* testing; see Section 5.6.2.

6.6.2 Three-class plan

The additional parameter in a three-class plan is as follows:

M = a quantity that is used to separate marginally acceptable from unacceptable. A value at or above M in any sample is unacceptable.

Hence, the three-class plan is where the food can be divided into three classes according to the concentration of micro-organisms detected:

- 'acceptable' if counts are below m
- 'marginally acceptable' if counts are above m but less than M
- 'unacceptable' (reject) if counts are greater than M

For example, a sampling plan for *Enterobacteriaceae* could be as follows:

$$n = 5, \quad c = 2, \quad m = 10, \quad M = 100$$

This means that two units from a sample number of 5 can contain between 10 and 100 *Enterobacteriaceae* and be acceptable. However, if three units contain *Enterobacteriaceae* between 10 and 100 or just one sample has greater than 100 *Enterobacteriaceae*, then the batch is unacceptable and the lot is rejected. Hence, the three-class sampling plan includes a tolerance value for the random distribution of microbes in foods.

The stringency of the sampling plan can be decided using the ICMSF (1986) concept based on the hazard potential of the food and the conditions in which a food is expected to be subject to before consumption (Table 6.1).

6.7 Principles

6.7.1 Defining a 'lot' of food

A lot is 'a quantity of food or food units produced and handled under uniform conditions'. This implies a homogeneity within a lot. However, in most instances, the distribution of micro-organisms within a 'lot' of food is heterogeneous. If a lot is in fact composed of different production batches, then the producer's risk (i.e. the risk that an acceptable lot will be rejected) can be high since the sample units analysed may by chance be those from a poor-quality batch. In contrast, by defining individual production batches as lots, a more precise identification of poor-quality (reject) food can be made.

6.7.2 Sample unit number

The number of sample units 'n' refers to the number of units that are chosen randomly. The samples should represent the composition of the lot from which it is taken. A sample unit can be an individual package or portions. The sample units must be taken in an unbiased fashion and must represent the food lot as well as possible. Micro-organisms in food are often heterogeneously distributed and this makes the interpretation of sample unit results difficult. Random choice of samples is required to try and avoid biased sampling; however, difficulties arise when the food is non-homogenous, such as a quiche.

The choice of n is usually a compromise between what is an ideal probability of assurance of consumer safety and the workload the laboratory can handle. It is important to first determine the nature of the hazard and then determine the appropriate probabilities of acceptance (Tables

Table 6.4 Probability of accepting (P_a%) a food lot; two-class plan, $c = 0$

Number of samples (n)	Probability of acceptance (P_a%)						
	Actual percentage of defective samples						
	2	5	10	20	30	40	50
3	94	86	73	51	34	22	13
5	90	77	59	33	17	8	3
10	82	60	35	11	3	1	(<0.5)
20	67	36	12	1	(<0.5)	(<0.5)	(<0.5)

Adapted from ICMSF 1986 and reprinted with permission from the University of Toronto Press.

6.4–6.6). It is uneconomical to test a large portion of a food lot. However, the stringency of a sampling plan for a hazardous micro-organism can be set using the relationship between the number of sample units analysed and the acceptance/rejection criteria (n and c values; Section 6.6.2).

6.7.3 Operating characteristic curve
It is possible when using a sample plan that a relatively poor lot of food will be accepted and a good lot is rejected. This is represented by the 'operating characteristic' curve. This is a plot of the following:
1 The probability of acceptance (P_a) on the 'y' axis, where P_a is the expected proportion of times a lot of this given quality is sampled for a decision.
2 Percentage of defective sample units comprising a lot (p) on the 'x' axis. This is also known as a measure of lot quality.
Figure 6.4 gives the operating characteristic curve for the sampling plan $n = 5$, $c = 3$.

The operating characteristic curve changes according to the values of 'n' and 'c'. Figure 6.5b is a selected area of Figure 6.5a to emphasise the high chance of accepting lots with up to 30%

Table 6.5 Probability of accepting (P_a%) a food lot; two-class plan, $c = 1$–3

Number of samples (n)	Value of c	Probability of acceptance (P_a%)						
		Actual percentage of defective samples						
		2	5	10	20	30	40	50
5	1	100	98	92	74	53	34	19
	2	100	100	99	94	84	68	50
	3	100	100	100	99	97	91	81
10	1	98	91	74	38	15	5	1
	2	100	99	93	68	38	17	5
	3	100	100	99	88	65	38	17
15	1	96	83	55	17	4	1	<0.5
	2	100	96	82	40	13	3	<0.5
	4	100	100	99	84	52	22	6
20	1	94	74	39	7	1	< 0.5	< 0.5
	4	100	100	96	63	24	5	1
	9	100	100	100	100	95	76	41

Adapted from ICMSF 1986 and reprinted with permission from the University of Toronto Press.

Table 6.6 Probability of accepting (P_a%) a food lot; three-class plan

Percentage defective (P_d%)	Value of c	Percentage marginal (P_m%)					
		10	20	30	50	70	90
Number of samples (n) = 5							
50	3	3	3	2	<0.5		
	2	3	2	1	<0.5		
	1	2	1	<0.5			
40	3	8	7	6	2	<0.5	
	2	8	6	4	<0.5		
	1	6	4	1	<0.5		
30	3	17	16	15	7	<0.5	
	2	16	14	11	2	<0.5	
	1	14	9	5	2	<0.5	
20	3	33	32	31	20	4	<0.5
	2	32	29	24	9	1	<0.5
	1	29	21	13	2	<0.5	
10	3	59	58	56	43	18	<0.5
	2	58	55	47	23	5	<0.5
	1	53	41	27	7	1	<0.5
5	3	77	77	75	60	31	2
	2	77	72	63	35	9	<0.5
	1	70	55	38	12	1	<0.5
0	3	100	99	97	81	47	8
	2	99	94	84	50	16	1
	1	92	74	53	19	3	<0.5
Number of samples (n) = 10							
40	3	1	<0.5				
	2	<0.5					
	1	<0.5					
30	3	3	2	1	<0.5		
	2	2	1	<0.5			
	1	2	<0.5				
20	3	10	8	5	<0.5		
	2	9	6	2	<0.5		
	1	7	3	1	<0.5		
10	3	34	29	20	3	<0.5	
	2	32	21	10	1	<0.5	
	1	24	11	4	1	<0.5	
5	3	59	51	36	20	8	2
	2	55	39	20	8	2	<0.5
	1	43	21	8	2	<0.5	
0	3	99	88	65	17	1	<0.5
	2	93	68	38	5	<0.5	
	1	74	38	15	5	1	<0.5

Adapted from ICMSF 1986 and reprinted with permission from the University of Toronto Press.

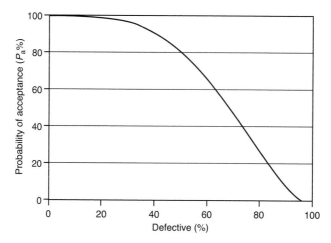

Figure 6.4 Operating characteristic curve for sample plan $n = 5, c = 3$.

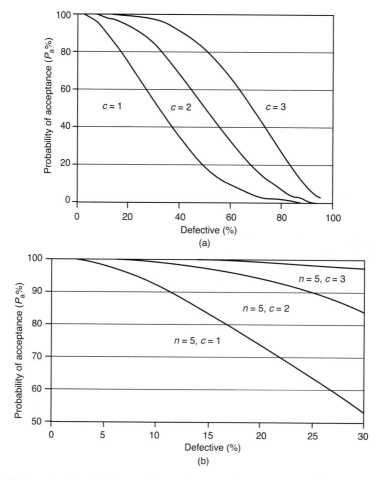

Figure 6.5 (a) Operating characteristic curve for $n = 5, c = 1$–3. (b) Enlarged area of Figure 6.5a.

defectives. If a producer sets a limit of 10% defectives (i.e. $p = 10\%$) using a two-class plan of $n = 5$, $c = 2$, then the probability of acceptance (P_a) is 99%. This means that on 99 of every 100 occasions when a 10% defective lot is sampled, one may expect to have two or fewer of the five tests showing the presence of the organism and thus calling for 'acceptance', while on 1 of every 100 times, there will be three or more positives, calling for non-acceptance. **Therefore, a sampling plan of $n = 5$, $c = 2$ will mean that 10% defective lots will be accepted on most (99%) sampling occasions!** Even increasing the number of samples to 10 ($n = 10$, $c = 2$) means that 10% defective batches will be accepted 93% of occasions. Hence, there is a need for the proactive approach of HACCP for assured food safety; see Chapter 8.

It is therefore apparent that no practical sampling plan can ensure the absence of the target micro-organism, and the concentration of the target organism may be greater than the set limit in parts of the food lot not sampled. The absence of a target organism in five randomly chosen samples only gives a 95% confidence that the food lot is less than 50% contaminated. If 30 samples had been analysed then the food lot is (with 95% confidence) contaminated at less than 10%. It requires 300 randomly taken samples giving the absence of the target micro-organism a 95% confidence that the food lot is less than 1% contaminated. Therefore, no sampling plan can guarantee the absence of a pathogen, unless every gram of the food was analysed leaving nothing for consumption.

6.7.4 Producer risk and consumer risk
It follows from the operating characteristic curve that it is possible that a 'bad' lot of food will on occasions be accepted, and conversely a 'good' lot will be rejected. This is known as the 'consumer's risk' and 'producer's risk', respectively. The 'consumer's risk' is considered to be the probability of accepting a lot whose actual microbial content is substandard as specified in the plan, even though the microbiological analysis of the sample units conform to acceptance (P_a). The 'producer's risk' is the converse, that is the producer rejects a batch of food due to unrepresentative poor microbiological results. It is expressed by '$1 - P_a$' in Figure 6.6.

6.7.5 Stringency of two- and three-class plans, setting n and c
The stringency of the two-class sampling plan depends upon the values chosen for n and c. Here, n is the number of samples units analysed and c is the maximum acceptable number

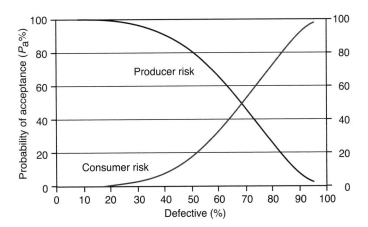

Figure 6.6 Producer's risk/consumer's risk curve.

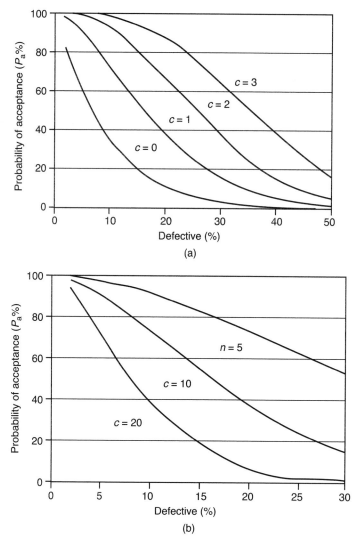

Figure 6.7 (a) Stringency of sampling plans, illustrated by $n = 10, c = 0$–3. (b) Stringency of sampling plans, illustrated by $n = 5$–20, $c = 1$.

of sample units that may exceed the value of m. Also, m is the maximum number of relevant bacteria/g. The lot is rejected if c is exceeded. If for a given value for c, the number of sample units (n) is increased, then the better in microbiological terms the food lot must be to have the same chance of being passed. Conversely, for a given sample size n, if c is increased, the sampling plan becomes more lenient as there is a higher probability of acceptance (P_a). This can be seen in Figure 6.7a where increasing $c = 0$–3 means the plan becomes more stringent, less chance of accepting a defective lot. However, in Figure 6.7b, where the number of samples is increased ($n = 5$–20) and c is fixed ($c = 1$), the plan becomes more stringent.

In a three-class plan, it is the values of n and c which determine the probability of acceptance, P_a, for a food lot of given microbiological quality.

The microbiological quality is given by determining the percent 'defective' proportions:

P_d = % defective; above M

P_m = % 'marginally acceptable'; m to M

P_a = % 'acceptable'; equal to or less than m

Since the three terms must equal 100%, then only the first two terms need to be determined. Probability values for three-class plans are given in Table 6.5.

Taking a lot of food of which 20% of the sample counts are marginally acceptable (P_m = 20%) and 10% 'defective' (P_d = 10%), the effects of n and c can be compared in Table 6.6. For $n = 5$, $c = 3$, the probability of acceptance (P_a) is 58% of occasions; if c is lowered to 1, then P_a decreases to 41%. However, if n is increased to 10 ($c = 3$), then P_a is 29% of occasions. The most stringent is $n = 10$, $c = 3$, where $P_a = 11\%$ of occasions.

This level of acceptance (11 in 100) for lots where 10% are defective and 20% are marginally acceptable reinforces the fact that microbiological hazards must be controlled in food using the proactive HACCP approach rather than a retrospective end-product testing approach.

Therefore, the setting of n and c varies with the desired stringency (probability of rejection). For stringent cases, n is high and c is low; for lenient cases, n is low and c is high. As n decreases, the chance of acceptance of bad lots increases.

6.7.6 Setting the values for *m* and *M*

The level of the target organism which is acceptable and attainable in the food is 'm'. It can be set from levels attained in GMP or, if the target organism is a pathogen, then 'm' may be set at zero for a given volume of sample (i.e. 25 g).

'M' is only used in three-class plans as the hazardous or unacceptable level of contamination caused by poor hygienic practice. There are three methods of setting the value of M:

1 As a utility (spoilage or shelf-life) index, where it relates to levels of bacteria causing detectable spoilage (odour, flavour) or to a decrease in shelf-life to an unacceptable short period; see Section 3.2.

2 As a general hygiene indicator, relating to levels of an indicator organism reflecting an unacceptable condition of hygiene.

3 As a health hazard where it relates to infectious dose. This value can be set using epidemiological and laboratory data, and similar sources of information; see Table 4.2.

Therefore, the values of m and M are independent of each other and have no set relationship.

6.8 Microbiological limits

6.8.1 Definitions

There are various terms used in reference to microbiological limits and these are defined as follows:

- 'Microbiological standards' refer to compulsory microbiological levels laid down in statue.
- 'Microbiological guidelines' refer to levels set out in guidance which does not have legal force.
- 'Microbiological criteria' may refer to either of the above or to levels in use by the food industry.
- 'Microbiological specifications' are agreed within or between companies and generally have no direct legal implications.

Thus, specifications can be prepared for raw materials supplied to a food processor, of foods at various stages of preparation and for final products. In the last case, the microbiological specifications may be those agreed as reasonable and attainable by the company or they may be standards imposed by or jointly agreed with an external agency. Specifications may include standards for total numbers of micro-organisms, food pathogens, indicator or spoilage organisms.

When compiling microbiological specifications for raw materials and final products, it is desirable to start with as wide a range of relevant test methods as is practicable so that comprehensive data on the background microbiology can be built up. These methods should give the highest recovery of organisms and reproducible results. Ideally, test samples should be exchanged between laboratories and analysed by the agreed methods to ensure that similar results are being obtained. Greater attention should be paid to raw materials and foods where erratic or unexpected results are obtained. Specifications should reflect what is attainable under GMP but should include tolerances to allow for sampling inaccuracies.

6.8.2 Limitations of microbiological testing
When intending to use microbiological criteria for testing food, a large number of problems need to be considered:
1 Cost of analysis with regard to trained personnel, equipment and consumables.
2 Sampling problems, obtaining 'representative' samples is very difficult.
3 Accepting a food lot which contains unacceptable levels of micro-organisms/toxins, simply because of presence at low levels and heterogenous distribution.
4 Variation in results, such as plate count which have 95% confidence limits of ±0.5 log cycles; see Table 5.1.
5 Destructive testing means that samples cannot be retested.
6 Time-consuming due to need for prolonged incubation periods.
7 Sensitivity and robustness of detection methods.

6.9 Examples of sampling plans

6.9.1 Egg products
Eggs are perishable since they are highly nutritious and hence subject to microbiological growth and spoilage. They are associated with certain foodborne pathogens, most notably *Salmonella* spp. Despite pasteurisation processes (liquid egg: 64.4°C, 2.5 minutes or 60°C, 3.5 minutes), eggs and egg products may be contaminated due to insufficient heat treatment or from post-pasteurisation contamination. *Salmonella* cells can subsequently multiply to an infectious dose due to temperature abuse after thawing or rehydration. *St. aureus* has been identified as a foodborne pathogen associated with pasta where it can grow and form toxic levels of enterotoxin. Sampling plans for eggs and egg products are given in Table 6.7.

6.9.2 Milk and milk products
Dairy products are highly nutritious, have a neutral pH and water activity conducive for the growth of foodborne pathogens. Dairy products are divided into two groups:
1 The more perishable (fresh) products such as milk, cream, flavoured milk and skim milk drinks, fresh cheese (cottage cheese) and fermented milks.

Table 6.7 Sampling plan for egg products

Target organism	Case	Plan cases	n	c	m	M
					Limit (/g)	
Aerobic plate count	2	3	5	2	5×10^4	10^6
Coliforms	5	3	5	2	10	10^3
Salmonella spp.	10	2	5	0	0	—
(general population)	11	2	10	0	0	—
	12	2	20	0	0	—
Salmonella spp.	10	2	15	0	0	—
(high-risk population)	11	2	30	0	0	—
	12	2	60	0	0	—

Adapted from ICMSF 1986 and reprinted with permission from the University of Toronto Press.

2 The relatively stable products having extended shelf-life under appropriate conditions of storage, such as hard cheese, butter, dried milk products, ice cream mixes, evaporated (canned) milk and sterilised or ultra high temperature (UHT) milk (for fluid consumption). Microbiological criteria cannot be effectively applied to group 1, since the products will probably have been consumed before the microbiological analysis has been completed. Milk products of group 2 which are associated with microbiological hazards (i.e. dried milk and ripened cheese) are usually microbiologically tested before distribution. Sampling plans proposed by ICMSF (1986) are given in Table 6.8.

6.9.3 Processed meats

Processed meats include a range of meat products that have been processed by heat treatment, curing, drying and fermenting. There are a number of microbiological hazards associated with meat products. The ICMSF (1986) sampling plans are given in Table 6.9. The main target organisms are *St. aureus*, *Cl. perfringens* and *Salmonella*.

Table 6.8 Sampling plan for dairy products

Product	Target organism	Case	Plan case	n	c	m	M
						Limit (/g)	
Dried milk	Aerobic plate count	2	3	5	2	3×10^4	3×10^5
	Coliforms	5	3	5	1	10	10^2
	Salmonella spp.	10	2	5	0	0	—
	(normal population)	11	2	10	0	0	—
		12	2	20	0	0	—
	Salmonella spp.	10	2	15	0	0	—
	(high-risk population)	11	2	30	0	0	—
		12	2	60	0	0	—
Cheese, hard and semi-soft types	*St. aureus*	8	2	5	0	10^4	—

Adapted from ICMSF 1986 and reprinted with permission from the University of Toronto Press.

Table 6.9 Sampling plan for processed meats

Product	Target organism	Case	Plan class	n	c	Limit (/g) m	M
Dried blood, plasma and gelatin	*St. aureus*	8	3	5	1	10^2	10^4
	Cl. perfringens	8	3	5	1	10^2	10^4
	Salmonella spp.	11	2	10	0	0	—
Roast beef and pâté	*Salmonella* spp.	12	2	20	0	0	—

Adapted from ICMSF 1986 and reprinted with permission from the University of Toronto Press.

6.9.4 Cereals and cereal products

A range of bakery products are covered by this sampling plan (Table 6.10). Moulds and the persistence of bacterial spore formers are important as the products are often dry (low water activity).

6.9.5 Cook-chill and cook-freeze products

Guidelines from the Department of Health (United Kingdom) are given in Table 6.11. The guidelines target five foodborne pathogens and the aerobic plate count as an indicator of microbial load.

6.9.6 Seafoods

The US Food and Drug Administration and Environmental Protection Agency have given guidance levels for seafood microbiological hazards (Table 6.12).

Table 6.10 Sampling plan for cereals and bakery products

Product	Target organism	Case	Plan case	n	c	Limit (/g) m	M
Cereals	Moulds	5	3	5	2	10^2–10^{4a}	10^5
Soya flour, concentrates and isolates	Moulds	5	3	5	2	10^2–10^4	10^5
	Salmonella spp.	10	2	5	0	0	—
Frozen bakery products (ready-to-eat) with low acid or high a_w fillings or toppings	*St. aureus*	9	3	5	1	10^2	10^4
	Salmonella spp.	12	2	20	0	0	—
Frozen bakery products (to be cooked) with low acid or high a_w fillings or toppings (e.g. meat pies, pizzas)	*St. aureus*	8	3	5	1	10^2	10^4
	Salmonella spp.	10	2	5	0	0	—
Frozen entrées containing rice or corn flour as a main ingredient	*B. cereus*	8	3	5	1	10^3	10^4
Frozen and dried products	*St. aureus*	8	3	5	1	10^2	10^4
	Salmonella spp.	10	2	5	0	0	—

[a]The exact value will vary with the type of grain.
Adapted from ICMSF 1986 and reprinted with permission from the University of Toronto Press.

Table 6.11 Sampling plan for cook-chill and cook-freeze foods at point of consumption

Target organism	Limit
Aerobic plate count	$<10^5$/g
E. coli	<10/g
St. aureus	<100/g
Cl. perfringens	<100/g
Salmonella spp.	Absent in 25 g
L. monocytogenes	Absent in 25 g

Adapted from Department of Health (UK) guidelines.

Table 6.12 US Food and Drug Administration and Environmental Protection Agency guidelines for seafood microbiology hazards

Product	Guideline/tolerance
Ready-to-eat fishery products (minimal cooking by consumer)	Enterotoxigenic *Escherichia coli* (ETEC) – 1×10^3 ETEC/g, LT or ST positive
Ready-to-eat fishery products (minimal cooking by consumer)	*Listeria monocytogenes* – presence of organism
All fish	*Salmonella* species – presence of organism
All fish	*Staphylococcus aureus* – (1) positive for staphylococcal enterotoxin, or (2) *Staphylococcus aureus* level is equal to or greater than 10^4/g (MPN)
Ready-to-eat fishery products (minimal cooking by consumer)	*Vibrio cholerae* – presence of toxigenic 01 or non-01
Ready-to-eat fishery products (minimal cooking by consumer)	*Vibrio parahaemolvticus* – levels equal to or greater than 1×10^4/g (Kanagawa positive or negative)
Ready-to-eat fishery products (minimal cooking by consumer)	*Vibrio vulnificus* – presence of pathogenic organism
All fish	*Clostridium botulinum* – (1) Presence of viable spores or vegetative cells in products that will support their growth, or (2) presence of toxin
Clams and oysters, and mussels fresh or frozen – imports	Microbiological – (1) *E. coli* – MPN of 230/100 g (average of subs or 3 or more of 5 subs); or (2) APC – 500 000/g (average of subs or 3 or more of 5 subs)
Clams, oysters, and mussels, fresh or frozen – domestic	Microbiological – (1) *E. coli* or faecal coliform – 1 or more of 5 subs exceeding MPN of 330/100 grams or 2 or more exceeding 230/100 grams; or (2) APC – 1 or more of 5 subs exceeding 1 500 000/g or 2 or more exceeding 500 000/g
Salt-cured, air-dried uneviscerated fish	Not permitted in commerce (*Note:* small fish exemption)
Tuna, mahi mahi and related fish	Histamine – 500 ppm set based on toxicity, 50 ppm set as defect action level, because histamine is generally not uniformly distributed in a decomposed fish. Therefore, if 50 ppm is found in one section, there is the possibility that other units may exceed 500 ppm

6.10 Implemented microbiological criteria

6.10.1 Microbiological criteria in the European Union

Most countries in the EU have implemented the Food Hygiene Directive (93/43/EEC; Section 11.6) into national law. Although most have restricted the implementation to the first five principles of HACCP, there are differences in implementation within Europe; for example, the Netherlands have a high emphasis on sampling programmes within inspections. Many businesses use microbiological criteria in contractual agreements with suppliers and customers and as a means of monitoring the hygiene of the production environment. These criteria are laid out in guidelines developed by individual companies of the relevant industry sector, and are generally based on GMP. Under the EC Food Hygiene Directive, it is possible to include microbiological guidelines in Industry Guides to Good Hygienic Practice. The EU microbiological criteria have recently been revised.

Table 6.13 Aerobic plate count categories for different types of ready-to-eat foods

Food group	Product	Category
Meat	Beefburgers and pork pies	1
	Poultry (unsliced)	2
	Pate, sliced meat (except ham and tongue)	3
	Tripe and other offal	4
	Salami and fermented meat products	5
Seafood	Herring and other pickled fish	1
	Crustaceans, seafood meals	3
	Shellfish (cooked), smoked fish, taramasalata	4
	Oysters (raw)	5
Dessert	Mousse/dessert	1
	Cakes, pastries, slices and desserts – without dairy cream	2
	Trifle	3
	Cakes, pastries, slices and desserts – with dairy cream	4
	Cheesecake	5
Savoury	Bhaji	1
	Cheese-based bakery products	2
	Spring rolls, satay	3
	Houmus, tzatziki and other dips	4
	Fermented foods, bean curd	5
Vegetable	Vegetables and vegetable meals (cooked)	2
	Fruits and vegetables (dried)	3
	Prepared mixed salads	4
	Fruits and vegetables (fresh)	5
Dairy	Ice cream (dairy and non-dairy)	2
	Cheese, yogurt	5
Ready-to-eat meals		2
Sandwiches and filled rolls	Without salad	3
	With salad	4

Adapted from Anon. 2000, with permission of The Health Protection Agency.

Table 6.14 Guidelines for ready-to-eat foods

Food category (see Table 6.13)	Criterion			
	Satisfactory	Acceptable	Unsatisfactory	Unacceptable/ potentially hazardous
Aerobic plate count (30°C; 48 ± 2 hours)				
1	$<10^3$	$10^3-<10^4$	$\geq 10^4$	N/A
2	$<10^4$	$10^4-<10^5$	$\geq 10^5$	N/A
3	$<10^5$	$10^5-<10^6$	$\geq 10^6$	N/A
4	$<10^6$	$10^6-<10^7$	$\geq 10^7$	N/A
5	—[a]	—[a]	—[a]	—[a]
Indicator organisms				
E. coli (total) 1–5	<20	$20-<100$	100	N/A
Enterobacteriaceae 1–5	<100	$10^2-<10^4$	$>10^4$	N/A
Listeria spp. (total) 1–5	<20	$20-<100^a$	>100	N/A
Pathogens				
Salmonella serovars 1–5	Not detected in 25 g			Present in 25 g
Campylobacter spp. 1–5	Not detected in 25 g			Present in 25 g
E. coli O157:H7 and other VTEC 1–5	Not detected in 25 g			Present in 25 g
V. parahaemolyticus – seafoods 1–5	<20	$20-<10^2$	$10^2-<10^3$	$\geq 10^3$
L. monocytogenes 1–5	<20	$20-<10^2$	N/A	$\geq 10^2$
S. aureus 1–5	<20	$20-100$	$100-<10^4$	$\geq 10^4$
C. perfringens 1–5	<20	$20-<100$	$100-<10^4$	$\geq 10^4$
B. cereus and B. subtilis group 1–5	$<10^3$	$10^3-<10^4$	$10^4-<10^5$	$\geq 10^5$

N/A denotes not applicable.

[a] Guidelines for aerobic plate counts may not apply to certain fermented foods, for example salami, soft cheese and unpasteurised yogurt. These foods fell into category 5. Acceptability is based on appearance, smell, texture and the levels or absence of pathogens.

Adapted from PHLS 2000, with permission of The Health Protection Agency.

The Food Hygiene Directive (93/34/EEC) provides:

'Without prejudice to more specific Community rules, microbiological criteria and temperature control criteria for certain classes of foodstuffs may be adopted in accordance with the procedure laid down in Article 14 and after consulting the Scientific Committee for Food set up by Decision 74/234/EEC.'

6.10.2 EU directives specifying microbiological standards for foods

Some microbiological criteria are standards which are specified in EC product-specific food hygiene directives. Some of these 'vertical' directives make provision for them to be introduced in the future, or, where standards have been set, there is scope for them to be revised or added. The EU may also lay down suitable laboratory methods. Such directives relate to food primarily during manufacturing, transport and wholesale storage but not usually at retail or catering level (except where small producers sell direct to the consumer).

Food safety in the EU is partially controlled through:

1 Vertical directives, which deal with products of animal origin: fresh meat, poultry, milk, fish, eggs. These apply at manufacture, storage and during transport.
2 Horizontal directives, which apply safety measures to all foodstuffs not covered by the vertical directives and when all foods enter the retail market.

See Section 11.6 for a fuller account of EU regulations.

6.11 UK guidelines for ready-to-eat foods

In the United Kingdom, microbiological guidelines for ready-to-eat foods were first published by the former 'Public Health Laboratory Service' (PHLS), now 'Health Protection Agency' (HPA) in 1992 and recently revised in 2009 (PHLS 1992, 1996, 2000; HPA 2009, see Web resources). They were collated to assist in the implementation of the UK Food Safety Act 1990 (Section 30) which provides for food examiners to examine samples of food submitted to them and issue certificates specifying the results of their examination. The guidelines are for application during the shelf-life of the product and are not at point of production or mandatory UK Government standards. Foods are divided into five categories (Table 6.13) for the aerobic colony count analysis, whereas the same criteria are used for all indictor organisms and foodborne pathogens. The guidelines revision in 2000 included new *Enterobacteriaceae* and *E. coli* criteria (Table 6.14).

Apart from setting proscriptive limits for certain pathogens, the guidelines recommend ranges of bacterial counts for a number of different types of food which allow the division of results into 'Satisfactory', 'Borderline – limit of acceptability', 'Unsatisfactory' or 'Unacceptable/potentially hazardous'. Although the guidelines have no formal status and refer only to 'ready-to-eat' food, they do reflect the opinions of experienced workers with access to a wealth of unpublished data collected over 50 years by the HPA (formerly PHLS). They are applied to single samples and therefore do not have the statistical validity of the ICMSF sampling plans. The application of the guidelines for surveillance purposes has already been covered in Section 1.12.7.

7 Hygienic production practices

7.1 Contribution of food handlers to foodborne illness

Food handlers are often associated with food poisoning outbreaks, estimates of the contribution range from 7% to 20%. Table 7.1 summarises the human carriage of foodborne pathogens, which would include workers in the food industry. A large review of 816 foodborne disease outbreaks (from 1927 to the first quarter of 2006) implicating food handlers included outbreaks caused by bacteria (*Salmonella*, *St. aureus*, *Shigella* and *Streptococcus*), viruses (Norovirus, hepatitis A) and protozoal parasites (*Cyclospora*, *Giardia* and *Cryptosporidium*) (Greig *et al.* 2007, Todd *et al.* 2007a, 2007b). Multiple foods and multi-ingredient foods are identified most frequently with outbreaks involving food handlers, probably due to the more frequent hand contact during preparation and serving. Within the 816 outbreaks, 11 involved >1,000 people, and 4 involved >3000 people.

Among the most frequently reported factors associated with the involvement of the infected worker in the large study above were bare hand contact with the food and inadequate hand washing. Many workers were asymptomatic spreaders of intestinal pathogens and therefore were not aware of the risk they represent, or had infected family members. Guzewich & Ross (1999) reported that hepatitis A virus (HAV) and Noroviruses probably accounted for >60% of foodborne outbreaks, and that infected food handlers are the most common source of contamination.

Infected food handlers may be:
- symptomatic and shed viruses during periods of illness;
- asymptomatic (including those recovered from illness) who continue to shed viruses, for example, Noroviruses may continue to be shed for at least 3 weeks after recovery;
- spreading viruses through contact with sick people (such as sick child at home).

The stability of human enteric viruses means that they can persist in food longer than bacterial contaminants, and they have low infectious doses (~1000 particles).

A HAV vaccine is available; however, it has not been used for food handlers. Instead, personal hygiene has been emphasised to prevent food contamination.

7.2 Personal hygiene and training

All applicants seeking employment in the food industry, who are likely to come into direct or indirect contact with foods, should be examined medically to ensure fitness for work. The examination should include answering a questionnaire recording the past medical history of the applicant, especially enteric infections, skin rashes, boils, discharges from the eyes, nose and ears (indicative of *St. aureus* infection), and recent travel in regions where intestinal disorders

Table 7.1 Human carriage of foodborne pathogens

Faeces	Salmonella spp., E. coli, Shigella spp., Norwalk-like virus, hepatitis A, Giardia lamblia	1 in 50 employees are highly infective and shedding 10^9 pathogens/g faeces
Vomit	Norwalk-like virus	10 viral particles are infectious
Skin, nose, boils and skin infections	St. aureus	60% of the population are carriers; there are 10^8 organisms per drop of pus
Throat and skin	Streptococcus group A	10^5 Strep. pyogenes in a cough

Reproduced from Snyder 1995, with kind permission of Springer Science and Business Media.

are prevalent or endemic. Stool sampling may be used to screen for carriers of enteric pathogens such as *Salmonella*, although sometimes their excretion can be intermittent. Obviously, medical examination and questionnaires should be applied to other employees on a regular basis. No one suffering from enteric infections should be allowed to work in the high-risk areas of food processing. As a matter of routine, all cuts, sores and other skin abrasions must be covered with approved waterproof dressings which are brightly coloured and incorporate a metal strip to detection if lost. It may not be practical to exclude carriers of *St. aureus* from food production areas, given the high percentage of human carriage ($\sim$30–40%). Therefore, minimising the spread of the organism by additional means is necessary, including wearing gloves and adopting good personal hygiene: no picking or wiping of nose with (gloved) hand.

All employees should be trained in the basic principles of food and personal hygiene. This should also include basic concepts of bacteriology, such as the ubiquitous nature of bacteria, their dissemination and rapid growth of bacteria under certain conditions. Hence, the role of employees in food poisoning cases and food spoilage. Complementing this should be methods for cleaning and disinfecting of equipment and contact surfaces, and personal hygiene.

In order for training to be effective, it must be continuous, which can take the form of one-to-one talks, discussion groups and refresher courses. Supporting material such as displaying posters, and photos of good and bad practices in suitable locations, and changing them to avoid indifference can provide a visual impact.

Staff should be trained as appropriate for their duties as given by ICMSF (1988). Table 7.2 shows the level of knowledge expected for line operators. To be effective, the management level in a food business management should be familiar with HACCP and its implementation.

Table 7.2 Level of knowledge expected for line operators

1	The major sources of micro-organisms in the product for which they are responsible
2	The role of micro-organisms in disease and food spoilage
3	Why good personal hygiene is required
4	The importance of reporting illness, lesions and cuts to supervisory staff
5	The nature of the control required at their point in the process
6	The proper procedures and frequency for cleaning equipment for which they are responsible
7	The procedures necessary to report deviations from control specifications
8	The characteristics of normal and abnormal product at their given step in the process
9	The importance of maintaining proper records
10	How to monitor the critical control point of operations within their responsibility

Reproduced from ICMSF 1988, with permission of Blackwell Publishing Ltd.

Table 7.3 Online food safety sites

Name	URL	Comment
Five keys to safer food	http://www.who.int/foodsafety/consumer/5keys/en/index.html	WHO global food hygiene message
Gateway to Government food safety information	http://www.foodsafety.gov	Gives access to a wide range of US and other governments food safety information sources
Food Safety Information Center (USDA)	http://foodsafety.nal.usda.gov	Provides food safety information for the general public, as well as industry, researchers and those in education
Foodlink	http://www.foodlink.org.uk	Contains a wide range of topics, and organises National Food Safety week (United Kingdom)
Eat well, be well	http://www.eatwell.gov.uk/keepingfoodsafe	Run by UK Food Standards Agency
Food safety web	http://www.foodsafetyweb.info/resources/NonEnglish.php	Food safety resources for non-English speakers

Management support of hygiene training programme could lead to increased productivity, and reduction in customer complaints which can be extremely costly (cf. Table 1.7).

There is a considerable amount of food safety training material available online. Some are from independent sources, and others from governmental regulators bodies (Table 7.3).

Food handlers, including seasonal workers, need to be educated specifically with regard to microbial hazards and personal hygiene. Most of the problems with foodborne viruses are due to contamination of food products that require manual handling and receive minimal processing (i.e. shellfish and salads). It has been recommended that food processing managers exclude symptomatic food handlers from high-risk areas until 48 hours after recovery (Cowdens *et al.* 1995). However, this is more applicable to bacterial than viral pathogens.

It is evident that the causes of gastroenteritis vary with age (Figure 1.1). Noroviruses and rotaviruses probably cause the majority of gastroenteritis in children under the age of 4 years, whereas bacteria (*Campylobacter* and *Salmonella* spp.) are the major cause of gastroenteritis in other age groups. Figure 1.2 indicates that males suffer from gastroenteritis more than females, except for the age group >74, probably because of the lower ratio of men to women in this age group. A possible reason for part of this difference is that fewer men than women wash their hands after using the lavatory: 33% compared with 60%; males taking an average of 47 seconds, compared with 79 seconds taken by females.

More than 1000 HAV particles (9.2% transfer rate) can easily be transferred from faecally contaminated fingers to foods and surfaces (Bidawid *et al.* 2000). Since the virus is likely to survive and remain infectious after any food processing steps, it is important to emphasise the need to implement Good Hygienic Practice (GHP), GMP and HACCP to prevent initial contamination.

The microbial flora of the hand is composed of the resident flora and transients. In the case of food equipment contact surfaces, it is possible to reduce the numbers of contaminating micro-organisms such that any remaining organisms will not undermine the quality of the processed food. Unfortunately, it is not possible to disinfect skin to the same degree and therefore bare hands are a potential means of transferring micro-organisms. Such distribution may involve the transfer of organisms from hands to food or their transfer from food to food via the hands. Special care must therefore be taken to ensure that these transmission routes are minimised.

Washing ones hands with soap and water principally removes the transient skin flora, which are mainly bacteria from the environment. Therefore, the numbers of enteric pathogens, including *E. coli* and *Salmonella*, which can pass through toilet paper onto the hands should be greatly reduced by thorough hand washing.

However, it is virtually impossible to remove the indigenous resident flora on the hands even when using bactericidal soaps and creams. This resident flora is a particular problem with employees who are carriers of *St. aureus* in their nasal passages. Many of these employees will also carry the organism on their hands as part of the established flora. Since it is impractical to prohibit ∼30−40% (the likely *St. aureus* carrier rate) of employees from contact with foods, other means of restricting the dissemination of *St. aureus* must be adopted.

Personnel must be encouraged to develop an attitude where hand washing becomes a virtually automatic response to certain situations. Thus, hands, fingernails and wrists should be thoroughly washed: (1) before starting work; (2) before and after lunch and tea breaks; (3) after using the toilet; (4) when leaving or returning to the processing area for any other reason; (5) when changing jobs within the process area; (6) when the hands become unexpectedly soiled or contaminated in any way such as after handling equipment or food which may be of a substandard quality. See Table 7.1 for details of pathogens carried by food handlers.

7.3 Cleaning

Having an effective cleaning and disinfection programme is a crucial step in hygienic food production. The main purpose is to reduce the number of pathogens in the environment and hence reduce potential food contamination. Subsequently, the efficiency of cleaning and disinfection regimes greatly affects final product quality and is essential in food production. However, the most effective sanitation programme cannot make up for poorly designed equipment and factory design. The terminology used in cleaning and disinfection technology is given in Table 7.4.

Cleaning must be performed at regular and frequent intervals, even continuously, so that a consistently good product quality is maintained. How this cleaning is done depends principally on:

1 the nature of the soil or contamination to be removed,
2 the type of surface to be cleaned,
3 the materials used for cleaning,
4 the degree of water hardness,
5 the standard of cleanliness required.

Cleaning must be the first step to remove organic material that would otherwise protect the pathogens from disinfection. The basic steps are as follows:

- Dry cleaning to remove gross soil or organic material.
- Wet cleaning with detergent to remove any residual material. This involves soaking, washing, rinsing and drying.

The final step is the disinfection process, which requires disinfectants to reduce or kill the remaining pathogens, rinsing off of these agents followed by drying. Since pathogens vary in their sensitivity to disinfectants, it is important to have prior knowledge of the pathogens that may be present.

Table 7.4 Definitions used concerning detergents and disinfectants

Term	Definition
Bactericide	A chemical agent which, under defined conditions, is capable of killing vegetative forms of bacteria but not necessarily bacterial spores
Bacteriostat	A chemical agent which, under defined conditions, is capable of preventing the growth of bacteria
Clean surface	One that is free from soil of whatever form and is odourless. Thus, it is one from which food debris, detergents and disinfectants have been removed. It will not contaminate foods in contact with it and has residual numbers of micro-organisms, if any, that could not undermine product quality during subsequent production. A clean surface is not necessarily sterile
Cleaning	Covers those processes concerned with the removal of soil from surfaces but not those concerned with sterilisation
Detergent	A substance which assists in cleaning when added to water
Disinfectant	Originally defined in medical terms as a chemical agent which destroys disease-producing organisms; now more correctly defined as an agent capable of destroying a very wide range of micro-organisms but not necessarily bacterial spores
Disinfection	Covers those processes concerned with the destruction of most of the micro-organisms, but not necessarily bacterial spores, on surfaces and in equipment. Any viable micro-organisms remaining are not capable of affecting the microbiological quality of foods coming into contact with the disinfected parts
Fungicide	A chemical agent which, under defined conditions, is capable of killing fungi including their spores
Sanitation	An all-embracing term covering those factors which assist in improving or maintaining man's physical well-being including the general cleanliness of his environment and the preservation of his health
Sanitiser	A substance that reduces the numbers of micro-organisms to an acceptable level (N.B.: This term is widely used in the United States and is virtually synonymous with the popular use of the term 'disinfectant')
Sanitising	See 'disinfection'
Soil	Any unwanted food residue, organic or inorganic matter remaining on equipment and other surfaces
Sterilisation	The process of destroying all forms of life, including microbial life
Steriliser	A chemical agent capable of destroying all forms of life

The type of soil to be removed varies according to the composition of the food and the nature of the process to which the food has been subjected. However, the food constituents themselves vary tremendously in terms of their clean ability (Tables 7.5 and 7.6) so that a wide choice of cleaning materials must be available for their removal. Food residues may be dry particulate, dried-on, cooked-on, sticky, fatty or slimy. Such residues may be best removed by physical means or by the use of hot or cold water almost invariably supplemented with detergents of one type or another. The length of time a food residue is left undisturbed also affects the case of cleaning. For example, fresh raw milk can be readily washed away, but if it is allowed to dry, greater difficulty will be experienced. This is due to the denaturation of the milk protein and the breakdown of the fat emulsion which results in the fat spreading over other milk particles, making them more difficult to remove.

The preliminary cleaning of smaller items of equipment may involve presoaking in warm (~45°C) or cold water to remove loosely adhering debris, followed by non-abrasive brushing.

Table 7.5 Soil characteristics

Food component	Water solubility	Removal	Affect of heating (i.e. cooking)
Protein	Alkaline conditions required, and slightly soluble under acidic conditions	Hard	Denaturation occurs which makes it more difficult to clean
Fat	Only under alkaline conditions	Hard	More difficult to clean
Sugar	Water soluble	Easy	Caramelisation occurs, making it more difficult to clean
Mineral salts	Water solubility varies	Varies from hard to easy	Generally insignificant

Steel wool and wire brushes should not be used since not only do they damage many surfaces, including stainless steel, but metal particles may also pass into foods and be a cause of customer complaints (cf. HACCP Section 8.5). Any cleaning aid causing damage to stainless steel and other food contact surfaces should be avoided since crevices can retain bacteria which will persist despite regular cleaning regimes (Holah & Thorpe 1990; cf. Section 7.5).

7.4 Detergents and disinfectants

Detergents must be capable of removing many different types of soil under a variety of conditions. They are not expected to possess bactericidal properties. However, during cleaning, detergents do physically remove a large number of bacteria. To achieve the above characteristics, various chemicals are blended to produce the correct traits for a particular application.

Detergents can be classified as follows:

1 *Inorganic alkalis*: Caustic and non-caustic.
2 Inorganic and organic acids.
3 *Surface active agents*: Anionic, non-ionic, cationic and amphoteric.
4 *Sequestering agents*: Inorganic and organic.

Table 7.6 Factors that influence the effectiveness of disinfectants

Factors affecting disinfectants	Aldehyde	Chlorine-releasing agent	Iodophor	Quaternary ammonium compound	Phenols and bis-phenols	Peroxygen
Organic matter	Inhibits effectiveness	Inhibits effectiveness	Some inhibition of effectiveness	Inhibits effectiveness	Slightly inhibits effectiveness	Inhibits effectiveness
Low temperature	Inhibits effectiveness	Slightly inhibits effectiveness	Slightly inhibits effectiveness	Inhibits effectiveness	Inhibits effectiveness	Slightly inhibits effectiveness
pH needed for maximum effectiveness	Alkaline	Acid	Acid	Acid	Alkaline or acid	Acid
Residual effects	Yes	No	Yes	No	Yes	No
Interaction with soaps	Compatible	Not compatible	Compatible	Not compatible	Compatible	Compatible

Table 7.7 Effectiveness of various disinfectants on microbial pathogens

Pathogen	Chlorhexadine	Chlorine-releasing agent	Iodophor	Phenol and bis-phenols	Single ammonium compound	Quaternary ammonium compound
C. jejuni	+	+++	+++	+++	+	+++
Salmonella serovars	+++	+++	+++	+++	+	+++
E. coli	+++	+++	+++	+++	+	+++
Cl. perfringens	−	++	++	−	−	−
St. aureus	−	+++	+++	+++	+++	+++
Aspergillus flavus	+	+++	+++	++	+/−	+++
Adenovirus	−	+++	+++	−	−	+++

Disinfectants for use on food contact surfaces should rapidly kill both Gram-positive and Gram-negative bacteria. The majority of mould spores should be killed and it would be advantageous if bacterial spores were inactivated as well. Disinfectants should be stable in the presence of organic residues and hard water salts. They should also be:

- non-corrosive and non-staining to plant surfaces of whatever type,
- odourless or have an inoffensive odour,
- non-toxic and non-irritating to the skin and eyes,
- readily soluble in water and readily rinsable,
- stable during prolonged storage in concentrated form and stable during short-term storage in dilute form,
- competitively priced and cost-effective in use.

Disinfectants used in the food industry are generally restricted to four groups:

1 Chlorine-releasing agents, that is sodium hypochlorite, chlorine dioxide, sodium dichloroisocyanurate and chloramines-T
2 Quaternary ammonium compounds
3 Iodophors, that is povidone-iodine and poloxamer-iodine
4 Amphoteric compounds

Disinfectants are effective against bacteria, viruses and fungi, but were not designed to be effective against parasites (Table 7.7). In general, the descending order of sensitivity is as follows: lipid-enveloped viruses, Gram-positive bacteria, non-enveloped viruses, fungi, Gram-negative bacteria and finally bacterial spores.

7.5 Microbial biofilms

In nature and food systems, micro-organisms attach to surfaces, grow into microcolonies, and produce biofilms (Denyer *et al.* 1993, Kumar & Anand 1998, Stickler 1999, Zottola & Sasahara 1994). A surface is conditioned first by the adsorption of organic molecules before bacterial colonisation. Rough surfaces provide areas for colonisation as they protect the bacteria from shear forces. The resulting build-up of micro-organisms is called a 'biofilm'. These biofilms, containing both spoilage and pathogenic organisms lead to an increased risk of microbial contamination of the final product. Consequently, there is a reduced shelf-life and increased risk of disease transmission. Biofilm formation can also impair heat transfer and contribute to metal corrosion. When biofilms are a nuisance, the term 'biofouling' or 'fouling' is used.

Biofilms are composed of viable and non-viable micro-organisms which become embedded together and to the surface by polyanionic extracellular polymeric substances (EPSs). The EPS is primarily composed of polysaccharides as well as other microbial-derived compounds: proteins, phospholipids, teichoic and nucleic acids. The EPS protects the microbes within the biofilm from biocides and desiccation and will also contribute to the persistence of micro-organisms within the biofilm.

Microbial cells embedded in EPS are phenotypically different from when they are growing in suspension (planktonic). One major difference is the increased resistance (orders of 10–100-fold) to antimicrobial agents (Druggan *et al.* 1993, Holmes & Evans 1989, Krysinski *et al.* 1992, Le Chevallier *et al.* 1988). A biofilm can less attractively be called 'slime' due to the build-up of EPS. Food can become contaminated with undesirable spoilage and pathogenic bacteria from sloughed portions of biofilms. Hence, biofilm formation leads to serious hygiene problems and economic losses due to food spoilage and the persistence of food pathogens. The major role proposed for the glycocalyx is that it protects attached bacteria from antibacterial agents. In the dairy industry, milk lines and gaskets can be colonised and are therefore a potential source of product contamination. Moulds may grow on the walls and ceilings of high-humidity food preparation areas and subsequently may infect the product. Many potentially pathogenic species have been isolated from biofilms associated with food preparation surfaces, including *St. aureus* and *L. monocytogenes*. Attached cells may not only exhibit the increased resistance to biocides expected from biofilm-associated organisms but also exhibit an increased resistance to heat treatment (Oh & Marshall 1995).

Biofilms form on any submerged surface where bacteria are present. They form in a sequence of events (Figure 7.1):

1 Nutrients from the food are adsorbed to the surface to form a conditioning film. This leads to a higher concentration of nutrients compared to the fluid phase and favours biofilm formation. The conditioning also affects the physico-chemical properties of the surface, that is surface free energy, changes in hydrophobicity and electrostatic charges which influence the microbial colonisation.

2 Micro-organisms attach to the conditioned surface. The adhesion of bacteria is initially reversible (van der Walls attraction forces, electrostatic forces and hydrophobic interactions) and subsequently irreversible (dipole–dipole interactions, hydrogen, ionic and covalent bonding and hydrophobic interactions). The bacterial appendages flagella, fimbriae and exopolysaccharide fibrils (also known as glycocalyx, slime or capsule) are involved in contact with the conditioning film. The exopolysaccharide is important in cell–cell and cell–surface adhesion and also protects the cells from dehydration.

3 The irreversibly attached bacteria grow and divide to form microcolonies which enlarge and subsequently coalesce to form a layer of cells covering the surface. During this phase, the cells produce more additional polymer which increases the cells anchorage and stabilises the colony from fluctuations of the environment.

4 The continuous attachment and growth of bacterial cells, and exopolysaccharide formation, leads to the biofilm formation. The biofilm layer may be several millimetres thick within a few days.

5 As the biofilm ages, sloughing occurs whereby detachment of relatively large particles of biomass from the biofilm occurs. The sloughed bacteria can either contaminate the food (note the subsequent non-homogenous presence of bacteria in food) or initiate a new biofilm further along the production line.

Conditioning film of food residues on work surface

Food residues (nutrient source)

Micro-organisms attached to conditioned surface

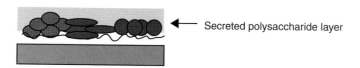

Micro-organisms divide and form micro-colonies. Polysaccharide formation stabilises the biofilm.

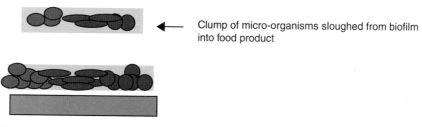

Secreted polysaccharide layer

Fragments of biofilm shed periodically

Clump of micro-organisms sloughed from biofilm
into food product

Figure 7.1 Biofilm formation.

Equipment design and plant layout are also important aspects of prevention and control strategies. Obviously, good design of equipment such as tanks, pipelines and joints is best to assist in the cleanability of the production line. The microtopography of a surface can complicate cleaning procedures when crevices and other surface imperfections shield attached cells from the rigours of cleaning. Stainless steel resists impact damage but is vulnerable to corrosion, while rubber surfaces are prone to deterioration and may develop surface cracks where bacteria can accumulate (Le Clercq-Perlat & Lalande 1994).

Generally, an effective cleaning and sanitation programme will inhibit biofilm formation. Mechanical treatment and chemical breakage of the polysaccharide matrix is very necessary for removal of biofilms. As can be seen in Figure 7.2, the adhered bacteria are covered with organic material (polysaccharide and food residues) which may inhibit the penetration of the

Mature biofilm

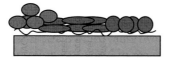

Detergent removal of surface (polysaccharide, food residues, etc.) material

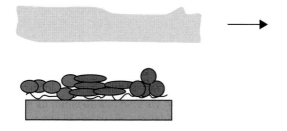

Disinfectant killing of micro-ogranisms

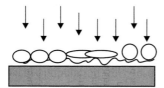

Rinse to remove dead microbial cells which otherwise could act as a nutrient source for a fresh biofilm

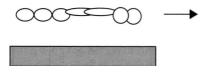

Figure 7.2 Biofilm removal.

disinfectant due to its lack of wetting properties. Therefore, a detergent activity is required to remove this outer layer prior to disinfection. The dead microbial biomass must be removed otherwise it may act as a conditioning film and nutrient source for further biofilm formation. New cleaning agents and enzyme treatments are being formulated for the effective removal of biofilms.

As previously stated, biofilms are problematic in the food industry. They are difficult to remove due to the combined issues of EPS recalcitrance and difficulties of cleaning complex processing equipment and production facilities. Therefore, biofilm control requires both the implementation of effective cleaning and sanitising procedures, along with well-designed processing equipment and the food processing environment which are amenable to routine and thorough cleaning. For rapid evaluation, ATP bioluminescence (Section 5.4.2) can be used to evaluate the effectiveness of cleaning regimes.

There are many problems associated with achieving efficient cleaning. These include (1) the faulty design and incorrect siting of food processing equipment; (2) recontamination of equipment that has been previously cleaned; (3) inadequate time for cleaning or cleaning too infrequently; (4) inadequate labour employed; (5) misuse of cleaning and disinfecting agents by gross variation from the recommended concentrations; and (6) lack of commitment by management and operatives.

Additional problems may be encountered if the factory hygiene officer does not enforce the recommended procedures, also if cleaning supervisors or operatives are absent and nobody is detailed to replace them. These problems are exacerbated in the food industry where there is a high turnover of labour resulting in new employees being given insufficient training.

With regard to the cleaning methods themselves, these are sometimes altered without full recognition of the possible long-term detrimental effects which the change may have on the physical state of the equipment or on the efficacy of action on the residual soils or micro-organisms. Promotion material for cleaning agents may well represent results that were obtained under test conditions, but are very different from the environment in which the product is to be used commercially. It is important, therefore, to always test the cleaning or disinfecting agent under typical commercial conditions before introducing it on a large scale in the factory. On the other hand, it should never be assumed that cleaning or disinfecting methods are effective because they always have been in the past. Checks should be made occasionally to confirm their continued reliability.

Finally, mechanical cleaning aids may break down and have to await repairs. When this happens, it puts additional pressure on the remaining working equipment which is, in turn, more liable to fail. Repairs may in consequence be hurried or inadequate and the whole cleaning programme may be jeopardised. This situation is often associated with a reluctance of senior management to spend sufficient money on cleaning aids.

7.6 Assessment of cleaning and disinfection efficiency

The efficiency of process line sanitation can be checked by visual inspection or by using microbiological techniques. Visual inspection is a simple but crude method which cannot determine the microbiological cleanliness achieved. Much depends on the care taken by the inspector. With experience, the inspector may know where to look for signs of inadequate cleaning but residual soils do vary in their visual detectability (e.g. some soils are not visible if there is a film of water on the cleaned surface), and high-intensity lighting must be directed on surfaces during inspection. In spite of these shortcomings, visual inspections are worthwhile provided they are performed assiduously. They can be carried out during or immediately after cleaning or even shortly before the start of the next production run. Inspections should be performed randomly so that employees involved in cleaning are unaware when they are due. A checklist of various items of equipment should be prepared and items, once cleaned and inspected, can be given a cleanliness rating which can be compared with earlier data. All the findings should be recorded in an inspection report. Additionally, records of the cleaning materials, dilutions and application times used should be kept where applicable. If equipment has not been adequately cleaned, there should be time available for corrective action before processing is restarted; particular care should be taken with the subsequent inspection of equipment given low cleanliness ratings.

The most commonly used surface cleaning assessment tests are by ATP bioluminescence, protein detection, as well as microbiological culturing and these have been discussed in Chapter 5. The ATP bioluminescence and protein tests determine the general hygienic status since they are not specific for microbial material, but will detect residual food material as well; see Sections 5.4.3 and 5.4.4. Microbiological culturing typically involves estimations from surface areas of the total numbers of viable bacteria, indicator organisms (i.e. *Enterobacteriaceae*, *E. coli*, etc.) and, where warranted, specific food spoilage, or food poisoning bacteria (i.e. *Listeria)*. In general, however, estimations are limited to 'total numbers' as this is the most sensitive guide to microbial hygienic status. The bacteria are removed from surfaces by means of sterile swabs, by rinsing with a known volume of sterile diluent or by agar contact method. Because microorganisms are often unevenly distributed on equipment surfaces, as large an area as possible should be sampled, but including less accessible points such as drains. For this reason, swabbing methods have found particular favour. The inherent problem with standard microbiological methods is that test results are normally unavailable until after 48-hour period due to the need for incubation of the test samples. Hence, the ATP bioluminescence and protein techniques have become test methods in the food industry as the results are obtained rapidly and the equipment has become more portable and simple to use. The ATP bioluminescence results can be collated, and trend analysis is applied to determine if the level of hygiene fluctuates during the week according to operatives, and so on.

The processed food may also be used to check cleaning efficiency. Foods first in contact with cleaned surfaces are most liable to pick up residual bacteria so that higher than expected counts on such foods may be indicative of inadequate cleaning. This principle is often used to monitor cleaning where liquid foods are being produced. Here samples are taken at regular intervals early in a production run, and if there is a gradual reduction in the bacterial count from the first sample onwards, rather than a fairly constant set of results, cleaning has been inadequate as the liquid is flushing out the residual bacteria.

In conclusion, it is pertinent to stress that when monitoring techniques show up areas where cleaning proves to be troublesome, it is vitally important to concentrate efforts in those areas and not to accept evidence of inefficient cleaning complacently.

8 **Food safety management tools**

8.1 **The manufacture of hygienic food**

Is there a way forward for the manufacture of food which is nutritious and appetising, and yet meets the expectations of the consumer regarding risk? A difficulty that arises in manufacturing 'safe' food is that the consumer is a mixed population with varying degrees of susceptibility and general lifestyle. Additionally, foods with 'high' levels of preservatives to reduce microbial growth are undesirable by the consumer and perceived as 'over processed' with 'chemical additives'. The consumer pressure is for greater varieties of fresh and minimally processed foods, natural preservatives with a **guarantee of absolute safety**.

The manufacture of safe food is the responsibility of everyone in the food chain, and food factory, from the operative on the conveyor belt to the higher management. It is not the sole responsibility of the food microbiologist. Nevertheless, the food microbiologist in industry will need to not only know which food pathogens are likely to occur in the ingredients, but also the effect of the food matrix and processing steps on cell survival in order to best advise on the most appropriate manufacturing regimes. The best methods for microbiological analysis are still being developed. It is obvious from the plethora of differing methods adopted by different countries that food poisoning statistics cannot be directly compared between countries due to the differing methods of analysis applied.

The production of safe food requires the following:
- Control at source
- Product design and process control
- Good Hygienic Practice during production, processing, handling and distribution, storage, sale, preparation and use
- A preventative approach because effectiveness of microbial end-product testing is limited
Taken from CAC ALINORM 97/13 (Codex Alimentarius Commission 1995).

An initial means of improving food safety is to prevent contamination of the raw materials in the first place. This has already been referred to with regard to water contamination by hepatitis A virus (HAV) and protozoal parasites (Sections 4.4 and 4.6). However, preventing microbial contamination is not always possible as microbes are part of our ecosystem. Soil contains bacteria, viruses and protozoa. Plants have microflora (Section 1.11.2), and microbes are part of the normal skin and intestinal flora of animals and fish. Many pathogens survive in the environment for long periods of time (Table 8.1) and can be transmitted to humans by a variety of routes (Figure 8.1). Inadequate temperature control during food production and preparation can increase the level of microbial pathogens and spoilage organisms prior to consumption.

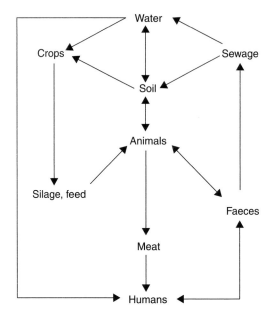

Figure 8.1 Routes of enteric pathogen transmission to humans.

Table 8.1 Survival of pathogenic micro-organisms in sewage sludge, soil and on vegetables

Organism	Conditions	Survival
Coliforms	Soil	30 days
Mycobacterium	Soil	Up to 2 years
tuberculosis	Soil	5–15 months
	Radish	3 months
Salmonella spp.	Soil	72 weeks
	Potatoes at soil surface	40 days
	Vegetables	7–40 days
	Beet leaves	21 days
	Carrots	10 days
	Cabbage/gooseberries	5 days
	Apple juice, pH 3.68	>30 days and multiplies
	Apple juice pH <3.4	2 days
S. Typhi	Soil	30 days
	Vegetables/fruit	1–69 days
	Water	7–30 days
Shigella spp.	Tomatoes	2–7 days
Vibrio cholerae	Spinach, lettuce, non-acid vegetables	2 days
Enterovirus group (polio, echo, coxsackie)	Soil	150–170 days
	Cucumber, tomato, lettuce, at 6–10°C	>15 days
	Radish	>2 months

Traditional ways to control microbial spoilage and safety hazards in foods include freezing, blanching, pasteurisation, sterilisation, canning, curing, syruping and the inclusion of preservatives.

Food additives such as preservatives are required to ensure that processed foods remain safe and unspoiled during its shelf-life. A range of preservatives are used in food manufacture including traditional foods. Many preservatives are effective under low-pH conditions: benzoic acid (<pH 4.0), propionic acid (<pH 5.0), sorbic acid (<pH 6.5), sulphites (<pH 4.5). The parabens (benzoic acid esters) are more effective at neutral pH conditions. For concise reference values on pH limits of microbial growth, the reader is directed to ICMSF (1996a) and related publications, as well as Table 2.8.

Food preservation technologies broadly comprise those which entail a physical treatment (e.g. heating or freezing) and those which use chemical additives (e.g. curing). With physical treatment, the food remains safe as long as any surviving pathogens are controlled and no additional contamination occurs after processing. In the latter, microbial contaminants are controlled for the duration of the chemical active lifetime. In order to design adequate treatment processes, an understanding of the factors affecting microbial growth is necessary. However, food is a chemically complex matrix, and predicting whether, or how fast, micro-organisms will grow in any given food is difficult. Most foods contain sufficient nutrients to support microbial growth, but other factors such as naturally occurring preservatives (benzoic acid in cranberries) and low pH will retard microbial growth.

Production processes can be very complicated. A general flow diagram for poultry processing is given in Figure 8.2. This is a relatively simple flow diagram with only one branching. It does not indicate the temperature and time at each step which affect microbial growth. Therefore, safe food production requires an all-encompassing approach involving the food operatives at the shop floor through to the management. Hence, a number of management safety tools such as Good Hygienic Practices (GHPs), Good Manufacturing Practice (GMP), Total Quality Management (TQM) and Hazard Analysis Critical Control Point (HACCP) need to be implemented.

Regarding legislation, the reader should always seek the appropriate regional authority. For information on hygienic food production with an emphasis on the factory layout, equipment design and staff training, the reader is recommended Forsythe & Hayes (1998) 'Food Hygiene, Microbiology and HACCP' and Shapton & Shapton (1991) 'Safe Processing of Foods'. Detailed information on individual micro-organisms can be found in the ICMSF book series, especially *Microorganisms in Foods 5: Characteristics of Microbial Pathogens* (1996a) and 'Microorganisms in Foods 6: Microbial Ecology of Food Commodities' (1998a) which should be regarded as essential requirements on the bookshelf (or library) of any food microbiologist.

The whole issue of safe food manufacturing comes within the umbrella of quality control and quality assurance. Hence, it requires the hygienic design of equipment and factory combined with managerial commitment to safety and quality. Diagrammatically, this is represented in Figure 8.3. The current issue concerning food safety is microbial risk assessment and the development of 'Food Safety Objectives' (ICMSF 1998c). These are governmental activities that eventually may decide the permissible level of foodborne pathogens (Figure 8.4). This issue is covered in Chapter 9.

There are some foods which are currently difficult to produce without a significant risk of foodborne infection. Outbreaks of *Salmonella* spp. and *E. coli* O157:H7 associated with raw seed sprouts have occurred in several countries, and currently the elderly, children and those with compromised immune systems are advised not to eat raw sprouts (such as alfalfa) until

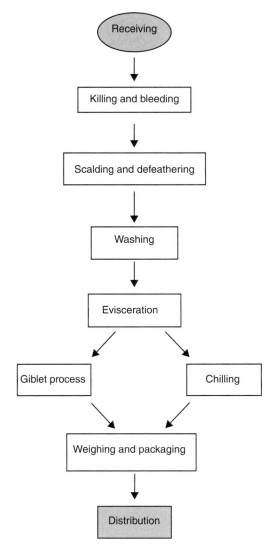

Figure 8.2 Flow sheet for poultry processing.

effective measures to prevent sprout-associated illness are identified (Taormina *et al.* 1999); see Section 1.11.2.

The globalisation of the food supply is recognised as a major trend contributing to food safety problems. Pathogenic micro-organisms are not contained within a single country's borders. Additionally, tourism and increased cultural interests may lead to new eating habits such as the consumption of Japanese sushi in Western countries. The continuous increase in international trade has been achievable partly through advances in food manufacturing and processing technologies as well as improvements in transportation. Regional trade arrangements and the overall impact of the Uruguay Round Agreements (Section 11.5) have reduced many tariff- and subsidy-related constraints to free trade, encouraging increased production and export from the

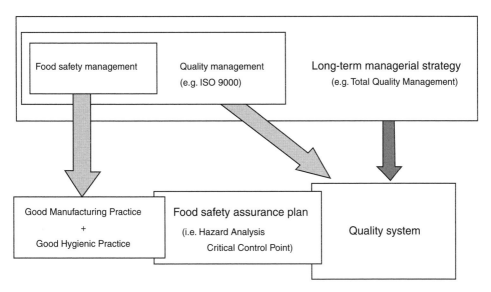

Figure 8.3 Food safety management tools. Adapted from Jouve *et al.* (1998), with kind permission of ILSI Europe.

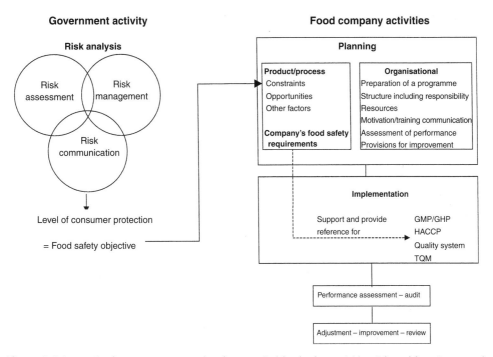

Figure 8.4 Interaction between goverments' and companies' food safety activities. Adapted from Jouve *et al.* (1998), with kind permission of ILSI Europe.

countries with the most cost-effective production means. However, many exporting countries do not have the infrastructure to ensure high levels of hygienic manufacture.

The continuing integration and consolidation of agriculture and food industries and the globalisation of food trade are changing the patterns of food production and distribution, as well as supply and demand. Production of raw materials is increasingly concentrated in fewer, specialised and larger production units. These changes may not only have many benefits, but also present new challenges to safe food production and may have widespread repercussions on health. The pressure to produce food for export is very significant in developing economies and can lead to improper agricultural practices. The consequences may include the following:

- Accidental or sporadic low-level microbial contamination of a single product can result in a major epidemic of foodborne illness
- High levels of mycotoxins, often resulting from poor storage and handling conditions
- High pesticide residues in food
- Industrial contamination of food with metals and chemicals such as polychlorinated biphenyls (PCBs) and dioxins

Changes in farming practices such as the introduction of new plant varieties and new crop rotation practices can introduce or increase the presence of hazards such as mycotoxins contamination of the crop. Because of the extensive distribution of food from a single source, the potential for many food consumers to be affected by a localised contamination has increased. This was demonstrated by the dioxin incident in Belgium and the BSE in England where a risk generated in one country had global implications (Table 1.8).

In direct contrast to the general concern on the presence of bacteria in food, there are in fact a number of foods which deliberately contain bacteria and fungi. These are the 'fermented' foods (Section 3.7) and these foods have been produced since the early era of civilisation. The taste, texture and flavour of the food are due to microbial metabolism. Hence, they are regarded as microbiologically safe. These foods now form the basis over the past 10–12 years for the development of 'functional foods'.

In Japan, and to a lesser extent in the United States, research into functional food has expanded greatly in recent years. Functional foods are expected to cause a health or physiological effect such as reduced risk to disease. Most functional foods currently approved contain either oligosaccharides or lactic acid bacteria for promoting intestinal health. In 1998, the Food and Drug Administration (FDA) recognised 11 foods or food components as showing correlation between intake and health benefits (Diplock *et al.* 1999). A significant portion of function foods is concerned with lactic acid bacteria ingestion (probiotics). Therefore, this book has reviewed probiotics as an extension from the age-old practice of fermented foods; see Section 3.8. The legislation regarding genetically modified foods will not be covered as there are more suitable texts available (FAO/WHO 1996, IFBC 1990, Jonas *et al.* 1996, Moseley 1999, OECD 1993, SCF 1997, Tomlinson 1998, WHO 1991).

Schlundt of the World Health Organisation (WHO) has described how approaches to manufacturing safe food have evolved in three stages:

1 GHPs in production and preparation to reduce the prevalence and concentration of the microbial hazard
2 HACCP and HACCP-like approach, which proactively identifies and controls the hazard
3 Risk analysis (including microbiological risk assessment), which focuses on the consequences to human of ingesting the microbial hazard, and the occurrence of the hazard in the whole food chain (from farm to fork)

These three steps are covered in this book as together they form the best means of food safety management. An overview of the development of food safety systems is given in Table 8.2.

Table 8.2 Important milestones in the development of food safety systems

Time	Activity
Distant past	Use of 'prohibition' principle to protect special groups within society against foodborne illnesses
1900 to present	Microbiological examination of food
1922	Introduction of process performance criteria by Esty and Meyer for canned, low-acid food products
1930–1960	Use of risk assessment (for different pathogenic organisms) in setting process performance criteria for heat pasteurisation of milk
1960	Introduction of good manufacturing practices
1971	Introduction of formal Hazard Analysis Critical Control Point system
ca. 1978	Start of predictive modelling of bacterial growth in food
1995	Introduction of formal quantitative risk analysis

Although industry and national regulators strive for production and processing systems which ensure that all food is 'safe and wholesome', a complete freedom from risks is an unattainable goal. Safety and wholesomeness are related to a level of risk that society regards as reasonable in the context, and is in comparison with other risks in everyday life.

The microbiological safety of foods is principally assured by the following:

- Control at the source
- Product design and process control
- The application of GHPs during production, processing (including labelling), handling, distribution, storage, sale, preparation and use
- The above, in conjunction with the application of the HACCP system. This preventative system offers more control than end-product testing, because the effectiveness of microbiological examination in assessing the safety of food is limited.

Consideration of safety needs to be applied to the complete food chain, from food production on the farm (or equivalent) through to the consumer. This is commonly known as the 'from farm to fork' approach. To achieve this, an integration of food safety tools is required (Forsythe & Hayes 1998; Figure 8.1):

- GMP
- GHP
- HACCP
- Microbial Risk Assessment (MRA)
- Quality management; ISO series
- TQM

These tools can be implemented worldwide which can ease communication with food distributors and regulatory authorities especially at port of entry.

8.2 Microbiological safety of food in world trade

It is important to understand that the ratification of the World Trade Organisation Agreement is a major factor in developing new hygiene measures for the international trade in food. There has been a noted requirement for quantitative data on the microbial risks associated with different classes of foods, and traditional GMP-based food hygiene requirements (i.e. end-product testing) are being challenged. Subsequently, risk assessment (Chapter 9) as a decision-making criterion for risk management will put more emphasis on predictive microbiology for the generation of exposure data and establishing critical limits for HACCP schemes.

The Final Act of the Uruguay Round of multilateral trade negotiations established the World Trade Organisation (WTO) to succeed the General Agreement on Tariffs and Trade (GATT, Section 11.5). The Final Act led to the 'Agreement on the Application of Sanitary and Phytosanitary Measures' (SPS Agreement) and the 'Agreement to Technical Barriers to Trade' (TBT Agreement). These are intended to facilitate the free movement of foods across borders, by ensuring that means established by countries to protect human health are scientifically justified and are not used as non-tariff barriers to trade in foodstuffs. The Agreement states that SPS measures based on appropriate standards, codes and guidelines developed by the Codex Alimentarius Commission (Section 11.4) are deemed to be necessary to protect human health and consistent with the relevant GATT provisions.

The SPS Agreement is of particular relevance to food safety. It provides a framework for the formulation and harmonisation of sanitary and phytosanitary measures. These measures must be based on science and implemented in an equivalent and transparent manner. They cannot be used as an unjustifiable barrier to trade by discriminating among foreign sources of supply or providing an unfair advantage to domestic producers. To facilitate safe food production for domestic and international markets, the SPS Agreement encourages governments to harmonise their national measures or base them on international standards, guidelines and recommendations developed by international standard-setting bodies.

The purpose of the TBT Agreement is to prevent the use of national or regional technical requirements, or standards in general, as unjustified technical barriers to trade. The agreement covers all types of standards including quality requirements for foods (except requirements related to sanitary and phytosanitary measures), and it includes numerous measures designed to protect the consumer against deception and economic fraud. The TBT Agreement also places emphasis on international standards. WTO members are obliged to use international standards or parts of them except where the international standard would be ineffective or inappropriate in the national situation.

The WTO Agreement also states that risk assessment should be used to provide the scientific basis for national food regulations on food safety and SPS measures, by taking into account risk-assessment techniques developed by international organisations.

Because of SPS and WHO, the Codex standards, guidelines and other recommendations have become the baseline for safe food production and consumer protection. Hence, the Codex Alimentarius Commission has become the reference for international food safety requirements.

8.3 Consumer pressure effect on food processing

In the production of food, it is crucial that proper measures are taken to ensure the safety and stability of the product during its whole shelf-life. In particular, modern consumer trends and food legislation have made the successful attainment of this objective much more of a challenge to the food industry. Firstly, consumers require more high quality, preservative-free, safe but mildly processed foods with extended shelf-life. For example, this may mean that foods have to be treated at mild pasteurisation rather than sterilisation temperatures. As acidity and sterilisation are two crucial factors in the control of outgrowth of pathogenic spore-forming bacteria, such as *Cl. botulinum*, addressing this consumer need calls for innovative approaches to ensure preservation of products (Gould 1995, Schellekens 1996, Peck 1997). Secondly, legislation has restricted the use and permitted levels of some currently accepted preservatives in different foods. This has created problems for the industry because the susceptibility of some micro-organisms to most currently used preservatives is falling. For example, work has identified that resistance to weak acid preservatives in spoilage yeast is mediated by a multidrug resistance protein (Piper *et al.* 1998). The consumer has demanded convenience foods which require minimal further processing before consumption. They also require less 'additives' and consequently less preservatives which affect product shelf-life. Therefore, new processing techniques are being introduced to increase product quality such as milder thermal processing, microwave heating, ohmic heating and high-pressure processing techniques; see Section 3.6. All these processes need to be fully evaluated regarding safe food production especially since 'mild' treatment might confer resistance to the stomach acid and hence lower the infectious dose; see Section 9.5.5. Additionally, there has been an increase in the consumption of food outside the home and an increase in the population of the vulnerable elderly.

There has been the dramatic appearance of emerging pathogens such as Shiga-toxin producing *E. coli* (STEC), formerly known as verotoxigenic *E. coli* (VTEC), multiantibiotic-resistant *S.* Typhimurium DT104 and a greater awareness of viral gastroenteritis. The link between variant CJD and BSE has been established and drastically affected abattoir procedures in certain countries and production methods.

8.4 The management of hazards in food which is in international trade

The management of microbiological hazards for foods in international trade can be divided into five steps (ICMSF 1997):
1 *Conduct a risk assessment*: The risk assessment and consequential risk management decisions provide a basis for determining the need to establish microbiological safety objectives.
2 *Establish food safety objectives (FSOs)*: A microbiological FSO is a statement of the maximum level of a microbiological hazard considered acceptable for consumer protection. These should be developed by governmental bodies with a view to obtaining consensus with respect to a food in international trade.
3 *Achievable FSO*: The FSO should be achievable throughout the food chain. This can be applied through the general principles of food hygiene and any product-specific codes and HACCP systems. The HACCP requirements must be developed by the food industry.
4 *Establish microbiological criteria, when appropriate*: This must be performed by an expert group of food microbiologists.
5 *Establish acceptance procedures for the food at port of entry*. A list of approved suppliers as determined by inspection of facilities and operations, certification, microbiological testing and/or other testing such as pH and water activity measurements.
Therefore, an understanding of HACCP (Section 8.7), Microbiological Criteria (Chapter 6), Microbiological Risk Assessment (Chapter 9) and Food Safety Objectives (Section 9.7) is required.

8.5 HACCP

Most food poisoning can be prevented by the application of the basic principles of food hygiene throughout the food chain. This is achievable through the following:
1 Education and training of food handlers and consumers in the application of safe food production practices.
2 Inspection of premises to ensure consistent hygienic practices are adhered to.
3 Microbiological testing for the presence or absence of foodborne pathogens and toxins.

Traditionally, the safety of food being produced was through end-product testing for the presence of food pathogens or their toxins. This retrospective approach, however, does not guarantee safe food for several reasons. Safe food production can be consistently achieved through the adoption of HACCP. This approach to safe food production is more fully explained later in this chapter and has been accepted worldwide. Therefore, factory inspection has changed in emphasis towards inspection of HACCP implementation.

The HACCP system for managing food safety was derived from two major developments:

- The HACCP system in 1960s, as pioneered by the Pillsbury Company, the United States Army and NASA as a collaborative development for the production of safe foods for the US space programme. NASA required 'zero defects' in food production to guarantee the astronauts' food (Bauman 1974).
- TQM systems which emphasise a total system approach to manufacturing that could improve quality while lowering costs.

HACCP is a scientifically based protocol. It is systematic, identifies specific hazards and measures for their control to ensure the safety of food. It is interactive, involving the food plant personnel. HACCP is a tool to assess hazards and establish control systems. Its focus is on prevention of problems occurring rather than a reliance on end-product testing. HACCP schemes can accommodate change, such as advances in equipment design, processing procedures or technological developments (Codex 1997a).

HACCP can be applied throughout the food chain from primary production to final consumption and its implementation should be guided by scientific evidence of risks to human health. As well as enhancing food safety, implementation of HACCP can provide other significant benefits. In addition, the application of HACCP systems can aid inspection by regulatory authorities and promote international trade by increasing confidence in food safety.

The successful application of HACCP requires the full commitment and involvement of management and the workforce. It also requires a multidisciplinary approach. The application of HACCP is compatible with the implementation of quality management systems, such as the ISO 9000 series, and is the system of choice in the management of food safety within such systems.

Guidance for the establishment of HACCP-based systems is detailed in *Hazard Analysis and Critical Control Point System and Guidelines for its Application* (CAC 1997a) and ICMSF 1988.

Table 8.3 outlines the implementation of HACCP in a food company. The seven steps in bold are known as the 'Seven Principles of HACCP' and are covered in the next section.

8.6 Prerequisite programme

Before HACCP can be properly implemented, there are a number of 'Prerequisite Programmes' or 'PRP' that are required. These have been defined by the WHO (1999) as 'Practices and conditions needed prior to and during the implementation of HACCP and which are essential for food safety, as described in Codex Alimentarius Commission's General Principles of Food Hygiene and other Codes of Practice'.

Prerequisite programmes include (but are not limited to) the following:

1 Building design and work flow pattern
2 Equipment design
3 Supplier control
4 Receiving, storage and distribution
5 Specifications

Table 8.3 The seven principles of the HACCP system and preliminary activities (CAC, Committee on Food Hygiene, 1997a, 1997b)

Decision by management to use the HACCP system

Training and formation of the HACCP team

Development of the HACCP plan document, including the following parts

Assemble the HACCP team

Describe the food product and its distribution

Identify the intended use and consumers

Develop and verify the flow diagram for the production process

On-site confirmation of the flow diagram

1. Conduct a hazard analysis	List all potential hazards associated with each step, conduct a hazard analysis and consider any measures to control identified hazards
2. Determine the critical control points (CCPs) (Figure 8.5)	Determine CCPs
3. Establish critical limit(s)	Establish critical limits for each CCP
4. Establish a system to monitor control of the CCP	Establish a system of monitoring for each of the CCP
5. Establish corrective actions	Establish the corrective action to be taken when monitoring indicates that a particular CCP is not under control
6. Establish verification procedures	Establish procedures for verification to confirm that the HACCP system is working effectively
7. Establish documentation and record	Establish documentation concerning all keeping procedures and records appropriate to these principles and their application

 6 Staff training and personal hygiene
 7 Cleaning and disinfection
 8 Pest control
 9 Chemical control
10 Traceability and recall
11 Waste management

This list is largely taken from the NACMCF 1997 HACCP System.

In America, the regulatory use of HACCP also added 'Sanitation Standard Operating Procedures', or SSOPs, as part of the necessary prerequisite programmes for the HACCP system; see Table 8.3. These are written procedures determining how a food processor will meet sanitation conditions and practices in a food plant. This has emphasised the need for adequate cleaning, and so on, in a food production plant where cross-contamination and post-processing contamination are potential hazards. Other countries have not adopted a separate SSOPs approach, but kept cleaning within the company's HACCP scheme. The key to good sanitary practice is that sanitisers should be applied to areas that are already visually clean. Otherwise, the organic material will neutralise the active agent in the sanitiser, rendering it ineffective.

8.7 Outline of HACCP

In order to produce a safe food product with negligible levels of foodborne food pathogens and toxins, three controlled stages must be established:

1 Prevent micro-organisms from contaminating food through hygienic production measures. This must include an examination of ingredients, premises, equipment, cleaning and disinfection protocols and personnel.
2 Prevent micro-organisms from growing or forming toxins in food. This can be achieved through chilling, freezing or other processes such as reduction of water activity or pH. These processes, however, do not destroy micro-organisms.
3 Eliminate any foodborne micro-organisms, for example, by using a time and temperature processing procedure or by the addition of suitable preservatives.

These principles are given in the Codex Alimentarius Commission (1993) and the National Advisory Committee on Microbiological Criteria for Foods (NACMCF 1992, 1997). Hence, it is an internationally recognised procedure. Differences arise however with the interpretation and implementation of these seven principles. The approach here will adhere to the Codex, WHO and NACMCF 1997 format. The Codex principles are given in bold lettering. It should be noted that the Codex document reverses principles 6 and 7.

8.7.1 Food hazards
A hazard is defined as follows:

'A biological, chemical, or physical property that may cause a food to be unsafe for consumption'.

Biological hazards are living organisms, including microbiological organisms: bacteria, viruses, fungi and parasites.

Chemical hazards are in two categories: naturally occurring poisons, chemicals or deleterious substances. These are natural constituents of foods and are not the result of environmental, agricultural, industrial or other contamination. Examples are aflatoxins and shellfish poisons. The second group are poisonous chemicals or deleterious substances which are intentionally or unintentionally added to foods at some point in the food chain. This group of chemicals can include pesticides, fungicides as well as lubricants and cleaners. The much publicised and tragic use of melamine in infant formula and other milk products was to increase the apparent nitrogen content, without consideration of the toxicological effects (Table 1.8).

A physical hazard is any physical material not normally found in food which causes illness or injury. Physical hazards include glass, wood, stones, metal which may cause illness and injury. Examples of hazards are given in Table 8.4.

8.7.2 Preparation for HACCP
Before the HACCP seven principles can be applied, there is the need to:

1 *Assemble HACCP team*: The food operation should assure that the appropriate product-specific knowledge and expertise are available for the development of an effective HACCP plan. Optimally, this may be accomplished by assembling a multidisciplinary team. Where such expertise is not available on-site, expert advice should be obtained from other sources. The scope of the food chain is involved and the general classes of hazards to be addressed (e.g. does it cover all classes of hazards or only selected classes).
2 *Describe the product*: A full description of the product should be drawn up, including relevant safety information such as composition, physical/chemical structure (including a_w, pH, etc.),

Table 8.4 Hazards associated with food

Biological	Chemical	Physical
Macrobiological	Veterinary residues: antibiotics, growth stimulants	Glass
Microbiological	Plasticisers and packaging migration: vinyl chloride, bisphenol A	Metal
Viruses	Chemical residues: pesticides (DDT), cleaning fluids	Stones
Pathogenic bacteria	Allergens	Wood
Spore forming	Toxic metals: lead, cadmium, arsenic, tin, mercury	Plastic
Non-spore forming	Food chemicals: preservatives, processing aids	Parts of pests
Bacterial toxins	Radiochemicals: ^{131}I, ^{127}Cs	Insulation material
Shellfish toxins: domoic acid, okadaic acid, NSP, PSP[a]	Dioxins, polychlorinated biphenyls (PCBs)	Bone
Parasites and protozoa	Prohibited substances	Fruit pits
Mycotoxins: ochratoxin, aflatoxins, fumonsins, patulin	Printing inks	

[a]NSP, neurotoxic shellfish poison; PSP, paralytic shellfish poison.
Adapted from Snyder (1995) and Forsythe (2000).

microcidal/static treatments (heat treatment, freezing, brining, smoking, etc.), packaging, durability and storage conditions and method of distribution.

3 *Identify intended use*: The intended use should be based on the expected uses of the product by the end user or consumer. In specific cases, vulnerable groups of the population, for example institutional feeding, may have to be considered.

4 *Construct flow diagram*: The flow diagram should be constructed by the HACCP team. The flow diagram should cover all steps in the operation. When applying HACCP to a given operation, consideration should be given to steps preceding and following the specified operation.

5 *On-site confirmation of flow diagram*: The HACCP team should confirm the processing operation against the flow diagram during all stages and hours of operation, and amend the flow diagram where appropriate.

8.7.3 Principle 1: hazard analysis

Conduct a hazard analysis. Prepare a list of steps in the process where significant hazards occur and describe the preventative measures.

The HACCP team should list all of the hazards that may be reasonably expected to occur at each step from primary production, processing, manufacture and distribution until the point of consumption. The evaluation of hazards should include the following:

- The likely occurrence of hazards and severity of their adverse health effects
- The qualitative and/or quantitative evaluation of the presence of hazards
- Survival or multiplication of micro-organisms of concern
- Production or persistence in foods of toxins, chemicals or physical agents
- Conditions leading to the above

Table 8.5 Example of a hazard analysis worksheet

Firm Name: _____   Product Description: _____

_____   _____

Firm Address: _____   Method of Storage and Distribution:_____

_____   _____

                                             Intended Use and Consumer: _____

_____   _____

(1)	(2)	(3)	(4)	(5)	(6)
Ingredient/ processing step	Identify potential hazards introduced, controlled or enhanced at this step (I)	Are any potential food-safety hazards significant? (Yes/No)	Justify your decisions for column 3	What preventative measures can be applied to prevent the significant hazards?	Is this step a critical control point? (Yes/No)
	Biological				
	Chemical				
	Physical				
	Biological				
	Chemical				
	Physical				
	Biological				
	Chemical				
	Physical				
	Biological				
	Chemical				
	Physical				

The hazard analysis should identify which hazards can be eliminated or reduced to acceptable levels as required for the production of safe food.

Table 8.5 is an example of a worksheet for use in documenting the identified hazards.

8.7.4 Principle 2: critical control points
Identify the critical control points (CCPs) in the process.

The HACCP team must identify the CCPs in the production process which are essential for the elimination or acceptable reduction of the hazards which were identified in Principle 1. These CCPs are identified through the use of decision trees such as that given in Figure 8.5 (NACMCF 1992). A series of questions are answered which lead to the decision as to whether the control point is a CCP. Other decision trees can be used if appropriate.

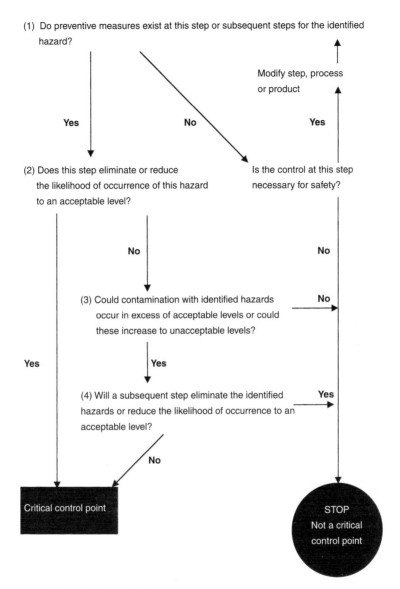

Figure 8.5 Critical control point decision tree.

A CCP must be a quantifiable procedure in order for measurable limits and monitoring to be achievable in Principles 3 and 4. If a hazard is identified for which there is no control measure in the flow diagram, then the product or process should be modified to include a control measure. In the past, some groups have differentiated CCPs into CCP1 as primary CCPs which eliminate hazards and CCP2 which only reduce hazards. This approach had the advantages of identifying which hazards were of crucial importance. For example, milk pasteurisation would have been CCP1, whereas assessment of raw milk on receipt would have been CCP2. This two-stage approach is, however, not in current practice.

The cooking step is an obvious CCP for which critical limits of temperature and time can be set, monitored and corrected; see Section 2.5. Non-temperature related control factors include water activity and pH; see Section 2.6.

8.7.5 Principle 3: critical limits

Establish critical limits for preventative measures associated with each identified CCP.

Critical limits must be specified and validated, if possible, for each CCP. The critical limit will describe the difference between safe and unsafe products at the CCP. Critical limit must be a quantifiable parameter: temperature, time, pH, moisture or a_w, salt concentration or titratable acidity, available chlorine.

8.7.6 Principle 4: CCP monitoring

Establish CCP monitoring requirements. Establish procedures from the results of monitoring to adjust the process and maintain control.

Monitoring is the scheduled measurement or observation of a CCP relative to its critical limits. The monitoring procedures must be able to detect loss of control at the CCP. Monitoring should provide the information in time (ideally online) for correcting the control measure. Ideally, by following the trend in measured values, the correction can take place before deviation from the critical limits.

The monitoring data must be evaluated by a designated person with knowledge and authority to carry out corrective actions when indicated. If monitoring is not continuous, then the amount or frequency of monitoring must be sufficient to guarantee that the CCP is in control. Most monitoring procedures for CCPs will need to be done rapidly because they relate to online processes and there will not be time for lengthy analytical testing. Physical and chemical measurements are often preferred to microbiological testing because they may be done rapidly and can often indicate the microbiological control of the product. Microbiological testing based on single samples or sampling plans will be of limited value in monitoring those processing steps which are CCPs. This is mainly because the conventional microbiological methods are too time-consuming for effective feedback. All records and documents associated with monitoring CCPs must be signed by the person(s) doing the monitoring and by a responsible reviewing official(s) of the company.

8.7.7 Principle 5: corrective actions

Establish corrective actions to be taken when monitoring indicates a deviation from an established critical limit.

Specific corrective actions must be developed for each CCP in order to deal with deviations from the critical limits. The remedial action must ensure that the CCP is under control and that the affected product is appropriately recycled or destroyed as appropriate.

8.7.8 Principle 6: verification

Establish procedures for verification that the HACCP system is working correctly.

Verification procedures must be established. These will ensure that the HACCP plan is effective for the current processing procedure. NACMCF (1992) gives four processes in the verification of HACCP:

1 Verification that critical limits at CCPs are satisfactory.
2 Ensure that the HACCP plan is functioning effectively.
3 Documented periodic revalidation, independent of audits or other verification procedures.

4 Government's regulatory responsibility to ensure that the HACCP system has been correctly implemented.

The frequency of verification should be sufficient to confirm that the HACCP system is working effectively.

The frequency of verification should be sufficient to confirm that the HACCP system is working effectively. Verification and auditing methods, procedures and tests, including random sampling and analysis, can be used to determine if the HACCP system is working correctly.

Examples of verification activities include the following:

- Review of the HACCP system and its records
- Review of deviations and product dispositions
- Confirmation that CCPs are kept under control
  Verification should be conducted:
- routinely, or on an unannounced basis, to assure CCPs are under control;
- when there are emerging concerns about the safety of the product;
- when foods have been implicated as a vehicle of foodborne disease;
- to confirm that changes have been implemented correctly after a HACCP plan has been modified;
- to assess whether a HACCP plan should be modified due to a change in the process, equipment, ingredients, and so on.

8.7.9 Principle 7: record keeping

Establish effective record-keeping procedures that document the HACCP system.

HACCP procedures should be documented. Records must be kept to demonstrate safe product manufacture and that appropriate action has been taken for any deviations from the critical limits.

Examples of documentation are as follows:

- Hazard analysis
- CCP determination
- Critical limit determination

Examples of records are as follows:

- CCP monitoring activities
- Deviations and associated corrective actions
- Modifications to the HACCP system

Table 8.6 is a useful worksheet for documenting the HACCP system.

8.8 Microbiological criteria and HACCP

Since HACCP is meant to be the internationally accepted means of assuring food safety, it may be argued that end-product testing is no longer required. However, food distributors (supermarket chains, etc.) have continued to insist on microbiological criteria (Chapter 6). The main role of end-product testing should be during the production of a new food and for verification of the HACCP scheme. In the context of HACCP, it must be emphasised that only those microbiological criteria which refer to foodborne hazards can be considered since HACCP systems are specifically designed to control hazards which are significant for food safety. Microbiological criteria which occur in legislation and are in the published literature may provide a reference point to assist food manufacturers in evaluating their data during hazard analysis; see Section 6.10.

Table 8.6 Example of an HACCP form

Firm Name: _____   Product Description: _____

Firm Address: _____   Method of Storage and Distribution: _____

_____   Intended Use and Consumer: _____

(1)	(2)	(3)	(4)	(5)	(6)	(7)	(8)	(9)	(10)
Critical Control Point (CCP)	Significant Hazard(s)	Critical Limits for Each Preventive Measure	Monitoring				Corrective Action(s)	Records	Verification
			What	How	Frequency	Who			

Signature of Company Official: _____   Date: _____

It should be noted that single sample analysis is limited as a HACCP verification procedure. Basically, one can only be 100% certain the food does not contain a hazard by analysing 100% of the food. The statistical confidence of sample analysis and sampling plans is covered in Section 6.4. The microbiological analysis of food using sampling plans result in the acceptance or rejection of a 'lot'. This includes the statistical probability that a lot may be falsely accepted or rejected, known as 'producer's risk' and 'consumer's risk'. Hence, statistical analysis demonstrates that microbiological testing is not a standalone verification tool for hazard analysis systems.

8.9 Microbiological hazards and their control

8.9.1 Sources of microbiological hazards

The common sources of foodborne pathogens are as follows:

1 *The raw ingredients*: The microbial flora of raw ingredients entering a food factory can be controlled by using reputable suppliers, certificates of quality, temperature monitoring on receipt, and so on. Ingredients may be rejected on receipt if they fail to comply with agreed standards.

2 *Personnel*: It has been estimated that approximately 1 in 50 employees is shedding 10^9 pathogens per gram faeces without showing any clinical symptoms (Snyder 1995; Table 7.1). Subsequently, poor personal hygiene such as failing to wash hands after going to the toilet can leave 10^7 pathogens under the fingernails. The movement of personnel in a factory needs to be controlled. Frequently, there will be a low-risk (or low-care) area partitioned from the high-risk (or high-care) area. Essentially, the low-risk area is where the ingredients are stored, weighed, mixed and cooked. After cooking, the food enters the high-risk area. Increased diligence is required since there will not be any further heat treatment to destroy any bacteria, and so on, from personnel or environmental contamination and also to prevent the growth of surviving organisms.

3 *The environment (air, water and equipment)*: The microbial quality of water needs to be frequently monitored as this can have severe repercussions if it is contaminated with potential foodborne pathogens. The accumulation of food residues can result in biofilm formation (Section 7.5) which requires physical removal as sanitisers will be neutralised by the organic material.

The severity of the illness caused by the organisms can be determined from standard texts, especially the ICMSF books, and is simplified in Table 4.3. The likely occurrence of the foodborne pathogen can also be determined from ICMSF and related literature; see Table 4.1. A list of web pages is given in the Appendix section where further information on foodborne pathogens can be obtained. A standard reference source is the FDA Bad Bug book. There are numerous websites giving information on current food poisoning outbreaks worldwide which may be accessed without any charges. National food poisoning statistics can be obtained from regulatory authorities. Useful websites can be accessed from the sites listed in the Web Resources section.

Details of the food's pH, a_w, heat treatment process, and so on, can be used to predict the micro-organisms of concern in the foodstuff (Sections 2.5 and 2.6 and Table 2.8). Although the infectious dose (Table 4.2, cf. Section 9.5.5) has been determined for a number of foodborne pathogens, this should be used only for indicative purposes since the susceptibility of the consumer will vary according to their immunocompetence, age and general health.

The microbial load of a product after processing can be predicted using storage tests and microbiological challenge testing, supplemented with predictive modelling (Sections 2.8). The

Table 8.7 Equivalent cooking time and temperature regimes

Temperature (°C)[a]	Time
60	45 minutes
65	10 minutes
70	2 minutes
75	30 seconds
80	6 seconds

[a]To convert to °F, use the equation: $°F = (9/5)°C + 32$. As a guidance: $60°C = 140°F$.
Data sourced from Department of Health, UK.

shelf-life of the product can be determined according to chemical, physical and microbiological parameters (Section 3.2).

8.9.2 Temperature control of microbiological hazards

For thousands of years, heat treatment has been the most effective and best method to control pathogenic and spoilage micro-organisms in food. Many food manufacturing processes include boiling, cooking or baking as heating steps. The resultant physico-chemical changes increase the digestibility of certain foods as well as improving the food's texture, taste, smell and appearance. The need for adequate cooking for food safety, however, is often overlooked. It is therefore not surprising that temperature control is one of the major identified reasons for food poisoning outbreaks; see Table 1.3.

The cooking step is an obvious CCP for which critical limits of temperature and time can be set, monitored and corrected. The time and temperature of the cooking process should be designed to give at least a 6 log kill of vegetative cells, that is 10^7 cells/g reduced to 10 cells/g; see Section 2.5.2. It will not kill spores and hence the time required to cool the food to a safe temperature needs to be monitored to prevent spore outgrowth. A list of time and temperature equivalencies is given in Table 8.7.

The objective of a 6 log kill is based upon two tenets:

1 Food pathogens are in the minority of the microbial flora on meat. Subsequently, any raw meat kept at temperatures that would allow food pathogens to multiply to 1×10^6/g would be rejected as spoilt. The cooking time and temperature regime are sufficient to kill 1×10^6/g of an enteric pathogen in the coldest area of the product. Since the number of food pathogens should be less than 1×10^6/g, the chance of vegetative cells of a pathogen surviving is reduced to a negligible level.

2 Since blood coagulates at $73.9°C$, visual observation can indicate if the correct temperature has been reached. Therefore, clear juices indicate that the blood proteins have been denatured and likewise bacterial vegetative cells. Hence, the phrases 'cook until well done' and 'cook until the juices run clear' are often quoted. However, bacterial spores do survive this time and temperature regime.

An example of the effect of cooking temperature on the survival of *E. coli* O157:H7 was given in Table 2.4. It can be seen that accurate temperature monitoring is necessary since the relationship of cell death to temperature is logarithmic. In other words, a small decrease in cooking temperature can result in considerable numbers of cells surviving the process.

The cooling period must be short enough to prevent spore outgrowth and germination from mesophilic *Bacillus* and *Clostridium* spp. Notably, the cooking process would create an anaerobic environment in the food which is ideal for *Clostridium* spp. The temperature growth range of *Cl. perfringens* is 10–52°C (Table 2.8); hence, cooling regimes should be designed to minimise the time the food is between these temperatures. A lower limit of 20°C is normally adopted since it only multiplies slowly below this value.

The control of the holding temperature of foods after processing and before consumption is crucial for safe food production. Generally recommended holding temperatures are as follows:

- Cold served foods <8°C
- Hot served foods >63°C

The microbiological 'Danger Zone' is often given as temperature between 8°C and 63°C. Ready to serve food should be at these temperatures for the minimum of time since it is proposed that any surviving mesophilic pathogens or post-processing contaminants are able to rapidly multiply at these temperatures to infectious levels. There is, however, no microbiological reason for the high (63°C) hot serve temperature since no foodborne pathogen grows between 53°C and 63°C. The best explanation is that hot servers do not maintain the temperature of food accurately and the extra 10°C is a safety barrier. These holding temperatures do vary between countries and where necessary local regulatory authorities should be consulted.

At cold serve temperatures (<8°C), *Clostridium* spp. are not a problem since they do not grow below 10°C. *St. aureus* produces toxins above 10°C, but is able to multiply down to 6.1°C. Subsequently, temperature abuse can lead to *St. aureus* growth and toxin production. Since *St. aureus* toxin is not inactivated by reheating to 72°C, *St. aureus* toxin production must be prevented. Chilled foods should be stored below 3.3°C to assure safety from spore germination and subsequent toxin production by *Cl. botulinum* E. In contrast, *Cl. botulinum* types A and B only multiply and produce toxin above 10°C. *L. monocytogenes* and *Y. enterocolitica* have a minimum growth temperature of −0.4°C and −1.3°C, respectively. Therefore, the cold holding time must be limited for foods which are not going to be reheated. It should be noted that different national regulatory authorities will have different hot and cold holding temperatures. This illustrates the need for risks to be assessed on a more worldwide scientific basis.

A *P* value (length of cooking period at 70°C; Section 2.5.1) of 30–60 minutes will give a shelf-life of at least 3 months depending upon the risk factors involved.

8.9.3 Non-temperature control of microbiological hazards

In addition to temperature, the water activity (a_w), pH and the presence of preservatives are important factors in the control of microbiological hazards. The limits of microbial growth are covered in Section 2.6. The pH and water activities of various foods are given in Table 2.8 and can be used to predict the relevant microbiological hazards.

8.10 HACCP plans

As explained in Section 8.7, HACCP plans require a team of company personnel who understand the details of the food's production. The following outlines are given as examples to be modified accordingly. HACCP has not been applied extensively in the retail trade. However, there is a source of HACCP–TQM technical guidelines available on the web; see Web Resources section.

8.10.1 Production of pasteurised milk

Figure 8.6 outlines the production of pasteurised milk. Receipt of raw milk is a CCP since it must be from certified tuberculosis-free herds and must be checked for total microbial load. The crucial step is the pasteurisation process (72°C, 15 seconds) which is designed to eliminate milk-borne pathogens and the cooling period (less than 6°C in 1.5 hours) which prevents the growth of surviving organisms. Therefore, pasteurisation is a CCP and the time and

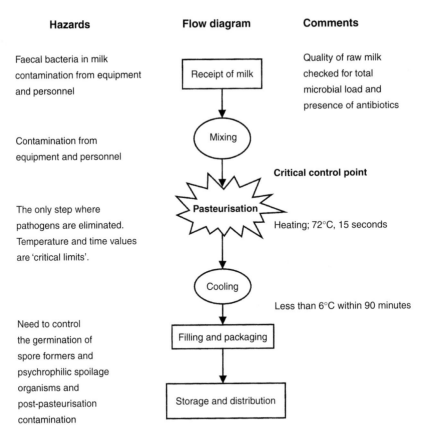

Figure 8.6 HACCP flow diagram for the pasteurisation of milk.

temperature are the critical limits. Since there are no further heat treatments, contamination of the pasteurised milk (especially from raw milk) must be avoided and therefore aseptic packaging is also a CCP.

8.10.2 Swine slaughter in the abattoir

Animal slaughter must be controlled to minimise the contamination of the meat with intestinal bacteria, which may include foodborne pathogens (Figure 8.7). Common pathogens associated with pigs are *C. jejuni, C. coli, Salmonella* spp. and *Y. enterocolitica*. Pathogenic micro-organisms commonly found in the processing environment are *L. monocytogenes, St. aureus* and *A. hydrophila*. Since the meat will subsequently be heat treated to eliminate pathogens, it is not necessary to produce a sterile animal carcass at the abattoir stage of food production. Therefore, the HACCP approach is to minimise pathogen contamination and not necessarily to eliminate it.

The key steps are enclosure of the rectum to prevent faecal contamination of the meat and disinfection of tools which can be contaminated with pathogenic organisms and vehicles of cross-contamination.

8.10.3 Chilled food manufacture

Chilled foods are typically multicomponent foods and therefore the HACCP flow diagram is complicated since each ingredient is represented. Figure 8.8 shows the flow diagram for a

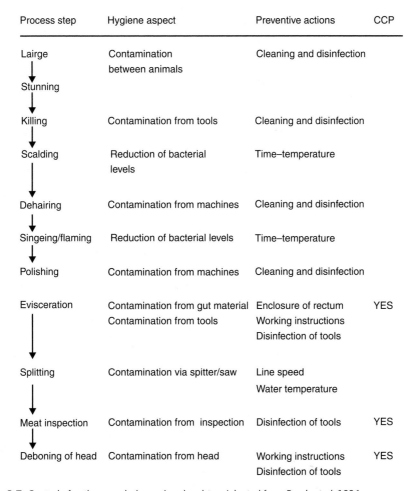

Process step	Hygiene aspect	Preventive actions	CCP
Lairge ↓ Stunning	Contamination between animals	Cleaning and disinfection	
↓ Killing	Contamination from tools	Cleaning and disinfection	
↓ Scalding	Reduction of bacterial levels	Time–temperature	
↓ Dehairing	Contamination from machines	Cleaning and disinfection	
↓ Singeing/flaming	Reduction of bacterial levels	Time–temperature	
↓ Polishing	Contamination from machines	Cleaning and disinfection	
↓ Evisceration	Contamination from gut material Contamination from tools	Enclosure of rectum Working instructions Disinfection of tools	YES
↓ Splitting	Contamination via spitter/saw	Line speed Water temperature	
↓ Meat inspection	Contamination from inspection	Disinfection of tools	YES
↓ Deboning of head	Contamination from head	Working instructions Disinfection of tools	YES

Figure 8.7 Control of pathogens during swine slaughter. Adapted from Borch *et al.* 1996.

chicken salad. There are nine identified CCPs. These are primarily after the mixing step and are more fully explained in Table 8.8. Since there is no heat treatment, careful temperature monitoring during storage is required to prevent significant multiplication of any foodborne pathogens present.

The process flow diagram for a bakery is reproduced in Figure 8.9. The monitoring of cleaning and fumigation, pest control and glass–perspex control are separate from the production process.

8.10.4 Generic models
HACCP implementation has mainly been product specific. However, a more amenable approach for food manufacturers with large numbers of products is generic HACCP. A number of generic HACCP models are available which have been produced by Canada, New Zealand and the United States; see Web Resources section. The generic models from the USDA for fresh squeezed orange juice and dried meats (beef jerky) are given in Tables 8.9 and 8.10.

Growing and harvesting(1)[1] — Raw material processing(2)   — Receiving of ingredients(3)[2]

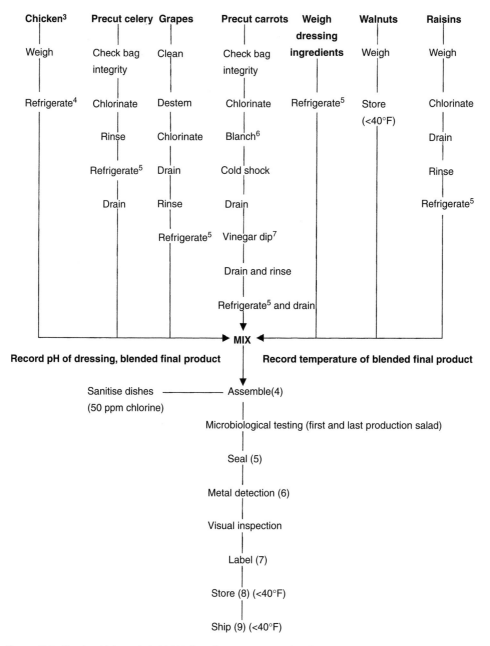

Chicken[3]	Precut celery	Grapes	Precut carrots	Weigh dressing ingredients	Walnuts	Raisins
Weigh	Check bag integrity	Clean	Check bag integrity	Refrigerate[5]	Weigh	Weigh
Refrigerate[4]	Chlorinate	Destem	Chlorinate		Store (<40°F)	Chlorinate
	Rinse	Chlorinate	Blanch[6]			Drain
	Refrigerate[5]	Drain	Cold shock			Rinse
	Drain	Rinse	Drain			Refrigerate[5]
		Refrigerate[5]	Vinegar dip[7]			
			Drain and rinse			
			Refrigerate[5] and drain			

➤ **MIX** ◄

Record pH of dressing, blended final product   **Record temperature of blended final product**

Sanitise dishes ——————— Assemble(4)
(50 ppm chlorine)

Microbiological testing (first and last production salad)

Seal (5)

Metal detection (6)

Visual inspection

Label (7)

Store (8) (<40°F)

Ship (9) (<40°F)

Figure 8.8 Chunky chicken salad HACCP flow diagram. Reprinted with permission from Anon. (1993c) in the *Journal of Food Protection*. Copyright held by the International Association for Food Protection, Des Moines, Iowa, US.

Table 8.8 HACCP worksheet for critical control points: chilled chicken salad

Item	Hazard	Control	Limit	Monitoring frequency/documentation	Action (for exceeding limit)	Personnel responsible
(1) Growing and harvesting	Antibiotic, chemical	Supplier compliance to raw material specifications	Regulatory approved chemicals and antibiotics; specified tolerances	Certificates of conformance for each lot. Random annual monitoring by QC	Reject lot	Shipping and receiving operator
(2) Raw material processing	Chemical, physical and microbiological	Certified supplier HACCP programme	Free of pathogens and foreign material	Monitor supplier HACCP programme	Reject as supplier	QC must monitor HACCP programme
(3) Raw ingredient storage temperatures	Microbiological	Compliance to raw material specifications	Precooked chicken 0–10°F. Fruits/vegetables ≤40°F (4.4°C). All others ≤80°F (24.4°C)	Check recorders in coolers daily	Report to shift supervisor and QC. Investigate time/temperature abuse and evaluate risk	Line operator
(4) Filling/ assembly	Microbiological	Temperature control specifications	Components/finished product temperature ≤ 40°F (4.4°C)	Check temperatures once/shift	Report to shift supervisor and QC. Adjust temperature according to specification and evaluate risk	Line operator
(5) Sealer	Microbiological	Proper sealer settings to ensure all trays hermetically sealed and have consistent seal bead	Upper limit tolerance of sealer	Sealer settings checked every 15 minutes. Visual inspection every 2 hours	Examine all packages since last check. Peel off membrane and reseal	Seal inspector
(6) Labeller	Incorrect code, dates, traceability	Legible and correct dates on all sleeves and shippers	Use proper labels	Each batch and at product changeover	Destroy incorrect/illegible code dated sleeves and shippers	Packaging operator

(Continued)

Table 8.8 (*Continued*)

Item	Hazard	Control	Limit	Monitoring frequency/ documentation	Action (for exceeding limit)	Personnel responsible
(7) Metal detector	Physical (metal)	Online metal detector	No metal detected	Calibrate before each batch or every 2 hours	Locate source of contamination. Destroy product	Packaging operator
(8) Plant storage of finished product	Microbiological	Product storage temperature ≤ 40°F (4.4°C)	≤40°F (4.4°C)	Check continuous recorders in coolers daily	Report to shift supervisor and QC. Place product on hold. Investigate time/temperature abuse and evaluate risk	Shipping and receiving operator
(9) Shipping temperature	Microbiological	Ship ≤ 40°F (4.4°C)	≤40°F throughout shipment	Each load temperature monitored by continuous recorder	Report to shift supervisor and QC. Place product on hold. Investigate time/temperature abuse and evaluate risk	Shipping operator/stock handler

Reprinted with permission from Anon. (1993c) in the *Journal of Food Protection*. Copyright held by the International Association for Food Protection, Des Moines, Iowa, US.

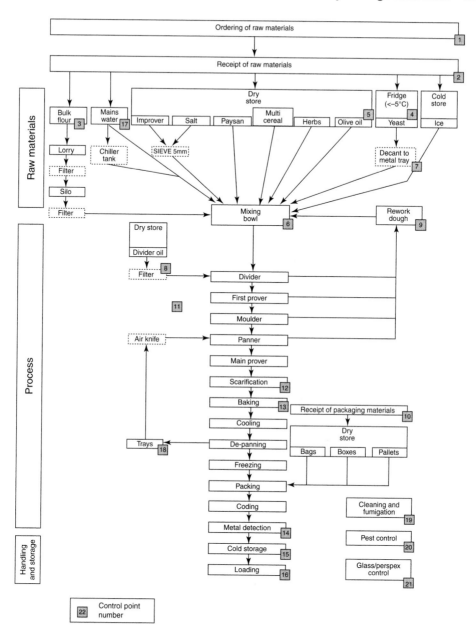

Figure 8.9 Process flow diagram for a bakery.

8.11 GMP and GHP

GMP covers the fundamental principles, procedures and means needed to design an environment suitable for the production of food of acceptable quality. GHP describes the basic hygienic measures which establishments should meet and which are the prerequisite(s) to other approaches, in particular HACCP. GMP/GHP requirements have been developed by governments,

Table 8.9 Generic HACCP plan for fresh squeezed orange juice

			HAZARD ANALYSIS WORKSHEET		
(1) Ingredient/processing step	(2) Identify potential hazards introduced, controlled or enhanced at this step	(3) Are any potential food safety hazards significant? (Yes/No)	(4) Justify your decision for column 3	(5) What preventative measure(s) can be applied to prevent the significant hazards?	(6) Is this step a critical control point? (Yes/No)
Receiving	BIOLOGICAL	Y	Environmental contamination	Control source	N
	Pathogens CHEMICAL Pesticides		Spraying	Letter of warrantee No raw fertilisation No drops. No harvesting before pre-harvest interval	
	PHYSICAL				
Grading	BIOLOGICAL	Y	Assumed to enter product	Wash and sanitise remove culls, decayed product, spills, and so on	N
	Pathogens CHEMICAL Pesticides PHYSICAL				
Wash/sanitising	BIOLOGICAL	Y	Assumed to be on product at receipt	Using effective sanitisation	Y
	Pathogens CHEMICAL Pesticides PHYSICAL				

Extraction	BIOLOGICAL	N	Controlled by sanitising	SSOP and GMP	N
	Pathogens				
	CHEMICAL				
	Pesticides				
	PHYSICAL				
Filling	BIOLOGICAL	N	Controlled by sanitising	SSOP and GMP	N
	Pathogens				
	CHEMICAL				
	Pesticides				
	PHYSICAL				
Chilling/holding	BIOLOGICAL	N	Need to maintain at refrigeration temperature (41°F)	SSOP and GMP	N
	Pathogens				
	CHEMICAL				
	Pesticides				
	PHYSICAL				

Data from FAMFES, web address in Appendix.

Table 8.10 Generic HACCP plan for dried meats (beef jerky)

Hazard description	Critical limits	Monitoring procedures	Deviation procedures	Verification procedure	HACCP records
Step: receiving; fresh and frozen meat					
Bacterial growth	Meat shall exhibit no unusual colours or odours	For each lot, meat receiver shall perform sensory evaluation of meat prior to unloading. He will inspect each combo of fresh meat or select two boxes of frozen meat per pallet for evaluation	Meat receiver: Fresh or frozen do not unload truck, notify Production Manager	HACCP coordinator will run a lab analysis on a sample of meat: once a week 3 samples (3 cases composite per sample). HACCP coordinator to verify once a week for proper monitoring at receiving (organoleptic inspection and temperature) (specification to be established)	Beef Jerky Meat Receiving Log Beef Jerky Processing Record
Bacterial growth	Meat temperature, shall be less than or equal to 4°C for fresh or solid frozen (frozen) microbiological criteria to be established by each plant	For each lot, meat receiver shall record temperature of meat at centre and on surface of each combo of fresh meat or freezing condition of two boxes of frozen meat per pallet	If meat temperature is 4–7°C; meat receiver shall notify QC, use second thermometer and retest. If meat temperature is >7°C, then hold lot and notify QC who will investigate possible causes, and decide on proper disposition	HACCP coordinator will run a lab analysis on a sample of meat: once a week 3 samples, (3 cases composite per sample). HACCP coordinator to verify once a week for proper monitoring at receiving (organoleptic inspection and temperature) (specification to be established)	Beef Jerky Meat Receiving Log

Foreign particles in meat	Product must be from listed supplier with contractual specification (supplier shall do an acceptable boneless meat reinspection programme)	Receiver shall ensure product is received from a listed supplier	If product is arriving from registered establishment that is not a QC-listed supplier, the lot is identified, inspected and is used for other products if found unacceptable	Inspect the cold storage to ensure all non-listed suppliers have been identified	Receiving Log
Step: receiving; dry ingredients					
Spices have an excessive bacterial load or contain foreign material	Specifications to be developed by the establishment. Each lot certified by supplier	Receiver shall monitor each lot received for supplier's certification (Lab analysis)	Receiver holds the lot until receiving appropriate certification and notify QC	QC samples one lot every 10 lots and runs analyses versus critical limits	Receiving records QC lot results
Step: receiving; packaging material					
Packaging material not of food quality	All packaging material must be 'approved' by AAFC to be received	Receiver ensures that product is approved before allowing and receiving	Return lot and notify QC	Once every 3 months, QC inspects stocks of packaging material to ensure all material is approved QC reviews receiving bills once a month	Receiving records

(Continued)

Table 8.10 (*Continued*)

Hazard description	Critical limits	Monitoring procedures	Deviation procedures	Verification procedure	HACCP records
Step: thawing of meat					
Bacterial pathogens growth	Thawing of meat shall be done in a room with temperature not exceeding 10°C	Designated employee records thawing at room temperature every 6 hours	Designated employee reports out of spec. room temperature to the Maintenance Foreperson and holds batch for QC evaluation	QC audits temperature records weekly. QC takes temperature and correlates results with records on a weekly basis. Audits employees work in monitoring time/temperature weekly. QC takes (monthly) samples to evaluate microbial acceptability of thawed product. Criteria to be developed by plant	Plant temperature record Beef Jerky Processing Record
Bacterial pathogens growth	Meat surface temperature maximum 7°C	Designated employee records meat temperature every 6 hours and before meat is removed from the room	If meat surface temperature > 7°C, hold the batch, submit samples for analysis. Process product, but hold until results are available. If APC > 5 × 10^5/g, test finished product, otherwise release product If *St. aureus* > 5 × 10^5/g, test for pathogens; if AAPC/HPB guidelines are not met, condemn product	QC takes temperature and correlate results weekly. Audit employees work in monitoring time/temperature weekly QC takes (monthly) samples to evaluate microbial acceptability of thawed product Criteria to be developed by plant	Meat temperature Record-Thaw Room Beef Jerky Processing Record

	Critical limits	Monitoring procedures	Corrective actions	Verification	Records
Step: weighing of nitrite					
Excessive amount may be toxic. Insufficient amount may allow for germination of *C. botulinum* spores	Minimum concentration is 100 ppm. Maximum concentration is 200 ppm	Designated employee keeps records on number of bags prepared and their use; write amount of nitrite in each bag	Designated employee notifies QC for follow-up and assessment	QC verifies operator's records, inventory of nitrite, weight of nitrite bags	Beef Jerky Ingredient record. Nitrite Concentration Out of Compliance Sheet Beef Jerky Complete Processing Record
Step: formulation mixing with spices and massaging					
Lack or excess Na/K nitrite	100–200 ppm	Foreperson ensures that operator checks off chart for ingredients. Correct ingredient and product identification tags	Product since last satisfactory monitoring is held by foreperson for rework	Foreperson to verify operator's records 'X' times/week. QC to verify operator's records and laboratory test records 'X' times/month. QC verifies once/week a sample for nitrite content	Spice room records. Operator's records. Foreperson verification records. Chemical lab records. QC verification records
Step: screening of spice					
Foreign material >2 mm	No contamination is acceptable	Foreperson ensures that the operator on duty screens spice for every batch and notifies QC of any finding. If foreign material is found, remove it from product, label sample bag with manufacturer's codes and send to QC	Foreperson re-instructs the employee on normal procedures and completes record. Quality team shall submit material to supplier. All remaining spice from the lot in question shall be returned to the manufacturer for rescreening or credit	QC audits employee practices weekly	Jerky Spice Check Sheet Beef Jerky Processing Record

(Continued)

Table 8.10 (Continued)

Hazard description	Critical limits	Monitoring procedures	Deviation procedures	Verification procedure	HACCP records
Step: smoking of spice					
Pathogen growth due to incorrect come-up time.	Exact cooking cycles to be defined here (come-up time) *House #A* 'X' hours at 71°C	Smokehouse operator on duty checks cooking cycles against critical limits and initials smokehouse chart recorders	Smokehouse operator on duty is to add more time if time is insufficient. If temperature is insufficient, hold batch. Take a sample of the product. Notify QC	Maintenance Foreman: Is to check accuracy of smokehouse thermometer using a mercury-based reference thermometer. Also check chart recorder accuracy (time and temperature) monthly. QC reviews and initials smokehouse charts weekly	Chart Recorders Beef Jerky Moisture Level Sheet Beef Jerky Processing Record
Bacterial survival due to improper time/temperature	*House #B* 'Y' hour at 70°C, 'Z' hours at 75°C				
Product not shelf-stable due to an excessive a_w	Final product have a_w less than or equal to 0.85	Foreperson checks at a frequency 'X' that the designated employee tests all lots prior to release	Hold batch, notify quality team Hold all suspect product until QC assessment adjust/reject lot QC shall submit samples for analysis. Process product, but hold until results are available. If results APC > 5 × 10⁵/g, test finished product. If *St. aureus* > 1 × 10⁴/g, test for toxin; if positive, condemn product; if finished product APC > 5 × 10³/g, test for pathogens; if AAPC/HPB guidelines are not met, condemn product	QC audits at a frequency 'X', test procedures, test tool calibration, log book. QC audits operator's practices for every smokehouse load of beef jerky for 5 minutes	a_w records

Step: metal detector

| Presence of undetected metal particles | No contamination is acceptable for ferrous and non-ferrous particles > 2 mm | Operator on duty sets up the metal detector using two dummy bags of jerky each containing a metal particle (one contains ferrous, the other, non-ferrous) 2 mm in diameter. The dummy bags shall be passed through the detector every hour to ensure that detector is functioning | Operator on duty sets rejected package aside. Pass rejected package through detector. If package fails test, set aside and notify quality team, open package and examine each piece of meat. QC will examine the identified metal to locate the most likely source of metal, check the equipment and take further action as needed | QC audits the operation by passing the dummy bags (same as for monitoring) through the detector once per shift | Metal detector Check Sheet Rejected Jerky Metallic Particle Record Beef Jerky Processing Record |

Date: Approved by:

Canadian Food Inspection Agency, web address in Appendix.
Canadian Food Inspection Agency. Reproduced with permission of the Minister of Public Works and Government Services, Canada (2009).

the Codex Alimentarius Committee on Food Hygiene (FAO/WHO) and the food industry, often in collaboration with other groups and food inspection and control authorities.

General GHP requirements usually cover the following:

1 The hygienic design and construction of food manufacturing premises
2 The hygienic design, construction and proper use of machinery
3 Cleaning and disinfection procedures (including pest control)
4 General hygienic and safety practices in food processing including:

 i The microbial quality of raw foods
 ii The hygienic operation of each process step
 iii The hygiene of personnel and their training in the hygiene and safety of food

GMP codes and the hygiene requirements they contain are the relevant boundary conditions for the hygienic manufacture of foods. They should always be applied and documented. *No food processing methods should be used to substitute for GMPs in food production and handling.*

8.12 Quality systems

The quality of a product may be defined as its measurement against a standard regarded as excellent at a particular price, which is satisfactory both to the producer and to the consumer. The aim of quality assurance is to ensure that a product conforms as closely as possible and consistently to that standard at all times. Quality can be measured in terms of the senses (e.g. taste panels), chemical composition, physical properties and the microbiological flora, both quantitative and qualitative.

An 'excellent' quality at a specific price can only be achieved by answering in the affirmative to the questions asked by quality assurance and quality control: 'Yes, we are doing the right thing and we are doing things right.' To answer thus means that a successful quality assurance scheme (and likewise a food hygiene scheme with all its ramifications) must be operating with the full and sincere support of top management and all those concerned with the implementation of the scheme. There must, of course, be full control over all aspects of production so that a consistency of product quality is maintained. This necessitates strict control over the initial quality of raw materials, over the process itself and over packaging and storage conditions. In microbiological terms, the build-up of bacteria during a process run must be monitored at critical processing steps.

8.13 TQM

TQM represents the 'cultural' approach of an organisation; it is centred on quality and based on the participation of all members of the organisation and the concept of continuous improvement. It aims at long-term success through customer satisfaction, benefits to the members of the organisation and benefits to society in general.

TQM is similar in emphasis to quality assurance and has been defined as 'a continual activity, led by management, in which everybody recognises personal responsibility for safety and quality' (Shapton & Shapton 1991). This requires the company, as a whole, achieving uniformity and quality of a product and thus safety is maintained. Hence, TQM is broader in scope than HACCP including quality and customer satisfaction in its objectives (Anon. 1992).

Quality systems cover organisational structure, responsibilities, procedures, processes and the resources needed to implement comprehensive quality management. They apply to, and interact with, all phases of a product cycle. They are intended to cover all quality elements.

A combination of HACCP, quality systems, TQM and business excellence provides a total systems approach to food production, which embraces quality, productivity and food safety. TQM and quality systems provide the philosophy, culture and discipline necessary to commit every member of an organisation to the achievement of all managerial objectives related to quality. Within this framework, the inclusion of HACCP as the key specific safety assurance plan provides the necessary confidence that products will conform to safety needs and that no unsafe or unsuitable product will leave the production site. Collectively, these tools provide a comprehensive and proactive approach to further reduce the risk of food safety problems.

In 1987, the International Organisation for Standardisation in Geneva, Switzerland, published the *ISO 9000* standards. They are equivalent to the European Standards EN29000 series and the British Standards *BS 5750:1987*. The ISO 9000 series is composed of five standards:

- *ISO 9000 (Quality management and quality assurance standards)*: Guidelines for selection and use.
- *ISO 9001 (Quality systems)*: Model for quality assurance in design/development, production, installation and servicing.
- *ISO 9002 (Quality systems)*: Model for quality assurance in production and installation.
- *ISO 9003 (Quality systems)*: Model for quality assurance in final inspection and test.
- *ISO 9004 (Quality management and quality system elements)*: Guidelines.

These standards can be used as a starting point for designing TQM programmes and should be used for managing the HACCP system.

9 Microbiological risk assessment

9.1 Risk analysis and microbiological risk assessment

Microbiological risk assessment is the stepwise analysis of hazards that may be associated with a particular type of food product, permitting an estimation of the probability of occurrence of adverse effects on health from consuming the product in question (Notermans & Mead 1996). It is sometimes referred to as 'Quantitative Microbial Risk Assessment' (Haas *et al.* 1999) and can be described as a methodology to organise and analyse scientific information to estimate the probability and severity of an adverse event (Cassin *et al.* 1998b). Its origin is therefore in the area of 'risk analysis'.

As described in Chapter 1, changes in food processing techniques, food distribution and the emergence of new foodborne pathogens can change the epidemiology of foodborne diseases. The increase in international trade in food has increased the risk from cross-border transmission of infectious agents and underscores the need to use international risk assessment to estimate the risk that microbial pathogens pose to human health. The globalisation and liberalisation of world food trade, while offering many benefits and opportunities, also present new risks. Because of the global nature of food production, manufacturing and marketing, infectious agents can be disseminated from the original point of processing and packaging to locations thousands of miles away. Therefore, new strategies are required for evaluating and managing food safety risks. Risk analysis generates models which will enable the changes in food processing, distribution and consumption to be assessed with regard to their influence on food poisoning potential. It is a management tool, initially for governmental bodies, to define an appropriate level of protection and establish guidelines to ensure the supply of safe foods. Nevertheless, it is also a tool for food companies with which they can assess the effect of processing changes on microbial risks (risk mitigation strategies).

Food safety must be ensured by the proper design of the food product and the production process. This means that the optimal interaction needs to be assured between intrinsic and extrinsic parameters (which are appropriate for the product's shelf life) as well as conditions for handling, storage, preparation and use. As previously outlined, the assured method of controlling hazards is HACCP (Chapter 8). Nevertheless, HACCP requires the identified (microbial) hazard to be eliminated or reduced to an 'acceptable level', but what is an 'acceptable level', does it vary with age, and so on, and who decides? To address these issues, one has to consider 'risk'.

Definition: Risk is a function of the probability of an adverse health effect and the severity of that effect, consequential to a hazard(s) in food.

Risk analysis is the 'third wave' of food safety (Schlundt 2002, WHO). The three being:
1 Good Hygienic Practices in production and preparation to reduce the prevalence and concentration of the microbial hazard.

338

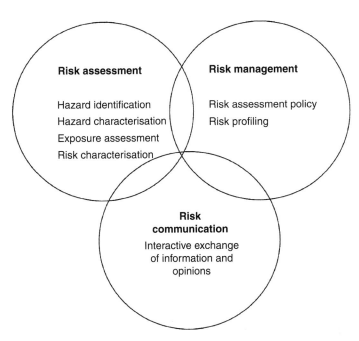

Figure 9.1 The general framework of microbiological risk assessment.

2 HACCP and HACCP-like approach which proactively identifies and controls the hazard.
3 Risk analysis which focuses on the consequences to humans of ingesting the microbial hazard, and the occurrence of the hazard in the whole food chain (from farm to fork).

As previously described, risk analysis consists of three components (Figure 9.1, FAO/WHO 1995).

Risk assessment identifies the risk and factors that influence it. It requires scientifically derived information and the application of established scientific procedures carried out in a transparent manner. However, sufficient scientific information is not always available and so an element of uncertainty must be associated with any decision.

Risk management shows how can the risk be controlled or prevented. This can be achieved through hygienic handling, processing and implementation of HACCP procedures and the setting of criteria/standards.

Risk communication informs others of the risk. It provides the public with the results of expert scientific review of food hazards and their risk to the general public or specific groups such as those who are immunodeficient, infants and the elderly. Risk communication provides industry and the consumer with information to reduce, prevent or avoid the food risk.

These three activities should be kept separate. However, there is an essential exchange of information between them which is represented by the circles overlapping (Figure 9.1).

9.2 Origin of microbiological risk assessment

On the international scene, the Uruguay Round Agreements (the SPS Agreement in particular) established the tenet that sanitary measures should be established on the basis of an assessment of the risk as appropriate to the circumstances. The purposes of the Sanitary and Phytosanitary

Table 9.1 Pathogen–commodity microbiological risk assessments

Hazard	Commodity
S. Enteritidis	Eggs and egg products
Salmonella spp.	Poultry, red meat, sprout, fish
C. jejuni	Poultry
Enterohaemorrhagic E. coli	Beef, sprouts
L. monocytogenes	Soft cheese, ready-to-eat products, smoked fish, minimally processed vegetables (i.e. salads and precooked frozen vegetables)
V. parahaemolyticus	Shellfish
Vibrio spp.	Seafood
Cyclospora	Fresh produce
Cryptosporidium	Drinking water
Rotavirus	Drinking water
Shigella spp.	Vegetables
B. cereus	Pasteurised milk
Cronobacter spp. and Salmonella	Powdered infant formula

measures (SPS) and Technical Barriers to Trade (TBT) agreements are to prevent the use of national or regional technical requirements, or standards in general, as unjustified TBT. The TBT agreement covers all types of standards including quality requirements for foods (except requirements related to SPS measures), and it includes numerous measures designed to protect the consumer against deception and economic fraud. The TBT agreement also places emphasis on international standards. World Trade Organisation (WTO) members are obliged to use international standards or parts of them, except where the international standard would be ineffective or inappropriate in the national situation. The WTO agreement also states that risk assessment should be used to provide the scientific basis for national food regulations on food safety and SPS measures, by taking into account risk assessment techniques developed by international organisations (Klapwijk *et al.* 2000).

Because of SPS and WHO, the Codex Alimentarius Commission standards, guidelines and other recommendations have become the baseline for safe food production and consumer protection. Hence, the Codex Alimentarius Commission has become the reference point for international food safety requirements. In turn, the Codex Alimentarius Commission has identified risk assessment of microbiological hazards in foods as a priority area of work. The Codex Committee on Food Hygiene (CCFH), within the Codex Alimentarius Commission, is responsible for risk management of food in international trade. They have the overall responsibility for all provisions on food hygiene prepared by Codex commodity committees. CCFH decides on the pathogen–commodity combinations for which expert risk assessment advice is required (Table 9.1). See Section 11.4 for more information on Codex Alimentarius Commission and its committees.

Initially, risk managers in the Codex Alimentarius Commission called for an expert advisory body regarding public health protection from foodborne hazard. Subsequently, a joint FAO/WHO consultation on risk management and food safety concluded that the work of CCFH would benefit from advice from an expert body on foodborne microbiological hazards for the purposes of risk management. Their report suggested that such a committee of experts could provide scientific advice on microbiological risk assessment similar to that provided by Joint FAO/WHO Expert Committee on Food Additives (JECFA) and Joint FAO/WHO Meeting on Pesticide Residues (JMPR) on food additives, contaminants, veterinary drug residues and pesticide residues. In response to the Codex Alimentarius Commission, the FAO and WHO

Table 9.2 Joint FAO/WHO expert meetings on microbiological risk assessment (JEMRA)

Topic (Year)	Microbiological risk assessment (MRA) series number[a]
Risk assessments of *Salmonella* in eggs and broiler chickens (2002)	1 & 2
Hazard characterisation for pathogens in food and water (2003)	3
Risk assessment of *Listeria monocytogenes* in ready-to-eat foods (2004)	4 & 5
Cronobacter spp. and other micro-organisms in powdered infant formula (2004)	6
Exposure assessment of microbiological hazards in food: Guidelines (2005)	7
Risk assessment of *Vibrio vulnificus* in raw oysters (2005)	8
Risk assessment of choleragenic *Vibrio cholerae* O1 and O139 in warm-water shrimp in international trade (2005)	9
Cronobacter spp. and *Salmonella* in powdered infant formula (2006)	10

[a]Accessible from http://www.who.int/foodsafety/micro/jemra/assessment.

established a Joint Expert Meeting on Microbiological Risk Assessment (JEMRA) which is an expert consultative process to collect, collate and evaluate risk assessment data for significant pathogens in food at the international level. This process involves the following:

1 Preparation of scientific descriptions of state-of-the-art knowledge.
2 Sharing this knowledge with all interested parties.
3 Interaction with risk managers to focus the scientific work towards areas where prevention is feasible.
4 Scientific scrutiny of the data presented.
5 Preparation of reports to enable an evaluation of the risk and an answer to the specific management question.

Hence, the JEMRA reports on risk assessment provide the scientific basis for the Codex committee CCFH on risk management.

An increasing number of completed studies are published or are in development under the aegis of the Joint FAO/WHO activities on risk assessment of microbiological hazards in foods. These include *Campylobacter*, *E. coli*, *Cronobacter* spp., *L. monocytogenes*, *Salmonella*, *Vibrio* and viruses. These are covered in greater detail in Chapter 10 and can be downloaded (pdf format) from the internet (see internet directory). A number of other related publications have been released and many can be downloaded in pdf format from the WHO website including those of JEMRA; see Table 9.2.

The WHO/FAO (2000a)preliminary report on hazard characterisation considered both food- and waterborne pathogens together because a significant number of 'food poisoning' outbreaks are caused by fruits and vegetables which have been irrigated with contaminated water. To date, the majority of quantitative risk assessments have been for bacterial hazards rather than viruses, toxigenic fungi and parasitic protozoa because of the greater volume of data available.

In 1997, the European Union Scientific Co-operation Task (SCOOP, coordinated by France) was established on microbiological risk assessment for foodborne pathogens and toxins. It ran for 2 years and assessed the collation of microbiological risk assessment data within member countries of the European countries (SCOOP 1998). The study was the first of its kind to determine what different countries were doing with respect to microbiological risk assessment according to the definitions of the Codex Alimentarius Commission (1999). The project

Table 9.3 Microbiological risk assessment studies (see also Table 9.2)

Micro-organism	References
Salmonella serovars	Kelly *et al.* 2000
Eggs and egg products	Todd 1996a, Whiting & Buchanan 1997, FAO/WHO (2002), FSIS 1998
Poultry industry	Brown *et al.* 1998, Oscar 1998a,b (see internet directory for Poultry FARM), Fazil *et al.* 2000
L. monocytogenes	Peeler & Bunning 1994, Farber *et al.* 1996, Van Schothorst 1996, 1997, Buchanan *et al.* 1997, Notermans *et al.* 1998, Bemrah *et al.* 1998 , FDA 1999, Lindqvist & Westöö 2000, Fazil *et al.* 2000, FDA 2001
E. coli O157:H7	Cassin *et al.* 1998a, Marks *et al.* 1998, Haas *et al.* 2000, Hoornstra & Notermans 2001
Bacillus cereus	Zwietering *et al.* 1996, Todd 1996b, Notermans *et al.* 1997, Notermans & Batt 1998, FSIS 1998, Carlin *et al.* 2000.
Campylobacter spp. fluoroquinolone resistance	Medema *et al.* 1996, Fazil *et al.* 2000b, FDA 2000a
V. parahaemolyticus	FDA 2000b
Sous-vide products	Barker *et al.* 1999
BSE	Gale 1998
Drinking water	Soker *et al.* 1999, Teunis & Havelaar 1999, FAO/WHO 2000b, Gale 2001

involved 39 scientists from 13 countries. The study concluded that although microbiological risk assessment was rapidly developing across Europe, very few complete risk assessments had been published.

Cassin *et al.* (1998a) introduced the term 'Process Risk Model' which integrated microbiological risk analysis with scenario analysis and predictive microbiology. In addition, their use of the term 'dose–response assessment' is effectively equivalent to 'hazard characterisation' in the Codex Alimentarius Commission and WHO (1995) terminology (cf. Potter 1996). The International Life Sciences Institute (ILSI 2000) have published a 'Revised framework for microbial risk assessment' which is a revision of their 'Food Safety Management Tools' monograph (ILSI 1998b). Originally, the ILSI risk assessment was more appropriate for waterborne microbial hazard. Although this is a very valuable document, the standard Codex Alimentarius Commission approach will be used in this book.

Therefore, microbiological risk assessment has only relatively recently been applied to microbial food safety issues. A list of various published pathogen–commodity studies is given in Tables 9.1–9.3 and will be considered in greater detail in Chapter 10.

9.3 Microbiological risk assessment – an overview

The risk analysis, as described above, starts with a brief profile on our current knowledge on the micro-organism. The profile should include the organism's incidence in the environment, domestic animals, foods and human implications. The risk profile will be used as the basis for the scientific risk assessment of the micro-organism using scientific literature. During this time, new research or calls for data into poorly understood areas may be initiated. Data on the prevalence and frequency of the micro-organism in the food chain and the infection outcomes will be analysed to determine the significance of food contamination to human illness and its severity. The risk assessment should also predict the most effective measures to be taken in order to control the micro-organism in food.

The risk analysis will provide the information to define a food safety objective (covered later in Section 9.7), that is either an acceptable level of the pathogenic micro-organism or an acceptable prevalence in a product, which is achievable by the implementation of HACCP plans, and so on.

Although many foodborne micro-organisms have been recognised as pathogenic (Table 1.2), not every ingestion of a pathogen results in an infection or subsequent illness. There is variation in the infectivity of the micro-organism as well as variation in the population susceptibility. Therefore, the risk of a foodborne disease is the combination of the likelihood of exposure to the pathogen through ingestion and the likelihood that the exposure will result in infection/intoxication and subsequently illness, of which there are varying degrees of illness including death. The risk can be quantified on a population basis to predict the likely number of infections, illnesses or deaths per 100 000 population per year, or per meal, and so on. It can also help to identify those stages from 'farm to fork' that contribute to an increase in risk of foodborne illness, and help focus on steps that most effectively reduce the risk of foodborne pathogens. These are termed 'risk mitigating strategies'. Where insufficient data are available, qualitative (descriptive) risk assessments can be carried out. A qualitative risk assessment may be first constructed prior to deciding if a more time-consuming, resource-demanding quantitative risk assessment is necessary. Therefore, risk assessment has the following two main objectives:

1 Quantify the risk to a defined population group from consumption of a defined product. If there is sufficient data, determine the risk from levels and frequency of contamination at the time of consumption, the amount of consumption (meal size and frequency) and an appropriate dose–response relationship to translate the exposure into public health outcomes.

2 Identify strategies and actions that can be used to decrease the level of health risk. This usually requires the modelling of production, processing and handling of the food and changes in the 'farm to fork' chain. Subsequently, it may identify the steps in food production that are critical to food safety, and those at which control actions or interventions would produce the greatest reduction in risk of foodborne illness. Hence, the potential use for critical control point identification in HACCP implementation.

Microbiological risk assessment is only one integral component in a series of steps leading to the management of microbiological hazards for foods in international trade. In order to complete an effective microbiological risk assessment, it is imperative that key information about the food be available concerning the technologies and handling practices used from production to consumption. Safety concepts need to be built into the development of food products, for example through HACCP implementation. Subsequently, these must be incorporated into Good Manufacturing Practices, Good Hygienic Practices and Total Quality Management (Figure 8.3). Hence, in the future, microbiological risk assessment should provide better information for the development of HACCP schemes, especially setting critical control points and company food safety activities (Figure 8.4). However, it can take several years for a formal risk analysis to be completed. In order to assist food companies, the WTO allows the use of internationally accepted criteria to be used. These criteria must be based on previous risk analysis procedures with Codex Alimentarius Commission and national governments as the risk managers. When the food safety objectives (Section 9.7) have been defined, the food companies will need to convert them into their own product or process criteria (Figure 8.4). Because of limited resources, food companies will take a simpler approach than governmental risk managers. They will focus on the prevalence and concentration of a recognised pathogen in their food ingredients and finished product (see Section 9.5.3 on exposure assessment) and can use predictive microbiology to determine the likely changes in prevalence and concentration during processing, preparation and storage. Subsequently, they can identify factors (i.e. initial microbial load) which could be controlled to reduce the associated risk (risk mitigation). See Section 2.8.5 for an example using *Shigella flexneri* (Figure 2.17).

Currently, the application of risk assessment to food microbiological safety is an evolving subject, where areas lacking sufficient information are being identified for further research and analysis. An increasing number of risk assessments have been completed and there has been particular focus on the use of MRA to establish and/or implement quantitative risk-based microbiological targets at the international level.

The FAO/WHO is investigating the role of MRA in developing quantitative risk-based microbiological targets or metrics (FAO/WHO 2006a). Public health goals are set by governments to improve the public health status and reduce disease burden. MRA can determine the scale of risk reduction required to achieve the public health goal. For example, one means of reducing salmonellosis' incidence from chicken from 50 to 10 per 100 000 consumers per year would require a reduction in the frequency of *Salmonella* in chicken by 20%. Therefore, 10 salmonellosis cases per 100 000 consumers per year would be the Appropriate Level of Protection (ALOP). A range of other scenarios of control measures and associated risk outcomes would also be considered during the MRA.

The Codex Alimentarius Commission (CAC) has defined three new 'intermediate' risk-based microbiological targets, namely, food safety objectives (FSOs), performance objectives (POs) and performance criteria (PCs).

- *FSO*: The maximum frequency and/or concentration of a hazard in a food at the time of consumption that provides or contributes to the ALOP (see also Section 9.7).
- *PO*: The maximum frequency and/or concentration of a hazard in a food at a specified step in the food chain before the time of consumption that provides or contributes to an FSO or ALOP, as applicable.
- *PC*: The effect in frequency and/or concentration of a hazard in a food that must be achieved by the application of one or more control measures to provide or contribute to a PO or an FSO.

The purpose of these intermediate targets is to communicate to the appropriate food industry the limits required at specific points in food supply chain in order to achieve a specified public health goal or level or protection.

There are two types of quantitative microbiological risk assessment (QMRA): deterministic and probabilistic. *Deterministic risk assessment* is based on single value inputs and outputs, and provides a relatively straightforward means of using MRA to develop metrics. However, the disadvantage is that the risk assessment is less accurate, for example limited insights into uncertainty and a tendency to focus on extreme situations such as worst-case scenarios. *Probabilistic risk assessment* differs in that it overcomes these disadvantages. The inputs and outputs of the probabilistic approach are a distribution of values, and this poses a challenge of how to express the outcome as a metric to be achieved by appropriate control measures. Work is ongoing to better elucidate how each of the above risk assessment approaches may be used to establish quantitative targets.

9.4 Microbiological risk assessment – structure

Compared with risk assessment of chemical hazards, the risk assessment of biological hazards is a 'new developing' science. Microbial risk assessment varies from chemical risk assessment due to the following reasons:

1 Micro-organisms may multiply or die in food. However, concentrations of chemicals do not change.
2 Microbiological risks are primarily the result of a single exposure, whereas chemical risks are often due to cumulative effects.
3 The level of microbial contamination is highly variable, whereas the administration of veterinary drugs can be controlled to minimise human exposure to residues.
4 Micro-organisms are rarely homogenously distributed in food.
5 Micro-organisms can be distributed via secondary transmission (i.e. person to person), in addition to direct ingestion of food.
6 The exposed population may exhibit short-term or long-term immunity. This will vary according to the hazardous micro-organism.
(ILSI 2000; see also Table 9.4).

Table 9.4 Comparison of microbiological risk assessment with chemical risk assessment

Component	Chemical risk assessment	Microbial risk assessment
Purpose	Determine: (a) If drug is genotoxic (b) Estimate safe levels in foods (c) Drug withholding times Provides for drug use that poses insignificant health risk	Characterise risk in terms of the probability of illness or death from naturally occurring microbial contaminants Assess impact of changes in food Production/processing on risk Develop food standards Eventually establish critical limits for HACCP
Hazard identification	Chemical structure of drug	Agent identified
	Evidence of toxicity in animal bioassay and no observed effect level (NOEL) (dose) derived	Evidence of causal role in foodborne disease gathered from outbreak investigation or epidemiological study
Exposure assessment	Assumptions made about consumption of foods derived from treated animals	Usually complex determinations of prevalence and concentrations of pathogen in food at consumption
	With ADI calculate maximum residue limits (MRL) in foods	Account for growth/death dynamics of microbes
	Withdrawal time set to ensure MRL not exceeded	Based on surveillance studies, simulation modelling
		Investigate scenarios to determine effects of processing changes on risk
Hazard characterisation Dose–response assessment	NOEL derived from bioassay with safety factor calculate acceptable daily intake (ADI)	Data from human volunteer studies, animal models, outbreak investigations, used to estimate effect of varying levels of contamination Usually involves complex modelling
Risk characterisation	Should be negligible risk if regulatory compliance achieved	Risk estimate expressed in terms of probability of illness or death, for example from one portion of food, expected number of cases/100 000 population, and so on, estimates for subgroups of population

A foundation for microbiological risk assessment was established through a series of consultations held by FAO and WHO and through the documents developed by the CAC. Since 1995, WHO, in collaboration with FAO, has been developing a formalised framework of risk analysis of foodborne hazards. The process has included a number of joint expert consultations:
1 The application of risk analysis to food standards and food safety issues which dealt principally with risk assessment (Geneva, 1995).
2 Risk management and food safety (Rome, 1997).
3 The application of risk communication to food standards and safety matters (Rome, 1998).
4 Risk assessment of microbiological hazards in foods (Geneva, 1999).
5 Risk management (Rome, 2000).
6 The interaction between assessors and managers (Kiel, 2000).

The principles and recommendations contained in the reports are intended to serve as guidelines for the Codex Committees to review the standards and advisory texts in their respective areas of responsibility and to provide a common framework to governments, industry and other parties wishing to develop risk analysis activities in the field of food safety. See Web Resources Section for a listing of URLs from which the documents can be downloaded.

Two major publications from Codex are as follows:
1 Principles and guidelines for the conduct of microbiological risk assessment, CAC/GL-30 (CAC 1999).
2 Proposed draft principles and guidelines for the conduct of microbiological risk management (at step 3 of the procedure) CX/FH 00/6 (CCFH 2000).

Although it is accepted that the formalised use of risk analysis in food microbiology is in its infancy, it is likely that in the near future, microbiological risk assessment will have a greater importance in the determination of the level of consumer protection that a government considers necessary and achievable.

9.4.1 Risk assessment
The goal of risk assessment is to provide an estimate of the level of illness from a pathogen in a given population. To achieve this, the process must be scientifically based consisting of the following steps:
1 Statement of purpose.
2 *Hazard identification*: The identification of the hazard, its nature, known or potential health effects associated with the hazard, and the individuals at risk from the hazard.
3 *Exposure assessment*: Describes the exposure pathways and considers the likely frequency and level of intake of food contaminated with the hazard.
4 *Hazard characterisation*: The effect of the hazard on both frequency and severity, and may include a dose–response assessment.
5 *Risk characterisation*: Identifies the likelihood that a population of individuals would experience an adverse health outcome from exposure to the food and might contain the pathogen. It also describes the variability and uncertainty of the risk and identifies data gaps in the assessment.
6 Production of a formal report.
  Figure 9.2 shows the sequence of steps.

Risk assessment requires a multidisciplinary approach including, for example, microbiologists, epidemiologists and statisticians. The same components of hazard identification, exposure assessment, hazard characterisation and risk characterisation are considered in both quantitative and qualitative risk assessments. Qualitative risk assessments may be chosen to

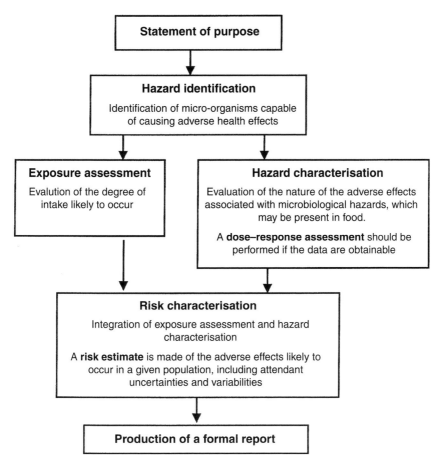

Figure 9.2 Risk assessment flowchart. Reprinted from Notermans, S., Mead, G.C. & Jouve, J.L. (1996) Food products and consumer protection: a conceptual approach and a glossary of terms. *Int. J. Food Microbiol.* **30**, 175–183; copyright 1996, with permission from Elsevier.

identify, describe and rank hazards associated with a food. Quantitative risk assessment may be chosen when substantive scientific data are required for analysis, and risk assessments almost always generate a numerical expression of risk. See Section 9.5 for more detail. Qualitative risk assessment is a useful method to determine which hazards are associated with a particular food, and when there are many gaps in the available data which limit the precision necessary for a quantitative risk assessment. For example, the probability of contamination, the extent of pathogen growth in the food and the amount of food consumed are often uncertain. Subsequently, qualitative risk assessment can be used to identify gaps in the data, and focusing research on aspects that would have the greatest impact of public health.

9.4.2 Risk management

Risk management is a distinct process from risk assessment. It entails comparing policy alternatives, in consultation with all interested parties, considering risk assessment and other factors relevant for the health protection of consumers and for the promotion of fair trade practices, and if needed, selecting appropriate prevention and control options. Currently, HACCP is a major means of hazard control. See Section 8.7 for more detail.

9.4.3 Risk communication

Risk communication is the interactive exchange of information and opinions throughout the risk analysis process concerning hazards and risks, risk-related factors and risk perceptions, among risk assessors, risk managers, consumers, industry, the academic community and other interested parties, including the explanation of risk assessment findings and the basis of risk management decisions.

9.5 Risk assessment

The goal of a risk assessment is to estimate the risk of illness in a given population from a pathogen and to understand the factors that influence it. Starting with a statement of purpose, the process (as defined by Codex) is shown in Figure 9.2. The definitions of these components are based on the Codex document (CAC 1999); there are 11 principles involved as given in Table 9.5.

Risk assessment collates information on foodborne hazards that enable decision-makers to identify interventions (risk mitigations) leading to improved public health. This may include regulatory action, voluntary activities and educational initiatives. Risk assessment can also be

Table 9.5 Principles of microbiological risk assessment (CAC 1999)

Principle

(1) Microbiological risk assessment should be soundly based on science
(2) There should be a functional separation between risk assessment and risk management
(3) Microbiological risk assessment should be conducted according to a structured approach that includes hazard identification, exposure assessment, hazard characterisation and risk characterisation

(4) A microbiological risk assessment of microbiological hazards should clearly state the purpose of the exercise, including the form of risk estimate that will be the output

(5) The conduct of a microbiological risk assessment should be transparent
(6) Any constraints that impact on the risk assessment such as cost, resources or time should be identified and their possible consequences described

(7) The risk estimate should contain a description of uncertainty and where the uncertainty arose during the risk assessment process

(8) Data should be such that uncertainty in the risk estimate can be determined; data and data collection systems should, as far as possible, be of sufficient quality and precision that uncertainty in the risk estimate is minimised

(9) A microbiological risk assessment should explicitly consider the dynamics of microbiological growth, survival and death in foods and the complexity of the interaction (including sequelae) between human and agent following consumption as well as the potential for further spread

(10) Wherever possible, risk estimates should be reassessed over time by comparison with independent human illness data

(11) A microbiological risk assessment may need re-evaluation, as new relevant information becomes available

used to identify data gaps and target research that should have the greatest value in terms of public health impact.

The knowledge in each step of the risk assessment is combined to represent a cause-and-effect chain from the prevalence and concentration of the pathogen (exposure assessment) to the probability and magnitude of health effects (risk characterisation; Lammerding & Paoli 1997). In risk assessment, 'risk' consists of both the probability and impact of disease. Therefore, risk reduction can be achieved either by reducing the probability of disease or by reducing its severity.

9.5.1 Statement of purpose

The specific purpose of the risk assessment should be clearly stated and the output form, and possible output alternatives, should be defined. This stage refers to problem formulation and is intended to form a practical framework and a structured approach either for a full-risk assessment or for a standalone process (such as hazard characterisation). During this stage, the cause of concern, the goals, breadth and focus of the risk assessment should be defined. The statement may also include data requirements, as they may vary depending on the focus and the use of the risk assessment and the questions relating to uncertainties that need resolving. Output might, for example, take the form of a risk estimate of an annual occurrence of illness, or an estimate of annual rate of illness per 100 000 population, or an estimate of the rate of human illness per eating occurrence.

9.5.2 Hazard identification

Definition: Hazard identification consists of the identification of biological, chemical and physical agents (micro-organisms and toxins) capable of causing adverse health effects which may be present in a particular food or group of foods.

Hazard identification consists of the identification of biological, chemical and physical agents (micro-organisms and toxins) capable of causing adverse health effects which may be present in a particular food or group of foods. It often involves evaluating epidemiological data linking foods and pathogens to human illness. Hazard identification can be used as a screening process to identify pathogen–commodity combinations of greatest concern to the risk managers. Information on potentially hazardous micro-organisms and toxins can be obtained from numerous sources such as government surveillance studies and various highly reputable organisations (i.e. ICMSF publications). The information may describe microbial growth and death conditions (pH, a_w, D values). Hazard identification is easier in microbial risk assessments than chemical risk assessments since the microbiological hazards are well recognised due to their short incubation period (days), whereas in chemical risk assessment, the adverse health effect may require a long time period (years) after exposure; see Table 9.4. Micro-organisms associated with foodborne illness are given in Tables 1.2 and 4.1.

The key to hazard identification is the availability of public health data and a preliminary estimate of the sources, frequency and amount of the hazard(s) under consideration. Although foodborne bacterial pathogens with certain foods are well recognised, surveillance data and epidemiological studies can reveal high-risk products and processes. The information collected is subsequently used in 'exposure assessment' where the effect of food processing, storing and distributing (covering from processing to consumption) on the number of foodborne pathogens is assessed.

9.5.3 Exposure assessment

Definition: Exposure assessment is the qualitative and/or quantitative evaluation of the likely intake of biological, chemical and physical agents via food as well as exposure from other sources if relevant.

Exposure assessment determines the likelihood of consumption and the likely dose of the pathogen that the consumers may be exposed to in a food. It should be in reference to a specified portion size of food at the time of consumption or a specified volume of water consumed per day. Overall, it describes the pathways through which a hazardous micro-organism enters the food chain and is subsequently distributed and challenged in the production, distribution and consumption of food. This may include an assessment of actual or anticipated human exposure. For foodborne microbiological hazards, exposure assessment might be based on the possible extent of food contamination by a particular hazard and on consumption patterns and habits. Exposure to foodborne pathogens is a function of the frequency and amount of food consumed and the frequency and level of contamination.

The steps in food production that affect human exposure to the target organism from primary production to consumption are described as the 'farm-to-fork' sequence. It is convenient in exposure assessment to divide the sequence into a series of modules as shown in Figure 9.3. The diagram emphasises the two sets of data required in a quantitative risk assessment: prevalence and concentration of the specified pathogen (hazard). Depending upon the scope of the risk assessment, exposure assessment can begin with either the pathogen prevalence in raw materials or with the description of the pathogen population at subsequent steps such as during processing. Where surveillance data are lacking or insufficient, the effect of processing on prevalence and concentration can be modelled using predictive microbiology (Section 2.8). Nevertheless, where possible such predicted values should be verified with surveillance data.

A flow diagram for exposure assessment is given in Figure 9.4. This should be combined with the activities described for hazard characterisation in Figure 9.5.

Exposure assessment is one of the most complex and uncertain aspects of microbial risk assessment. Hence, modelling and simulation studies are required. A large emphasis is placed on estimating the effects of a number of factors on the microbial population. These factors include the following:

- The microbial ecology of the food
- Microbial growth requirements (intrinsic, extrinsic parameters, Table 2.8)
- The initial contamination of the raw materials
- Prevalence of infection in food animals
- The effect of the production, processing, cooking, handling, storing, distribution steps and preparation by the final consumer on the microbial agent (i.e. the impact of each step on the level of the pathogenic agent of concern)
- The variability in processes involved and the level of process control
- The level of sanitation, slaughter practices, rates of animal–animal transmission
- The potential for (re)contamination (e.g. cross-contamination from other foods and recontamination after a heat treatment)
- The methods or conditions of packaging, distribution and storage of the food (e.g. temperature of storage, relative humidity of the environment, gaseous composition of the atmosphere), the characteristics of the food that may influence the potential for growth of the pathogen (and/or toxin production) in the food under various conditions, including abuse (e.g. pH, moisture content or water activity, nutrient content, presence of antimicrobial substances, competitive flora).

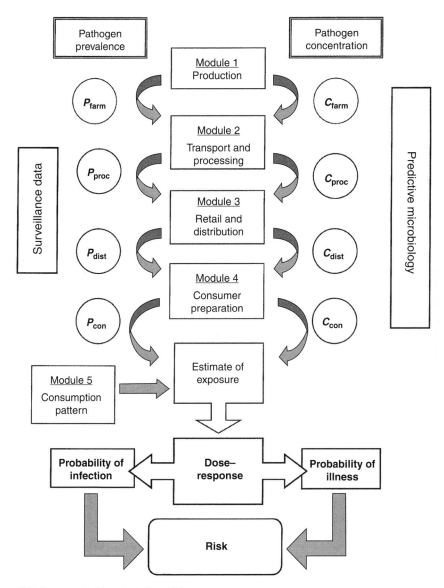

Figure 9.3 Framework of farm to fork modules.

Since preformed microbial toxins and viruses do not grow in food, their exposure assessment is simpler than that of bacteria which may multiply, die or adapt during the farm-to-fork stages.

The factors related to the food matrix are principally those that may influence the survival of the pathogen through the hostile environment of the stomach. They may include the composition and structure of the food matrix (e.g. highly buffered foods):

- Entrapment of bacteria in lipid droplets
- The processing conditions (e.g. increased acid tolerance of bacteria following pre-exposure to moderately acid conditions)
- Conditions of ingestion (e.g. initial rapid transit of liquids through an empty stomach)

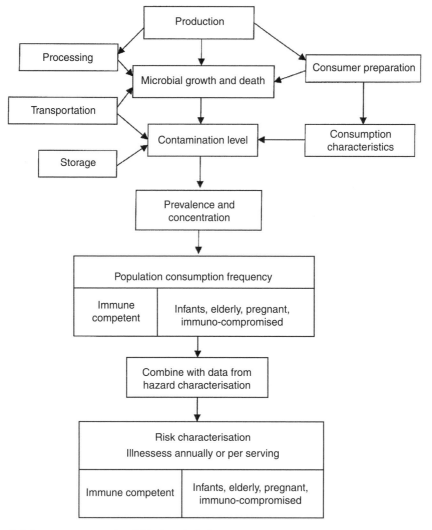

Figure 9.4 Exposure assessment and risk characterisation.

Data on microbial survival and growth in foods can be obtained from food poisoning out-breaks, storage tests, historical performance data of a food process, microbiological challenge tests and predictive microbiology (Section 2.8). These tests provide information on the likely numbers of organisms (or quantity of toxin) present in a food at the point of consumption.

Information on consumption patterns and habits may include the following:

- Socio-economic and cultural background, ethnicity
- Consumer preferences and behaviour as they influence the choice and the amount of the food intake (e.g. frequent consumption of high-risk foods)
- Average serving size and distribution of sizes
- Amount of food consumed over a year considering seasonality and regional differences

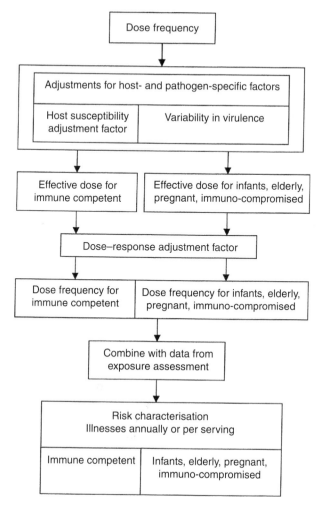

Figure 9.5 Hazard characterisation and risk characterisation.

- Food preparation practices (e.g. cooking habits and/or cooking time, temperature used, extent of home storage and conditions, including abuse)
- Demographics and size of exposed population(s) (e.g. age distribution, susceptible groups)

Bettcher *et al.* (2000) and Ruthven (2000) published a survey of food consumption for various regions of the world. Consumption values varied; for example the consumption of chicken meat ranged from 11.5 g (Far Eastern diet) to 44.0 g (European diet).

There are several factors that strongly affect consumer exposure, but which are poorly understood. These factors are as follows:

- Incidence of different pathogens in raw materials
- Effect of processing conditions (including alternative technologies)
- Effect of distribution conditions (e.g. chilled chain)
- Recontamination during handling
- Abuse by the consumer

- Heterogeneous distribution of micro-organisms within a food
- Person-to-person transmission
- *Host effects*: Age, pregnancy, nutritional status, concurrent or recent infections, use of medication, immunological status
- Food vehicle effects

9.5.4 Hazard characterisation

Definition: Hazard characterisation is the qualitative and/or quantitative evaluation of the nature of the adverse effects associated with biological, chemical and physical agents that may be present in food. If data are available, then a dose–response assessment should be performed.

Hazard characterisation provides an estimate of the nature, severity and duration of the adverse effects following ingestion of the hazard; that is for a given number of micro-organisms consumed at a sitting, what is the probability of illness? If sufficient data are available, then a dose–response relationship is determined (see below). This step, like exposure assessment, is very complex. Factors important to consider relate to the micro-organism, food and the host. A flow diagram for hazard characterisation is given in Figure 9.5; this should be combined with the activities described for exposure assessment in Figure 9.4. A preliminary guideline document on hazard characterisation has been released by WHO/FAO (2000a, see Web Resources section for URL).

Hazard characterisation can be a standalone process as well as a component of risk assessment. The hazard characterisation must be transparent (assumptions and variables well documented) such that risk managers can combine the information with an appropriate exposure assessment. This could even occur by combining the two components across different countries. A hazard characterisation developed for water exposure may be adapted to a food exposure scenario with modification for the food matrix effects. In general, hazard characterisations are fairly adaptable between risk assessments for the same pathogen. This contrasts with exposure assessments since those are highly specific to the production, processing and consumption patterns within a country or region.

Ingestion of a pathogen does not necessarily mean the person will become infected, nor that illness or death will occur. As shown in Figure 9.6, there are a number of barriers to infection and illness. These barriers can be compromised due to host and food matrix factors. The response (infection, illness, death) to pathogen ingestion will vary according to pathogen, food and host factors, commonly known as the 'infectious disease triangle' (Figure 9.7).

Pathogen factors

In determining the hazard characterisation for foodborne microbial pathogens, the biological aspects of the pathogen must be considered. Infection can be seen as resulting from the successful passage of multiple barriers in the host (Figure 9.6). These barriers are not equally effective in eliminating or inactivating pathogens. Each individual pathogen has some particular probability or relative frequency to overcome a barrier, which is conditional on the previous step(s) being completed successfully.

The infectivity of microbial pathogens is dependent upon many factors. These can be intrinsic to the pathogen (phenotypic and genetic characteristics) as well as host specificity. The stress response to temperature, drying, acidity, and so on, may affect the virulence of the pathogen (Section 2.7). Figure 9.7 lists the various pathogen factors which must be considered in hazard characterisation.

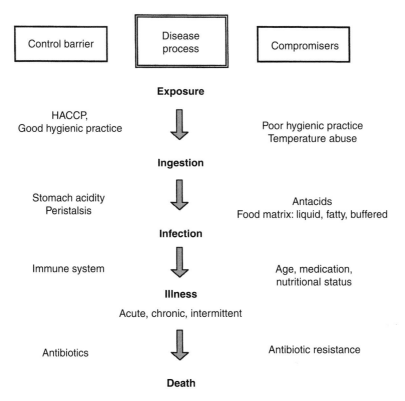

Figure 9.6 Barriers to infectious disease.

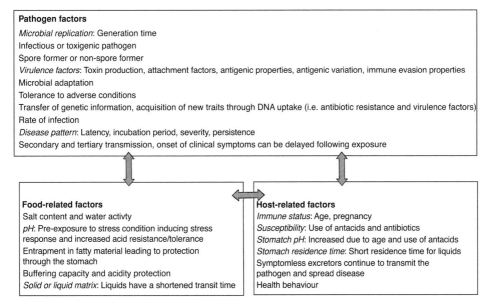

Figure 9.7 Factors affecting the infectivity of microbial pathogens.

Food-related factors

The food matrix can affect the survival of the microbial pathogen. For example, a food high in fat content can protect the organism from stomach acid, and hence increase the chances that it will survive and cause infection. The food processing can give the organism a heat shock which may cause microbial adaptation, including acid tolerance and hence increased survival through the stomach. Exposure to short-chain fatty acids can induce acid resistance in *Salmonella* serovars. Additionally, if the food is highly proteinaceous, it may act as a buffer and protect the pathogen from the stomach's acidity.

Host-related factors

As shown in Figure 9.7, there are a variety of host-related factors that influence the disease response. The well-recognised factors are age and immune status. It has been estimated that 20% of the total population may be immunoimpaired and hence are separately described in Figure 9.7 (Gerba *et al.* 1996). Prior exposure is of limited importance for foodborne pathogens since many do not invade the host body, but may be important for the protozoal parasite *Cyclospora cayetanensis*. See Section 10.1 for a description of human feeding trials with *Salmonella* serovars.

Hazard characterisation must consider the range of biological responses to pathogen ingestion. This ranges from asymptomatic infections, illness (acute, chronic and intermittent) to death; see Figure 9.6. Severity of the risk may be expressed as duration of the illness, the proportion of the population affected or as mortality rate and should identify the 'at-risk' groups. For pathogens that cause chronic sequelae (i.e. Guillain–Barré syndrome; Section 1.8), the effect on quality of life may be included in the hazard characterisation.

Figures 9.6 and 9.7 show the various host-related factors that affect susceptibility to microbial infections and severity of illness. The important aspect of hazard characterisation is to provide information on who is at risk and the associated severity for the susceptible sub-populations. In Section 1.7, the cost of foodborne disease in the United States was given as between $6.6 and $37.1 billion. The economic and social cost is so extensive that the WHO has identified improved food safety as one of its aims for the twenty-first century.

9.5.5 Dose–response assessment

The goal of a dose–response assessment is to determine the relationship between the magnitude of exposure (dose) to the pathogen and the severity and/or frequency of adverse health effects (response). Sources of information include the following:
1 Human volunteer studies
2 Population health statistics
3 Outbreak data
4 Animal trials

Because of the variety of disease response (Figure 9.6), the end point must be clearly delineated. Essentially, there are four possible dose responses:
1 Probability of infection following ingestion
2 Probability of illness (morbidity) following infection
3 Probability of chronic sequelae following illness
4 Probability of death (mortality)

It is generally assumed that the effects of foodborne pathogens are dose dependent but not cumulative (unlike many chemical hazards). Hence, the frequency of consumption must be determined since multiple exposures to low doses may not represent the same risk as a single exposure to a large dose.

Table 9.6 Aspects of the dose–response relationship (WHO/FAO 2000a)

1	Organism type and strain
2	Route of exposure
3	Level of exposure (the dose)
4	Adverse effect considered (the response)
5	Characteristics of the exposed population
6	Duration – multiplicity of exposure

The dose–response relationship is complex and in many cases may not be feasible. For example, Medema *et al.* (1996) reported that although the rate of *Campylobacter jejuni* infection was dose related, the rate of illness was not. This contrasts with *Salmonella* infections where higher doses are reported to result in greater frequency of severe illness (Coleman & Marks 1998).

Currently, there is a lack of data concerning pathogen-specific responses, the effect of the host's immunocompetence on the pathogen-specific responses, translation of infection into illness and translation of illness into different outcomes. Also, as given in Figure 9.7, there are numerous variable factors involved:

- Physiology, virulence and pathogenicity of the micro-organisms
- Variation in host susceptibility
- Food matrix

It is therefore very important that the dose–response analysis identifies the information which has been used and its source. In addition, variability (due to known factors such as amount of food consumed, population susceptibility) and uncertainty (insufficient experimental data or lack of knowledge of the pathogen/host/food being studied) in the data should be thoroughly described for the risk assessment to be transparent (Nauta 2000).

An important aspect is whether the effect on human health is a threshold, or cumulative action of the pathogen or toxin. Information on the biological mechanism is important when extrapolating from laboratory animal or *in vitro* studies to assess the relevance to human health. Aspects to be considered are outlined in Table 9.6.

Until relatively recently, it had been assumed that there was a threshold level of pathogens that must be ingested in order for the micro-organism to produce an infection or disease (the minimum infectious dose; see Table 4.2). This approach has been largely superseded by the proposal that infection may result from the survival of a single, viable, infectious pathogenic organism (single-hit concept). Thus, regardless of how low the dose, there is always a non-zero effect of infection and illness. It should be noted that the accuracy of the infectious dose is debatable. For example, it is commonly cited that the infectious dose for *C. jejuni* is as low as 500 bacteria. However, this value can be traced back to the article by Robinson (1981) who performed the investigation on himself. After a light breakfast, he drank a glass of milk that contained 500 bacteria, which consequently made him ill. In contrast, Martin *et al.* (1995) gave the probability for developing at least light symptoms which was estimated at 24% by consuming *C. jejuni* at the level of 10^2 cells compared with the probability of developing at least light symptoms by ingesting 10^8 cells which was 32%. The 'tolerable' level for *B. cereus* are accepted to be less than 10^4 cells (Section 10.5).

Very little is known about the effect that the food matrix has on microbial survival in the stomach and intestines, and its virulence. It is plausible that the lipid content of foods such as cheese and chocolate can protect *Salmonella* from the acid in the stomach. Some salmonellas are more acid resistant than others. The use of antacids by the host can affect the apparent virulence of *Salmonella* and *L. monocytogenes*.

Concurrently to the analysis of clinical and epidemiological information, mathematical modelling can assist in developing a dose–response relationship. This is particularly useful when extrapolating to low doses. An active area of research is the development of more appropriate mathematical models for dose–response assessment.

9.5.6 Dose–response models

Food is frequently contaminated with smaller numbers of microbial pathogens than those used in laboratory trials of human feeding studies and animal models. Therefore, mathematical models are needed to extrapolate low-dose responses from the high-dose data. Various dose–response models have been proposed to describe the relation between ingestion of a certain number (N) of a pathogenic micro-organism and the possible outcomes. The main models are exponential and beta-Poisson (Holcomb *et al.* 1999).

One approach assumes that each micro-organism has an inherent minimal infective dose, that is there is a threshold value, below which no response (depending upon the end point) is seen. The value of the minimal dose in the population may be assumed to follow different distributions. The alternative approach is that the actions of individual cells of pathogenic micro-organisms are independent and that a single micro-organism has the potential to infect and provoke a response in the individual, that is a single-hit, non-threshold model (Haas 1983). The exponential model assumes that the probability of a single cell causing infection is independent of dose. In contrast, the beta-Poisson model assumes that infectivity is dose dependent. Different microbial pathogens appear to fit different dose–response models.

Data for protozoan parasites can be well described by the exponential models described in the following sections.

Exponential model

$$P_i = 1 - \exp^{-r*N}$$

where
  P_i = probability of infection
  r = host/micro-organism interaction probability
  N = ingested dose of micro-organisms

This has been used for *Cryptosporidium parvum* and *Giardia lamblia* (Teunis *et al.* 1996), though Holcomb *et al.* (1999) modified it slightly into the simple exponential model and flexible exponential model (cf. Rose *et al.* 1991).

Simple exponential model

$$P_i = 1 - \exp^{-r*\log 10^{(N)}}$$

Flexible exponential model

$$P_i = \beta * \left[1 - p * \exp^{-\varepsilon \left\{ \log_{10}(N) - X \right\}} \right]$$

where

   P_i = probability of infection
   r = host/micro-organism interaction probability
   N = ingested dose of micro-organisms
   β = asymptomatic value of probability of infection as dose approaches $\beta = 1$ (Holcomb *et al.* 1999)
   χ = predicted dose at specified value of p where $p = 1 - Pr$
   ε = curve rate value affecting spread of curve along dose axis

   In contrast, the bacterial infection data are generally well described using beta-Poisson models (Haas 1983, Teunis *et al.* 1997, 1999) and the Weibull-gamma model (Holcomb *et al.* 1999, Todd & Harwig 1996).

Beta-Poisson model

$$P_i = [1 - (1 + N/\beta)]^{-\alpha}$$

where

   P_i = probability of infection
   N = ingested dose of micro-organisms
   α and β are parameters affecting the shape of curve that are specific to the pathogen (cf. Vose 1998)

   The beta-Poisson model is frequently used for describing dose–response relationships when assessing low levels of bacterial pathogens. It generates a sigmoid dose–response relationship that assumes no threshold value for infection (see Figures 9.8 and 9.9 for dose–response curves for *Salmonella* and *C. jejuni*, respectively). Instead, it assumes that there is a small but finite risk that an individual can become infected after exposure to a single cell of a bacterial pathogen (single-hit concept).

   Marks *et al.* (1998) compared two beta-Poisson models for a risk assessment of *E. coli* O157:H7 in hamburgers, one of which had a threshold value of three bacteria. The differences between the models were only significant in the low-dose range and the resulting estimates of risk were 100- to 1000-fold larger when using the non-threshold model, depending upon the cooking temperature. They concluded that the two-parameter beta-Poisson model seemed

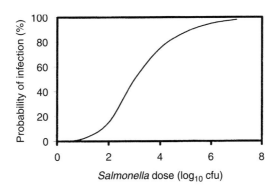

Figure 9.8 Beta-Poisson dose–response curve for *Salmonella* serovars.

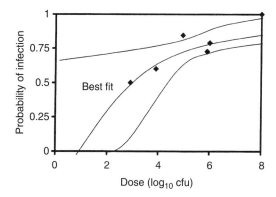

Figure 9.9 *C. jejuni* dose–response in human feeding trials (Black *et al.* 1988; Fazil *et al.* 2000).

inadequate as a default model for describing the complexity of dose–response interactions especially in cooked foods.

Weibull-gamma model

$$P_i = 1 - \left[1 + (N)^b/\beta\right]^{-\alpha}$$

where

P_i = probability of infection

N = ingested dose of micro-organisms

α, β and b are parameters affecting the shape of curve that are specific to the pathogen

The Weibull-gamma model assumes that the probability that any individual cell can cause an infection is distributed as a gamma function. Hence, this model is very flexible in that its shape depends on the parameters selected and has more recently been used for dose responses (Farber *et al.* 1996).

Beta-binomial model

Cassin *et al.* (1998a) developed a beta-binomial dose–response model for *E. coli* O157:H7 in hamburgers which yields variability for probability of illness from a particular dose in contrast to the original model which only specifies a mean population risk (Section 10.4.1).

$$P = 1 - (1 - P_i(1))^N$$

where

$P_i(1)$ = probability of illness from ingestion of one micro-organism, and this probability was assumed to be beta-distributed with parameters α and β.

By fitting the model to data from human feeding studies with *Sh. dysenteria* and *Sh. flexneri* (Crockett *et al.* 1996), a dose–response curve was generated which showed the estimated uncertainty in the average probability of illness versus the ingested dose. The variability between feeding studies was used for the uncertainty in the parameters α and β.

The same model may not be equally effective for all biological end points caused by the pathogen. For example, the exponential model did not fit *Listeria monocytogenes* infection of mice data (isolation from spleen and liver), but was a good model for describing the relationship

Table 9.7 Definitions of 'dose'

Measured dose	The dose as estimated by the mean concentration in the delivery matrix. Note that the dose unit (e.g. colony forming units) may contain one or more discrete organisms
Functional dose	The measured dose corrected for the sensitivity and specificity of the measurement method to detect viable, infectious agents
Administered dose	The number of viable, infectious agents actually administered, whether orally (ingestion, gavage, nasal-gastric intubation), by injection (intraperitoneal or intravenous), or by other means. Although this is the dose that is actually given to an individual, it can only be estimated
Effective dose	The number of viable, infectious agents that actually reach the site of infection

between dose and the frequency of death (FDA 2000a). The Gompertz equation gave the best fit for frequency of infection (Coleman & Marks 1998).

Gompertz model

$$P_i = 1 - \exp\left[-\exp(a + bf(x))\right]$$

where

a = model (intercept) parameter

b = model (slope) parameter

$f(n)$ = function of dose

It is important to note that unit dose and biological dose are almost always different due to the non-homogenous distribution of micro-organisms in food; see Table 9.7. A range of α and β values are give in Table 9.8. Infection values are different from the minimum infectious dose values commonly given; see Table 4.2.

Table 9.8 Dose–response parameters for food- and water-borne pathogens, where α and β are beta-Poisson parameters and N_{50} represents the ID_{50}[a]

Micro-organism	Model	Model parameters	Reference
Non-typhi *Salmonella*	beta-Poisson	$\alpha = 0.4059$ $\beta = 5308$	Fazil *et al.* 2000
E. coli	beta-Poisson	$\alpha = 0.1705$ $\beta = 1.61 \times 10^6$	
Echovirus 12	beta-Poisson	$\alpha = 0.374$ $\beta = 186.69$	Schiff *et al.* 1984
Rotavirus	beta-Poisson	$\alpha = 0.26$ $\beta = 0.42$ $\alpha = 0.265$ $N_{50} = 0.42$	Ward *et al.* 1986; Gale 2001
Cryptosporidium	Exponential	$r = 0.004191$	
Giardia lamblia	Exponential	$r = 0.02$	

[a]ID_{50} is the dose causing 50% infection.

Other models
It is probable that alternative dose–response models will be required according to whether the
microbial pathogen is infectious or toxigenic, and whether the organism produces the toxin
whilst passing through the intestinal tract (*Clostridium perfringens*) or is ingested preformed in
the food (*Staphylococcus aureus*). Dose–response modelling is further complicated by micro-
organisms such as *B. cereus* which cause two different illnesses: emetic and diarrhoeal (Section
4.3.10). Some organisms may tend to remain as colonies or clumps following ingestion and
hence infection due to a single cell may not frequently occur. Virulence mechanisms such
as attachment and effacement of enterocytes as found in pathogenic *E. coli* may make the
intestinal wall more susceptible to further infection and illness in contrast to the biological
response to toxins (i.e. *St. aureus* enterotoxins and aflatoxins) which are adsorbed through the
intestinal wall without damaging it. For an extensive treatment on mathematical modelling of
dose response and data fitting, the reader is referred to Haas *et al.* (1999).

9.5.7 Dose and infection
In general, dose–response models for foodborne pathogens should consider their discrete
(particulate) nature, and should be based on the concept of infection from one or more
'survivors' from an initial dose. There are however different definitions of dose and hence this
must be clearly stated in any study; see Table 9.7. The measured dose may need to be corrected
according to the sensitivity and/or specificity of the detection method which may not be
specific for the viable, infectious organism. Therefore, the functional dose is defined as the
corrected measured dose. It should be recognised that the standard agar plate viability count
gives 'colony forming units per g'; however, a colony may have been due to one or more initial
bacterial cells. Hence, the method will tend to underestimate the number of infectious bacterial
cells. An additional complication in estimating the dose is the non-random distribution of
micro-organisms in food.

The functional dose describes the average number of viable, infectious units in the inoculum.
Every individual in a population will consume a subsample containing a discrete, but unknown
number of units. This ingested dose can be characterised by a frequency distribution such as
a Poisson distribution for random distribution. It is assumed that each individual cell in the
ingested dose has a distinct probability of survival before colonisation. It is this relationship
between the actual number of surviving organisms (the effective dose) and the probability of
colonisation of the host that is a key concept in the derivation of dose–response models.

Infection is the result of a pathogen overcoming a number of barriers (Figure 9.6). The risk of
infection is increased when these barriers are compromised. Infections may be asymptomatic.
In this condition, the host does not develop any symptoms to the infection, and clears the
pathogens within a limited period of time. Since the person is unaware of their infection, they
can act as symptomless carriers of infection. The probability of sequelae and/or mortality for
a given illness depends on both the pathogen and the host. Host factors are commonly age
and immune status, but genetic determinants are increasingly being recognised as important
factors in infection susceptibility.

The most direct means of determining dose–response relationships for foodborne pathogens
would be to expose humans to the organisms under controlled conditions. Shocking though
this may seem, there have been a limited number of human feedings trials using volunteers,
and most of these have been in conjunction with vaccine trials. From the human volunteer
studies, 'Probability of infection' (P_i) values have been determined for a number of food- and
waterborne pathogens (Table 9.9). For example, there is a 1 in 2000 chance of an individual

Table 9.9 Probability of infection for food- and water-borne pathogens (Bennett *et al.* 1987, Notermans & Mead 1996)

Enteric pathogen	Probability of infection (P_i)	Fatality/case (%)
Campylobacter jejuni	7×10^{-3}	0.1
Salmonella serovars	2×10^{-3}	0.1
Shigella spp.	1×10^{-3}	0.2
V. cholerae (classical)	7×10^{-6}	1.0
V. cholerae El Tor	1.5×10^{-5}	4.0
Rotavirus	3×10^{-1}	0.01
Giardia spp.	2×10^{-2}	nd

nd, not determined.

becoming infected from a single *Salmonella* cell compared to a 1 in 7 million chance from *V. cholerae*. Medema *et al.* (1996) used the human feeding trial data of Black *et al.* (1988) to determine the dose response with a beta-Poisson model (Figure 9.9; Table 9.10) for infection with *C. jejuni*. The occurrence of symptoms did not follow a similar dose-related trend however. The beta-Poisson model for *E. coli* O157:H7 (Section 10.4.1) showed considerable variability. Since no comparable data were available for *E. coli* O157:H7, this study used the α and β values for *Sh. dysenteriae* due to the similarity in virulence

There are a number of limitations associated with the use of human feed trials:

1 The main problem is that they are almost always conducted with healthy, young (18–50 years old) adults, usually men. However, the most vulnerable members of society are the elderly, pregnant and very young.

2 Ethically, pathogens that are life-threatening (such as *E. coli* O157) or that cause disease only in high-risk sub-populations (such as *L. monocytogenes* serotype 4b) are not amenable to such volunteer studies.

3 Human feed trials usually only use a small number of volunteers per dose and a small number of doses. The average is six volunteers per dose, though the range is from 4 to 193; see Table 9.10. Since the studies are often for vaccine trials, the dose ranges are typically high to ensure a significant portion of the test population respond. Therefore, the doses are often not in the range of interest to dose–response modellers for foodborne infections.

4 The pathogen is often fed to the volunteer in a non-food matrix after neutralising the stomach acidity. Hence, the size of dose reaching the intestines is probably different to normal ingestion patterns (Kothary & Babu 2001).

Table 9.10 Human trial data for *C. jejuni* infection (Black *et al.* 1988, Medema *et al.* 1996). See also Figure 9.9

Dose (cfu)	Number of volunteers	Number infected	Number showing symptoms
8×10^2	10	5	1
8×10^3	10	6	1
9×10^4	13	11	6
8×10^5	11	8	1
1×10^6	19	15	2
1×10^8	5	5	0

However, obtaining human volunteer data does have an advantage that inter-species conversions, as required from animal models, are not necessary.

Animal models depend on the selection of appropriate animal(s) showing the same disease response and a conversion factor to human response. The major advantage is that a larger number of replicates and dose range can be used and the animals can be kept under more environmentally controlled conditions than human volunteers. Aside from ethical issues on animal experimentation, the animal models do have inherent limitations. The animals are often similar in age, weight and immune status. Hence, similar criticisms can be used as have also been applied to human feeding trials. Additionally, laboratory animals have very limited genetic variation, unlike humans. The use of animal models for *L. monocytogenes* risk assessment is discussed in Section 10.3.

Many national governments and several international organisations compile health statistics for infectious diseases including those that are transmitted by foods and water. Such data are crucial for the characterisation of microbiological hazards. In addition, surveillance-based data have been used along with food survey data to estimate dose–response relationships. Epidemiological studies are important as a means of verification of dose–response models. The effectiveness of dose–response models is typically assessed by combining them with exposure estimates and determining if they approximate the annual disease incidence for the organism.

Studies of outbreaks of foodborne illness may be very useful to dose–response modellers as they can provide data on the levels of infectious organism in the food, amount consumed and susceptible population (Mintz *et al*. 1994). The Minnesota Department of Health investigated two *Salmonella* outbreaks. The highest levels of *Salmonella* contamination were 4.3 organisms 100 g^{-1} of cheese and 6.0 organisms 65 g^{-1} (one-half cup) serving of ice cream. This low level had also been reported in previous outbreak investigations involving cheese and chocolate in the United States and Canada. The subsequent estimated infective doses derived from these epidemiological studies are several log orders lower than estimates of minimum infective doses from clinical trials using a limited number of volunteers. Unfortunately, data from epidemiological investigations of outbreaks may not always be totally reliable because the information was not collected according to a standardised format or procedure. Estimates of attack rate may be overestimated or underestimated as they may be based on symptoms rather than laboratory-confirmed cases where the causative organism was recovered. According to WHO/FAO (2000a), determination of the exposure dose in outbreak scenarios may be inaccurate because:

1 representative samples of the contaminated food or water were not obtained;
2 detection methods may not be sufficiently accurate (e.g. *Cryptosporidium* oocysts in water);
3 water or food consumption levels may be inaccurate estimates.

Section 10.1.2 provides the extensive JEMRA report on hazard identification and hazard characterisation of *Salmonella* in broilers and eggs and compares the dose–response models between feeding trials and outbreaks.

Buchanan *et al*. (2000) encouraged a more mechanistic approach to dose–response modelling that takes into account the limitations of extrapolating from human feeding trials of healthy (male) individuals to the general, diverse population. They proposed that there were three stages to be considered:

1 Gastric acidity barrier
2 Attachment/infectivity
3 Morbidity/mortality

Gastric acidity barrier
The infectivity of ingested pathogens depends initially upon their survival through the stomach contents. This depends upon the kill effect of the stomach acidity and the rate of gastric emptying. The rate of death can be expressed as a D value (Section 2.5.2) according to the equation:

$Log_{10} D = (0.554 \times pH) - 1.429$

The rate of emptying is given by:

$\%R = 100.4 \text{ minutes} + (-0.429 \times t)$

where $\%R$ = percent retention and t = time (minutes)

A combination of these equations predicts that at pH 2.2 (normal pH of gastric juice), only 1–2 cells/100 will survive, whereas at pH 4.0, there would be approximately 50% survival. Buchanan *et al.* (2000) found these predicted results correlated with experimentally derived values by Peterson *et al.* (1989).

Attachment/infectivity
The second stage in the infection model is the ability of the ingested pathogen to overcome the wash-out effect due to the (relative to microbial size) rapid flow of intestinal contents. Hence, the pathogen must attach to and colonise the intestinal epithelium. The attachment site varies according to the pathogen (i.e. microvilli via mannose-specific type 1 fimbrae in *Salmonella* serovars). Due to the lack of appropriate data, Buchanan *et al.* (2000) used a value that 1 in 100 surviving pathogens were able to attach and colonise the epithelial layer. Obviously, further research is required on this topic.

Morbidity/mortality
The final stage in the infection model is the likelihood of the ingested pathogen to cause symptoms and even death. The progress of the infection will depend upon the virulence mechanisms of the pathogen and the host defence (primarily immune) system. The immune system deteriorates with age and the defence system may be weakened due to inadequate diet and the use of prescription drugs such as antacids.

Hence, the mechanistic model implies that the rate of infection is dependent upon the stomach pH, intestinal attachment, virulence mechanisms and host immune status. Buchanan *et al.* (2000) used the model to simulate the effect of *Salmonella* ingestion on two populations: elderly (>65 years old) of which 30% suffered from achlorhydria (reduced gastric acid secretion), and adults between 20 and 65 years old, of which only 1% suffered from achlorhydria. The population sizes were set at 100 000 and the ingested dose was 100 *Salmonella* cells. Those with achlorhydria had stomach pH values of 4.0 compared with the normal pH 2.2 (Forsythe *et al.* 1988).

Based on a series of assumptions, such as rates of morbidity and mortality and proportion of each population becoming infected and dying, Table 9.11 was constructed (summarised from Buchanan *et al.* 2000). The model predicts the considerably greater number of deaths for the elderly.

In this example, the mechanistic model has only been applied to one dose, but could be further utilised for a range of exposure levels.

9.5.8 Risk characterisation
Definition: Risk characterisation is the integration of the three previous steps (hazard identification, exposure assessment, hazard characterisation) to obtain a **risk estimate** of the likelihood, and the severity of the adverse effects in a given population with attendant uncertainties.

Table 9.11 Predicted frequency of *Salmonella* infection following ingestion of 100 *Salmonella* cells by two 100 000 populations using a mechanistic model (summarised from Buchanan *et al.* 2000)

	Immune status	Adults <65 years old	Adult >65 years old
Number of individuals infected	Immunocompetent	3297	13 910
	Immunoimpaired	33	2669
	Total	3330	16 579
Number of symptomatic individuals	Immunocompetent	165	696
	Immunoimpaired	3	267
	Total	168	963
Number of deaths	Immunocompetent	3.3	13.9
	Immunoimpaired	0.2	21.3
	Total	3.5	35.2

Risk characterisation is the final stage of risk assessment. This can be qualitative (low, medium, high) or quantitative (number of human infections, illnesses or deaths per year or per 100 000 population) depending upon the exposure assessment. The degree of confidence in the risk estimate depends on the amount of knowledge in the previous steps, for example the variation in the human (sub)population susceptibility. Unfortunately, quantitative estimates of risk are difficult to obtain due to limitations in expertise, time, data and methodology. In risk characterisation, the variation is divided into '**uncertainty**' which reflects where important data are not available and '**variability**' where the data are not constant due to recognised factors such as variable amount of food eaten and population susceptibility. These must be distinguished in modelling (Nauta 2000).

The overall probability, or risk estimate, is determined from:

$$\text{Risk estimate} = \text{Dose response assessment} \times \text{Exposure assessment}$$

This is schematically shown in Figure 9.10, where the exposure assessment is the input value(s) for the dose–response relationship and result is the risk estimate, that is the probability of an adverse effect.

The risk characterisation should not only determine the relative risk that a hazard will have on a population but also its severity. Hence, the use of multiple end points, such as infection, illness (morbidity) and death (mortality).

Unfortunately, scientific data are frequently unavailable for parts of the risk assessment and so any uncertainties and variations should be noted in order to assist in making the process transparent. Additionally, any changes in the process would require a re-assessment of the risk. Hence, accurate risk communication (including uncertainties) is a very important aspect of risk analysis.

Risk characterisation estimates can be assessed by comparison with independent epidemiological data that relate hazards to disease prevalence. Using probabilistic models, Monte Carlo simulations can be used to modify previous assumptions and values to ascertain their relative importance. Important variables are frequently illustrated through tornado chart presentation.

A risk management strategy can then be formulated from the risk characterisation.

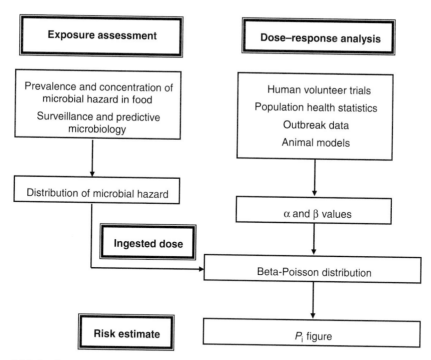

Figure 9.10 Predicting the probability of infection.

9.5.9 Production of a formal report

The risk assessment should be fully and systematically documented. To ensure transparency, the final report should indicate in particular any constraints and assumptions relative to the risk assessment. The report should be made available to independent parties on request.

9.5.10 Triangular distributions and Monte Carlo simulation

Risk assessments need to separate 'uncertainty' (due to lack of knowledge) from 'variability' (due to known factors such as biological variation) and must be described in a transparent fashion. There are two types of quantitative risk assessments: 'deterministic' and 'stochastic'. Deterministic models use single point estimates as input data whereas stochastic models use a distribution range of data values. To determine the risk estimate, the input data may be point estimates of the average value or worst case (95th percentile). In contrast to point estimates, probability distributions describe the relative weightings of each possible value and are characterised by a number of parameters: minimum, mostly likely and maximum. By describing these three values, a triangular distribution is generated which can be further analysed using Monte Carlo simulation (see below). An example of triangular distribution (−1.0, 0.5, 2.5) is given in Figure 9.11 for the effect of evisceration of poultry on microbial load. There is no log reduction in microbial numbers and the most likely effect was a 0.5 $\log_{10}$ increase. The problem with constructing a triangular distribution (as shown in Figure 9.11) is that it does not fully reflect the 'crude data' distribution. Nevertheless, the probability distribution produces a risk distribution that characterises the range of risk of a population. The stochastic method is also known as the 'probabilistic approach'.

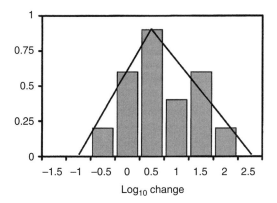

Figure 9.11 Triangular distribution of *C. jejuni* on poultry carcasses after evisceration (Fazil *et al.* 2000b).

The Monte Carlo process is a procedure that generates values of a random variable based on one or more probability distributions (Vose 1996). Monte Carlo simulation is a model that uses the Monte Carlo process to calculate a model output value many times with different input values. The purpose is to get a complete range of all possible scenarios. In microbiological risk assessment, there may be two or more variables, such as prevalence of the pathogen and the concentration of pathogen, which are multiplied together and hence generate a further probability range for further calculations (Vose 1997,1998).

Figure 9.12 shows the frequency distribution for three variables. There are three ways of assessing the data:

1 Determine the mean of each distribution and multiply them together.
2 Take the highest value for each data set to determine the worst-case scenario.
3 Use Monte Carlo simulation to take random values from each data set; after repeated sampling (several thousand times), determine a distribution curve of likely results.

The third approach is more representative of the situation.

A convenient means of Monte Carlo simulation is to first enter the variable ranges in Excel^{TM} (Microsoft Corp.) spreadsheet files and then use either @RISK (Palisaide Corp.), Crystal Ball 2000 (Decisioneering) or Analytica (Lumina Decision Systems, Inc.) which are risk analysis add-in tools to calculate the resultant distribution. See Web Resources section for contact details.

An example of illustrating risk mitigation factors using a tornado chart (so called due to its shape) is given in Figure 9.13 for *V. parahaemolyticus* (FDA 2000b; Section 10.6.1). The purpose is to easily communicate the major factors which might be altered in risk mitigating strategies.

9.6 Risk management

Risk management is required when epidemiological and surveillance data demonstrate that specific foods are a possible hazard factor towards consumer health due to the presence of hazardous micro-organisms or microbial toxins. Governmental risk managers must decide on appropriate control options to manage this risk. In order to understand the risk for consumers more explicitly, the risk managers may initiate a microbiological risk assessment. This

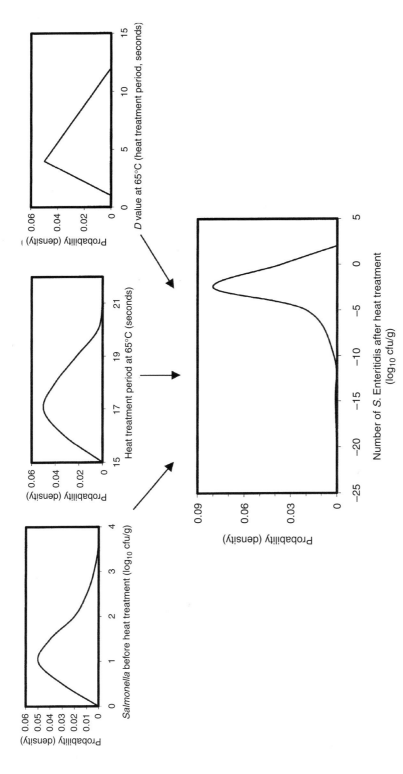

Figure 9.12 Monte Carlo simulation of *S.* Enteritidis distribution after heat treatment.

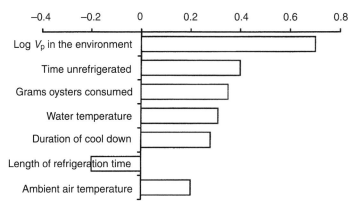

Figure 9.13 Risk factors for *V. parahaemolyticus* in raw molluscan shellfish (FDA 2000b).

assessment leads to an estimate of human risk along with associated uncertainty and variability limits. Risk managers must be aware of these limitations when considering risk management options. A separation between risk management and risk assessment activities must be maintained in order for the assessment to be transparent.

Following risk assessment, appropriate risk management steps should result in safe handling procedures and practices, food processing quality and safety assurance controls, and food quality and safety standards and criteria. If required, risk management may select and implement appropriate regulatory measures. A guidance document on the interaction between assessors and managers of microbiological hazards in foods has been developed (WHO/FAO 2000b). The international risk manager for food is the CAC.

A diagram of risk management is given in Figure 9.14.

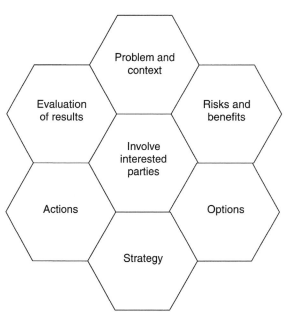

Figure 9.14 Risk management activities.

Risk management can be divided into four aspects:

1 Risk evaluation; initial risk management activities
2 Risk management option assessment
3 Implementation and management of decisions
4 Monitoring and review

The general principles are as follows:

1 Risk management should follow a structured approach.
2 Protection of human health should be the primary consideration in risk management decisions.
3 Risk management decisions and practices should be transparent.
4 Determination of risk assessment policy should be a specific component of risk management.
5 Risk management should ensure the scientific integrity of the risk assessment process by maintaining a functional separation of risk management and risk assessment.
6 Risk management decisions should take into account the uncertainty in the output of the risk assessment.
7 Risk management should include clear, interactive communication with consumers and other interested parties in all aspects of the process.
8 Risk management should be a continuing process that takes into account all newly generated data in the evaluation and review of risk management decisions.

Risk limitation has two principles:

1 The individual risk resulting from a risk source should not exceed the maximum permissible level.
2 The risk is to be reduced 'as low as reasonably achievable' and correct implementation of optimisation must be demonstrated.

A specific aspect of this approach relates to the criteria utilised and the related values. With regard to the maximum permissible level and to the negligible level, where this approach has been applied, the criterion utilised is frequently the lifelong risk of death. The figures range from 10^{-8} to 4.10^{-3} though most figures converge between 10^{-6} and 10^{-4}. For the food industry, the maximum permissible level of risk may also be translated into an expression of the maximum level and/or frequency of a hazard, for example in a given product and termed as FSO; see Section 9.7.

The SPS Agreement (Section 11.5) refers to the 'appropriate level of sanitary or phytosanitary protection' (ALOP). This is defined as the level of protection deemed appropriate by the WTO member for establishing a sanitary or phytosanitary measure to protect human, animal or plant life or health within its territory. This assumes a threshold value dividing an unacceptable risk from an acceptable risk. The CCFH (2000) proposes a 'tolerable risk' again with a dividing threshold value. The ICMSF defines an FSO as a statement of the frequency or maximum concentration of a microbiological hazard in a food considered acceptable for consumer protection (ICMSF 1996b,1998b). 'As low as reasonably achievable' (ALARA) is a concept for risk management which does not have an absolute value dividing acceptable and unacceptable (tolerable) risk. Instead, risk is categorised into three bands: intolerable, tolerable and acceptable (Figure 9.15). Intolerable (unacceptable) risks are managed by regulations and bans. The goal of risk management is to use strategies to reduce the risk to ALARA and may be a more appropriate approach than defining a threshold value between acceptable and unacceptable.

HACCP is a risk management system previously developed on the basis of qualitative risk assessment. Two key advantages that now come from using a quantitative approach are the ability to link the HACCP plan to an estimate of public health impact and a measure of the level of confidence that the evaluators have in their results.

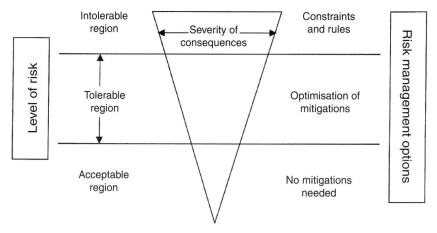

Figure 9.15 As low as reasonably achievable (ALARA) approach to risk mangement (Jouve 2001).

Risk assessment for microbiological hazards is partially derived from chemical risk assessment. However, microbial contaminants are a significantly more complex group of hazards than chemicals. The analysis must estimate microbial growth rates under plausible conditions of food processing and storage. This is focused on processing steps that either lead to microbial contamination or allow microbial growth. At its present state of development, the HACCP approach of risk management does not use microbial dose–response information. Instead, it can use FSOs (numbers of microbes per gram of food; described in Section 9.7).

9.6.1 Risk assessment policy
Guidelines for value judgement and policy choices which may need to be applied at specific decision points in the risk assessment process are known as the risk assessment policy. Risk assessment policy setting is risk management responsibility, which should be carried out in full collaboration with risk assessors, and which serves to protect the scientific integrity of the risk assessment. The guidelines should be documented so as to ensure consistency and transparency. Examples of risk assessment policy setting are establishing the population(s) at risk, establishing criteria for ranking hazards and guidelines for application of safety factors.

9.6.2 Risk profiling
Risk profiling is the process of describing a food safety problem and its context, in order to identify those elements of the hazard or risk relevant to various risk management decisions. The risk profile would include identifying aspects of hazards relevant to prioritising and setting the risk assessment policy and aspects of the risk relevant to the choice of safety standards and management options. A typical risk profile might include:
- a brief description of the situation, product or commodity involved;
- identification of what is at risk, for example human health, economic concerns, potential consequences, consumer perception of the risks;
- description of risks and benefits.

A semi-quantitative risk profile approach has been proposed (CCFRA 2000) which offers considerable help in microbiological risk assessment (see also Voysey & Brown 2000). The Danish Veterinary and Food Administration constructed a risk profile of *C. jejuni* (Section

10.2.2) which was further utilised in the construction of a risk assessment of human illness to *C. jejuni* in broilers (Section 10.2.3).

9.7 Food safety objectives

The ICMSF (2002) introduced the concept of FSOs. This is a statement of the maximum frequency and/or concentration of a microbiological hazard in a food considered acceptable for consumer protection. The FSO is a risk management tool linking risk assessment and effective measures to control identified risks. For practical implementation in specific sectors of the food chain, it is the responsibility of governmental authorities to translate the output of risk analysis into FSOs. Such objectives delineate the specific target(s) that any food operator concerned should endeavour to achieve through appropriate interventions. Although it is recognised that with regard to food microbiology the formalised approach to risk analysis is in its infancy, it is likely that in the near future, MRA will have a greater importance in the determination of the level of consumer protection that a government considers necessary and achievable. It offers a practical means to convert public health goals into values or targets that can be used by regulatory agencies and industry.

FSOs are a statement of the maximum level of a microbiological hazard in a food considered acceptable for human consumption and hence are the 'translation' of microbiological risk. Whenever possible, FSOs should be quantitative and verifiable in order to be incorporated into the principles of both HACCP and GHP. FSOs give the scientific basis for authorities to develop their own inspection procedures of food industry, quantify equivalent inspection procedures of differing countries and represent the minimum target on which food operators base their own approach; cf. Figure 8.4. Food companies may adopt the inspecting authorities' FSOs as their own food safety requirements or establish more demanding food safety requirements depending upon commercial issues. FSOs subsequently can direct product and process planning, design and implementation of food safety management programmes covering GMP, GHP, HACCP and quality assurance systems; see Chapter 8.

FSOs are a statement of the maximum level of a microbiological hazard in a food considered acceptable for human consumption. It should:
- be technically feasible
- include quantitative values
- be verifiable
- be developed by governmental bodies with a view to reaching consensus for a food in international trade

The FSO approach combines scientific data from risk assessment in order to set quantifiable standards that address specific health outcomes. Since the FSO must be met at the point of consumption, it is necessary to determine the growth of the foodborne pathogen during storage and distribution. The processing safety objective is the FSO which takes into account any expected pathogen growth. For example, if the FSO is less than 100 cfu/g of *L. monocytogenes* and 1 log cycle of growth is expected, then the processing safety objective is no more than 10 cfu *L. monocytogenes*/g. Obviously, if no pathogen growth is predicted, the processing safety objective is the same as the FSO. The PCs are the reduction necessary during processing to achieve the processing safety objective, that is a 5-log reduction.

Microbiological criteria (Chapter 6) such as FSOs could be based upon established methods of certification, inspection and/or microbiological testing. The establishment of microbiological criteria should consider:

- evidence of actual or potential hazards to health;
- microbiology of the raw materials;
- effects of processing;
- likelihood and consequences of contamination and growth during handling, storage and use;
- category of consumers at risk;
- the distribution system and potential for consumer abuse;
- the reliability of the methods used to determine product safety;
- cost/benefit ratio of the application;
- intended use of the food.

Microbiological criteria are used to ensure safety of food, adherence to GMPs, the keeping quality of certain perishable foods and/or the suitability of a food or ingredient for a particular purpose. Microbiological criteria when appropriately applied can be a useful means for ensuring safety and quality of foods, which in turn elevates consumer confidence. It can also provide the food industry and regulatory agencies with guidelines for control of food processing systems. Internationally accepted criteria can advance free trade through standardisation of food safety and quality requirements. However, it should be recognised that sampling plans have inherent risks associated with them; see Figures 6.4–6.6, Sections 6.7.2 and 6.7.3.

9.8 Risk communication

Risk communication is an exchange of information and opinions throughout the risk analysis process concerning risk, risk-related factors and risk perception, among all interested parties, to explain risk assessment findings and the basis of risk management decisions.

Goals of risk communication:

- Promote awareness and understanding of the specific issues under consideration during the risk analysis process by all participants.
- Promote consistency and transparency in arriving at and implementing risk management decisions.
- Provide a sound basis for understanding the risk management decisions proposed or implemented.
- Improve the overall effectiveness and efficiency of the risk analysis process.
- Contribute to the development and delivery of effective information and education programs, when they are selected as risk management options.
- Strengthen the working relations and mutual respect among participants.
- Promote the appropriate involvement of all interested parties in the risk communication process.
- Exchange information on the knowledge, attitudes, values, practices and perceptions of interested parties concerning risk associated with food and related topics.
- Foster public trust and confidence in the safety of the food supply chain.

Due to numerous food scares and well-publicised food poisoning outbreaks, the public have become increasingly concerned about the risks associated with food (Table 1.8). Hence, effective risk communication with consumers is both important and necessary. Risk communication is required to adequately address and respond to needs for criteria, hazards, risks, safety and general concerns about food. Risk communication provides the public with results of expert scientific review of food hazard identification and assessment of risk to the general population or target group. It also provides the private and public sectors with information necessary to prevent, reduce and minimise food risks through systems of quality and safety. Additionally, it is essential that risk communication provides sufficient information for populations with greatest risk in terms of any particular hazard to exercise their own options to achieve protection.

Aspects of risk communication:

1 Nature of the risk:
  • Characteristics and importance of hazard of concern
  • Magnitude and severity of risk
  • Urgency of situation and trend
  • Probability of exposure
  • Amount of exposure that constitutes a significant risk
  • Population at risk

2 Nature of the benefits
  • Benefits associated with risk
  • Who benefits and in what way
  • Balance point between risk and benefit
  • Magnitude and importance of benefit
  • Total benefit to all affected populations

3 Uncertainties in risk assessment
  • Methods used to assess risk
  • Importance of each of uncertainties
  • (In)accuracy of available data
  • Assumptions on which the estimates are based
  • Effect of changes in the estimates on risk management decisions

4 Risk management options
  • Action(s) taken to control/manage the risk
  • Actions individuals may take to reduce individual risk
  • Justification for choosing a specific risk management option
  • Benefit(s) of a specific option
  • Who benefits?
  • Cost of managing a risk – who pays?
  • Risks that remain after a risk management option is implemented

The differences in the public's perception of 'risk' and 'benefit' need to be understood and how these perceptions vary between countries and social groups. For example, there are marked differences in the acceptance of genetically modified food ingredients between United Kingdom and United States, partly due to public distrust of reassurances from politicians and food experts since the emergence of BSE-vCJD.

The FAO/WHO (1998) described the principles of risk communication:

1 Know the audience
2 Involve the scientific experts
3 Establish expertise in communication
4 Be a credible source of information
5 Share responsibility
6 Differentiate between science and value judgment
7 Assure transparency
8 Put the risk in perspective

9.9 Future developments in microbiological risk assessment

The area of microbiological risk assessment has developed very rapidly from its early applications in the 1990s. It is therefore difficult to publish a predictive list which will not age too rapidly. Hence, this section concentrates on topics requiring further investigation.

If microbiological risk assessment proves to be effective in facilitating the worldwide distribution of 'safe' food products, then it could subsequently enhance world trade in food. Additionally, microbiological risk assessment could lead to improved HACCP schedules due to the establishment of critical control points which reduce the microbial hazard to a scientifically justifiable 'acceptable' level.

9.9.1 International methodology and guidelines

In order to facilitate the international adoption of microbiological risk assessment, it will be necessary to:
1 produce an internationally agreed methodology;
2 produce guidelines on hazard characterisation, exposure assessment and risk characterisation to give detailed guidance in information and data requirements and how to evaluate such information;
3 agree on dose–response models.

In order to facilitate risk assessment experience, collaborative projects involving countries with risk assessment experience and inexperienced countries in a joint risk assessment are required. A means of presenting and disseminating the results of preliminary and completed microbiological risk assessments would support these initiatives.

More national microbiological risk assessments are required in contrast to the current ones which largely rely on international data (cf. Denmark *C. jejuni* Section 10.2.3).

9.9.2 Data

A database of international risk assessment information should be established. This database should include data from developing countries. It is expected that time and temperatures of storage, preparation and cooking practices differ between countries. Countries from all regions of the world need to report on local practices of storage and handling of food in the home by consumers and in food-service establishments, including storage temperatures and times. Consumption details, such as size of portion and frequency of consumption, are also needed. The system of data input from industry should be such that there would be no punitive action by government nor use by competing companies.

Adoption of microbiological risk assessment is likely to increase the need for data on the presence and concentration of micro-organisms in food products to validate the risk assessment model(s). As already shown, risk assessments of *Salmonella* (and *L. monocytogenes*) are restricted due to the standard practice of detection criteria being the presence or absence of one cell in 25 g food, rather than enumeration. Quantitative data on the concentration of pathogens, such as *Salmonella* on poultry carcasses, are required from all regions of the world. Cross-contamination determination in the processing and preparation stages is needed for many microbial pathogens in the exposure assessment step.

Surveillance systems need to be established to support epidemiological and outbreaks investigations. These data are needed for dose–response assessment models with populations of different susceptibilities. Sentinel studies are required to gain a more accurate estimation of the number of people with foodborne illness each year.

Predictive microbiology has to date largely focused on microbial growth (temperature abuse, etc.) and death (pasteurisation, cooking). There is a need to develop further predictive microbiology as a core facility and enable different countries to apply the models to their own farm to fork processing conditions.

Predictive microbiology needs to be extended to account for the following:
- Processes with fluctuating time and temperature regimes
- Bacterial survival at chill and frozen temperatures
- Bacterial adaptation during processing; see Section 2.7

In the future, the dose–response models should represent the following:

1 The probability of infection following exposure
2 The probability of illness following infection
3 The probability of sequelae and/or mortality following illness

The dose–response curves for non-Typhi *Salmonella* in the next chapter (Section 10.1) demonstrates the diversity of mathematical models which currently can be applied to the same set of data. Hence, an internationally agreed methodology for dose–response analysis is needed in the future (albeit different models for different microbial hazards), so that one could predict the likely number of people becoming infected or ill from the number of bacteria (etc.) in food (FSO/microbiological criteria). The other important part of this will be to find new ways of obtaining relevant human data to guide our models, a process in which improved outbreak investigations will lead the way.

FSOs and microbiological criteria need to be set which incorporate the risk associated with operating characteristic curves; see Section 6.7.2.

9.9.3 Training courses and use of resources

Risk assessment requires a multidisciplinary approach involving, for example risk assessors, statisticians, microbiologists and epidemiologists. However, such skilled, experienced personnel are not distributed worldwide. On a global scale, the creation of networks via the internet would be valuable for the transfer of information, technical advice and collaborative studies. The JEMRA has already been established by the FAO/WHO for the provision of expert advice on microbiological risk assessment.

These need to be further developed with working examples of microbiological risk assessment. It should be recognised however that the internet is not available worldwide and alternative effective communication methods must be made available to all who need it, especially in developing countries.

10 Application of microbiological risk assessment

There are a growing number of published microbiological risk assessments, some of which have been focused on one aspect, that is hazard characterisation (see Tables 9.2 and 9.3). Of particular importance are those by the Joint Expert Meeting on Microbiological Risk Assessment (JEMRA) (Section 9.2) which contain an extensive literature review of the subject matter, including unpublished data.

The microbiological risk assessments published to date vary considerably in the depth of assessment and also in structure. The reader should appreciate that these examples do not always use the Codex structure and definitions for risk assessment, and that this is an evolving scientific discipline. The following cases are not an exhaustive survey of published microbiological risk assessments but are abridged examples to give an indication of the processes involved. For a 'hazard identification' of each organism, see the relevant section in Chapter 4.

MRA can predict pathogen behaviour and transmission along a defined food chain, and can assess the effects of different control measures on public health risk to consumers (i.e. risk per servings). This *direct* application of MRA has been demonstrated by a number of risk assessments at both the national and international level, and is widely recognised as one of the strengths of MRA (Tables 9.2 and 9.3). For example, the two FAO/WHO (2004, 2006b) risk assessments on *Cronobacter* spp. (*Enterobacter sakazakii)* and *Salmonella* in powdered infant formula have resulted in a web-based risk model (JEMRA 2008; Section 10.7).

10.1 *Salmonella* risk assessments

10.1.1 *S.* Enteritidis in shell eggs and egg products

The US Food Safety and Inspection Service (FSIS 1998) completed a 2-year comprehensive risk assessment of *S.* Enteritidis in shell eggs which can be accessed from the web (see Table 9.3 and Web Resources section for the URL). The *S.* Enteritidis risk assessment (SERA) model can also be downloaded. It requires Excel^{TM} (Microsoft Corp.) and @RISK (Palisade Corp.) to run. There is an online study guide to the SERA model by Wachsmuth; see Web Resources section. The model is in many ways the archetypal risk assessment for foodborne pathogens and is frequently referred to by the other *Salmonella* risk assessments. The consequence of risk models, such as FSIS (1998), is the action plan in the United States to eliminate *S.* Enteritidis illness (Anon. 2009).

The objectives of the risk assessment were to:
- identify and evaluate potential risk reduction strategies;
- identify data needs;
- prioritise future data collection efforts.

The risk assessment model consists of five modules (Figure 10.1):

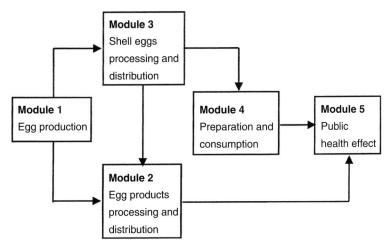

Figure 10.1 Farm to table risk assessment model for eggs and egg products (FSIS 1998).

Egg production module
This module estimates the number of eggs produced that are infected (or internally contaminated) with *S*. Enteritidis.

Shell egg processing and distribution module
This module follows the shell eggs from collection on the farm through processing, transportation and storage. The eggs remain intact throughout this module. Therefore, the primary factors affecting the *S*. Enteritidis are the cumulative temperatures and times of the various processing, transportation and storage stages. The two important modelling components are the time until the yolk membrane loses its integrity (and hence barrier to *S*. Enteritidis) and the subsequent growth rate of *S*. Enteritidis in eggs.

The lag period before yolk membrane breakdown time (YMT) was estimated from the equation:

$$\mathrm{Log}_{10}\,YM = \{(2.08 - 0.04257^{*}T) \pm (2.042^{*}0.15245)[(1/32) + ((T - 21.6)^{2}/(32^{*}43.2))]^{0.5}\}$$

The subsequent growth rate was estimated from:

$$\mathrm{Growth\ rate\ (log}_{10}\ \mathrm{cfu/h)} = -0.1434 + 0.026^{*}\mathrm{internal\ egg\ temperature\ (^{\circ}C)}$$

Egg products processing and distribution module
This module tracks the change in numbers of *S*. Enteritidis in egg processing plants from receiving through pasteurisation. The death rate of *S*. Enteritidis in whole eggs and yolk during pasteurisation is determined from experimentally derived *D* values (Figure 2.10). There are two sources of *S*. Enteritidis in egg products: from the internal contents of eggs and from cross-contamination during breaking.

Preparation and consumption module
This module estimates the increase or decrease in the number of *S*. Enteritidis organisms in eggs or egg products as they pass through storage, transportation, processing and preparation.

Table 10.1 FSIS (1998) public health module results

	Category	Mean
Normal population	Exposed	1 889 200
	Ill	448 803
	Recover w/no treatment	425 389
	Physician visit and recovery	21 717
	Hospitalised and recovered	1 574
	Death	123
	Reactive arthritis	13 578
Susceptible population	Exposed	521 705
	Ill	212 830
	Recover w/no treatment	196 295
	Physician visit and recovery	14 491
	Hospitalised and recovered	1 776
	Death	269
	Reactive arthritis	6 416
Total population	Exposed	2 410 904
	Ill	661 633
	Recover w/no treatment	621 684
	Physician visit and recovery	36 208
	Hospitalised and recovered	3 350
	Death	391
	Reactive arthritis	19 994

The public health module

This module calculates the incidences of illnesses and four clinical outcomes (recovery without treatment, recovery after treatment by a physician, hospitalisation and mortality) as well as the cases of reactive arthritis associated with consuming S. Enteritidis positive eggs (Table 10.1).

The results of the baseline model predicts the following:
- Average production of 46.8 billion shell eggs per year (in the United States)
- 2.3 million eggs contain S. Enteritidis
- This results in 661 633 human illnesses per year of which:
    94% of illnesses recover without medical care
    5% visit a physician
    0.5% are hospitalised
    0.05% of cases result in death
- Twenty percent of the population are considered to be at a higher risk: infants, elderly, transplant patients, pregnant women and individuals with certain diseases.

The output of the module was validated using egg culturing data from California to predict that there were 2.2 million S. Enteritidis contaminated eggs in the United States each year. The model was also validated using public health surveillance data.

The beta-Poisson model (Section 9.5.6) from *Salmonella* human volunteer feeding trials (1930–1973) estimates a probability of infection of 0.2 from ingesting 10^4 *Salmonella* cells (Figure 9.8). Because an infectious dose does not necessarily lead to illness, the probability of infection is greater than the probability of illness. These data were obtained using serotypes other than S. Enteritidis and hence were accepted as not being totally appropriate. It is further discussed by Fazil *et al.* (2000).

Table 10.2 USDA minimum time and temperature requirements for three egg products

Liquid egg product	Minimum temperature requirements (°F)	Minimum holding time requirements (minutes)
Albumen	134	3.5
	132	6.2
Whole egg	140	3.5
Plain yolk	142	3.5
	140	6.2

The baseline egg products model predicts that the probability is low that any cases of *S.* Enteritidis will result from the consumption of pasteurised egg products. However, the current FSIS time and temperature regulations do not provide sufficient guidance to egg products industry for the large range of products the industry produces (Table 10.2). Time and temperature standards based on the amount of bacteria in the raw product, how the raw product will be processed and the intended use of the final product will provide greater protection to the consumers of egg products.

The percent reduction for total human illnesses was calculated for two scenarios differing from current practice within the shell egg processing and distribution module. The first scenario was that if all eggs were immediately cooled after lay to an internal temperature of 7.2°C (45°F), then maintained at this temperature, then a 12% reduction in human illnesses would be the result. Similarly, an 8% reduction in human illnesses would be the result if eggs were maintained at an ambient (i.e. air) temperature of 7.2°C (45°F) throughout shell egg processing and distribution.

10.1.2 Hazard identification and hazard characterisation of *Salmonella* in broilers and eggs

In the JEMRA report (2003) entitled 'Risk Assessments of *Salmonella* in Eggs and Broiler Chickens', an extensive amount of information (as common with other JEMRA reports) is complied and can be downloaded from the internet; see Web Resources section for the URL. It contains an overview of *Salmonella* outbreaks, summarised in Table 10.3. The first section describes the public health outcomes, pathogen characteristics, host characteristics and food-related factors that may affect the survival of *Salmonella* in the human intestinal tract. The second section reviewed three dose–response models for salmonellosis and compared with 33 sets of outbreak data (Section 9.5.6). Where possible, differences in the dose response was characterised for susceptible and normal subgroups of the population.

The three dose–response models reviewed were as follows:
1 *S.* Enteritidis in eggs from FSIS (1998). This is a beta-Poisson model derived using the results of feeding studies of *Shigella dysenteriae*, with illness as the biological endpoint.
2 Health Canada (2000) model based on Weibull function (Section 9.5.6) which was derived from human feeding studies for several different bacterial pathogens and data from two *Salmonella* outbreaks (Paoli 2000, Ross 2000). The model was taken from an unpublished report at the time of writing.
3 Beta-Poisson model derived from human feeding study data of prisoners (McCullough & Eisele 1951a–c).

Table 10.3 Population sizes associated with *Salmonella* outbreaks (JEMRA 2003)

Salmonella serovar	Food vehicle	Population	Dose (log$_{10}$)	Number of people exposed	Number of people ill
S. Cubana	Dye	Susceptible	4.57	17	12
S. Enteritidis	Beef	Normal	5.41	5	3
S. Enteritidis	Beef	Normal	2.97	3517	967
S. Enteritidis	Cake	Normal	2.65	5102	1371
S. Enteritidis	Cake	Normal	5.8	13	11
S. Enteritidis	Chicken	Normal	3.63	16	3
		Susceptible		133	53
S. Enteritidis	Egg	Normal	6.3	114	63
		Normal	3.8	884	558
		Susceptible	1.4	156	42
S. Enteritidis	Ice cream	Normal	2.09	452	30
S. Enteritidis	Peanut	Normal	1.72	3990	644
S. Enteritidis	Sauce	Normal	4.74	39	39
S. Enteritidis	Soup	Normal	6.31	123	113
S. Heidelberg	Cheese	Normal	2.22	205	68
S. Infantis	Ham	Normal	6.46	8	8
S. Newport	Hamburger	Normal	1.23	1813	19
S. Oranienburg	Soup	Normal	9.9	11	11
S. Typhimurium	Ice cream	Normal	3.79	1400	770
		Normal	8.7	7	7
		Susceptible		1	1
S. Typhimurium	Water	Normal	2.31	7572	805
		Susceptible		1216	230

The feeding trail data from McCullough & Eisele (1951a, 1951b) which used S. Anatum, S. Bareilly, S. Derby, S. Meleagridis and S. Newport are shown in Figure 10.2. The data have been corrected (as per JEMRA 2003) for subjects who had received multiple doses and hence may have acquired some immunity. The appropriate beta-Poisson fit is also shown.

The epidemiological models of FSIS (1998) and Health Canada were subdivided into normal and susceptible subgroups and are shown in Figure 10.3. The beta-Poisson fit from Figure 10.2 is also included for comparison.

Data from 33 outbreaks were compiled and used to compare the previous dose–response curves. The Japanese data was used where large-scale facilities (>750 meals/day or >300 dishes of a single menu) had saved 50 g portions for a minimum of 2 weeks at −20°C for possible future examination. Since S. Enteritidis is the major cause of *Salmonella* food poisoning, the majority of outbreaks were from this serovar, in contrast to the feeding data where a number of serovars were tested. The outbreak data are shown in Figure 10.4. The data were fitted to the three models previously described. Neither models best fitted the outbreak data over the full dose range; the 'naïve beta-Poisson model' in particular significantly underestimated the probability of illness.

Overall, it was estimated that children under 5 years of age were 1.8–2.3 times more likely to become ill with the same ingested dose as the normal population. There was no evidence to indicate that S. Enteritidis was more infectious than another serovar at the same ingested dose. Hence, a common dose–response model for all non-typhoid and non-paratyphoid *Salmonella*

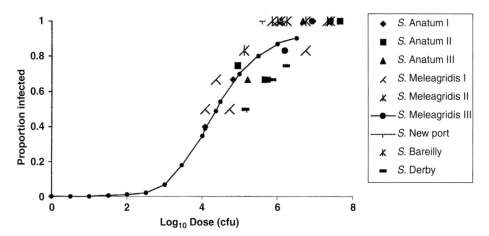

Figure 10.2 Human volunteer feeding trials for *Salmonella* serovars. From JEMRA (2003) using the data of McCullough & Eisele (1951a, 1951b).

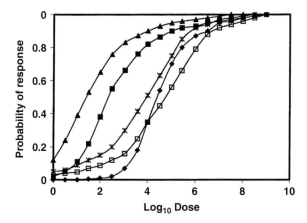

Figure 10.3 Epidemiological models of FSIS (1998) and Health Canada (JEMRA 2003). Health Canada: □, normal; *, susceptible. FSIS: ■, normal; ▲, susceptible. ◇, Naïve beta-Poisson (Figure 10.2).

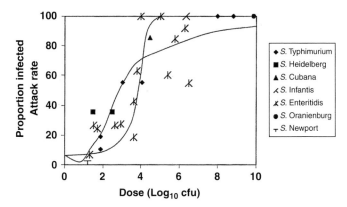

Figure 10.4 Outbreak dose–response curve (FAO/WHO 2002).

spp. can be used. The report did not consider secondary transmission, chronic sequelae such as reactive arthritis nor the effect of the food matrix.

10.1.3 Exposure assessment of *Salmonella* spp. in broilers

Kelly *et al.* (2000) released the fullest exposure assessment of *Salmonella* spp. in broilers to date. This includes products that are fresh, frozen and further processed. The preliminary report is available for downloading (see Web Resources section for URL), and contains an extensive literature review (104 pages in length). The report offers a model framework for future risk assessments and highlights current limitations in the information available. As previously stated, although *Salmonella* is a well-studied food poisoning bacterium, current approved microbiological criteria require enumeration on presence or absence basis (less than one *Salmonella* cell 25/g); hence, there is a lack of quantitative data on numbers of *Salmonella* in the primary source (i.e. poultry) and in food. There is an overlap in approach with the *Campylobacter jejuni* in fresh chicken risk assessment (Section 10.2.1). However, *C. jejuni* is more temperature sensitive than *Salmonella*, does not grow at temperatures below 30°C, but does have a reported higher infectivity (Table 2.8, Section 10.2).

The framework is similar to that shown in Figure 9.3, though the fourth module has a slightly different name.

Production module: This module estimates the prevalence of *Salmonella* positive broilers leaving the farm. Data required include the source of infection, prevalence of *Salmonella* in the flock, the number of *Salmonella* per positive bird and sampling methodology (see Section 9.5.7 for dose estimate). Various epidemiological and farm management factors will influence these values, and currently, there is little quantitative data on the number of *Salmonella* per bird.

Transport and processing module: This module ultimately aims to estimate the prevalence and concentration of *Salmonella* at the end of processing. Therefore, it needs estimations of prevalence and concentration at each step of processing (cf. *C. jejuni*, Section 10.2.1). An estimation of cross-contamination during transportation and processing needs to be included, though currently there are little data on this topic (see Christensen *et al.* 2001, Section 10.2.3).

Retail, distribution and storage module: This module estimates the changes in prevalence and concentration between processing and consumer preparation. Periods when the temperature supports microbial multiplication on contaminated meat need to be determined in order to estimate the growth and persistence of the pathogen. Temperature abuse can occur in the retail sector as well the consumer aspect of this module. Future studies collecting relevant data for predictive microbiology models are required to improve this module; see Section 2.8.

Preparation module: The changes in *Salmonella* numbers due to preparation, including cross-contamination, are considered in this module. Improved predictive models are required with regard to the thawing (thermal profiles) of frozen contaminated carcasses as well as death rates (*D* values) due to cooking (Section 2.5, Figure 2.10c). The outputs from this module are an estimate of the prevalence of contaminated products and the number of ingested *Salmonella* cells.

Consumption pattern module: This module requires data on the consumers. This does not only include age, sex, immunological status but also behaviour which can be age and nationality related. Most information available is based on 'average consumption per day' and does not describe portion size nor frequency of consumption. The amount ingested by different sub-populations (i.e. normal and susceptible) is estimated and combined with the output from the preparation module to generate an overall estimate of exposure.

The estimate of exposure can be used through dose–response analysis (hazard characterisation) to determine both the probability of infection and illness (Section 9.5.6).

10.1.4 *Salmonella* spp. in cooked chicken

Buchanan and Whiting (1996) published a three-stage risk assessment example concerning *Salmonella* spp. in cooked chicken. This early study was not intended to follow the Codex approach, but was an illustration of the use of predictive microbiology. The production process is that raw chicken is stored at 10°C for 48 hours before being cooked at 60°C for 3 minutes and then stored at 10°C for 72 hours before consumption. The 10°C stored temperature is in the 'Danger Zone' of microbial growth and represents mild temperature abuse.

Stage 1: Number of Salmonella spp. in raw chicken before cooking. The number of *Salmonella* cells on raw chicken will vary; however, an expected level of contamination is given in Figure 10.5. The contamination range varies from no *Salmonella* cells in 75% of samples to 1% containing 100 cells per gram of meat. The amount of *Salmonella* growth at 10°C for 48 hours before cooking can be determined using growth models (Section 2.8) by assuming the meat pH is 7.0 and the sodium chloride level is 0.5%.

Stage 2: Effect of cooking (60°C, 3 minutes) on Salmonella numbers in chicken. The decimal reduction time (D value) at 60°C is 0.4 minutes. The effect of heat treatment on *Salmonella* numbers can be calculated using the equation:

$$\log(N) = \log(N_0) - (t/D)$$

where
N is the number of micro-organisms (cfu/g) after the heat treatment
N_0 is the initial number of bacteria (cfu/g)
D is the D value, log(cfu/g)/min
t is the duration of the heat treatment (minutes)
Note, for simplicity, that no effect on *Salmonella* numbers is taken into consideration for the time period during warming the food to 60°C and cooling afterwards.

This equation gives the number of surviving *Salmonella* after the cooking process and is designed to give a 7-D kill (see Section 2.5.2).

Stage 3: Salmonella cell numbers following storage at 10°C, 72 hours before consumption. As before in stage 1, the growth curve for *Salmonella* can be predicted to estimate the number of *Salmonella* in cooked chicken after storage but before consumption.

By determining the survival number and subsequent growth for each initial population level of *Salmonella* gives an estimate of the numbers of *Salmonella* that a population of consumers is likely to ingest. In this example, 1% of the chicken samples contained 100 *Salmonella* cells/g gave a probability of infection (P_i) of 4.1×10^{-8}/g food consumed (see Section 9.5.6 for an explanation of P_i). This means there was less than one cell surviving for every 10 000 g of food. Hence, the *Salmonella* risk associated with cooked chicken under these conditions of storage is minimal.

The above example can be used as a template to determine the effect of changing the cooking regime and storage conditions. For example, raising the initial storage temperature to 15°C and reducing the cooking time to 2 minutes causes the probability of infectivity (P_i) to be unacceptably high; see Figure 10.5 right-hand side.

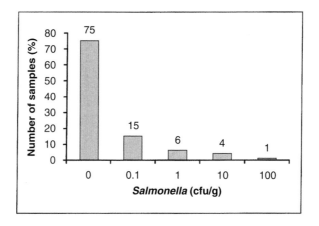

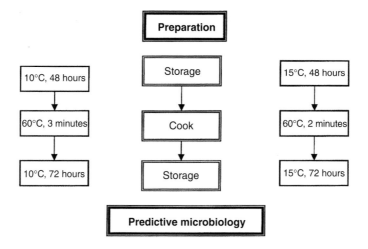

Salmonella (%)									
Safe process					Temperature abuse				
75	15	6	4	1	75	15	6	4	1
P_i 0	8.8×10^{-11}	6.5×10^{-10}	5.1×10^{-5}	4.1×10^{-8}	0	1.1×10^{-1}	4.1×10^{-1}	7.0×10^{-1}	8.6×10^{-1}

Figure 10.5 Probability of *Salmonella* infection per gram of cooked chicken after temperature abuse. From Buchanan & Whiting (1996). Reprinted with permission from the *Journal of Food Protection.* Copyright held by the International Association for Food Protection, Des Moines, Iowa, US.

10.1.5 *Salmonella* spp. in cooked patty

Whiting (1997) published an early microbiological risk assessment for *Salmonella* spp. in cooked patty that was in five steps:

1 *Initial distribution*: The initial microbial load was taken from published data (Surkiewicz *et al.* 1969), where 3.5% of the samples have *Salmonella* present at levels greater than 0.44 cfu/g (Figure 10.6).

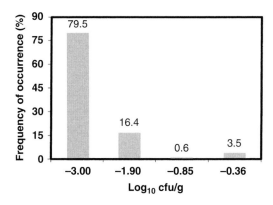

Figure 10.6 Frequency of *Salmonella* in poultry patty. Reprinted from Whiting 1997, with permission of the IFT.

2 *Storage*: Conditions chosen were 21°C, 5 hours and the growth rate predicted from a *Salmonella* model (Gibson *et al.* 1988).
3 *Cooking*: Published *D* values were used to determine the extent of heat treatment at 60°C for 6 minutes.
4 *Consumption*: A typical serving of 100 g was assumed.
5 Infectious dose determination.

The model calculated that one *Salmonella* cell has the mean probability of $10^{-4.6}$ of being an infectious dose (Table 10.4). However, 3% of the predictions gave risks greater than 10^{-3}. This was due to the small number of initial samples with high *Salmonella* contamination. The usefulness of the model is in determining the effect of altering the variables, such as cooking temperature to 61°C, which reduces the median probability of $10^{-7.4}$.

The approach of Whiting (1997) is very similar to that of Whiting's paper for *Listeria monocytogenes* (Miller *et al.* 1997; Section 10.3.2).

10.1.6 Poultry FARM
A series of predictive models for *Salmonella* and *C. jejuni* infections from chicken called 'Poultry Food Assess Risk Model' (Poultry FARM) can be downloaded from the internet (see internet directory). The models use @RISK™ (Palisade Corp.) and Excel™ (Microsoft Soft) spreadsheets to predict the change in microbial load of 100 000 servings of chicken

Table 10.4 Risk assessment model for *Salmonella* in a cooked poultry patty

Stage	Statistics
Initial distribution	−2.7 log cfu/g
Storage (21°C for 5 hours)	0.17 log cfu/g
Cooking (60°C for 6 minutes)	−4.42 log cfu/g
Consumption (100 g)	−2.42 log cfu/g
Infectious dose (probability)	$10^{-4.6}$

Adapted from Whiting 1997.

between packaging and processing. Additionally, the model predicts the number of cases of severe outcomes per 100 000 servings and the overall public health impact of the chicken. The internet site includes a full (58 pages long) explanation of the Poultry FARM model.

10.1.7 Domestic and sporadic human salmonellosis

Hald *et al.* (2001) produced an alternative quantitative risk assessment method using 'Bayesian Monte Carlo' which combined Bayesian inference with Monte Carlo simulation. The method was applied to quantifying the contribution of animals to domestic and sporadic human salmonellosis. Data from the 1999 Danish national surveillance of *Salmonella* in animals, foods and humans were used to demonstrate the method as an alternative to 'stable to table'.

The number of domestic and sporadic cases (caused by different *Salmonella* serovars) was estimated from the registered number of cases. This was then compared with the prevalence of the *Salmonella* serovars isolated from different animal sources, weighted according to the consumption pattern and the association of particular *Salmonella* serovars with specific food sources. The probability of observing the actual number of human cases was determined using a Poisson likelihood function from the data of *Salmonella* prevalence in the various food types and amount ingested. The formula used was:

$$\lambda_{ij} = M_j * p_{ij} * q_i * a_j * (\text{non} - \text{se})$$

where

λ_{ij} = the expected number of cases per year of type i from source j
M_j = the amount of source j available for consumption per year
p_i = the prevalence of type i in source j
q_i = the bacteria dependent factor for type i
a_j = the food source dependent factor for source j
non-se = serotype $\neq$ *S*. Enteritidis and source = eggs, else non-se = 1

The 'Bayesian Monte Carlo' technique was then used to determine the distribution of cases with food sources. The most important source was estimated to be eggs (54%), pork (%) and poultry (8%). A fuller mathematical account is given in the original article.

10.2 *Campylobacter* risk assessments

10.2.1 *C. jejuni* risk from fresh chicken

FAO/WHO (2002) have produced a quantitative model for the risk from *C. jejuni* on fresh chicken which can be accessed from the web; see Web Resources section. The model determines the fate of the organism through the food chain ('farm to fork') using a slightly different risk assessment framework: hazard identification, exposure analysis, dose response and risk characterisation. This was termed the 'Process Risk Model' (see also *Escherichia coli* O157:H7, Section 10.4). The objectives of the study were to generate a model of the production of chickens, gain a better understanding of chicken processing operations and identify important steps in the process that influence risk. Subsequently, the study would generate a 'tool' to assist in decision-making in order to reduce the risk to the consumer from foodborne pathogens. The study did not include long-term illnesses such as GBS.

Figure 10.7 summarises the food chain for *C. jejuni* in chicken and includes the variation in data obtained from published literature and can be compared with the framework model of Figure 9.3. The *Campylobacter* prevalence going into the process (P_F) was also described

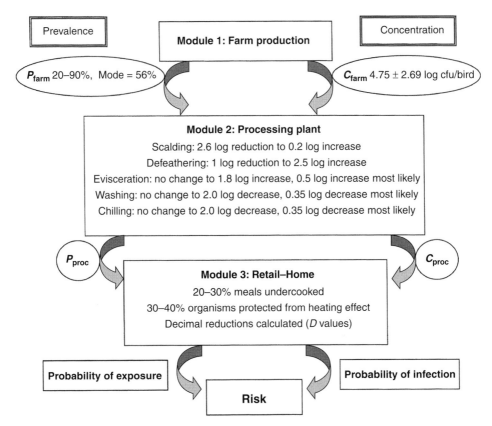

Prevalence

Module 1: Farm production

Concentration

P_{farm} 20–90%, Mode = 56%

C_{farm} 4.75 ± 2.69 log cfu/bird

Module 2: Processing plant

Scalding: 2.6 log reduction to 0.2 log increase
Defeathering: 1 log reduction to 2.5 log increase
Evisceration: no change to 1.8 log increase, 0.5 log increase most likely
Washing: no change to 2.0 log decrease, 0.35 log decrease most likely
Chilling: no change to 2.0 log decrease, 0.35 log decrease most likely

P_{proc}

C_{proc}

Module 3: Retail–Home

20–30% meals undercooked
30–40% organisms protected from heating effect
Decimal reductions calculated (D values)

Probability of exposure

Probability of infection

Risk

Figure 10.7 *C. jejuni* in chicken. Adapted from Fazil *et al.* 2000; also see Figure 9.3.

using a beta distribution (mode = 56%). The effect of processing on *C. jejuni* prevalence and concentration was determined for the five stages: scalding, defeathering, evisceration, washing and chilling. Due to considerable uncertainty in the literature, the numbers of *C. jejuni* could either increase or decrease. A triangular distribution (Section 9.5.10; Figure 9.11) was used due to the absence of complete information which gave minimum, maximum and most likely parameters. After processing, it was assumed, due to the well-known physiology of the organism (Table 2.8), that no multiplication would occur during transit from the 'processing plant' and the home. It has previously been determined that 20–30% of meals are undercooked. In addition, given the temperature sensitivity of the organism (Table 2.8), it was assumed that 30–40% of *C. jejuni* on contaminated chickens would be in areas that were protected from the direct heat. Previously published *D* values were used to determine the effect of cooking on the *C. jejuni* viable count. These calculations subsequently generate the probability of exposure and the dose likely to be ingested by the consumer. Using dose–response analysis, the probability that the consumer will be infected at the given ingested dose can be determined. The dose–response analysis of Medema *et al.* (1996) using human feeding trial data from Black *et al.* (1988) has previously been given in Figure 9.9 (see Table 9.10 for original data).

The prevalence of contaminated carcasses leaving the processing plant and the concentration of *C. jejuni* on the carcasses was determined using combined @RISK™ (Palisade

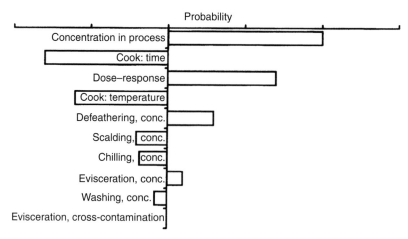

Figure 10.8 *Campylobacter jejuni* importance analysis.

Corp.)-Microsoft Excel™ (Microsoft Corp.) software to incorporate the Monte Carlo analysis. The model was run for 10 000 iterations. Furthermore, the data were used to determine the probability of illness and number of illnesses. The variation in values at each stage illustrates the need for mathematical simulations to determine the variation and uncertainty of the estimates. See Section 9.5.10 for an explanation of Monte Carlo analysis and related topics. The most likely prevalence of *C. jejuni* on chickens was 65–85%, with an average of $\log_{10}$ 3.8 *C. jejuni* cells per carcass. The distribution of probability of infection per $\frac{1}{4}$ chicken serving was 2.23×10^{-4} probability. These values can be converted into an estimate of number of illnesses per year (in the United States) by using the risk per serving, the number of chicken servings consumed in a year and the size of the population consuming the chickens. The predicted number of illnesses is in the order of 2.7 million per year. The microbiological risk assessment also predicts the parameters ('importance analysis') that are important in influencing the risk (Tornado chart, Figure 10.8) either because of their uncertainty or variability. Uncertain parameters require future research, whereas variable factors could be important control points to reduce the risk.

10.2.2 Risk profile for pathogenic species of *Campylobacter* in Denmark

In 1998, the Danish Veterinary and Food Administration started to collate a risk profile of *C. jejuni*. Afterwards, the study was used to produce a risk assessment of human illness from *C. jejuni* in broilers (Christensen *et al.* 2001). The studies collated information from a wide range of studies within Denmark and have been summarised below to demonstrate a country-specific microbiological risk assessment. Part of the reasoning for this work is the general trend of governments to reduce the prevalence of pathogens such as *Campylobacter* in poultry. The Dutch Government stated that the prevalence of *Campylobacter* in poultry must be reduced to 0%. Similarly, the Netherlands aims to reduce the level of *Campylobacter* in poultry meat to below 15%.

The rate of incidence of campylobacter gastroenteritis in Denmark was approximately 50 reported cases/100 000 population. However, it is probable that the actual number of cases was approximately 20 times greater due to under-reporting (Christensen *et al.* 2001). In Denmark, the number of sporadic cases peaks in the months of summer season, whereas the number of outbreaks culminates in the months of May and October. The cases were more prevalent in the 10–19-year old age group. A case–control investigation of foodborne risk factors for sporadic campylobacteriosis in Denmark (May 1996 to September 1997) used 227 cases and 250 control persons. On the basis of the established risk factors, the ethiological fraction was able to explain approximately 50% of the human cases: 5–8% due to insufficiently cooked poultry, 15–20% to meat prepared by grill, 5–8% to drinking water and 15–20% was due to overseas travel.

The incidence of *Campylobacter* spp. in farm animals was studied in pigs, cattle, broilers and turkeys. *Campylobacter coli* and *C. jejuni* were detected in 95% and 0.3% of pig slurry samples, respectively. It has also been shown that 66% of pig carcasses before freezing contained *Campylobacter* compared with 14% after freezing. The prevalence of *Campylobacter* in cattle was 51%, with *C. jejuni* as the most frequently isolated species. The *Campylobacter* prevalence in broilers was 37% with a distinct seasonal variation with the most frequent incidence being in the summer half. *C. jejuni* was the most frequently isolated species. The incidence in turkeys was assumed to be similar to that of chickens. Fresh poultry showed a high prevalence of *Campylobacter* (20–30%) compared with only 1% of beef and pork samples.

Human exposure to *Campylobacter* is not only from food, but also from the environment (such as bathing water), wild animals and pets. *Campylobacter* was detected in 29% of faeces samples from dogs less than 5 months old; 16 of the 21 isolates being *C. jejuni*. *Campylobacter* was detected in 2/42 of faeces samples from cats less than 5 months old. Both isolates were *Campylobacter upsaliensis*.

Ongoing research programs include determining *Campylobacter* prevalence in environmental reservoirs (such as private wells and bathing water) and unpasteurised milk, the incidence and significance of the 'viable but non-culturable form', dose–response relations, the exact incidence in the population and the significance of chronic sequelae. This study provided data to initiate a risk assessment of *Campylobacter* from broiler chickens outlined below.

10.2.3 Risk assessment of C. *jejuni* in broilers

As described above, the Danish Veterinary and Food Administration produced a risk profile of *C. jejuni* which was subsequently used towards producing a risk assessment of *C. jejuni* in broilers (Christensen *et al.* 2001). The prevalence of *C. jejuni* in flock and their concentration after bleeding was the starting point of the assessment model which had two subsequent modules: (1) slaughter and processing, and (2) preparation and consumption. Cross-contamination between positive and negative contaminated flocks was modelled, but not cross-contamination within flocks.

The prevalence and concentration of *C. jejuni* in each module was estimated and used to describe a dose response with associated probability of infection and probability of illness. The probability of illness associated with chilled chickens and frozen chickens was 1 case per 6000 meals and 1 case per 26 000 meals, respectively. The probability of illness was comparable with the 1999 registered number of campylobacter enteritis cases in Denmark (Figure 10.9).

As before, risk assessment enables predictions on the effect of risk mitigation strategies. In this case, a 25-fold reduction in the number of human illnesses could be obtained by the following:

1 A 100-fold reduction in the concentration of *C. jejuni* on the broiler chickens (i.e. 1000 to 10 cfu/g)
2 A 25-fold reduction in the flock prevalence (i.e. 60 reduced to 2.4%)
3 A 25-fold increase in 'safe' consumer behaviour during preparation of a chicken meal.

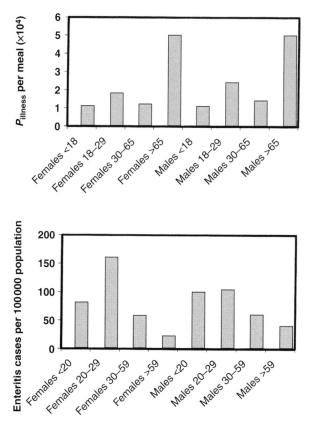

Figure 10.9 Probability of illness and *Campylobacter* enteritis in Denmark (Christensen *et al.* 2001).

10.2.4 *Campylobacter* fluoroquinolone resistance

Antimicrobial agents are not only used for treatment of human infections, but are also added to animal feed to prevent infections or growth promotion. This use of antimicrobials in agriculture is an issue that is of considerable concern due to the increasing resistance in human pathogens to medically important antibiotics (Section 1.12.7).

A significant proportion of resistance is due to the overuse and misuse of medically important antibiotics. However, some antibiotic resistant bacteria in meat are transmissible to humans. This has been reported for the *Salmonella* and *Campylobacter* and the commensal bacterium *Enterococcus*. The rise of ciprofloxacin (a medically important fluoroquinolone) resistance is linked to the use of the veterinary antibiotic enrofloxacin, which is also a fluoroquinolone. The mechanism of resistance is through the mutation of DNA gyrase which confers cross-resistance to ciprofloxacin. It is plausible that the prolonged use of enrofloxacin has exerted a selective pressure for spontaneous mutations of DNA gyrase, the target enzyme of both antibiotics.

In the United Kingdom, MRA was used to assess the risk of illness from fluoroquinolone-resistant *Campylobacter* attributable to various pathways: food, the environment and human sources. Human sources referred to foreign travel and the human use of ciprofloxacin (VLA 2005). A farm-to-consumption risk model was developed to assess and compare the risk of infection from conventional, free-range, organic and non-UK chicken. For the conventional chicken model, the exposure assessment quantified scenarios where fluoroquinolone-resistant *Campylobacter* colonisation or contamination may occur during on-farm broiler chicken production as well as during transport. The impacts of processing, cross-contamination events and undercooking during preparation in the consumer's home were also simulated. Due to the better quality of data, these aspects were also considered from the point of retail in the free-range, organic and non-UK chicken risk assessments.

The risk estimation for each source included the risk of illness, the number of fluoroquinolone-resistant *Campylobacter* cases and the number of excess illness days due to patient treatment failures because of pathogen resistance. A number of risk management strategies were modelled to estimate potential risk reductions.

The US Centre for Veterinary Medicine (FDA 2000a) have also produced a risk assessment on the human health impact of fluoroquinolone-resistant *Campylobacter* associated with the consumption of chicken. This is available on the web; see internet directory under FDA. This risk assessment also includes two models: one which uses Excel™ (Microsoft Corp.) and hence readily available, the other uses @Risk (Palisade Corp) and can be modified for Crystal Ball users (Decisioneering). The risk values obtained are related to 1998 input data only.

The model is summarised in Figure 10.10. It is divided into five modules:

1 *Campylobacter* culture-confirmed cases observable in US population
2 Total number of *Campylobacter* infections in a year in US population
3 Number of those with fluoroquinolone resistance from chickens and administered fluoro-quinolone
4 Quantity of fluoroquinolone-resistant *Campylobacter* contaminated chicken meat consumed in year
5 (i) Using the model to manage risk
  (ii) Measure the level of risk
  (iii) Controlling the risk

The reader should note that for clarity, the author has used the term 'module', whereas the original document uses the term 'section'. Table 10.5 summarises the results.

Module 1 explained the process of extrapolating the number of culture-confirmed cases reported to the US Communicable Disease Surveillance Centre to the total number of culture-confirmed cases in the United States, and subdivided the number according to whether the infection was invasive or enteric (with and without bloody diarrhoea).

Module 2 uses the values calculated in Module 1 to estimate the predicted total number of *Campylobacter* cases in the United States. This gives a mean estimate of 1.92 million cases per year (1.6–2.6 million, 90% CI).

Module 3 showed that about 5000 people (2585–8595, 90% CI) were given fluoroquinolone treatment, who were in fact infected with fluoroquinolone-resistant *C. jejuni* from consuming contaminated chicken.

Module 4 estimated that 1 450 000 000 lb (967 000 000–2 000 000 000, 90% CI) of chicken (boneless) contaminated with fluoroquinolone-resistant *C. jejuni* was consumed.

Module 5 determined the final estimates of risk. For the average US person, only 1 in 61 093 were affected by the use of fluoroquinolone, whereas for the target sub-population (person with campylobacter enteritis, seeking care and prescribed fluoroquinolone), the risk increased to 1 in 30.

The model can be used to determine the maximum prevalence of fluoroquinolone-resistant *C. jejuni* that is permissible before there is an unacceptable human health impact.

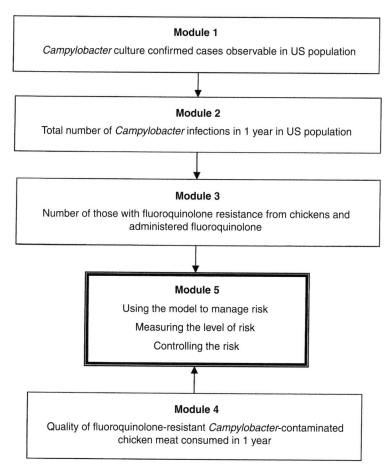

Figure 10.10 Risk assessment of fluoroquinolone use in poultry (FDA 2000a).

On September 12, 2005, the US FDA withdrawal of its approval for use of fluoroquinolones in treating poultry took effect. This was a consequence of an earlier proposal in October 2000 by the FDA's Centre for Veterinary Medicine to withdraw their use based on evidence that the use of fluoroquinolones in poultry production was causing an increase in human *Campylobacter* infections by strains that were resistant to fluoroquinolones such as ciprofloxacin. The withdrawal of use applies only to use in poultry and not other approved veterinary uses. For further information, see http://www.fda.gov/oc/antimicrobial/baytril.pdf.

10.3 *L. monocytogenes* risk assessment

10.3.1 *L. monocytogenes* hazard identification and hazard characterisation in ready-to-eat foods

JEMRA have published a report on the hazard identification and hazard characterisation of *L. monocytogenes* in ready-to-eat (RTE) foods as part of the ongoing program of microbiological risk assessments (Tables 9.2 and 9.3; Buchanan & Linqvist 2000). The objectives were to

Table 10.5 Risk assessment on the human health impact of fluoroquinolone (FDA 2000a)

Module	Step	Value		
1	US population	270 298 524		
	Catchment site population	20 723 982		
	Observed FoodNet invasive cases	43		
	Observed FoodNet enteric cases	3 985		
	Estimated mean population invasive infections	5 621		
	Estimated mean population enteric infections	51 976		
	Estimated mean culture-confirmed cases	Enteric		Invasive
		Non-bloody	Bloody	
		28 077	23 898	561
2	Proportion seeking care	12%	26.7%	100%
	Proportion submitting stool sample	19%	55.4%	100%
	Proportion of samples tested in laboratory	94.5%	94.5%	100%
	Porportion of cultures confirmed	75%	75%	100%
	Illness in population	1 702 043	228 040	561
	Total number of cases	1 930 644		
3	Number of fluoroquinolone-resistant infections from chickens (59%)	1 004 205	134 543	331
	Proportion seeking care	12.2%	26.7%	100%
	Number seeking care	122 078	35 878	331
	Proportion treated with antibiotic	47.9%	63.7%	100%
	Number treated	58 450	22 854	331
	Proportion receiving fluoroquinolone treatment	55.08%	55.08%	55.08%
	Number of chicken-related cases treated with fluoroquinolone	32 195	12 588	182
	Proportion of *Campylobacter* infections from chicken that are FQ resistant	10.4%		
	Number of fluoroquinolone-resistant infections from chicken seeking care, receiving fluoroquinolone	3 352	1 311	19
	Total number of fluoroquinolone-resistant infections from chicken seeking care, receiving fluoroquinolone	4 682		
4	Total prevalence of *Campylobacter*	88.1%		
	Prevalence of fluoroquinolone-resistant *Campylobacter* among *Campylobacter* isolates from slaughter plant	11.8%		
	Estimated prevalence of fluoroquinolone-resistant *Campylobacter* in broiler carcasses	10.4%		
	Chicken consumption per head	51.4 lb		
	Total consumption in the United States	1.39×10^{10} lb		
	Total consumption of chicken with fluoroquinolone-resistant *Campylobacter* in the United States	1.45×10^9 lb		

quantitatively evaluate the nature of the adverse health effects associated with *L. monocytogenes* in RTE foods, and to assess the relationship between magnitude of the dose and the frequency of these health effects. The entire document (prepared by Buchanan & Linqvist) was part of the JEMRA 2000 meeting (Table 9.2) and can be downloaded from the internet; see Web Resources section for URL. RTE foods are considered by Codex Alimentarius to be any food or beverage that is normally consumed in its raw state or is any that is handled, processed, mixed, cooked or otherwise prepared into a form in which it is normally consumed without further processing.

The report has an extensive literature survey and various dose–response models. Due to the lack of human feeding trials and surrogate pathogens to determine the probability of infection (Section 9.5.6), data from epidemiological studies and animal models were used. Various dose–response models were assessed which had different end points (infection, morbidity and mortality). The dose–response relationship which best described the interaction between *L. monocytogenes* and humans was not resolved, though the highly variable response is known to be dependent upon the combined interaction of the host, pathogen and food matrix. Subsequently, it was recommended that several multiple dose–response models should be used in developing the risk assessment. Data from animal studies that modelled lethality or severe invasive listeriosis were more related to human disease than modelling infection (Section 9.5.6).

A summary of selected dose–response models that were reviewed is given in Table 10.6, and Figures 10.11 and 10.12 (murine model) show the dose–response models for frequency of infection (Gompertz-log) and mortality (exponential). The mathematical models are described in Section 9.5.6. A comparison of the FDA models for general population, neonates and elderly with dose–response curves from only epidemiological data showed a lower median probability of response at a specified dose (Figure 10.11). The difference was probably due to the FDA models being based on mortality not morbidity and that other models were based on the use of highly virulent strains of *L. monocytogenes*. The predicted risk of serious listeriosis was determined to be five times the risk of mortality.

10.3.2 *L. monocytogenes* exposure assessment in RTE foods

JEMRA (Table 9.2, Ross *et al.* 2000) have published a report on the exposure assessment *L. monocytogenes* in RTE foods as part of the ongoing program of microbiological risk assessments. The report both reviews (extensively) the relevant current publications and further develops exposure assessment for *L. monocytogenes*. In addition to reviewing 11 risk assessments, the report gave seven new examples of exposure assessments in RTE foods. Raw and unpasteurised milk, ice cream and soft mould-ripened cheese were modelled from retail to point of consumption. Minimally processed vegetables, smoked salmon and semi-fermented meats were modelled from production to point of consumption. The risk of 100 cfu/g *L. monocytogenes* at point of consumption was compared with the effect of the 'zero tolerance' policy. The aim of these examples was to illustrate the effect on exposure of processing, low contamination levels in products that do not permit growth of *L. monocytogenes*, long-term storage on increase or decrease of *L. monocytogenes* concentration and consumption frequency and meal size. Additionally, the report covered the use of predictive microbiology in modelling risk assessments (Section 2.7).

A generic model was developed which emphasised the need to monitor changes in the prevalence and concentration of *L. monocytogenes* in RTE foods (Figure 10.13). An example exposure assessment predicts that consumers at normal risk consume between 4 and 22 servings of soft cheese per year, and that for consumers at high risk consume between 3 and 17 servings per year. Of those servings, 4% (median) are predicted to be contaminated.

A problem with exposure assessment for *L. monocytogenes* (as well as *Salmonella* serotypes) is that current data often lacks information on the organism's concentration in food since many authorities have a 'zero tolerance' policy. Hence, laboratory procedures are often designed to determine only the presence or absence of the organism in a 25 g sample. The lack of sufficient

Table 10.6 Dose–response models for *L. monocytogenes* (Buchanan & Linqvist 2000)

Model/study	Biological end point	Model/parameters	Comments
Buchanan *et al.* (1997)	Serious listeriosis Based on annual statistics and food survey data	Exponential $P_i = 1.18 \times 10^{-10}$	Based on immuno-compromised individuals. Predicted morbidity$_{50}$ = 5.9×10^9 cfu
Lindqvist & Westöö (2000)	Serious listeriosis Based on annual statistics and food survey data	Exponential $P_i = 5.6 \times 10^{-10}$	Based on immuno-compromised individuals. Predicted morbidity$_{50}$ = 1.2×10^9 cfu
Chocolate milk Buchanan & Linqvist (2000)	Febrile gastroenteritis Outbreak data	Exponential $P_i = 5.8 \times 10^{-8}$	Based on chocolate milk outbreak and limited to immuno-compromised individuals
Farber *et al.* (1996)	Serious infection	Weibull-Gamma $\alpha = 0.25$ $\beta_{high\ risk} = 10^{11}$ $b = 2.14$	Estimated dose for 50% population to be infected: High risk = 4.8×10^5 cfu Low risk = 4.8×10^7 cfu
Butter Buchanan & Linqvist (2000) FDA (2001)	Serious listeriosis Outbreak data	Exponential $P_i = 1.02 \times 10^{-5}$	Outbreak data from Finland for immuno-compromised individuals. Predicted morbidity$_{50}$ = 6.8×10^4 cfu
Mexican-style cheese Buchanan & Linqvist 2000 FDA (2000)	Morbidity Outbreak data	Exponential $P_i = 3.7 \times 10^{-7}$	Outbreak data from the United States Predicted morbidity$_{50}$ = 1.9×10^6 cfu
FDA-General FDA-Neonates FDA-Elderly FDA (2000)	Mortality Combined human and surrogate animal (murine) data (Golnazariuan *et al.* 1989)	Five models compared At 10^{12} dose: P_i(general) = 8.5×10^{-16} P_i(neonates) = 5.0×10^{-14} P_i(elderly) = 8.4×10^{-15}	Gompertz-log equation best fit for frequency of infection, whereas exponential model was best for mortality

Reproduced with kind permission of the Food and Agriculture Organisation of the United Nations. Sourced from http://www.who.int/foodsafety/publications/mirco/en/mra4.pdf.

incidence and prevalence data at the point of consumption means that currently predictive microbiology models have to be used. These models will need to be further validated in products of similar microbial ecology and extrinsic parameters.

The 'zero tolerance' policy was not shown to provide a greater level of public health than other less stringent criteria such as <100 *L. monocytogenes* cells/g food.

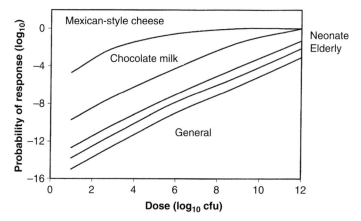

Figure 10.11 *Listeria monocytogenes* dose–response curves for various age groups and sources (Buchanan & Linqvist 2000). Reproduced with kind permission of the Food and Agriculture Organisation of the United Nations. Sourced from http://www.who.int/foodsafety/publications/mirco/en/mra4.pdf.

Note from FDA website: 'During the course of the expert consultation that was held on 30 April to 4 May 2001 to finalise the risk assessments on *L. monocytogenes* in RTE foods and *Salmonella* spp. in broilers and eggs, an error was discovered in the simulation model used in the *Listeria* risk assessment which affected the risk estimates for *Listeria*. This error partially related to work that was undertaken in the exposure assessment component of the risk assessment. Therefore, this document is currently under review and revision and a new version will be made available at a later date. If you have already downloaded the exposure assessment document from this website, please be advised that it is a preliminary document and the final version may have substantial changes.'

10.3.3 Relative risk of *L. monocytogenes* in selected RTE foods

A draft risk assessment of the relative risk to public health from foodborne *L. monocytogenes* in selected RTE foods was released by the USDA in January 2001 (FDA 2001) for further comment.

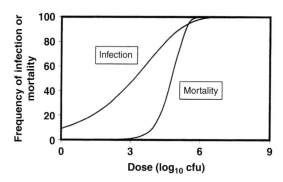

Figure 10.12 Dose–response curves for illness and mortality following administration of *L. monocytogenes* to mice (Buchanan & Linqvist 2000). Reproduced with kind permission of the Food and Agriculture Organisation of the United Nations. Sourced from http://www.who.int/foodsafety/publications/mirco/en/mra4.pdf.

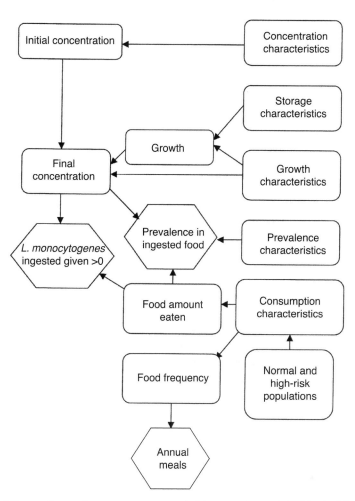

Figure 10.13 Influence diagram for exposure assessment of *L. monocytogenes* in RTE foods. Adapted from Ross *et al.* 2000.

The risk assessment was separated into three sub-populations: perinatal (fetuses and newborns <30 days after birth), elderly and 'intermediate-age' which was the remaining population both healthy and susceptible. The risk assessment was divided into the four categories as according to Codex (CAC 1999). The dose–response relationship was based on a mouse model (see Section 10.3.1; Buchanan & Linqvist 2000) with murine death as the end point as opposed to infection. A dose of 1 billion *L. monocytogenes* was chosen to compare the responses of the three age groups. The model predicted that for every 100 million servings (each containing 1 billion *L. monocytogenes*), the most likely number of deaths were as follows:

- Intermediate age group 103; range 1–1190.
- Perinatal 14 000; range 3125–781 250.
- Elderly 332; range 1–2350.

The considerable range was due to variability and uncertainty in the data used in the model. The risk characterisation used 300 Monte Carlo simulations (30 000 iterations per simulation).

The risk characterisation also developed a 'relative risk ranking' for 20 food categories based on the predicted number of listeriosis cases per 100 million servings of the food category. The highest predicted risk of RTE foods was pâté and meat spreads, smoked seafood, fresh soft cheese, frankfurters and some foods from delicatessen counters. The exposure assessment predicted that five factors affected consumer exposure to *L. monocytogenes*:
1 Amounts and frequency of food consumed
2 Frequency and level of *L. monocytogenes* in RTE food
3 Potential to support microbial growth during refrigeration
4 Refrigeration storage temperature
5 Duration of refrigeration storage before consumption

10.3.4 *L. monocytogenes* in EU trade
The absence of agreed reference values for *L. monocytogenes* (except for dairy products) has led to controversy, especially intra-community trade of the EU. The lack of microbiological reference values has led to food products being declared unfit for human consumption because of the non-quantified (zero tolerance) demonstration of *L. monocytogenes* contamination. Therefore, microbiological risk assessment leading to the setting of food safety objectives for the EU was required and has been reported (Anon. 1999a).

Although human listeriosis is mainly caused by a few serovars (4b and 1/2 a, b) it was concluded that a wide range of strains might cause serious disease. Additionally, since none of the typing methods discriminates pathogenic from non-pathogenic or less virulent strains, all *L. monocytogenes* were regarded as potentially pathogenic.

The EU risk assessment of *L. monocytogenes* enables six food groupings relative top processing control of the organism (Table 10.7). Examples of products are as follows:
• *Groups B and D*: Meat products such as cooked ham, wiener sausages or hot smoked fish, soft cheese made from pasteurised milk.
• *Groups C and E*: Cold smoked or gravid fish and meat, cheese made from unpasteurised milk.
• *Group F*: Tartar, sliced vegetables and sprouts.
Groups B and D, and C and E are separated according to the technology used.

A concentration of 100 *L. monocytogenes* cells/g of food at the point of consumption was considered a low risk to consumers. However, due to the uncertainties related to this risk, levels lower than 100 cells/g may be required for those foods in which listeria growth may occur.

Table 10.7 Grouping of RTE food commodities relative to the control potential for *L. monocytogenes*

A	Foods heat-treated to a listericidal level in the final package
B	Heat-treated products that are handled after heat treatment. The products support growth of *L. monocytogenes* during the shelf-life at the stipulated storage temperature
C	Lightly preserved products, not heat treated. The products support growth of *L. monocytogenes* during the shelf-life at the stipulated storage temperature
D	Heat-treated products that are handled after heat treatment. The products are stabilised against growth of *L. monocytogenes* during the shelf-life at the stipulated storage temperature
E	Lightly preserved products, not heat treated. The products are stabilised against growth of *L. monocytogenes* during the shelf-life at the stipulated storage temperature
F	Raw, RTE foods

Table 10.8 US population distribution with respect to vulnerable groups

Population category (year)	% US population
Pregnant women (1988)	2.5
Children <5 years old (1992)	7.7
Elderly >65 years (1992)	12.7
Residents in nursing and related care facilities (1991)	0.7
Cancer cases under care (1992)	1.6
Organ transplant patients (1992)	0.02
Total HIV (AIDS) infections (January 1993)	0.03–0.04

Reprinted from Miller *et al.* 1997, with permission of the IFT.

L. monocytogenes levels above 100 cfu/g can be achieved after in-food growth. Therefore, risk management should be focused on those foods which support *L. monocytogenes* growth.

Suggested levels of *L. monocytogenes* were as follows:
- *Food groups D, E and F*: <100 cfu/g throughout the shelf-life and at point of consumption.
- *Food groups A, B and C*: Not detectable in 25 g at time of production.

The food safety objective (Section 9.7) should be to keep the concentration of *L. monocytogenes* in food below 100 cfu/g and to reduce the fraction of foods with a concentration above 100 *L. monocytogenes* per gram significantly. In risk communication, special attention should be addressed to consumer groups at increased risk (immuno-compromised) which represent a considerable and growing section of the total population.

10.3.5 *L. monocytogenes* in meatballs

Miller *et al.* (1997) published a quantitative risk assessment for *L. monocytogenes* in meatballs in order to identify control points for HACCP implementation. A 'farm to fork' approach was used as illustrated in Figure 10.14. The risk assessment used the 1994 US census data to identify the proportion of susceptible people in the general population (Table 10.8) and combined the survey data with predictive microbiology (Pathogen Modelling Program, Section 2.8) and D values to simulate the changes in *L. monocytogenes* numbers (Table 2.9). A working morbidity threshold of ca. 100 *L. monocytogenes* cells/g was used.

The risk assessment model estimated that eating 100 g meatballs (processed as in Figure 10.14 would result in the average ingestion of 995 *L. monocytogenes* cells. This was below the target of less than 100 cells/g (10^4 total dose) and hence appeared as a 'safe' process. However, there is considerable range in the prevalence and concentration of *L. monocytogenes* in the raw materials and hence it is plausible that a high initial number of *L. monocytogenes* cells might be reduced to a dose greater than the 100 cells/g target. Monte Carlo analysis (Section 9.5.10) was used to consider the frequency of samples that could exceed the target value. It was determined that 7.3% would exceed the target dose and hence the process was not as 'safe' as previous, more simple estimations (Figure 10.14).

The advantage of risk assessment models is their use to predict the effect of processing changes, and so on, on the risk. Miller *et al.* (1997) used the model to determine the effect of reducing the initial counts to 71% 10^{-3}, 24% 10^{-2}, 5% 10^{-1} and 0% 10^{0} cfu/g (see distribution in Figure 10.14). The model subsequently predicted that only 0.94% of the ingestion doses would be greater than 100 cells/g.

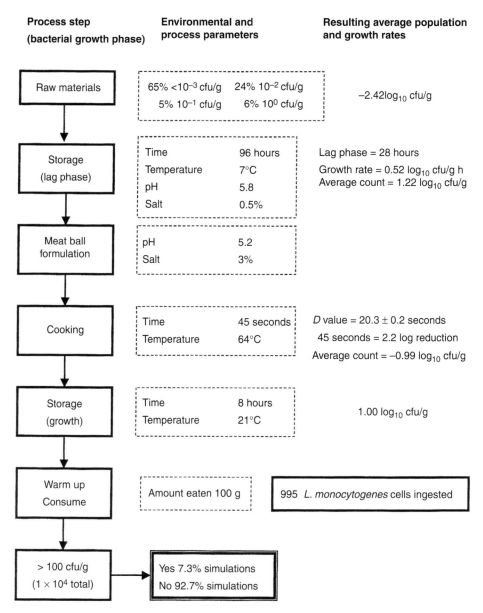

| Process step (bacterial growth phase) | Environmental and process parameters | Resulting average population and growth rates |

Raw materials — 65% <10^{-3} cfu/g   24% 10^{-2} cfu/g / 5% 10^{-1} cfu/g   6% 10^0 cfu/g — $-2.42\log_{10}$ cfu/g

Storage (lag phase) — Time 96 hours / Temperature 7°C / pH 5.8 / Salt 0.5% — Lag phase = 28 hours / Growth rate = 0.52 $\log_{10}$ cfu/g h / Average count = 1.22 $\log_{10}$ cfu/g

Meat ball formulation — pH 5.2 / Salt 3%

Cooking — Time 45 seconds / Temperature 64°C — D value = 20.3 ± 0.2 seconds / 45 seconds = 2.2 log reduction / Average count = -0.99 $\log_{10}$ cfu/g

Storage (growth) — Time 8 hours / Temperature 21°C — 1.00 $\log_{10}$ cfu/g

Warm up Consume — Amount eaten 100 g — 995 *L. monocytogenes* cells ingested

> 100 cfu/g (1×10^4 total) — Yes 7.3% simulations / No 92.7% simulations

Figure 10.14 *L. monocytogenes* in meatballs. Adapted from Miller *et al.* 1997, with permission of the IFT.

10.3.6 Listeriosis from RTE meat products

Lake *et al.* (2002) conducted a MRA for RTE meat products in New Zealand. There had been cases of invasive and non-invasive infection by *L. monocytogenes*, where RTE meats had been identified as the vehicle of infection. The food surveys revealed that the prevalence of *L. monocytogenes* in these products was similar to other countries. These foods had a high level of consumption in terms of the number of servings and serving sizes. Limited prevalence data indicated that contamination occurred across all types of meat products. Due to similarities in dietary habits, no quantitative MRA was undertaken in New Zealand. Instead, using information from the US quantitative MRA study (FDA/FSIS 2003), pâté, meat spreads and deli meats were regarded as representing the highest relative risks in the ready-to-eat meat group. However, fermented meat products had a lower relative risk.

10.4 *E. coli* O157 risk assessment

10.4.1 *E. coli* O157:H7 in ground beef

A model of *E. coli* O157:H7 in ground beef was constructed by Cassin *et al.* (1998a) to assess the impact of different control strategies. The model described the pathogen population from carcass processing through to consumer cooking and consumption. The paper also introduced the term 'Process Risk Model' which combined quantitative risk assessment with scenario analysis and predictive microbiology. This approach was also used in the *C. jejuni* microbiological risk assessment (Section 10.2.1).

The study used two mathematical models:

1 Describe the behaviour of the pathogen from the production of the food through processing, handling and consumption to predict human exposure.
2 The exposure estimate from (1) was then used in a dose–response model to estimate the human health risk associated with consuming food from the process.

The prevalence and concentration of the pathogen was determined for each stage of the food chain ('farm to fork'). The effect of each stage on the prevalence and concentration of *E. coli* O157 is shown in Figure 10.15. The figure starts with the prevalence (range 0/1131 to 188/11881) and concentration (range <2.0 to 5.0 $\log_{10}$ cfu/g) data for *E. coli* O157:H7 in faeces.

The effect of processing and grinding on the prevalence and numbers of *E. coli* O157:H7 was then determined (see Figure 10.15). For example, cross-contamination could increase the prevalence threefold, whereas spray washing could reduce the microbial numbers by 2.6–4.3 log orders. The predictive microbiology software Food Micromodel™ was used to simulate the growth of *E. coli* O157:H7 (modified Gompertz equation). The amount of growth (parameter C) was assumed to be no greater than 7–9 $\log_{10}$ cfu/g, the maximum rate of growth (parameter B) and length of time until maximum growth occurs (parameter M) were determined under the conditions: temperature = 10–15°C, pH 5.1 and 6.1 and water activity = 0.99 to 1.00. The inactivation of *E. coli* O157 during cooking was estimated according to the range of consumer preferences: rare to well done (54.4–68.3°C internal temperature, respectively). Because of the range of estimates at each stage, Monte Carlo simulation was used (25 000 iterations for each simulation, Section 9.5.10) to generate a representative distribution of risk.

The ingested dose varied with age group since adults consume almost double that of children (83 g compared with 42 g). The associated probability of exposure was calculated and a dose–response assessment constructed. The dose–response model was based on a modified beta-Poisson model for infection termed the beta-binomial model (see Section 9.5.6). It did not assume any threshold level and was based on α and β parameters similar to those of *Sh. dysenteriae* (Crockett *et al.* 1996). As can be seen in Figure 10.16, there was considerable uncertainty in the probability of illness for a particular dose. From the dose–response analysis, the probability of illness, HUS and mortality was determined. The probability of HUS was 5–10% of the probability of illness cases and the probability of mortality following HUS was assumed to be 12% of HUS cases (Figure 10.15).

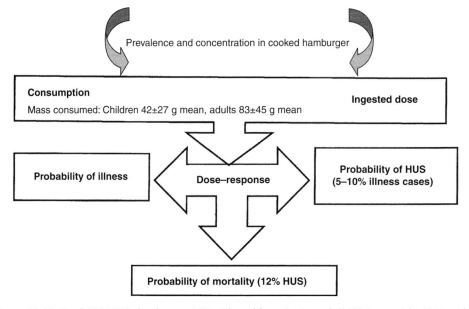

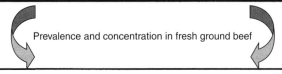

Processing and grinding

Mass of package of fresh ground beef: 5 kg   Mass of package of trimming : 5 kg

Cross-contamination factor: three-fold increase   Surface area of trimming: 0.25–1 cm²/g

Number of trimmings in package: 10–50   Faecal dilution factor: –5.1log₁₀

Reduction due to spray washing: 2.6–4.3 log₁₀ cfu/g   Reduction due to trimming: 1.4–2.5log₁₀ cfu/g

Growth during processing: –2 reduction to 5 generation increase, mode 0 generations

Prevalence and concentration in fresh ground beef

Storage   **Cooking**

Time in storage   Cooking preference: rare to well done

Maximum temperature: 4–15°C, mode 10°C   Final internal temperature: 56.1–74.4°C

Maximum lag time   Thermal inactivation regression parameters

Lag time   Maximum population density

Predictive microbiology

Gompertz parameter C= 7–9log₁₀ cfu/g, parameters B and M determined under the following conditions:

Temperature = 10–15°C, pH 5.1 and 6.1 and water activity = 0.99–1.00

Prevalence and concentration in cooked hamburger

Consumption   **Ingested dose**

Mass consumed: Children 42±27 g mean, adults 83±45 g mean

Probability of illness   Dose–response   **Probability of HUS**
                                                            (5–10% illness cases)

Probability of mortality (12% HUS)

Figure 10.15 *E. coli* O157:H7 in hamburgers MRA. Adapted from Cassin *et al.* (1998a), copyright 1998, with permission from Elsevier.

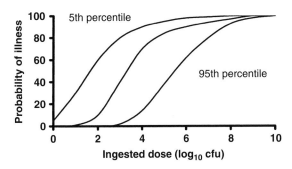

Figure 10.16 Beta-Poisson dose–response model for *E. coli* O157:H7. Adapted from Cassin *et al.* (1998a), copyright 1998, with permission from Elsevier.

The prevalence of packages containing *E. coli* O157 was estimated to be 2.9% and was comparable with surveillance data. The probability of illness per single hamburger meal ranges from 10^{-22} to 10^{-2} with a central tendency at 10^{-12}. Hence, the majority of hamburger meals are predicted to have a very small risk to the consumer, but is not negligible. The average probability of illness from a single meal was 5.1×10^{-5} for adults and 3.7×10^{-5} for children. Consequently, the model predicted a HUS probability of 3.7×10^{-6} and a mortality probability of 1.9×10^{-7} per meal for the very young. These values were deemed as representative of home-prepared hamburgers, and high for commercial production.

Spearman rank correlation coefficient (Figure 10.17) was used to show that risk was most sensitive to the concentration of *E. coli* O157:H7 in faeces and hence indicates the use of screening animals pre-slaughter as a control point.

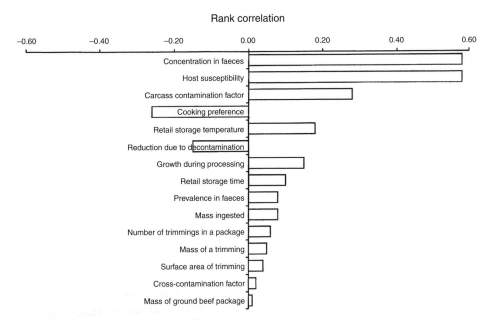

Figure 10.17 Spearman rank correlation of risk factors for *E. coli* O157:H7. Adapted from Cassin *et al.* (1998a), copyright 1998, with permission from Elsevier.

Table 10.9 Risk mitigation strategy, percent reduction per meal illness from *E. coli* O157:H7 following assumed compliance

	Strategy	Control variable	Predicted reduction in illness
1.	Storage temperature	Maximum storage temperature; 8°C mode, 13°C maximum	80%
2.	Pre-slaughter screening	Concentration of *E. coli* O157:H7 in faeces reduced 4 log orders	46%
3.	Hamburger cooking.	Cooking temperature, increase thorough cooking	16%
	Consumer information program on cooking hamburgers		

Reprinted from Cassin *et al.* (1998a), copyright 1998, with permission from Elsevier.

The model further enables changes in health risk associated with changes in control strategies to be predicted. The average probability of illness was predicted to be reduced by 80% by reducing microbial growth via reducing the storage temperature (Table 10.9). This was modelled on an average storage temperature of 8°C and a temperature abuse of no more than 13°C (compared with previous values of 10°C and 15°C, respectively). This risk reduction approach was predicted to be more effective than the reduction in the concentration of the *E. coli* O157:H7 in cattle faeces and even encouraging thorough cooking by the consumer.

An online explanation of the FSIS model of *E. coli* in ground beef is available; see Web Resources section for URL.

10.5 *Bacillus cereus* risk assessment

10.5.1 *B. cereus* risk assessment

Notermans *et al.* (1997) and Notermans & Batt (1998) described a risk assessment approach for *B. cereus*. The added difficulty in hazard characterisation and dose–response assessment of this organism is that it can cause two different illnesses according to toxin(s) production. Epidemiological studies by Kramer & Gilbert (1989) showed that the number of *B. cereus* in food causing diarrhoeal and emetic food poisoning varied from 1.2×10^3 to 10^8 and 1.0×10^3 to 5.0×10^{10} cfu/g, respectively. The median for both was approximately 1×10^7 cfu/g. Human volunteer studies used pasteurised milk naturally contaminated with *B. cereus* (Langeveld *et al.* 1996). Symptoms were observed with ingested doses greater than 10^8 cells. The dose–response curve as given by Notermans & Mead (1996) is reproduced (generalised) in Figure 10.18. There is probably considerable variation within *B. cereus* strains with regard to toxin production and the amount of toxin produced may be food related.

Since consumption of 10^4 *B. cereus* cells does not appear to be harmful (GMP and RTE levels are approximately 10^3 cfu/g (Anon. 1996)), risk assessment needs to be applied with foods with levels greater than this value. This contrasts with the zero tolerance approach of SERA. Surveillance data show that a number of RTE products such as boiled rice, cream, herbs and spices and even pasteurised milk may have levels of *B. cereus* greater than 10^4/g (te Giffel *et al.* 1997). Notermans *et al.* (1997) showed that 11% of milk consumed in the Netherlands contained $>1 \times 10^4$ *B. cereus* cells/mL and that about approximately 10^9–10^{10} portions of

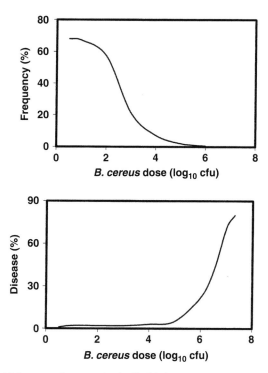

Figure 10.18 *B. cereus*: (a) frequency in pasteurised milk; (b) dose–response curve.

pasteurised milk were consumed annually. Although to date there is insufficient distribution frequency data for a detailed quantitative risk assessment for *B. cereus*, Zwietering *et al.* (1996) have constructed a predictive growth model for the organism.

10.6 *Vibrio parahaemolyticus* risk assessment

10.6.1 Public health impact of *V. parahaemolyticus* in raw molluscan shellfish

The FDA (2000b) have released a risk assessment of *V. parahaemolyticus* in raw molluscan shellfish. The document can be downloaded; see Web Resources section for URL. The report is sectioned according to the four risk assessment steps (hazard identification, exposure assessment, hazard characterisation and risk characterisation), and risk mitigation factors are also presented. The report states the various assumptions which have been made in order to elicit useful feedback. Dose–response analysis was included in both hazard characterisation and exposure assessment (post-harvest module).

An outline of the risk assessment is given in Figure 10.19. The FDA risk assessment was in response to four outbreaks in 1997–1998 which totalled over 700 cases. The process of exposure assessment is summarised in Figure 10.20. Information and data were gathered for the three modules: harvest, post-harvest and public heath. The harvest module simulated the variation in total and pathogenic *V. parahaemolyticus* densities as a function of environmental conditions and showed that salinity was not an important parameter. Subsequently, the growth of the pathogen was modelled using only water temperature. The post-harvest model simulated the

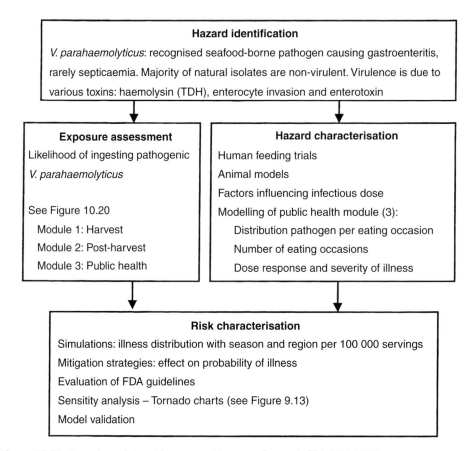

Hazard identification

V. parahaemolyticus: recognised seafood-borne pathogen causing gastroenteritis, rarely septicaemia. Majority of natural isolates are non-virulent. Virulence is due to various toxins: haemolysin (TDH), enterocyte invasion and enterotoxin

Exposure assessment

Likelihood of ingesting pathogenic

V. parahaemolyticus

See Figure 10.20

  Module 1: Harvest

  Module 2: Post-harvest

  Module 3: Public health

Hazard characterisation

Human feeding trials

Animal models

Factors influencing infectious dose

Modelling of public health module (3):

  Distribution pathogen per eating occasion

  Number of eating occasions

  Dose response and severity of illness

Risk characterisation

Simulations: illness distribution with season and region per 100 000 servings

Mitigation strategies: effect on probability of illness

Evaluation of FDA guidelines

Sensity analysis – Tornado charts (see Figure 9.13)

Model validation

Figure 10.19 *V. parahaemolyticus* risk assessment in raw molluscan shellfish (FDA 2001).

effect of current handling practices on the pathogen prevalence and concentration at time of consumption. The public health module estimated the distribution of the probable number of illnesses within one of five regions studied and season. The dose–response relationship was modelled using beta-Poisson, Gompertz and Probit relations (Section 9.5.6). The output from the three models was used to determine the risk of illness. For the Gulf Coast region, the average number of illnesses predicted was 25, 1200, 3000 and 400 in the winter, spring, summer and in the autumn (fall), respectively. Average nationwide risk of illness was 4750 cases with a range of 1000–16 000.

The risk assessment model enabled an evaluation of the current FDA criteria of $<10^4$/g *V. parahaemolyticus* shellfish. It was estimated that excluding all oysters with *V. parahaemolyticus* contamination $>10^4$ cells/g would reduce associated illness by 15% and cause a 5% loss of harvest.

Risk mitigating strategies were (see Figure 9.13) as follows:

- Cooling oysters immediately after harvest and being kept refrigerated, during which time the viability slowly decreased.
- Mild heat treatment (5 minutes, 50°C) giving a >4.5 log decrease in viability and almost eliminated the likelihood of illness.

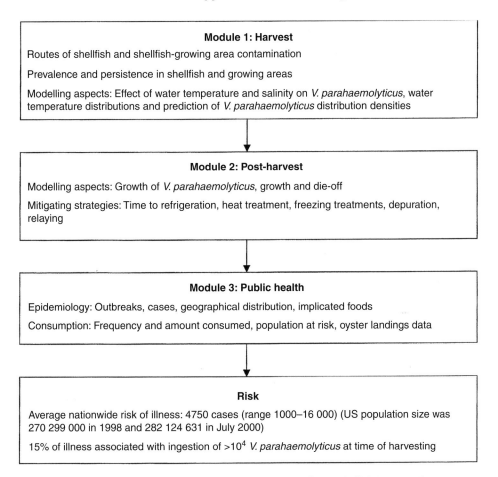

Module 1: Harvest

Routes of shellfish and shellfish-growing area contamination

Prevalence and persistence in shellfish and growing areas

Modelling aspects: Effect of water temperature and salinity on *V. parahaemolyticus*, water temperature distributions and prediction of *V. parahaemolyticus* distribution densities

Module 2: Post-harvest

Modelling aspects: Growth of *V. parahaemolyticus*, growth and die-off

Mitigating strategies: Time to refrigeration, heat treatment, freezing treatments, depuration, relaying

Module 3: Public health

Epidemiology: Outbreaks, cases, geographical distribution, implicated foods

Consumption: Frequency and amount consumed, population at risk, oyster landings data

Risk

Average nationwide risk of illness: 4750 cases (range 1000–16 000) (US population size was 270 299 000 in 1998 and 282 124 631 in July 2000)

15% of illness associated with ingestion of >10^4 *V. parahaemolyticus* at time of harvesting

Figure 10.20 Exposure assessment of *V. parahaemolyticus* in raw molluscan shellfish (FDA 2001).

- Quick freezing and frozen storage giving a 1–2 log decrease in viability and hence reducing likelihood of illness.

The model can be downloaded; see Web Resources section for URL.

10.7 *Cronobacter* spp. (*Enterobacter sakazakii*) and *Salmonella* in powdered infant formula

The term 'powdered infant formula' (PIF) is a generic term to cover a range of breast milk fortifiers and substitutes given to infants aged 0–12 months in age. These are not manufactured as sterile products, but conform to the Codex microbiological specifications (CAC 2008). Due to the raised awareness of neonatal infections due to 'Ent. sakazakii' through PIF, these specifications were revised from the former 1979 version; see Section 4.8.2. This bacterium is able to survive long periods of desiccation in infant formula, and can cause severe infections including necrotising enterocolitis, septicaemia and meningitis (Forsythe 2005). Risk management specifically for *Cronobacter* spp. and *Salmonella* came under consideration by Codex

Committee on Food Hygiene (FAO/WHO 2006b). In 2007, the WHO released the 'Safe preparation, storage and handling of powdered infant formula guidelines' and specific documents for different care settings (WHO 2007).

In response to international concern over the safety of infant formula, the FAO/WHO have undertaken two risk assessments (FAO/WHO 2004, 2006b; Table 9.2). One of the first outcomes was the recognition of three categories of organisms associated with neonatal infections and presence in PIF. These are as follows:

Category A. Clear evidence of causality: *Salmonella* serovars and *Cronobacter* species (*Ent. sakazakii*)

Category B. Causality plausible, but not yet demonstrated: *Enterobacter cloacae, Citrobacter koseri, C. freundii, Klebsiella oxytoca, Klebsiella pneumoniae, Pantoea agglomerans, Escherichia vulneris, E. coli, Hafnia alvei, Serratia* spp. and *Acinetobacter* spp.

Category C. Causality less plausible or not yet demonstrated: *B. cereus, Clostridium difficile, Clostridium perfringens, Clostridium botulinum, Staphylococcus aureus, L. monocytogenes* and coagulase negative staphylococci.

Category A means that there have been cases where contaminated reconstituted PIF has been the source of *Salmonella* and *Cronobacter* spp. infections, as demonstrated by molecular fingerprinting of clinical and formula isolates. Some *Salmonella* outbreaks have been caused by rare lactose-fermenting *Salmonella* which would not be detected using routine *Salmonella* methods (Anon. 1997a).

Both FAO/WHO meetings (2004, 2006b) stressed that multiplication of *Cronobacter* spp. (*Ent. sakazakii*) following reconstitution equated to increased risk of infection, and recommend reconstitution with water at temperatures greater than 70°C. Therefore, appropriate advice based on the effect of methods of reconstitution at high temperature and subsequent handling would be of use in reducing microbial risks. Since none of the organisms in Category B are spore formers, this practice would be expected to reduce the risk of infection from those organisms as well.

In 2007, JEMRA introduced a risk model using various conditions of PIF preparation which mimicked hygienic practices, and the subsequent risks were determined (JEMRA 2008) (Plate 22). The model has three components: concentration in powder supply, total PIF consumption and dose per serving, leading to an estimated number of cases and 'relative risk'. The user is able to select or input from a range of contamination levels, reconstitution temperatures, holding temperatures, and so on. Given that *Cronobacter* spp. levels have never reportedly exceeded 1 cfu/g PIF, it is important to stress the potential temperature abuse of reconstituted PIF which may enable low levels of opportunistic pathogens to multiply and cause an infection. Bacterial contaminants may be intrinsic to the powder or from environmental sources during preparation. The majority of cases are low birth weight neonates who have poorly developed immune systems, and limited competing intestinal flora. Hence, ingestion of formula with raised levels of an opportunistic pathogen should be minimised. The reader is encouraged to consult Farber and Forsythe (2008) for more information on this topic.

10.8 Viral risk assessments

In general, compared with most bacterial pathogens, the food- and waterborne viral pathogens are less understood and more fundamental research is required into detection methods and surveillance programmes to support risk assessments. The use of hepatitis A vaccine for food handlers (as practiced in some US cities) needs further investigation with regard to cost-effectiveness. Since viral contamination of food can occur anywhere in the production process, they should be considered more fully in HACCP schemes. The detection of viral contaminants needs to be improved to establish a baseline for future intervention and prevention programs; studies are needed to estimate the burden of illness and cost of illness due to foodborne viral infections.

10.8.1 Viral contamination of shellfish and coastal waters

Rose and Sobsey (1993) completed a four-stage (Codex format) quantitative risk assessment of viral contamination of shellfish and coastal waters. The hazard identification stage summarised the viral hazards associated with contaminated shellfish: polivirus, echovirus and rotavirus. The dose–response assessment (part of hazard characterisation) used previous human feeding trial data (see Table 9.8 for echovirus 12 (low infectivity) and rotavirus (high infectivity)). Exposure assessment was determined by multiplying the number of viral particles per gram of shellfish by the amount ingested per year. The average number of viral particles per gram of shellfish (range 0.2–31 pfu/100 g) and the arithmetic average for each sampling site were calculated. The amount of shellfish consumed per person per year in the United States ingested was 250 g clams, 74 g oysters and 53 g other shellfish. It was assumed that individuals consumed an equal amount of shellfish through the year. An average serving was determined to be 60–240 g. Hence, exposures to a single serving ranged from 0.11–18.6 pfu (average 6 pfu) for a 60 g serving to 0.43–74.4 pfu for a 240 g serving. The risk characterisation determined risk estimates using the beta-Poisson probability models of echovirus 12 and rotavirus, virus contamination level and the two levels of consumption to represent low and high exposures. Illness and death rates were compared over a range of exposure doses and were multiplied by the probability of infection to determine the risk of illness and death. Using echovirus 12 probability model for the consumption of 60 g raw shellfish, the individual risks ranged from 2.2×10^{-4} to 3.5×10^{-2}. The risk being four times greater for 240 g servings. Using average virus levels, this equates to approximately 1 in 100 chance of being infected following ingestion of raw shellfish from approved waters. However, using the rotavirus model, the risk increased to 3.1×10^{-1} due to its greater infectivity with an average virus exposure of 6 pfu/60 g shellfish.

This early risk assessment understandably does not use the distribution range of viruses in a Monte Carlo simulation. It is also plausible that the use of improved detection methods would increase the virus incidence and level in shellfish compared to that used in this study.

11 International control of microbiological hazards in foods: regulations and authorities

As emphasised in earlier chapters, national and international surveillance is required these days in order to reduce the risk of worldwide foodborne disease, improve production processes and global economies. A number of national and international surveillance systems are covered in Section 1.11. This chapter is more focused on organisations with international roles in food safety.

11.1 World Health Organisation, global food security from accidental and deliberation contamination

The International Health Regulations (IHR) were agreed by the international community in 1969 and adopted by the World Health Organisation (WHO) in the same year. They are a regulatory framework for global public health security based on preventing the international spread of infectious diseases. The initial IHR required Member States to notify the WHO of all cases of cholera, plague and yellow fever. However, the revised IHR (2005), which came into force on 15 June 2007, includes food contamination and foodborne disease events. The IHR require the WHO to be notified of all public health emergencies of international concern, and includes guidelines for surveillance procedures and response to public health emergencies.

The WHO Food Safety Department can provide guidance to Member States. It is also the focal point for WHO collaboration with FAO in the Codex Alimentarius Commission. The Joint FAO/WHO Expert Committee on Food Additives (JECFA) and the Joint FAO/WHO Meetings on Pesticide Residues (JMPR) assess the risks associated with chemicals in food, for use by Member States and by the Codex Alimentarius Commission. Risk assessment of microbiological agents in food is through the Joint FAO/WHO Expert Meetings on Microbiological Risk Assessment (JEMRA). The WHO Food Safety Department can therefore provide expert advice on specific chemical and microbiological threats in food. Risk assessment and technical advice are needed to assess any food terrorism threat and ensure that there are appropriate levels of international preparation and emergency responses.

The increase in international trade in food has increased the risk from cross-border transmission of infectious agents, and underscores the need for international risk assessment to estimate the risk that microbial pathogens pose to human health. The globalisation and liberalisation of world food trade, while offering many benefits and opportunities, also present new risks. Because of the global nature of food production, manufacturing and marketing, infectious agents can be disseminated from the original point of processing and packaging to locations thousands of miles away. Foodborne disease is of considerable importance as a global public health issue, and was recognised as a priority area for WHO at its fifty-third session of

Box 11.1 The WHO has given five keys to safer food

1 Keep clean
2 Separate raw and cooked
3 Cook thoroughly
4 Keep food at safe temperatures
5 Use safe water and raw materials.

the World Health Assembly held in Geneva in May 2000. The Director-General of the World Health Organisation, Dr Gro Harlem Brundtland (WHO 2001), proposed that there are three major challenges to protect the health of the consumer:

1 Re-establish consumer confidence from the farm to the table by re-assessing and improving existing food safety systems.
2 Ensure reasonable food safety standards to apply throughout the world and assist all countries to reach those standards.
3 Develop global standards for pre-market approval systems of genetically modified food to ensure that these new products are not only safe, but also beneficial for consumers and more efficient than existing products (Box 11.1).

The WHO and the Food and Agriculture Organisation of the United Nations (FAO) have been in the forefront of the development of risk-based approaches for the management of public health risks for hazards in food. Since risk analysis is now well established for chemical hazards, the WHO and FAO are using their expertise for risk analysis of microbiological hazards.

Conventional food safety systems, typified by the pasteurisation and sterilisation procedures of the dairy industry, have been improved by the adoption of HACCP. Yet despite the increasing implementation of HACCP, the increased number of reported 'food poisoning' cases in many countries has emphasised the need for further food safety systems with an emphasis being on assessing the direct microbiological risk to humans. This new approach, microbiological risk assessment, requires an improved epidemiological knowledge of foodborne diseases throughout the food chain. Since the food is in a global supply chain, this knowledge needs to be built up on a global perspective and hence involves governments as well as companies. Hence, the activities of the European Food Safety Authority and outbreak monitoring networks such as Enter-Net and Salm-Surv; see Web Resources section for URLs.

The 'farm to fork' approach to food safety has highlighted that it is easier to keep food products free from microbial (etc.) contamination in the supply chain if one can ensure contamination-free animals and poultry on the farm. An example is the approach in Sweden where nearly *Salmonella*-free poultry production has been achieved. It costs about $8 million per year to implement. But this amount is small compared to the cost of medical treatment which is estimated at $28 million.

Antibiotics have been added to animal feeds to inhibit the growth of microbial pathogens. Unfortunately, this has, most likely, caused the selection of antibiotic strains which can also be resistant to medically important antibiotics with life-threatening consequences. Hence, the use of antibiotics needs to be more carefully controlled (Sections 1.9 and 1.12.6).

To ensure global food safety, developing and developed countries need to participate together in establishing food safety systems. WHO and its Member States have recognised food safety as a worldwide challenge and the World Health Assembly passed a food safety resolution in May 2000 with reference to food safety as an essential public health issue. The resolution focuses on the need to develop sustainable, integrated food safety systems for the reduction of health risk along the entire food chain. Joint FAO/FAO activities are outlined in Section 11.3.

The WHO have given their aims for food safety in the twenty-first century as follows:
- Strengthening national food safety policies and infrastructures
- Advising on food legislation and enforcement
- Evaluating and promoting safe food technologies
- Educating food handlers, health professionals and consumers in food safety
- Encouraging food safety in urban settings
- Promoting food safety in tourism
- Establishing epidemiological surveillance of foodborne diseases
- Monitoring chemical contamination of food
- Developing international food safety standards
- Assessing foodborne hazards and risks

The WHO aims to decrease the burden of microbiological foodborne diseases by taking into consideration the whole food chain and applying a holistic approach. This strategy comprises the following actions:
- Development of a strategy for microbiological monitoring in food (development of national capacities, investigation of potentials for database registration); see FERG below.
- Microbiological risk assessment methodology (development of international agreement on methodology via guidance documents)
- Microbiological risk assessment advice (to Codex and member countries via joint experts consultations)
- Development and strengthening of risk communication methodology (guidelines for studying risk perception, methods for efficient communication, feasibility of a rapid alert system)
- New preventive strategies (potential of new food production, inspection or investigation methods to contribute to food safety, expert consultation on emerging pathogens with priority given to foodborne viruses)
- Strengthening and coordination of global efforts on the surveillance of foodborne diseases and outbreak response (guidelines and standards for surveillance, laboratory and other capacities, networking).

The WHO has responded to concerns of its Member States that food might be used as a vehicle for disseminating biological, chemical and radionuclear agents by support to the Member States to increase the response capacity of their national health systems. Their guidance document (WHO 2002) states:

> 'Food terrorism is defined as an act or threat of deliberate contamination of food for human consumption with chemical, biological or radionuclear agents for the purpose of causing injury or death to civilian populations and/or disrupting social, economic or political stability (WHO). The chemical agents in question are man-made or natural toxins, and the biological agents referred to are communicably infectious or non-infectious pathogenic microorganisms, including viruses, bacteria and parasites. Radionuclear agents are defined in this context as radioactive chemicals capable of causing injury when present at unacceptable levels.'

The WHO document covers the security of food and bottled water preparation and not threats to animal or plant health or to the provision of sufficient quantities of food for a population. The key points are as follows:

- Prevention, for example by reducing access to agents of use in food terrorism, and control at port of entry.
- Surveillance, strengthening national and international networks such as Enter-Net.
- Preparedness, integrating food terrorist response with existing emergency response resources, and assessing vulnerability. For example, the CDC has a method for assessing potential threats; 'Bioterrorism Preparedness and Response Office' (http://www.bt.cdc.gov).

11.2 The foodborne disease burden epidemiology reference group (FERG)

In November 2007, the FERG was established by the WHO (Kuchenmüller *et al.* 2009). Its objective is to determine the global burden of foodborne disease. This is the incidence and prevalence of morbidity, disability and mortality associated with acute and chronic infection. Initially, FERG will consider microbial, parasitic, zoonotic and chemical contamination of food. Of particular interest will be assessing the chemical and parasitic causes of foodborne disease burden since little work has been done on these topics to date. See Box 11.2 for a list of action points.

The burden of foodborne diseases estimates will be measured in DALYs (Disability-Adjusted Life Years) as they are useful as a comparative measure of disease burden and can form a basis for more detailed cost estimates related to the economic impact. The DALY measure combines the years of life lost due to premature death and the years lived with disability for varying degrees of severity, making time itself the common metric for death and disability. One DALY is a health gap measure, equating to 1 year of healthy life lost.

FERG will collaborate closely with the Food and Agriculture Organisation of the United Nations (FAO) and others to undertake regional consultations to discuss the regional specific profiles of foodborne syndromes and etiologic agent for future burden estimation.

Box 11.2 FERG action points

Acute infectious diseases
- Examine pathogen-specific global burden in children
- Develop detailed analysis of cause of death data on WHO mortality database
- Conduct or commission relevant burden work
- Develop cause attribution models and estimate percentage foodborne
- Recommend and/or conduct intervention studies to increase data availability

Chronic infectious diseases
- Evaluate the known microbial and chemical causes of foodborne disease
- Develop cause attribution models and estimate percentage foodborne

Acute and chronic chemicals
- Identify major causes, particularly for developing countries, and commission-relevant burden work
- Develop cause attribution models and estimate percentage foodborne

11.3 Regulations in international trade of food

Currently, food safety measures are not consistent around the world and such differences can lead to trade disagreements among countries. This is particularly true if microbiological requirements are not justified scientifically. A useful website giving food and agricultural import regulations and standards can be found in the Web Resources section.

The standards, guidelines and recommendations adopted by the Codex Alimentarius Commission (Section 11.4) and international trade agreements, such as those administered by the World Trade Organisation (WTO), are playing an increasingly important role in protecting the health of consumers and ensuring fair practices in trade. In 1962, the Joint FAO/WHO Food Standards Programme was created with the Codex Alimentarius Commission as its executive organisation. The Codex Alimentarius, or the food code, is a collection of international food standards that have been adopted by the Codex Alimentarius Commission. Codex standards cover all the main foods, whether processed, semi-processed, or raw. The principal objectives of Codex Alimentarius Commission are to protect the health of consumers and ensure fair practices in the food trade.

In the case of microbiological hazards, Codex has elaborated standards, guidelines and recommendations that describe processes and procedures for the safe preparation of food. The application of these standards, guidelines and recommendations is intended to prevent or eliminate hazards in foods or reduce them to acceptable levels.

The WTO SPS Agreement entered into force in 1995 and applies to all sanitary and phytosanitary measures, which may, directly or indirectly, affect international trade. It provides basic rights and obligations for WTO Members and directs them to harmonise sanitary, guidelines and recommendations. For food safety, the standards, guidelines and recommendations established by the CAC relating to food additives, veterinary drug and pesticide residues, contaminants, sampling and methods of analysis, and codes and guidelines of hygienic practice are recognised as the basis for harmonisation of sanitary measures.

WTO Members may introduce or maintain measures that result in a higher level of sanitary or phytosanitary protection than that would be achieved by measures based on international standards, guidelines and recommendations. In this regard, WTO Members are required to ensure that their sanitary and phytosanitary measures are based on an assessment, as appropriate to the circumstances, of the risks to human, animal or plant life or health, taking into account the risk assessment techniques developed by the relevant international organisations. Article 5 of the SPS Agreement provides an impetus for the development of microbiological risk assessment to support the elaboration of standards, guidelines and recommendations related to food safety.

The first Joint FAO/WHO Expert Consultation on the Application of Risk Analysis to Food Standards Issues was held in 1995. It delineated the basic terminology and principles of risk assessment and concluded that the analysis of risks associated with microbiological hazards presents unique challenges. The report of the Joint FAO/WHO Expert Consultation on Risk Management and Safety held in 1997 identified a risk management framework and the elements of risk management for food safety (FAO/WHO 1997). The Joint FAO/WHO Expert Consultation on the Application of Risk Communication to Food Standards and Safety Matter held in 1998 identified elements and guiding principles of risk communication and strategies for effective risk communication (FAO/WHO 1998). In addition to the foundation provided by the series of Joint FAO/WHO Expert Consultations, the Codex Committee on Food Hygiene has elaborated principles and guidelines for microbiological risk assessment.

Table 11.1 Codex committees

Subject	Host country
Food labelling	Canada
Food additives and contaminants	The Netherlands
Food hygiene	United States
Pesticide residues	The Netherlands
Veterinary drugs in foods	United States
Methods of analysis and sampling	Hungary
Food import and export inspection and certification systems	Australia
General principles	France
Nutrition and foods for special dietary uses	Germany

The 'Draft Principles and Guidelines for the Conduct of Microbiological Risk Assessment' were adopted by the twenty-third session of the Codex Commission in June 1999 (CAC 1999).

11.4 Codex Alimentarius Commission

The Codex Alimentarius Commission was established in 1962 to implement the Joint Food and Agricultural Organisation (FAP) and World Health Organisation (WHO) Food Standards Programme. The aim of the programme is to set minimum standards to protect the consumer and provide a framework to ensure fair trade across international borders. Codex also plays a role in promoting coordination on all food standards work between international governmental and non-governmental organisations, which includes helping to set priorities for work in this area (Table 11.1).

The Codex Alimentarius (Latin for food law/code) is a collection of internationally accepted food standards presented in a uniform manner. The aim of these food standards is to protect consumers' health and to ensure fair practices in the food trade. Codex also publishes advisory tests in the form of codes of practice, guidelines and other recommended measures to assist in achieving its objective.

The Codex machine operates through a labyrinthine structure of committees which feed recommendations up to the main Codex Alimentarius Commission. There are a number of commodity-type committees dealing with hygiene and quality issues; examples are Codex Committee for Food Hygiene, Codex Committee for Milk and Milk Products, and Codex Committee for Nutrition and Foods for Special Dietary Uses. There are also a number of other committees dealing with executive, procedural and regional issues, for example Codex Committee on General Principles, and Codex Coordinating Committee for Europe.

11.5 Sanitary and phytosanitary measures (SPS), technical barriers to trade (TBT) and WHO

The Final Act of the Uruguay Round of multilateral trade negotiations established the WTO to succeed the General Agreement on Tariffs and Trade (GATT). The Final Act led to the 'Agreement on the Application of Sanitary and Phytosanitary Measures' (SPS Agreement) and the 'Agreement to Technical Barriers to Trade' (TBT Agreement). The ratification of the WTO Agreement is a major factor in developing new hygiene measures for the international trade in food.

The GATT decision on SPS reaffirmed that no member should be prevented from adopting or enforcing measures necessary to protect human, animal or plant life or health (GATT 1994). The SPS (Article 5) provisions of the World Trade Agreement encourage 'harmonisation' of standards for food safety:

> 'Members shall ensure that their sanitary or phytosanitary measures are based on an assessment, as appropriate to the circumstances, of the risks to human, animal or plant health, taking into account risk assessment techniques as described by relevant international organisations.'

These are intended to facilitate the free movement of foods across borders, by ensuring that means established by countries to protect human health are scientifically justified and are not used as non-tariff barriers to trade in foodstuffs. The agreement states that SPS measures based on appropriate standards, codes and guidelines developed by the Codex Alimentarius Commission are deemed to be necessary to protect human health and consistent with the relevant GATT provisions. The WTO SPS Agreement also recognises the Office Internationale des Epizooties (OIE) and the International Plant Protection Convention (IPPC) as international standard-setting bodies for animal and plant health, respectively. Codex, OIE and IPPC should coordinate their standard-setting activities, as appropriate, to ensure that international food safety standards adequately consider and incorporate factors relevant to the impact of animal and plant health issues on food safety, for example bovine spongiform encephalopathy (BSE) as a cause of illness (vCJD) in humans.

The Agreement of Sanitary and Phytosanitary Measures came into force in 1995, and in support of the SPS Agreement, the CAC now has a comprehensive action plan to incorporate risk analysis in its activities wherever appropriate. The SPS Agreement recognises the right of governments to protect its population's health from hazards that might be introduced through imported food by imposing sanitary measures, even though these might mean trade restrictions. However, such sanitary measures must be based on risk assessment in order to avoid unjustifiable, protective trade measures.

The management of foodborne hazards requires a transparent scientific risk assessment. Risk assessment needs to be undertaken by people that have scientific and public credibility and retain confidence in their conclusions. Both the risk assessment and risk management processes and decisions must be transparent and accompanied by effective communication activities. Many developing countries, however, are poorly equipped to effectively manage existing and emerging risks from food. As well as improved methods for the long-distance distribution of food, there have also been improvements in detection methods for pathogen food contaminants: both chemical and biological. However, the detected presence of a compound may not necessarily make it a public health hazard. Therefore, risk assessment is necessary to assess the significance of the 'contaminant' and inform the public appropriately. Decisions must be made in accordance with internationally recognised standards which are scientifically and not politically based. Hence, in the food industry, the Codex standards are internationally recognised and compliant with SPS provisions. Conventional Good Manufacturing Practice based food hygiene requirements (i.e. end-product testing) are being improved through the implementation of Hazard Analysis Critical Control Point (HACCP). Subsequently, risk assessment will put more emphasis on predictive microbiology for the generation of exposure data. In turn, this will assist in establishing critical limits for HACCP schemes (Buchanan 1995, Notermans et al. 1999, Seera et al. 1999; Chapter 8).

The adoption of food safety systems such as HACCP has been encouraged by various international bodies such as the Food and Agriculture Organisation of the United Nations (FAO), WHO and the FAO/WHO Codex Alimentarius Commission. Implementation of HACCP has changed the emphasis from reactive control of problems following end-product analysis to identifying hazards in the production process, and ensuring the steps which control the hazards are effective. Subsequently, HACCP has been adopted into law to different extents. However, HACCP does not assess the risk associated with food consumption as required by the SPS Agreement (Article 5).

Hence, Article 5 of the SPS Agreement provides an impetus for the development of microbiological risk assessment to support the elaboration of standards, guidelines and recommendations related to food safety. It provides a framework for the formulation and harmonisation of sanitary and phytosanitary measures. These measures must be based on science and implemented in an equivalent and transparent manner. They cannot be used as an unjustifiable barrier to trade by discriminating among foreign sources of supply or providing an unfair advantage to domestic producers. To facilitate safe food production for domestic and international markets, the SPS Agreement encourages governments to harmonise their national measures or base them on international standards, guidelines and recommendations developed by international standard-setting bodies.

11.6 European Union legislation

There are three types of legislation within the EU:

1 *Regulation*: A legal act which has general applications and is binding in its entirety and directly applicable to the citizens, courts and governments of all Member States. Regulations do not, therefore, have to be transferred into domestic laws and are chiefly designed to ensure uniformity of law across the community.

2 *Directive*: A binding law directed to one or more Member States. The law states objectives which the Member States(s) are required to confirm within a specified time. A directive has to be implemented by Member States by amendment of their domestic laws to comply with the stated objectives. This process is known as 'approximation of laws' or 'harmonisation' since it involves the alignment of domestic policy throughout the community.

3 *Decision*: An Act which is directed at specific individuals, companies or Member States, which is binding in its entirety. Decisions addressed to Member States are directly applicable in the same way as directives.

Food hygiene law and technical legislation such as additives or labelling are mostly 'Single Act Measures'; this means they are part of the progress towards the single market. Single Act Measures are those which are essential for the free flow of goods and services in a truly common market; they are subject to majority voting.

The aim of EU food legislation is to ensure a high standard of public health protection and that the consumer is adequately informed of the nature and, where appropriate, the origin of the product. In the context of a developing internal market, the Commission has adopted a particular approach to the food sector in order to establish a large market without barriers, and at the same time ensuring consumer safety with the widest possible choice. The Commission has combined the principles of mutual recognition of national standards and rules, contained within Articles 30–36 of the Treaty of Rome – this approach works alongside 100a which resulted from the amendment to the treaty by the Single European Act. The primary source for European Union information is the 'Official Journal'.

The 1985 European Commission White Paper 'Completing the Internal Market' catalogued the measures necessary to allow for the free movement of goods (including food services, capital and labour) which would lead to the removal of all physical, technical and fiscal barriers between Member States. Since 1 January 1993, food has moved freely within the EU with the minimum of inspection at land or sea frontiers. Harmonised rules have been adopted, applicable to all food produced in the EU, underpinned by the principle of mutual recognition of national standards and regulations for matters that do not require EC legislation. Specific directives were in place for meat and meat products, live bivalve molluscs, fishery products, milk and milk products, and eggs. Foods entering the EC from countries outside the EU will be subject to EU hygiene standards.

11.6.1 Food hygiene directive (93/43/EEC)

One of the most significant EU Directive for the food industry was the adoption of the Food Hygiene Directive (93/43/EEC) on the Hygiene of Foodstuffs (Anon. 1993a). It is extremely significant in the development of EU food law and will form the basis for food hygiene control across Europe for years to come. The Directive deals with general rules of hygiene for foodstuffs and the procedures for verification of compliance with the rules. 'Food hygiene' means all measures necessary to ensure the safety and wholesomeness of foodstuffs. 'Wholesome food' means food which is fit for human consumption. 'Food business' means any undertaking whether public or private, and whether operating for profit or not. The measures cover all stages after primary production (harvesting, slaughter and milking), preparation, processing, manufacturing, packaging, storing, transportation, distribution, handling and offering for sale or supply to the consumer. Food business operators must identify any step in their activities which is critical to ensuring food safety and ensure that adequate safety procedures are identified, implemented, maintained and reviewed on the basis of the principles used to develop HACCP (Section 8.5). The Directive is a horizontal Directive and therefore applies across the whole of the food industry. It covers producers, manufacturers, distributors, wholesalers, retailers and caterers. In essence, this Directive combines the proactive approach of food safety by HACCP implementation and codes of Good Hygienic Practice into food law.

11.7 Food safety agencies

There are numerous national food safety agencies established around the world; the websites for some of them are given in the 'Food Safety Resources on the World Wide Web' section. The Danish National Food Agency takes a 'farm to fork' approach and has a broad remit. The Swedish National Food Administration (SNFA) also has a broad remit, has a proactive health role and has powers to legislate. In Germany, the enforcement responsibilities are within the regional governments, the German Federal Institute for Consumer Health and Veterinary Medicine (BgVV) and the Ministry of Health. The Australian and New Zealand Food Authority (ANZFA) has a narrower remit and currently focuses on the development of food standards and codes of practice to protect public health and promote fair trade. Its role is to make recommendations and has industry representation. This is in contrast to the Food Authority in Ireland where commercial interests have been deliberately excluded in order to establish public confidence in its independence. The Canadian Food Inspection Agency (CFIA) has responsibility for human, plant and animal health and has enforcement powers. The FDA, however, does not have responsibility for meat; this is the responsibility of the US Department of Agriculture.

Many agencies restructured in response to the considerable public concern over food safety. The European Food Safety Authority (EFSA) was formed in part due to the BSE/vCJD situation in the UK and dioxin contamination in Belgium. The Scientific Committee on Food responsibilities are the '*scientific and technical questions concerning consumer health and food safety associated with the consumption of food products and in particular questions relating to toxicology and hygiene in the entire food production chain, nutrition, and applications of agrifood technologies, as well as those relating to materials coming into contact with foodstuffs, such as packaging*'; EU website, see Web Resources for URL address.

11.7.1 Food authorities in the United States
US Department of Health and Human Services
This includes the Food and Drug Administration (FDA) and the Centers for Disease Control and Prevention (CDC).

The FDA's responsibilities are for the following:
- All domestic and imported food sold in interstate commerce, including shell eggs, but not meat and poultry
- Bottled water
- Wine beverages with less than 7% alcohol

Its role is to enforce food safety laws concerning domestic and imported food, except meat and poultry. The FDA Food Code is endorsed by USDA's Food Safety and Inspection Service (FSIS) and CDC. It provides a template by which food safety rules concerning food safety in restaurants, grocery stores, nursing homes and other institutional and retail settings. The Food Code is neither federal law nor regulation. However, it is used by more than 3000 state and local regulatory agencies across the United States and aims to achieve consistency across the various regulatory jurisdictions. The Food Code is updated every 2 years and the URL is given in the Web Resources section.

The CDC has responsibility for all foods. Their role is to:
- investigate sources of foodborne disease outbreaks;
- maintain a nationwide system of foodborne disease surveillance;
- develop and advocate public health policies to prevent foodborne diseases;
- conduct research to help prevent foodborne illness;
- train local and state safety personnel.

US Department of Agriculture (USDA)
The USDA is composed of the FSIS, Co-operative State Research, Education, and Extension Service and the National Agricultural Library USDA/FDA Foodborne Illness Education Information Center.

The FSIS enforces food safety laws governing domestic and imported eat and poultry products. It has responsibility for:
- domestic and imported meat and poultry and related products such as meat- or poultry-containing stews, pizzas and frozen foods;
- processed egg products (generally liquid, frozen and dried pasteurised egg products).

The Co-operative State Research, Education and Extension Service is responsible for all domestic foods and some imported foods. Its food safety role is with US colleges and universities to develop research and education programmes on food safety for farmers and consumers.

The National Agricultural Library (NAL) USDA/FDA Foodborne Illness Education Information Center maintains a database of computer software, audiovisuals, posters, games, teachers' guides and other educational materials on preventing foodborne illness. It also helps educators, food service trainers and consumers to locate educational materials on preventing foodborne illness.

US Environmental Protection Agency (EPA)

The US EPA oversees drinking water and has a food safety role with regard to foods made from plants, seafood, meat and poultry. It:

- establishes safe drinking water standards;
- regulates toxic substances and wastes to prevent their entry into the environment and food chain;
- assists states in monitoring the equality of drinking water and finding ways to prevent contamination of drinking water;
- determines the safety of new pesticides, sets tolerance levels for pesticides in foods and publishes directions on safe use of pesticides.

Food Outbreak Response Coordinating Group (FORC-G)

The FORC-G was formed in 1997 to:

- increase coordination and communication among federal, state and local food safety agencies
- guide efficient use of resources and expertise during an outbreak
- prepare for new and emerging threats to the US food supply

The group is formed from the Department of Health and Human Services (which includes the FDA), the US Department of Agriculture and the EPA.

Glossary of terms

Acceptable daily intake (ADI)   An estimate of the quantity of a particular chemical in food or drinking water, expressed on a body mass basis (usually mg/kg body weight) which, it is believed, can be consumed on a daily basis over a lifetime without appreciable health risk, that is the practical certainty that no harm will result, even after a lifetime's exposure (assuming a standard body mass of 60 kg).

Adverse effect   Change in morphology, physiology, growth, development or lifespan of an organism which results in impairment of functional capacity or impairment of capacity to compensate for additional stress or increase in susceptibility to the harmful effects of other environmental influences. Decisions on whether or not any effect is adverse require expert judgement.

Assumption   An expert judgement made on the basis of incomplete information, which therefore has uncertainty associated with it.

CCP decision tree   A sequence of questions to assist in determining whether a control point is a critical control point.

Cleaning   The removal of soil, food residue, dirt, grease or other objectionable matter.

Clean in place (CIP)   A system used to clean process piping, bins, tanks, mixing equipment or larger pieces of equipment without disassembly, where interior product zones are fully exposed and soil can be readily washed away by the flow of the cleaning solution.

Clean out of place (COP)   A system (e.g. cleaning tanks) used to clean equipment parts, piping, and so on, after disassembly.

Cluster/outbreak/epidemic   'Cluster' is used to describe a group of cases linked by time or place, but with no identified common food or other source. In the context of foodborne disease, 'outbreak' refers to two or more cases resulting from ingestion of a common food. The term 'epidemic' is often reserved for crises or situations involving larger numbers of people over a wide geographical area.

Contaminant   Any biological or chemical agent, foreign matter, or other substances not intentionally added to food which may compromise food safety or suitability.

Contamination   The introduction or occurrence of a contaminant in food or food environment.

Control measure   Any action or activity that can be used to prevent, eliminate or reduce a food safety hazard to an acceptable level.

Control (noun)   The state wherein correct procedures are being followed and criteria are being met.

Control point   Any step at which biological, chemical or physical factors can be controlled.

Control (verb)   To take all necessary actions to ensure and maintain compliance with criteria established in the HACCP plan.

Corrective action   Any action to be taken when the results of monitoring at the CCP indicate a loss of control.

Criterion   A requirement on which a judgement or decision can be based.

Critical control point (CCP)   A step at which control can be applied and is essential to prevent or eliminate a food safety hazard or reduce it to an acceptable level.

Critical food-contact surface   A surface that contacts food or a surface from which drainage onto the food or onto surfaces that contact food ordinarily occurs during the normal course of operations after the food is subjected to a bactericidal control process, or when the food is not subjected to any bactericidal control process at the processor's facility.

Critical limit   A maximum and/or minimum value to which a biological, chemical or physical parameter must be controlled at a CCP to prevent, eliminate or reduce to an acceptable level the occurrence of a food safety hazard. A criterion which separates acceptability from unacceptability.

Critical non-food-contact surface or area   A surface (other than a food-contact surface) or area that could, through the action of man or equipment, contaminate a food that will not be subjected to a bactericidal control measure after the exposure of food or a food-contact surface to the surface or area. Critical non-food-contact surfaces and areas include equipment, vents, fixtures, drains, walls, floors, employee clothing, shoes, and accessories, and other surfaces in the plant that do not (or are not intended to) contact food.

Critical surfaces and areas   Critical food-contact surfaces and critical non-food-contact surfaces and areas.

Deviation   Failure to meet a critical limit.

Disinfection   The reduction, by means of chemical agents and/or physical methods, of the number of micro-organisms in the environment to a level that does not compromise food safety or suitability.

Dose–response assessment   The determination of the relationship between the magnitude of exposure (dose) to a biological, chemical or physical agent and the severity and/or frequency of associated adverse health effects (response).

D value (decimal reduction time)   90% (= 1 log) loss of viability due to a lethal process such as heat (cooking), acidity or irradiation. See also Z value.

Enterotoxins   Substances that are toxic to the intestinal tract causing vomiting, diarrhoea, and so on; most common enterotoxins are produced by bacteria.

Establishment   Any building or area in which food is handled and the surroundings under the control of the same management.

Exposure assessment   The qualitative and/or quantitative evaluation of the likely intake of biological, chemical and physical agents via food as well as exposures from other sources, if relevant.

Finished RF-RTE food   A refrigerated or frozen ready-to-eat food that is packaged.

Flow diagram   A systematic representation of the sequence of steps or operations used in the production or manufacture of a particular food item.

Food   Any substance, whether processed, semi-processed or raw which is intended for human consumption, including drinks, chewing gum and any substance which has been used in the manufacture, preparation or treatment of "food" but excluding cosmetics, tobacco and substances used only as drugs.

Foodborne disease   Any disease of an infectious or toxic nature caused by consumption of food.

Foodborne disease outbreak   Two definitions are used: (a) The observed number of cases of a particular disease exceeds the expected number. (b) The occurrence of two or more cases of a similar foodborne disease resulting from the ingestion of a common food.

Food handler   Any person who directly handles packaged or unpackaged food, food equipment and utensils, or food-contact surfaces and is therefore expected to comply with food hygiene requirements.

Food hygiene   All conditions and measures necessary to ensure the safety and suitability of food at all stages of the food chain.

Food safety   Assurance that food will not cause harm to the consumer when it is prepared and/or eaten according to its intended use.

Food safety objective (FSO)   A government target considered necessary to protect the health of consumers (this may apply to raw materials, a process or finished products).

Food safety requirement   A company-defined target considered necessary to comply with a FSO.

Food suitability   Assurance that food is acceptable for human consumption according to its intended use.

Genome   The total DNA content of an organism; chromosomal and extra-chromosomal elements such as plasmids.

HACCP   A systematic approach to the identification, evaluation and control of food safety hazards.

HACCP plan   The document which is based upon the principles of HACCP and which delineates the procedures to be followed.

HACCP system   The result of the implementation of the HACCP plan.

HACCP team   The group of people who are responsible for developing, implementing and maintaining the HACCP system.

Hazard   A biological, chemical or physical agent in a food, or condition of a food, with the potential to cause an adverse health effect.

Hazard analysis (HA)   The process of collecting and evaluating information on hazards associated with the food under consideration to decide which are significant and must be addressed in the HACCP plan.

Hazard characterisation   The qualitative and/or quantitative evaluation of the nature of the adverse health effects associated with biological, chemical and physical agents which may be present in food. For the purpose of microbiological risk assessment, the concerns relate to micro-organisms and/or their toxins. For biological agents, a dose–response assessment should be performed if the data are obtainable.

Hazard identification   The identification of biological, chemical and physical agents capable of causing adverse health effects and which may be present in a particular food, or group of foods.

Lot   A quantity of food or food units produced and handled under uniform conditions.

Microbiological risk   A risk that is related to the presence of a microbiological hazard (such as bacteria, viruses, yeast, moulds, algae, parasitic protozoa and helminths). This includes the chemical hazards they may produce (toxins and metabolites).

Monitor   Means to conduct a planned sequence of observations or measurements to assess whether a process or procedure is under control.

Monitoring   The act of conducting a planned sequence of observations or measurements of control parameters to assess whether a CCP is under control.

Pathogen   A micro-organism capable of causing human illness or injury.

Prerequisite programs   Procedures, including Good Manufacturing Practices, that address operational conditions providing the foundation for the HACCP system.

Primary production   Those steps in the food chain up to and including, for example, harvesting, slaughter, milking, fishing.

Process or processing   Any activity that is directly related to the production of a food, including any packaging activity.

Processor   Any person engaged in commercial, custom or institutional processing of a food.

Proteome   The protein complement of a genome.

Qualitative risk assessment   A risk assessment based on data which, whilst forming an inadequate basis for numerical risk estimate, nonetheless, when conditioned by prior expert knowledge and identification of attendant uncertainties, permits risk ranking (comparison) or separation into descriptive categories of risk.

Quality   The totality of characteristics of an entity that bear on its ability to satisfy stated or implied needs.

Quality assurance (QA)   All the planned and systematic activities implemented within the quality system, and demonstrated as needed, to provide adequate confidence that an entity will fulfil requirements for quality.

Quality control (QC)   The operational techniques and activities used to fulfil requirements of quality.

Quality management   All activities of the overall management function that determine the quality policy, objectives and responsibilities that implement them by means such as quality planning, quality control, QA and quality improvement with the quality system.

Quality system   The organisational structure, procedures, processes and resources needed to implement quality management.

Quantitative risk assessment   A risk assessment that provides numerical expression of risk and indication of the attendant uncertainties.

Ready-to-eat (RTE) food   A food that is customarily consumed without cooking by the consumer, or that reasonably appears to be suitable for consumption without cooking by the consumer.

Risk   A function of the probability of an adverse health effect and the severity of that effect, consequential to a hazard(s) in food.

Risk analysis   A process consisting of three components: risk assessment, risk management and risk communication.

Risk assessment   A scientifically based process consisting of the following steps: (1) hazard identification, (2) hazard characterisation, (3) exposure assessment and (4) risk characterisation. The definition includes quantitative risk assessment, which emphasises reliance on numerical expressions of risk, and also qualitative expressions of risk, as well as an indication of the attendant uncertainties.

Risk assessment policy   This consists of documented guidelines for scientific judgement and policy choices to be applied at appropriate decision points during risk assessment.

Risk characterisation   The qualitative and/or quantitative estimation, including attendant uncertainties, of the probability of occurrence and severity of known or potential adverse health effects in a given population based on hazard characterisation and exposure assessment.

Risk communication   The interactive exchange of information and opinions throughout the risk analysis process concerning hazards and risks, risk-related factors and risk perceptions, among risk assessors, risk managers, consumers, industry, the academic community and other interested parties, including the explanation of risk assessment findings and the basis of risk management decisions.

Risk estimate   Output of risk characterisation.

Risk management   The process, distinct from risk assessment, of weighing policy alternatives, in consultation with all relevant parties, considering risk assessment and other factors relevant for the health protection of consumers and for the promotion of fair trade practices, and, if needed, selecting appropriate prevention and control options.

Safety policy   The overall intentions and direction of an organisation with regard to safety as formally expressed by top management.

Sanitation Standard Operating Procedures (SSOPs)   Procedures established by an operator for the day-to-day sanitation activities involved in the production of safe and wholesome food.

Scenario set   A construct characterising the range of likely pathways affecting the safety of the food product. This may include consideration of processing, inspection, storage, distribution and consumer practices. Probability and severity values are applied to each scenario.

Sensitivity analysis   A method used to examine the behaviour of a model by measuring the variation in its outputs resulting from changes to its inputs.

Severity   The seriousness of the effect(s) of a hazard.

Sporadic case   A case that cannot be linked epidemiologically to other cases of the same illness.

Standard Operating Procedures (SOPs)   Procedures established by an operator for the day-to-day activities involved in the production of safe and wholesome food.

Step   A point, procedure, operation or stage in the food chain including raw materials, from primary production to final consumption.

Surveillance   The systematic collection, analysis and interpretation of data essential to the planning, implementation and evaluation of public health practice, and the timely dissemination of this information for public health action.

Synteny   Co-localisation of genes on chromosomes of related species.

Threshold   Dose of a substance or exposure concentration below which a stated effect is not observed or expected to occur.

Total Quality Management   An organisation's management approach centred on quality, based on the participation of all its members and aimed at long-term success through customer satisfaction and benefits to the members of the organisation and to society.

Transparent   Characteristics of a process where the rationale, the logic of development, constraints, assumptions, value judgements, decisions, limitations and uncertainties of the expressed determination are fully and systematically stated, documented and accessible for review.

Uncertainty   Lack of sufficient or reliable data or knowledge.

Uncertainty analysis   A method used to estimate the uncertainty associated with model inputs, assumptions and structure/form.

Validation   The element of verification focused on collecting and evaluating scientific and technical information to determine if the HACCP plan, when properly implemented, will effectively control the hazards.

Variability   Distribution of values due to known variables such as biological variation, seasonal changes and amount of food eaten.

Verification   Those activities, other than monitoring, that determine the validity of the HACCP plan and that the system is operating according to the plan.

Z value   Temperature increase required to increase the death rate tenfold. In other words, the temperature increase reduces the D value tenfold ($1\,\log_{10}$ unit).

List of abbreviations

ADI	Acceptable Daily Intake
ALARA	As Low As Reasonably Achievable
ALOP	Appropriate Level Of Protection
CAC	Codex Alimentarius Commission
CCFAC	Codex Commission for Food Additives and Contaminants
CCFH	Codex Commission for Food Hygiene
CCP	Critical Control Point
CCPR	Codex Commission for Pesticide Residues
CCRVDF	Codex Commission for Residues of Veterinary Drugs in Food
FAO	Food and Agriculture Organisation of the United Nations
FDA	Food and Drug Administration
FSO	Food Safety Objective
GAP	Good Agricultural Practice
GMP	Good Manufacturing Practice
HACCP	Hazard Analysis Critical Control Points
HTST	High-Temperature, Short-Time
IAFP	International Association of Food Protection
IAMFES	International Association of Milk, Food and Environmental Sanitarians
ICMSF	International Commission on Microbiological Specification for Foods
JECFA	Joint FAO/WHO Expert Committee on Food Additives
JMPR	Joint FAO/WHO Meeting on Pesticide Residues
LEE	Locus of Enterocycte Effacement
MRA	Microbiological Risk Assessment
MRL	Maximal Residue Level
NACMCF	US National Advisory Committee on Microbiological Criteria for Foods
PMTDI	Provisional Maximal Tolerable Daily Intake
PTWI	Provisional Tolerable Weekly Intake
QRA	Quantitative Risk Analysis
SPS	WTO Agreement on the Application of Sanitary and Agreement Phytosanitary Measures
WHO	World Health Organisation
WTO	World Trade Organisation

Food safety resources on the world wide web

Since websites have a habit of moving, it may be necessary to use a search engine with the Organisation name or topic to relocate the URL. The prefix 'http' has been omited from all addresses.

Topic	URL
Author's homepage. 'The Microbiology of Safe Food' companion site	http://www.wiley.com/go/forsythe; also http://www.theagarplate.com
Agreement of the Application of Sanitary and Phytosanitary Measures	http://www.wto.org/english/docs_e/legal_e/15-sps.pdf
Bacterial growth and survival predictors	
GInaFIT	http://cit.kuleuven.be/biotec/downloads/downloads.htm
Growth Predictor (Combase)	http://www.ifr.ac.uk/Safety/GrowthPredictor/
Food Micromodel	http://www.arrowscientific.com.au/predictive_micro_sw.html
Microfit	http://www.ifr.bbsrc.ac.uk/MicroFit/default.html
Pathogen Modeling Program	http//ars.usda.gov/Services/docs.htm?docid=6784
Seafood Spoilage Predictor	http://www.dfu.min.dk/micro/sssp/Home/Home.aspx
Sym'Previus	http://www.symprevius.net
Bacterial genomics	
Bacterial genomes	http://www.ncbi.nlm.nih.gov/genomes/lproks.cgi
BAGEL (Bacteriocin genome mining tool)	http://bioinformatics.biol.rug.nl/websoftware/bagel/bagel_start.php
ClustalW	http://www.ebi.ac.uk/Tools/clustalw2/index.html
J. Craig Venter Institute (formerly TIGR)	http://www.tigr.org
Human microbiome project	http://nihroadmap.nih.gov/hmp and http://www.hmpdacc.org/bacterial_strains.php
MetaHIT project	http://www.metahit.eu
National Centre for Biotechnology Information (NCBI)	http://www.ncbi.nlm.nih.gov
Ribosomal Database Project II	http://rdp.cme.msu.edu
Bioinformatics tutorials	http://www.wiley.com/go/forsythe
WebACT (Artemis comparison tool)	http://www.webact.org/WebACT/home
CAST (Council for Agricultural Science and Technology)	http://www.cast-science.org
CAC – Principles of microbiological risk analysis	http://www.who.int/fsf/mbriskassess/pdf/draftpr.pdf

(Continued)

Topic	URL
CAC Principles and Guidelines for the Conduct of Microbiological Risk Assessment. CAC/GL-30 (1999b)	http://www.codexalimentarius.net/download/standards/357/CXG_030e.pdf
CDC Foodborne disease outbreaks search facility	http://wwwn.cdc.gov/foodborneoutbreaks/
CDC Epi Info™	http://www.cdc.gov/epiinfo/
CDC automated surveillance outbreak detection algorithm (SODA)	http://www.cdc.gov/ncidod/dbmd/phlisdata
Codex Alimentarius Commission (CAC)	http://www.codexalimentarius.net/
Codex Alimentarius Commission. Procedural Manual. 16th edn	ftp://ftp.fao.org/codex/Publications/ProcManuals/Manual_16e.pdf
Codex – food hygiene	http://www.fao.org/docrep/W4982E/w4982e09.htm
Detection methods (approved) Association of Official Analytical Chemists (AOAC) International	http://www.aoac.org
Health Canada: Compendium of Methods for the Microbiological Examination of Foods	http://www.hc-sc.gc.ca/fn-an/res-rech/analy-meth/microbio/index_e.html
FDA: Bacteriological Analytical Manual	http://www.cfsan.fda.gov/~ebam/bam-toc.html
International Standardisation Organisation (ISO)	http://www.iso.ch
USDA Microbiology Laboratory Guidebook	http://www.fsis.usda.gov/Science/Microbiological_Lab_Guidebook/index.asp
European Commission (EC) EU microbiological criteria	http://www.europa.eu.int/comm/dg24/health/sc/scv/out26_en.html
European Commission (1999) Risk assessment and 93/43/EEC	http://europa.eu.int/comm/dg24/health/sc/oldcomm7/out07_en.html
EC harmonization of risk assessment	http://europa.eu.int/comm/food/fs/sc/ssc/out82_en.html
Salmonella Enteritidis Risk Assessment	http://www.europa.eu.int/comm/dg24/health/sc/scv/out26_en.html
Eurosurveillance (weekly and monthly reports)	http://www.eurosurveillance.org
FAO/WHO documents Assesors and managers of microbiological hazards	ftp://ftp.fao.org/docrep/nonfao/ae586e/ae586e00.pdf
FAO Microbiological Risk Assessment repository	http://www.fao.org/documents/advanced_s_result.asp?FORM_C=AND&SERIES=314
Development of practical risk management strategies based on microbiological risk assessment outputs	http://www.fao.org/ag/agn/agns/jemra/Ecoli.pdf
FAO/WHO. Food safety risk analysis – a guide for national food safety authorities (2006)	http://www.who.int/foodsafety/publications/micro/riskanalysis06/en
General – Food Control	http://www.fao.org/docrep/w8088e/w8088e04.htm
Hazard characterization for pathogens in food and water	ftp://ftp.fao.org/docrep/fao/006/y4666e/y4666e00.pdf
JECFA	http://www.fao.org/ag/agn/agns/jecfa.index.en.asp

JEMRA	http://www.fao.org/ag/agn/agns//micro_en.asp
Microbiological risk assessments	http://www.fao.org/ag/agn/agns/jemra_riskassessment.en.asp
Microbial risk assessment	http://www.who.int/foodsafety/micro/jemra/assessment/en
Principles and guidelines for incorporating microbiological risk assessment in the development of food safety standards, guidelines and related texts (2002)	http://www.who.int/foodsafety/publications/micro/march2002/en
Risk analysis, Geneva, 1995	http://www.who.int/fsf/mbriskassess/applicara/index.htm
Risk assessment, Geneva, 1999	http://www.who.int/fsf/mbriskassess/Consultation99/reporam.pdf
Risk assessment of *Salmonella* in eggs and broiler chickens	http://www.who.int/foodsafety/publications/micro/salmonella/en
RA on *Salmonella* in eggs and broiler chickens	http://www.who.int/foodsafety/micro/jemra/assessment/salmonella/en/
RA of *Salmonella* in eggs and broiler chickens – 2	http://www.who.int/foodsafety/publications/micro/salmonella/en
Risk communication, Rome, 1998	http://www.fao.org/docrep/005/x1271e/X1271E00.HTM
Risk management and food safety, Rome 1997	http://www.who.int/foodsafety/publications/micro/jan1997/en/index.html
Risk management, 2000	http://www.who.int/foodsafety/publications/micro/march2000/en/index.html
The interaction between assessors and managers of microbiological hazards in food (2000)	http://www.who.int/foodsafety/publications/micro/march2000/en
FDA	
Bad bug book	http://www.fda.gov/Food/FoodSafety/FoodborneIllness/FoodborneIllnessFoodbornePathogensNaturalToxins/BadBugBook/default.htm
Bacteriological analytical method (BAM)	http://www.cfsan.fda.gov/~ebam/bam-toc.html
Baytril decision	http://www.fda.gov/oc/antimicrobial/baytril.pdf
Guide to minimize microbial food safety hazards of fresh-cut fruits and vegetables	http://www.cfsan.fda.gov/~dms/guidance.html
Fluoroquinolone-resistant *Campylobacter*	http://www.fda.gov/downloads/AnimalVeterinary/SafetyHealth/IRecallsWithdrawals/UCM152308.pdf
V. parahaemolyticus in shellfish model	http://www.fda.gov/Food/ScienceResearch/ResearchAreas/RiskAssessmentSafetyAssessment/ucm050421.htm
Food and agricultural import regulations and standards	http://www.fas.usda.gov/itp/ofsts/us.html
Food irradiation	http://www.foodsafety.org/sf/sf057.htm
Foodrisk.org	http://www.foodrisk.org/

(Continued)

Topic	URL
Food Safety and Inspection Service (FSIS)	http://www.fsis.usda.gov/Science/Risk_Assessments/index.asp
E. coli O157 in beef risk assessment	http://www.fsis.usda.gov/OPHS/ecolrisk/prelim.htm
E. coli O157:H7	http://www.fsis.usda.gov/ophs/ecolrisk/pubmeet/index.htm
Salmonella in shell eggs risk assessment	http://www.fsis.usda.gov/ophs/risk/contents.htm
S. Enteritidis risk assessment model	http://www.fsis.usda.gov/ophs/risk/semodel.htm
Food Safety Risk Analysis Clearinghouse	http://www.foodrisk.org
Hazard Analysis Critical Control Point (HACCP)	
HACCP–TQM in retail	http://www.hi-tm.com
International HACCP Alliance	http://aceis.agr.ca
NACMCF HACCP plan	http://www.foodrisk.org
International Commission on Microbiological Specifications for Foods (ICMSF)	http://www.ICMSF.org
International Life Sciences Institute (ILSI)	http://www.ilsi.org
ILSI revised framework for microbial risk assessment	http://www.ilsi.org/file/mrabook.pdf
Microbial growth media and detection kits (OXOID Ltd)	http://www.oxoid.co.uk
Online training courses and model explanations	
Poultry FARM (Oscar)	https://apps.who.int/fsf/Micro/farmhp/PFARM1-HP.PDF
FAO/WHO initiative on microbial risk assessment	http://www.who.int/fsf/mbriskassess/IAFP_meeting_01/index.htm
Risk assessment frameworks	http://www.gov.on.ca/omafra/english/research/risk/frameworks/index.html
Risk assessment and Monte Carlo simulation tools	https://apps.who.int/fsf/Micro/farmhp/PFARM1-HP.PDF
Analytica (Monte Carlo simulation software)	http://www.lumina.com
Crystal Ball	http://www.decisioneering.com/crystal_ball/index.html
@RISK	http://www.palisade.com
RiskWorld	http://www.riskworld.com
Society for Risk Analysis	http://www.sra.org/index.php
Surveillance programmes	
Enter-Net (ECDC)	http://ecdc.europa.eu
Foodborne viruses in Europe	http://www.Eufoodborneviruses.co.uk
FoodNet (CDC)	http://www.cdc.gov/foodnet
Global Salmonella surveillance (Global Salm-Surv)	http://www.who.int/salmsurv/en/ and
	http://www.who.int/emc/diseases/zoo/SALM-SURV/SlideShow
Health Protection Agency (UK), communicable disease reports	http://www.hpa.org,uk/web/home
PulseNet	http://www.cdc.gov/pulsenet/
USDA Economic Research Service foodborne illness cost calculator	http://www.ers.usda.gov/data/foodborneillness

USDA risk analysis bibliography	http://www.nal.usda.gov/fnic/foodborne/risk.htm
WHO	http://who.int/fsf/index.htm
Activities, reports, news and events related to *Salmonella*	http://www.who.int/topics/salmonella/en/index.html
Antimicrobial resistance	http://www.who.int/emc/diseases/zoo/antimicrobial.html
Ensuring food safety in the aftermath of natural disasters	http://www.who.int/foodsafety/foodborne_disease/emergency/en/
Environmental health in emergencies and disasters	http://www.who.int/water_sanitation_health/hygiene/emergencies/emergencies2002/en/
Foodborne disease outbreaks: guidelines for investigation and control	http://www.who.int/foodsafety/publications/foodborne_disease/fdbmanual/en/index.html
Microbiological risk assessment	http://www.who.int/fsf/mbriskassess/index.htm
Salmonella in eggs and broiler chickens	http://www.who.int/foodsafety/micro/jemra/assessment/salmonella/en/
Study course	http://www.who.int/fsf/mbriskassess/studycourse/index.html
National Governmental Food Departments and Agencies (or equivalents)	
Austria. Rechtsinformationssystem	http://www.ris.bka.gv.at
Austria. Federal Environment Agency	http://udk.ubavie.gv.at
Australia. Department of Agriculture, Fish and Forestry	http://www.affa.gov.au
Australia and New Zealand Food Authority	http://www.foodstandards.gov.au
Canada. Food Inspection Agency	http://www.inspection.gc.ca/english/toce.shtml
Denmark. Fødevareministeriet	http://www.fvm.dk
European Union	
Rapid Alert System for Food and Feed (RASFF)	http://ec.europa.eu/food/food/rapidalert/index_en.htm
EU	http://ec.europa.eu
EU food hygiene	http://ec.europa.eu/food/food/biosafety/hygienelegislation/index_en.htm
Finland. Maa- ja Metsätalousministeriö	http://www.evira.fi/portal/en
France. Ministry of Agriculture	http://www.agriculture.gouv.fr
Germany. Bundes., Ernah., Landw., Forsten	http://www.bml.de
Greece. Hellenic Republic Ministry of Agriculture	http://www.minagric.gr
India. Ministry of Food Processing	http://www.allindia.com/gov/ministry/fpi/policy.htm
Ireland. Dept. of Agriculture and Food	http://www.irlgov.ie/daff
Ireland, Food Safety Authority	http://www.fsai.ie/
Italy. Istituto Nazionale di Economia Agraria	http://www.inea.it
Japan. Ministry of Agriculture Fisheries and Food	http://www.maff.go.jp/eindex.html

(*Continued*)

Topic	URL
Korea. Food and Drug Administration	http://www.kfda.go.kr/english/index.html
Netherlands. Ministry of Agric., Nat. Man. Fish.	http://www.minlnv.nl/international/
Portugal. Minist. Agric.Desen. rural e das Pescas	http://www.min-agricultura.pt
Portugal. Minist. Equip.Plane. Admin.	http://www.min-plan.pt
Russia. Ministry of Agriculture and Food	http://www.aris.ru/N/WIN_R/PARTNER
Scotland	http://www.food.gov.uk/scotland
Spain. Agritel – Minist. de Agric. Pesc. Aliment	http://www.aesan.msc.es/AESAN/web/evaluacion_riesgos/detalle/contaminante.shtml
Sweden. Jordbruksdepartementet	http://www.slv.se/Default.aspx?epslanguage=SV
United Kingdom Food Standards agency	http://www.foodstandards.gov.uk
Health Protection Agency (HPA)	http://www.hpa.org.uk/cdr
HPA Ready to eat food guidelines (2009)	http://www.hpa.org.uk/web/HPAwebFile/HPAweb_C/1259151921557
United Nations	
Codex Alimentarius Commission	http://www.codexalimentarius.net
Food and Agriculture Organization	http://www.fao.org
USA	
Centres for Disease Control	http://ftp.cdc.gov/pub/mmwr/MMWRweekly
Department of Agriculture	http://www.usda.gov
Food Safety and Inspection Service	http://www.fsis.usda.gov
Food and Drug Adminsitration (FDA)	http://www.fda.gov
World Health Organisation (WHO)	http://www.who.int
Foodborne Disease Burden Epidemiology Reference Group (FERG)	http://www.who.int/foodsafety/foodborne_disease/ferg/en/index.html

References

Abee, T. & Wouters, J.A. (1999) Microbial stress response in minimal processing. *Int. J. Food Microbiol.* **50**, 65–91.

Acheson, D. (2001) An alternative perspective on the role of *Mycobacterium paratuberculosis* in the etiology of Crohn's disease. *Food Control* **12**, 335–338.

ACMSF (2005) ACMSF second report on *Campylobacter*. HMSO, London.

Adams, M.R., Little, C.L. & Easter, M.C. (1991) Modelling the effect of pH, acidulant and temperature on the growth rate of *Yersinia enterocolitica*. *J. Appl. Bacteriol.* **71**, 65.

Adams, M.R. & Marteau, P. (1995) On the safety of lactic acid bacteria from food. *Int. J. Food Microbiol.* **27**, 263–264.

Adams, M. & Motarjemi, Y. (1999) *Basic Food Safety for Health Workers*. WHO/SDE/PHE/FOS/99.1. World Health Organization, Geneva.

Agbodaze, D. (1999) Verocytotoxins (Shiga-like toxins) produced by *Escherichia coli*: a mini review of their classification, clinical presentations and management of a heterogenous family of cytotoxins. *Comp. Immunol. Microbiol.* **22**, 221–230.

Allos, B.M. (1998) *Campylobacter jejuni* infection as a cause of the Guillain–Barré syndrome. *Emerg. Infect. Dis.* **12**, 173–184.

Alocilja, E.C. & Radke, S.M. (2003) Market analysis of biosensors for food safety. *Biosens. Bioelectron.* **18**, 841–846.

Amann, R.I., Ludwig, W. & Scheifer, K.H. (1995) Phylogenetic identification and *in situ* detection of individual microbial cells without cultivation. *Microbiol. Rev.* **59**, 143–169.

Anon. (1992) HACCP and Total Quality Management – winning concepts for the 90's: a review. *J. Food Prot.* **55**, 459–462.

Anon. (1993a) Council Directive 93/43/EEC on the hygiene of foodstuffs. *Offic. J. Eur. Comm.* No. L 175/1.

Anon. (1993b) *Listing of Codes of Practice Applicable to Foods*. Institute of Food Science and Technology, London.

Anon. (1993c) HACCP implementation: A generic model for chilled foods. *J. Food Protect.* **56**, 1077–1084.

Anon. (1996) Microbiological guidelines for some ready-to-eat foods sampled at the point of sale: an expert opinion from the Public Health Laboratory Service (PHLS). *PHLS Microbiol. Dig.* **13**, 41–43.

Anon. (1997a) Preliminary report of an outbreak of *Salmonella anatum* infection linked to infant formula milk. *Euro Surveill.* **2**, 22–24.

Anon. (1997b) Surveillance of enterohaemorrhagic *E. coli* (EHEC) infections and haemolytic uraemic syndrome (HUS) in Europe. *Euroserv.* **2**, 91–96.

Anon. (1999a) Opinion of the scientific committee on veterinary measures relating to public health on the evaluation of microbiological criteria for food products of animal origin for human consumption. Available from: http://europa.eu.int/comm/food/fs/sc/scv/out26_en.pdf.

Anon. (1999b) Where have all the gastrointestinal infections gone? *Euro Surveill Wkly Rep.* **3**, 990114.

Anon. (2000) EN 13783. Brussels, European Committee for Standardization.

Anon. (2004) EN14569. Brussels, European Committee for Standardization.

Anon. (2009) Egg safety, from production to consumption: an action plan to eliminate *Salmonella* Enteritidis illness due to eggs, President's Council on Food Safety, December 10, USA. Available from: http://www.fda.gov/Food/FoodSafety/Product-SpecificInformation/EggSafety/EggSafetyActionPlan/ ucm170615.htm.

ANZFA (1999) *Food Safety Standards – Costs and Benefits.* Australia New Zealand Food Authority. Available from: http://www.anzfa.gov.au.

Atlas, R.M. (1999) Probiotics – snake oil for the new millennium? *Environ. Microbiol.* **1**, 377–380.

Atmar, R. & Estes, M. (2001) Diagnosis of non-cultivatable gastroenteritis viruses, the human caliciviruses. *Clin. Microbiol. Rev.* **14**, 15–37.

Baik, H.S., Bearson, S., Dunbar, S. & Foster, J.W. (1996) The acid tolerance response of *Salmonella typhimurium* provides protection against organic acids. *Microbiology* **142**, 3195–3200.

Baker, D.A. & Genigeorgis, C. (1990) Predicting the safe storage of fresh fish under modified atmospheres with respect to *Clostridium botulinum* toxigenesis by modeling length of the lag phase of growth. *J. Food Protect* **53**, 131–140.

Baker, D.A. (1995) Application of modelling in HACCP plan development. *Int. J. Food Microbiol.* **25**, 251–261.

Banerjee, P. & Bhunia, A.K. (2009) Mammalian cell-based biosensors for pathogens and toxins. *TIBTECH* **27**, 179–188.

Baranyi, J. & Roberts, T.A. (1994) A dynamic approach to predicting bacterial growth in food. *Int. J. Food Microbiol.* **23**, 277–294.

Barer, M.R. (1997) Viable but non-culturable and dormant bacteria: time to resolve an oxymoron and a misnomer? *J. Med. Microbiol.* **46**, 629–631.

Barer, M.R., Kaprelyants, A.S., Weichart, D.H., Harwood, C.R. & Kell, D.B. (1998) Microbial stress and culturability: conceptual and operational domains. *Microbiology* **144**, 2009–2010.

Barker, G.C., Talbot, N.L.X. & Peck, M.W. (1999) Microbial risk assessment for sous-vide foods. In: *Proceedings of Third European Symposium on Sous Vide.* pp. 37–46. Alma Sous Vide Centre, Belgium.

Barsotti, L. Merle, P. & Cheftel, J.C. (1999) Food processing by pulsed electric fields (Part I and II), *Food. Rev.Int.* **15**, 163–213.

Bauman, H.E. (1974) The HACCP concept and microbiological hazard categories. *Food Technol.* **28**, 30–34 and 74.

Bearson, S., Bearson, B. & Foster, J.W. (1997) Acid stress responses in enterobacteria. *FEMS Microbiol. Lett.* **147**, 173–180.

Bemrah, N., Sana, M., Cassin, M.H., Griffiths, M.W. & Cerf, O. (1998) Quantitative risk assessment of human listeriosis from consumption of soft cheese made from raw milk. *Prev. Vet. Med.* **37**, 129–145.

Bennett, J.V., Homberg, S.D., Rogers, M.F. & Solomon, S.L. (1987) Infectious and parasitic diseases. *Am. J. Prev. Med.* **55**, 102–114.

Berg, D., Kohn, M., Farley, T. & McFarland, L. (2000) Multistate outbreaks of acute gastroenteritis traced to fecal-contaminated oysters harvested in Louisiana. *J. Infect. Dis.* **181**, S381–S386.

Berg, R.D. (1998) Probiotics, prebiotics or 'conbiotics'? *Trends Microbiol.* **6**, 89–92.

Best, E.L., Fox, A.J., Frost, J.A. & Bolton, F.J. (2005) Real-time single-nucleotide polymorphism profiling using Taqman technology for rapid recognition of *Campylobacter jejuni* clonal complexes. *J. Med. Microbiol.* **54**, 919–925.

Bettcher, D.W., Yach, D. & Guindon, G.E. (2000) Global trade and health: key linkages and future challenges. *Bull. WHO.* **78**, 521–534.

Beuret, C., Kohler, D. & Luthi, T. (2000) Norwalk-like virus sequences detected by reverse transcription polymerase chanin reaction in mineral waters imported into or bottled in Switzerland. *J. Food Prot.* **63**, 1576–1582.

Bidawid, S., Farber, J.M. & Sattar, S.A. (2000) Contamination of foods by foodhandlers: experiments on hepatitis A virus transfer to food and its interruption. *Appl. Environ. Microbiol.* **66**, 2759–2763.

Bidol, S.A., Daly, E.R., Rickert, R.E., *et al.* (2007) Multistate outbreaks of *Salmonella* infections associated with raw tomatoes eaten in restaurants – United States, 2005–2006. *MMWR* **56**, 909–911.

Biller, J.A, Katz, A.J., Flores, A.F., Buie, T.M. & Gorbach, S.L. (1995) Treatment of recurrent *Clostridium difficile* colitis with *Lactobacillus* GG. *J. Pediatr. Gastroenterology* **21**, 224–226.

Biziagos, E., Passagot, J., Crance, J.M. & Deloince, R. (1988) Long-term survival of hepatitis A virus and poliovirus type 1 in mineral water. *Appl. Environ. Microbiol.* **54**, 2705–2710.

Black, R.E., Levine, M.M., Clements, M.L., Highes, T.P. & Blaster, M.J. (1988) Experimental *Campylobacter jejuni* infection in humans. *J. Infect. Dis.* **157**, 472–479.

Blackstock, W.P. & Weir, M.P. (1999) Proteomics: quantitative and physical mapping of cellular proteins. *TIBTECH* **17**, 121–127.

Bloomfield, S.F., Stewart, G.S.A.B., Dodd, C.E.R., Booth, I.R. & Power, E.G.M. (1998) The viable but non-culturable phenomenon explained? *Microbiology* **144**, 1–3.

Booth, I.R., Pourkomailian, B., McLaggan, D. & Koo, S.-P. (1994) Mechanisms controlling compatible solute accumulation: a consideration of the genetics and physiology of bacterial osmoregulation. *J. Food Eng.* **22**, 381–397.

Boquet, P., Munro, P., Fiorentini, C. & Just, I. (1998) Toxins from anaerobic bacteria: specifically and molecular mechanisms of action. *Curr. Opin. Microbiol.* **1**, 66–74.

Borch, E. & Wallentin, C. (1993) Conductance measurement for data generation in predictive modelling. *J. Ind. Microbiol.* **12**, 286.

Borch, E., Nesbakken, T. & Christensen, H. (1996) Hazard identification in swine slaughter with respect to foodborne bacteria. *Int. J. Food Microbiol.* **30**, 9–25.

Bouwmeester, H., Dekkers, S., Noordam, M., *et al.* (2009) Review of health safety aspects of nanotechnologies in food production. *Regul. Toxicol Pharmacol.* **53**, 52–62.

Bowen, A.B. & Braden, C.R. (2006) Invasive *Enterobacter sakazakii* disease in infants. *Emerg. Infect. Dis.* Available from: http://www.cdc.gov/ncidod/EID/vol12no08/05-1509.htm.

Bower, C.K. & Daeschel, M.A. (1999) Resistance responses of microorganisms in food environments. *Int. J. Food Microbiol.* **50**, 33–44.

Bowers, J., Brown, B., Springer, J., Tollefson, L., Lorentzen, R. & Henry, S. (1993) Risk assessment for aflatoxin: an evaluation based on the multistage model. *Risk Anal.* **13**, 637–642.

Boyd, E.F., Wang, F.S., Whitham, T.S. & Selander, R.K. (1996) Molecular relationship of the salmonellae. *Appl. Environ. Microbiol.* **62**, 804–808.

Bracey, D., Holyoak, C.D. & Coote, P.J. (1998) Comparison of the inhibitory effect of sorbic acid and amphotericin B on *Saccharomyces cerevisiae*: is growth inhibition dependent on reduced intracellular pH? *J. Appl. Microbiol.* **85**, 1056–1066.

Bradbury, J. (1998) Nanobacteria may lie at the heart of kidney stones. *Lancet*, **352**(9122) 121.

Brenner, D.J. (1984) Facultatively anaerobic Gram-negative rods. In: *Bergey's Manual of Systematic Bacteriology*, Vol. 1 (eds N.R. Krieg & J.C. Holt, J.C.), pp. 408–516. Williams and Wilkins, Baltimore.

Broughall, J. & Brown, C. (1984) Hazard analysis applied to microbial growth in foods: development and application of three-dimensional models to predict bacterial growth. *J. Food Microbiol.* **1**, 13–22.

Broughall, J.W., Anslow, P.A. & Kilsby, D.C. (1983) Hazard analysis applied to microbial growth in foods: development of mathematical models describing the effect of water activity. *J. Appl. Bacteriol.* **55**, 101–110.

Brown, M.H., Davies, K.W., Billon, C.M.P., Adair, C. & McClure, P.J. (1998) Quantitative microbiological risk assessment: principles applied to determining the comparative risk of salmonellosis from chicken products. *J. Food Prot.* **61**, 1446–1453.

Brown, P., Will, R.G., Bradley, R., Asher, D. & Detwiler, L. (2001) Bovine spongiform encephalopathy and variant Creutzfeldt–Jakob disease: background, evolution and current concerns. *Emerg. Inf. Dis.* **7**, 6–16.

Brown, W.L. (1991) Designing *Listeria monocytogenes* thermal inactivation studies for extended shelf-life refrigerated foods. *Food Technol.* **45**, 152–153.

Bruce, M.E., Will, R.G., Ironside, J.W., *et al.* (1997) Transmissions to mice indicate that 'new variant' CJD is caused by the BSE agent. *Nature* **389**, 498–501.

Brul, S., Coote, P. (1999) Preservative agents in foods. Mode of action and microbial resistance mechanisms. *Int. J. Food Microbiol.* **50**, 1–17.

Brul, S. & Klis, F.M. (1999) Review: mechanistic and mathematical inactivation studies of food spoilage fungi. *Fungal Genet. Biol.* **27**, 199–208.

Buchanan, R.L. (1995) The role of microbiological criteria and risk assessment in HACCP. *Food Microbiol. (London)* **12**, 421–424.

Buchanan, R.L. & Whiting, R. (1996) Risk assessment and predictive microbiology. *J. Food Prot.* **59**, Suppl., 31–36.

Buchanan, R.L., Damert, W.G., Whiting, R.C. & Van Schothorst, M. (1997) Use of epidemiological and food survey data to estimate a purposefully conservative dose–response relationship for *Listeria monocytogenes* levels and incidence of listeriosis. *J. Food Prot.* **60**, 918–922.

Buchanan, R.L. & Edelson, S.G. (1999) Effect of pH-dependent, stationary phase acid resistance on the thermal tolerance of *Escherichia coli* O157:H7. *Food Microbiol.* **16**, 447–458.

Buchanan, R.L. & Linqvist, R. (2000) Preliminary report: hazard identification and hazard characterization of *Listeria monocytogenes* in ready-to-eat foods. JEMRA. Available from: http://www.fao.org/ag/agn/agns/jemra_riskassessment_listeria_en.asp.

Buchanan, R.L., Smith, J.L. & Long, W. (2000) Microbial risk assessment: dose–response relations and risk characterization. *Int. J. Food Microbiol.* **58**, 159–172.

Bückenhuskes, H.J. (1997) Fermented vegetables. In: *Food Microbiology: Fundamentals and Frontiers* (eds M.P. Doyle, L.R. Beuchat & T.J. Montville), pp. 595–609. ASM Press, Washington, DC.

Büllte, M., Klien, G. & Reuter, G. (1992) Pig slaughter. Is the meat contaminated by *Yersinia enterocolitica* strains pathogenic to man? *Fleischwirtschaft* **72**, 1267–1270.

Bunning, V.K., Lindsay, J.A. & Archer, D.L. (1997) Chronic health effects of foodborne microbial disease. *World Health Stat. Q.* **50**, 51–56.

Butzler, J.P., Dekeyser, P., Detrain, M. & Dehaen, F. (1973) Related vibrio in stools. *J. Pediatr.* **82**, 493–495.

Buzby, J.C. & Roberts, T. (1997a) Economic costs and trade implications of microbial foodborne illness. *World Health Stat. Q.* **50**, 57–66.

Buzby, J.C. & Roberts, T. (1997b) Guillain–Barré syndrome increases foodborne disease costs. *Food Rev.* **20**, 36–42.

Buzby, J.C. & Roberts, T. (2009) The economics of enteric infections: human foodborne disease costs. *Gastroenterology* **136**, 1851–1862.

Calvert, R.M., Hopkins, H.C., Reilly, M.J. & Forsythe, S.J. (2000) Caged ATP – internal calibration method for ATP bioluminescence assays. *Lett. Appl. Microbiol.* **30**, 223–227.

Campbell, G.A. & Mutharasan, R. (2007) A method of measuring *Escherichia coli* O157:H7 at 1 cell/mL in 1 liter sample using antibody functionalized piezoelectric-excited millimeter-sized cantilever sensor. *Environ. Sci. Technol.* **41**, 1668–1674.

Caplice, E. & Fitzgerald, G.F. (1999) Food fermentations: role of microorganisms in food production and preservation. *Int. J. Food Microbiol.* **50**, 131–149.

Carlin, F., Girardin, H., Peck, M.W., *et al.* (2000) Research on factors allowing a risk assessment of spore-forming pathogenic bacteria in cooked chilled foods containing vegetables: a FAIR collaborative project. *Int. J. Food Microbiol.* **60**, 117–135.

Cassin, M.H., Lammerding, A.M., Todd, E.C.D, Ross, W. & McColl, R.S. (1998a) Quantitative risk assessment for *Escherichia coli* O157:H7 in ground beef hamburgers. *Int. J. Food Microbiol.* **41**, 21–44.

Cassin, M.H., Paoli, G.M. & Lammerding, A.M. (1998b) Simulation modeling for microbial risk assessment. *J. Food Prot.* **61**, 1560–1566.

Caubilla-Barron, J. & Forsythe, S. (2007) Dry stress and survival time of *Enterobacter sakazakii* and other *Enterobacteriaceae J. Food Prot.* **70**, 2111–2117.

Caubilla-Barron, J., Hurrell, E., Townsend, S., *et al.* (2007) Genotypic and phenotypic analysis of *Enterobacter sakazakii* strains from an outbreak resulting in fatalities in a neonatal intensive care unit in France. *J. Clin. Microbiol.* **45**, 3979–3985.

CCFH (1999a) Discussion paper on recommendations for the management of microbiological hazards for food in international trade. CX/FX/98/10. Codex Committee on Food Hygiene. FAO, Rome.

CCFH (1999b) Management of *Listeria monocytogenes* in foods. Joint FAO/WHO Food Standards Programme, Codex Committee on Food Hygiene. 32nd Session. CX/FH 99/10 p. 27. FAO, Rome.

CCFH (2000) *Proposed Draft Principles and Guidelines for the Conduct of Microbiological Risk Management at Step 3.* Joint FAO/WHO Food Standards Programme, Codex Committee on Food Hygiene, July 2000. CX/FH 00/06. FAO, Rome.

CCFRA (2000) *An Introduction to the Practice of Microbiological Risk Assessment for Food Industry Applications.* Guideline 28. Campden & Chorleywood Food Research Association Group, Leatherhead, UK.

Centers for Disease Control (CDC) (1993) Multistate outbreak of *Escherichia coli* O157:H7 infections from hamburgers – Western United States, 1992–1993. *MMWR* **42**, 258–263.

Centers for Disease Control (CDC) (1999) Outbreaks of *Shigella sonnei* infection associated with eating fresh parsley: United States and Canada. *MMWR* **48**, 285–289.

Çetinkaya, B., Egan, K. & Morgan, K.L. (1996) A practice-based survey of the frequency of Johne's disease in south west England. *Vet. Rec.* **134**, 494–497.

Chapman, P.A. & Siddons, C.A. (1996) A comparison of immunomagnetic separation and direct culture for the isolation of verocytotoxin-producing *Escherichia coli* O157 from cases of bloody diarrhoea, non-bloody diarrhoea and asymptomatic contact. *J. Med. Microbiol.* **44**, 267–271.

Chiodini, R.J. & Hermon-Taylor, J. (1993) The thermal resistance of *Mycobacterium paratuberculosis* in raw milk under conditions simulating pasteurisation. *J. Vet. Diagn. Invest.* **5**, 629–631.

Chirife, J. & del Pilarbuera, M. (1996) Water activity, water glass & dynamics, and the control of microbiological growth in foods. *Crit. Rev. Food Sci. & Nutr.* **36**, 465–513.

Christensen, B., Rosenquist, H., Sommer, H. & Nielsen, N. (2001) Quantitative risk assessment of human illness associated with *Campylobacter jejuni* in broilers. *Campylobacter, Helicobacter* and related organisms (CHRO). 11th Workshop. Freiberg. Germany. Available from: http://www.lst.min.dk/publikationer/publikationer/publicationer/campuk/cameng_ref.doc.

Ciarlet, M. & Estes, M.K. (2001) Rotavirus and calicivirus infections of the gastrointestinal tract. *Curr. Opin. Gastroentrol.* **17**, 10–16.

Ciftcioglu, N., Bjorklund, M., Kuroikoski, S., Bergstrom, K. & Kajander, E.O. (1999) Nanobacteria: an infectious cause for kidney stone formation. *Kidney Int.* **56**, 1893–1898.

Cisar, J.O., De Qi, X.U., Thompson, J., Swaim, W, Hu, L. & Kopecko, D.L. (2000) An alternative interpretation of nanobacteria-induced biomineralization. *Proc. Natl. Acad. Sci. U. S. A.* **97**, 11511–11515.

Claesson, B.E.B., Holmlund, D.E.W, Linghagen, C.A. & Matzsch, T.W. (1994) *Plesiomonas shigelloides* in acute cholecystitis: a case report. *J. Clin. Microbiol.* **20**, 985–987.

Claesson, M.J., van Sinderen, D. & O'Toole, P.W. (2008) *Lactobacillus* phylogenomics – towards a reclassification of the genus. *Int. J. Syst. Evol. Microbiol.* **58**, 2945–2954.

Codex Alimentarius Commission (CAC) (1993) *Codex Guidelines for the Application of the Hazard Analysis Critical Control Point (HACCP) System.* Joint FAO/WHO Codex Committee on Food Hygiene. WHO/FNU/FOS/93.3 Annex II.

Codex Alimentarius Commission (CAC) (1995) *Hazard Analysis Critical Control Point (HACCP) System and Guidelines for Its Application.* Alinorm 97/13, Annex to Appendix II.

Codex Alimentarius Commission (CAC) (1997a) *Hazard Analysis Critical Control Point (HACCP) System and Guidelines for Its Application.* Annex to CAC/RCP 1–1969, Rev. 3. 1997.

Codex Alimentarius Commission (CAC) (1997b) *Principles for the Development of Microbiological Criteria for Animal Products and Products of Animal Origin Intended for Human Consumption.* European Commission, Luxembourg.

Codex Alimentarius Commission (CAC) (1997c) *Principles for the Establishment and Application of Microbiological Criteria for Foods.* CAC/GL 21 – 1997.

Codex Alimentarius Commission (1999) *Principle and Guidelines for the Conduct of Microbiological Risk Assessment.* CAC/GL 30.

Codex Alimentarius Commission (CAC) (2008) Code of hygienic practice for powdered formulae for infants and young children. CAC/RCP 66-2008. Available from: http://www.codexalimentarius.net/download/standards/11026/CXP_066e.pdf.

Coeyn, T. & Vandamme, P. (2003) Intragenomic heterogeneity between multiple 16S ribosomal RNA operons in sequenced bacterial genomes. *FEMS Microbiol. Lett.* **228**, 45–48.

Coghlan, A. (1998) Deadly *E. coli* strains may have come form South America. *New Scientist*, 10 January, p. 12.

Coleman, M. & Marks, H. (1998) Topics in dose–response modelling. *J. Food Prot.* **61**, 1550–1559.

Collinge, J., Sidle, K.C.L., Meads, J., Ironside, J. & Hill, A.F. (1996) Molecular analysis of prion strain variation and the aetiology of 'new variant' CJD. *Nature* **383**, 685–690.

Colwell, R.R., Brayton, P., Herrington, D., Tall, B., Huq, A. & Levine, M.M. (1996) Viable but nonculturable *Vibrio cholerae* O1 revert to a cultivable state in the human intestine. *World J. Microbiol. Biotechnol.* **12**, 28–31.

Cone, L.A., Voodard, D.R., Schievert, P.M. & Tomory, G.S. (1987) Clinical and bacteriological observations of a toxic-shock-like syndrome due to *Streptococcus pyogenes*. *N. Engl. J. Med.* **317**, 146–149.

Corlett, D.A. (1998) HACCP User's Manual. A Chapman & Hall Food Science Title. An Aspen Publication. Aspen Publishers, Gaithersburg, Maryland, USA.

Corry, J., Mansfield, L.P. & Forsythe, S.J. (2002) Culture media for the detection of *Campylobacter* and related organisms. In: *Culture Media for Food Microbiology* (eds J.E.L. Corry, G.D.W. Curtis & R. Baird). Elsevier Science, The Netherlands.

Corthier, G., Delorme, C., Ehrlich, S.D. & Renault, P. (1998) Use of luciferase genes as biosensors to study bacterial physiology in the digestive tract. *Appl. Environ. Microbiol.* **64**, 2721–2722.

Cousens, S.N., Vynnycky, E., Zeidler, M., Will, R.G. & Smith, P.G. (1997) Predicting the CJD epidemic in humans. *Nature* **385**, 197–198.

Cowden, J.M., Wall, P.G., Adak, G., Evans, H., Le Baigue, S. & Ross, D. (1995) Outbreaks of foodborne infectious intestinal disease in England and Wales: 1992 and 1993. *CDR Rev.* **5**, R109–R117.

Cowell, N.D. & Morisetti, M.D. (1969) Microbiological techniques – some statistical aspects. *J. Sci. Food Agric.* **20**, 573–579.

Crockett, C.S., Haas, C.N., Fazil, A., Rose, J.B. & Gerba, C.P. (1996) Prevalence of shigellosis in the US: consistency with dose–response information. *Int. J. Food Microbiol.* **30**, 87–99.

Cronquist, A., Wedel, S., Albanese, B., *et al.* (2006) Multistate outbreak of *Salmonella* Typhimurium infections associated with eating ground beef – United States, 2004. *MMWR* **55**, 180–182.

Crutchfield, S.R., Buzby, J.C., Roberts, T. & Ollinger, M. (1999) Assessing the costs and benefits of pathogen reduction. *Food Rev.* **22**, 6–9.

Cuthbert, J. (2001) Hepatitis A: old and new. *Clin. Microbiol. Rev.* **14**, 38–58.

D'Aoust, J.-Y. (1994) *Salmonella* and the international food trade. *Int. J. Food Microbiol.* **24**, 11–31.

Dainty, R.H. (1996) Chemical/biochemical detection of spoilage. *Int. J. Food Microbiol.* **33**, 19–33.

Dalgaard, P., Gram, L. & Huss, H.H. (1993) Spoilage and shelf-life of cod fillets packed in vacuum or modified atmospheres. *Int. J. Food Microbiol.* **19**, 283–294.

Davidson, P.M. (1997) Chemical preservatives and natural antimicrobial compounds. In: *Food Microbiology: Fundamental and Frontiers* (eds M.P. Doyle, L.R. Beuchat & T.J. Montville), pp. 520–556. ASM Press. Washington, DC.

de Boer, E., Tilburg, J.J.H.C., Woodward, D.L., Lior, H. & Johnson, W.M. (1996) A selective medium for the isolation of *Arcobacter* from meats. *Lett. Appl. Microbiol.* **23**, 64–66.

de Vos, W.M. (1999) Safe and sustainable systems for food-grade fermentations by genetically modified lactic acid bacteria. *Int. Dairy J.* **9**, 3–10.

de Wit, J.N. & van Hooydonk, A.C.M. (1996) Structure, functions and applications of lactoperoxidase in natural antimicrobial systems. *Neth. Milk Dairy J.* **50**, 227–244.

de Wit, M.A.S., Koopmans, M.P.G., Kortbeek, L.M., van Leeuwen, N.J., Bartelds, A.I.M. & van Duyn-hoven, T.H.P. (2001) Gastroenteritis in sentinel general practices, the Netherlands. *Emerg. Infect. Dis.* **7**, 82–91.

Denyer, S.P., Gorman, S.P. & Sussman, M. (1993) *Microbial Biofilms*. The Society for Applied Bacteriology Technical Series No. 30. Blackwell Scientific Publications, London, Edinburgh, Boston, Melbourne, Paris, Berlin, Vienna.

Desselberger, U. (1998) Viral gastroenteritis. *Curr. Opin. Infect. Dis.* **11**, 565–575.

Dingle, K.E., Colles, F.M., Ure, R., *et al.* (2002) Molecular characterization of *Campylobacter jejuni* clones: a basis for epidemiologic investigation. *Emerg. Infect. Dis.* **8**, 949–955.

Diplock, A.T., Aggett, P.J., Ashwell, M., Bornet, F., Fern, E.B. & Roberfroid, M.B. (1999) Scientific concepts of functional foods in Europe: consensus document. *Br. J. Nutr.* **81**, S1–S27.

Dodd, C.E.R., Sharman, R.L., Bloomfield, S.F., Booth, I.R. & Stewart, G.S.A.B. (1997) Inimical processes: bacterial self-destruction and sub-lethal injury. *Trends Food Sci. Technol.* **8**, 238–241.

Druggan, P., Forsythe, S.J. & Silley, P. (1993) Indirect impedance for microbial screening in the food and beverage industries. In: *New Techniques in Food and Beverage Microbiology* (eds R.G. Kroll, A. Gilmour, M. Sussman). Society for Applied Bacteriology, Technical Series No. 31. Blackwell Science, Oxford.

Earnshaw, R.G., Appleyard, J. & Hurst, R.M. (1995) Understanding physical inactivation processes: combined preservation opportunities using heat, ultrasound and pressure. *Int. J. Food Microbiol.* **28**, 197–219.

EFSA (2007) Report of the task force on zoonoses data collection on the analysis of the baseline study on the prevalence of *Salmonella* in holdings of laying hen flocks of *Gallus gallus. Parma, Italy:* EFSA, 21 February 2007. Available from: http://www.efsa.europa.eu/en/science/monitoring_zoonoses/reports/report_finlayinghens.html.

Eklund, T. (1985) The effect of sorbic acid and esters of para-hydroxybenzoic acid on the proton motive force in *Escherichia coli* membrane vesicle. *J. Gen. Microbiol.* **131**, 73–76.

Entis, P. & Lerner, I. (1996) 24-hour presumptive enumeration of *Escherichia coli* O157:H7 in foods by using the ISO-GRID® method with SD-39 agar. *J. Food Prot.* **60**, 883–890.

Entis, P. & Lerner, I. (2000) Twenty-four-hour direct presumptive enumeration of *Listeria monocytogenes* in food and environmental samples using the ISO-GRID method with LM-137 agar. *J. Food Prot.* **63**, 354–363.

Eurosurveillance (2002) Laboratory capability in Europe for foodborne viruses. *Eurosurveillance* **7**(4), 323. Available from: http://www.eurosurveillance.org/ViewArticle.aspx?ArticleId=323.

Eurosurveillance (2008) Special double issue on antimicrobial resistance. *Eurosurveillance* **13**(46 & 47), 19039, 19043. Available from: http://www.eurosurveillance.org/images/dynamic/EE/V13N46/V13N46.pdf

Ewing, W.H. (1986) The taxonomy of Enterobacteriaceae, isolation of Enterobacteriaceae and preliminary identification. The genus *Salmonella*. In: *Identification of Enterobacteriaceae* (eds P. Edwards & W.H. Ewing), 4th edn, pp. 1–91, 181–318. Elsevier, New York.

Falik, E., Aharoni, Y., Grinberg, S., Copel, A. & Klein, J.D. (1994) Postharvest hydrogen peroxide treatment inhibits decay in eggplant and sweet red pepper. *Crop Prot.* **13**, 451–454.

Fankhauser, R.I., Noel, J.S., Monroe, S.S., Ando, T. & Glass, R.I. (1998) Molecular epidemiology of 'Norwalk-like virus' in outbreaks of gastroenteritis in the United States. *J. Infect. Dis.* **178**, 1571–1578.

Fanning, S. & Forsythe, S.J. (2008) Isolation and identification of *Enterobacter sakazakii*. Chapter 2. In: *Emerging Issues in Food Safety: Enterobacter Sakazakii* (eds J. Farber & S.J. Forsythe). ASM Press, Washington, DC.

FAO/WHO (1995) *Application of risk analysis to food standards issues.* Report of the Joint FAO/WHO Expert Consultation. WHO, Geneva. WHO/FNU/FOS/95.3.

FAO/WHO (1996) *Biotechnology and Food Safety.* Report of a joint FAO/WHO consultation. UN Food and Agriculture Organisation, Rome.

FAO/WHO (1997) *Risk Management and Food Safety.* Report of a Joint FAO/WHO Expert Consultation, Rome, Italy, 1997. FAO Food and Nutrition Paper, No 65.

FAO/WHO (1998) *The Application of Risk Communication to Food Standards and Safety Matters.* Report of a Joint FAO/WHO Expert Consultation, Rome, Italy, 1998. FAO Food and Nutrition Paper, No 70.

FAO/WHO (2000a) Activities on risk assessment of microbiological hazards in foods. Preliminary document: *WHO/FAO Guidelines on hazard characterization for pathogens in food and water.* Available from: http://www.ftp.fao.org/docrep/fao/006/y4666e/y4666e00.pdf.

FAO/WHO (2000b) Report of the Joint FAO/WHO expert consultation on risk assessment of microbiological hazards in foods, Rome, 17–21 July 2000. Available from: http://www.fao.org.ES/ESN/rskpage.htm.

FAO/WHO (2002) Risk assessment of *Salmonella* in eggs and broiler chickens. Microbiological Risk Assessment Series, No. 2. Available from: http://www.who.int/foodsafety/publications/micro/salmonella/en.

FAO/WHO (2004) *Enterobacter sakazakii* and other microorganisms in powdered infant formula. Meeting Report. Microbiological Risk Assessment Series, No. 6. Available from: http://www.who.int/foodsafety/publications/micro/mra6/en.

FAO/WHO (2006a) Development of practical risk management strategies based on microbiological risk assessment outputs. Report of a Joint FAO/WHO Consultation. Available from: http://www.who.int/foodsafety/micro/jemra/meetings/2005/en.

FAO/WHO (2006b) *Enterobacter sakazakii* and *Salmonella* in powdered infant formula: Meeting report, Microbiological Risk Assessment Series No.10. Available from: http://www.who.int/foodsafety/publications/micro/mra10/en.

Farber, J. & Forsythe, S.J. (2008) *Emerging Issues in Food Safety: Enterobacter sakazakii.* ASM Press, Washington, DC.

Farber, J.M. & Peterkin, P.I. (1991) *Listeria monocytogenes*: A food-borne pathogen. *Microbiol. Rev.* **55**, 476–511.

Farber, J.M., Ross, W.H. & Harwig, J. (1996) Health risk assessment of *Listeria monocytogenes* in Canada. *Int. J. Food Microbiol.* **31**, 145–156.

Farmer, J.J. III, Asbury, M.A., Hickman, F.W., Brenner, D.J. and The *Enterobacteriaceae* study group (1980) *Enterobacter sakazakii*: a new species of "*Enterobacteriaceae*" isolated from clinical specimens. *Intl. J. System. Bacteriol.* **30**, 569–584.

Fasoli, S., Marzotoo, M., Rizzotti, L., Rossi, F., Dellaglio, F. & Torriani, S. (2003) Bacterial composition of commercial probiotic products as evaluated by PCR-DGGE analysis. *Int. J. Food Microbiol.* **882**, 59–70.

Fazil, A.M., Lammerding, A. & Ellis, A. (2000) A quantitative risk assessment model for *Campylobacter jejuni* on chicken. Available from: http://www.who.int/fsf/mbriskassess/studycourse/index.html.

Feng, P., Lampel, K., Karch, H. & Whittam, T. (1998) Genotypic and phenotypic changes in the emergence of *E. coli.* O157:H7. *J. Infect. Dis.* **177**, 1750–1753.

Ferguson, N.M., Donnelly, C.A., Ghani, A.C. & Anderson, R.M. (1999) Predicting the size of the epidemic of the new variant of Creutzfeldt–Jakob disease. *Br. Food J.* **101**, 86–98.

Fischetti, V.A., Medaglini, D., Oggioni, M. & Pozzi, G. (1993) Expression of foreign proteins on Gram-positive commensal bacteria for mucosal vaccine delivery. *Curr. Opin. Biotechnol.* **4**, 603–610.

Fleet, G.H., Heiskanen, P., Reid, I. & Buckle, K.A. (2000) Foodborne viral illness – status in Australia. *Int. J. Food Microbiol.* **59**, 127–136.

Foegeding, P.M. & Busta, F.F. (1991) Chemical food preservatives. In: *Disinfection, Sterilization and Preservation* (ed. S. Block), pp. 802–832. Lea & Febiger, Philadelphia.

Folk, R.L. (1999) Nanobacteria and the precipitation of carbonate in unusual environments. *Sedimentary Geology*, **126**, 1–4.

Food and Drug Administration (FDA) (1999) Structure and initial data survey for the risk assessment of the public health impact of foodborne *Listeria monocytogenes*. Preliminary information available from: http://vm.cfsan.fda.gov/~dms/listrisk.html.

Food and Drug Administration (FDA) (2000a) Draft risk assessment on the human health impact of fluoroquinolone resistant campylobacter associated with the consumption of chicken. Revised 9 February 2000. Available from: http://www.fda.gov/cvm/Risk_asses.htm.

Food and Drug Administration (FDA) (2000b) Draft risk assessment on the public health impact of *Vibrio parahaemolyticus* in raw molluscan shellfish. Available from: http://www.who.int/foodsafety/publications/micro/mra8/en/index.html.

Food and Drug Administration (FDA) (2001) Draft assessment of the public health impact of foodborne *Listeria monocytogenes among selected categories of ready-to-eat foods*. Center for Food Safety and Applied Nutrition (FDA) and Food Safety Inspection Service (FSIS, USDA). Available from: http://www.who.int/foodsafety/publications/micro/mra_listeria/en/index.html.

Food and Drug Administration (FDA) Online (2008). Guidance for industry: guide to minimize food safety hazards for fresh fruits and vegetables. Available from: http://www.fda.gov/Food/GuidanceCompliance RegulatoryInformation/GuidanceDocuments/ProduceandPlanProducts/ucm064574.htm.

Food and Drug Administration/Food Safety Inspection Service (FDA/FSIS) (2003) Quantitative assessment of the relative risk to public health from foodborne *Listeria monocytogenes* among selected categories of ready-to-eat foods. US Department of Agriculture. Available from: http://www.fda.gov/Food/ScienceResearch/ResearchAreas/RiskAssessmentSafetyAssessment/default.htm.

Food Safety Inspection Service (FSIS) (1998) *Salmonella* Enteritidis Risk Assessment. Shell Eggs and Egg Products. Available from: http://www.europa.eu.int/comm/dg24/health/sc/scv/out26_en.html.

Food Standards Agency (FSA) (2001) A review of the evidence for a link between exposure to *Mycobacterium paratuberculosis* (MAP) and Crohn's Disease (CD) in humans. A report of the Food Standards Agency, London.

FoodNet (2009) Preliminary FoodNet data on the incidence of infection with pathogens transmitted commonly through food – 10 States, 2008. *MMWR* **58**, 333–337.

Forsythe, S.J. (2000) *The Microbiology of Safe Food*. Blackwell Science, Oxford. Companion available from: http://www.wiley.com/go/forsythe.

Forsythe, S. (2005) *Enterobacter sakazakii* and other bacteria in powdered infant milk formula. *Matern Child Nutr* **1**, 44–50.

Forsythe, S.J., Dolby, J.M., Webster, A.D.B. & Cole, J.A. (1988) Nitrate- and nitrite-reducing bacteria in the achlorhydric stomach. *J. Med. Microbiol.* **25**, 253–259.

Forsythe, S.J. & Hayes, P.R. (1998) *Food Hygiene, Microbiology and HACCP*. A Chapman & Hall Food Science Book. Aspen Publishers, Gaithersburg, MD.

Forsythe, S.J. (2006) *Arcobacter*. In: *Emerging Foodborne Pathogens* (eds Y. Motarjemi & M. Adams). Woodhead Publishing, Cambridge.

Fouts, D.E., Mongodin, E.F., Mandrell, R.E., *et al.* (2005) Major structural differences and novel potential virulence mechanisms from the genomes of multiple *Campylobacter* species. *PLoS Biol.* **3**, 72–85.

Frenzen, P.D. (2004) Deaths due to unknown foodborne agents. *Emerg. Infect. Dis.* **10**, 1536–1543.

Frenzen, P.D., Drake, A., Angulo, F.J. & the emerging infections program foodnet working group (2005). Economic cost of illness due to *Escherichia coli* O157 infections in the United States. *J. Food Prot.* **68**, 2623–2630.

FSIS (Food Safety Inspection Service) (1998) *Salmonella* Enteritidis Risk Assessment. Shell Eggs and Egg Products. Available from: http://www.fsis.usda.gov/ophs/risk/contents.htm.

Fuller, R. (1989) Probiotics in man and animals. *J. Appl. Bacteriol.* **66**, 365–378.

Galanis, E., Wong, D., Patrick, M., *et al.* (2006) Web-based surveillance and global *Salmonella* distribution, 2000–2002. *Emerg. Infect. Dis.* **12**, 381–388.

Gale, P., Young, C., Stanfield, G. & Oakes, D. (1998) A review: development of a risk assessment for BSE in the aquatic environment. *J. Appl. Microbiol.* **84**, 467–477.

Gale, P. (2001) A review: developments in microbiological risk assessment for drinking water. *J. Appl. Microbiol.* **91**, 191–205.

GATT (1994) The application of the Uruguay round of multilateral trade negotiations: the legal texts, World Trade Organization. ISBN: 92–870-1121–4.

Geeraerd, A.H., Valdramidis, V.P. & Van Impe, J.F. (2005) GInaFiT, a freeware tool to assess non-log-linear microbial survivor curves. *Int. J. Food Microbiol.* **102**, 95–105.

Gehring, A.G., Albin, D.M., Reed, S.A., Tu, S. & Brewster, J.D. (2008) An antibody microarray, in multiwell plate format, for multiplex screening of foodborne pathogenic bacteria and biomolecules *Anal. Bioanal. Chem.* **391**, 497–506.

Gerba, C.P., Rose, J.B. & Haas, C.N. (1996) Sensitive populations: who is at the greatest risk? *Int. J. Food Microbiol.* **30**, 87–99.

German, B., Schiffrin, E.J., Reniero, R., *et al.* (1999) The development of functional foods: lessons from the gut. *Trends Biotechnol.* **17**, 492–499.

Gibson, A.M., Bratchell, N. & Roberts, T.A. (1988) Predicting microbial growth: growth responses of salmonellae in a laboratory medium as affected by pH, sodium chloride, and storage temperature. *Int. J. Food Microbiol.* **6**, 155–178.

Gill, C.O. & Phillips, D.M. (1985) The effect of media composition on the relationship between temperature and growth rate of *Escherichia coli*. *Food Microbiol.* **2**, 285.

Golnazariuan, C.A., Donnelly, C.W., Pintauro, S.J., *et al.* (1989) Comparison *of* infectious dose of *Listeria monocytogenes* F5817 as determined for normal versus compromised C57B1/6J mice. *J. Food Prot.* **52**, 696–701.

Gould, G.W. (ed.) (1995) *New Methods of Food Preservation.* Chapman and Hall, London.

Gould, G.W. (1996) Methods for preservation and extension of shelf life. *Int. J. Food Microbiol.* **33**, 51–64.

Graham, A.F. & Lund, B.M. (1986) The effect of citric acid on growth of proteolytic strains of *Clostridium botulinum. J. Appl. Bacteriol.* **61**, 39–49.

Grant, I.R., Ball, H.J., Neill, S.D. & Rowe, M.T. (1996) Inactivation of *Mycobacterium paratuberculosis* in cow's milk at pasteurisation temperatures. *Appl. Environ. Microbiol.* **62**, 631–636.

Granum, A.F. & Lund, B.M. (1997a) The effect of citric acid on growth of proteolytic strains of *Clostridium botulinum. J. Appl. Bacteriol.* **61**, 39–49.

Granum, P.E. & Lund, T. (1997b) MiniReview. *Bacillus cereus* and its food poisoning toxins. *FEMS Microbiol. Lett.* **157**, 223–228.

Grau, F.H. & Vanderlinde, P.B. (1992) Aerobic growth of *Listeria monocytogenes* on beef lean and fatty tissue: equations describing the effects of temperature and pH. *J. Food Prot.* **55**, 4.

Graves, D.J. (1999) Powerful tools for genetic analysis come of age. *TIBTECH* **17**, 127–134.

Graves, L.M. & Swaminathan, B. (2001) PulseNet standardized protocol for subtyping *Listeria monocytogenes* by macrorestriction and pulsed-field gel electrophoresis. *Int. J. Food Microbiol.* **65**, 55–62.

Green, D.H., Wakeley, P.R., Page, A., *et al.* (1999) Characterization of two *Bacillus* probiotics. *Appl. Environ. Microbiol.* **65**, 4288–4291.

Green, K.Y., Ando, T., Balayan, M.S., *et al.* (2000) Taxonomy of the caliciviruses. *J. Infect. Dis.* **181** (Suppl. 2), S322–S330.

Greig, J.D., Todd, E.C.D., Bartleson, C.A. & Michaels, B.S. (2007) Outbreaks where food workers have been implicated in the spread of foodborne disease. Part 1. Description of the problem, methods, and agents involved. *J. Food Prot.* **70**, 1752–1761.

Guarner, F. & Schaafsma, G.J. (1998) Probiotics. *Int. J. Food Microbiol.* **39**, 237–238.

Guzewich, J. & Ross, M.P. (1999) Evaluation of risks related to microbiological contamination of ready-to-eat food by food preparation workers and the effectiveness of interventions to minimize those risks. Available from: http://www.cfsan.fda.gov/~ear/rterisk.html.

Haas, C. (1999) On modeling correlated random variables in risk assessment. *Risk Anal.* **19**, 1205–1213.

Haas, C.N. (1983) Estimation of the risk due to low doses of microorganisms: a comparison of alternative methodologies. *Am. J. Epidemiol.* **118**, 573–582.

Haas, C.N., Rose, J.B. & Gerba, C.P. (1999) Quantitative Microbial Risk Assessment. John Wiley & Sons, Inc., New York, Chichester, Weinheim, Brisbane, Singapore & Toronto.

Haas, C.N., Thayyar-Madabusi, A., Rose, J.B. & Gerba, C.P. (2000) Development of a dose–response relationship for *Escherichia coli* 0157:H7. *Int. J. Food Microbiol.* **56**, 153–159.

Hain, T., Chatterjee, S.S., Ghai, R., *et al.* (2007) Pathogenomics of *Listeria* spp. *Int. J. Med. Microbiol.* **297**, 541–557.

Haire, D.L., Chen, G.M., Janzen, E.G., Fraser, L. & Lynch, J.A. (1997) Identification of irradiated foodstuffs: a review of the recent literature. *Food Res. Int.* **30**, 249–264.

Hald, T., Vose, D. & Wegener, H.C. (2001) Quantifying the contribution of animal-food sources to human salmonellosis in Denmark in 1999. Available from: http://www.lst.min.dk/publikationer/publikationer/publikationer/campuk/cameng_ref.doc.

Haldenwang, W.G. (1995) The sigma factors of *Bacillus subtilis. Microbiol. Rev.* **59**, 1–30.

Halliday, M.L., Kang, L.Y. & Zhou, T.K. (1991) An epidemic of hepatitis A attributable to the ingestion of raw clams in Shanghai, China. *J. Infect. Dis.* **164**, 852–859.

Hamilton-Miller, J.M.T., Shah, S. & Winkler, J.T. (1999) Public health issues arising from microbiological and labelling quality of foods and supplements containing probiotic organisms. *Public Health Nutr.* **2**, 223–229.

Harrigan, W.F. & Park, R.A. (1991) *Making Safe Food. A Management Guide for Microbiological Quality.* Academic Press, London.

Hart, C.A. & Cunliffe, N.A. (1999) Viral gastroenteritis. *Curr. Opin. Infect. Dis.* **12**, 447–457.

Hartman, P.A. (1997) The evolution of food microbiology. In: *Food Microbiology: Fundamentals and Frontiers* (eds M.P. Doyle, L.R. Beuchat & T.J. Montville), pp. 3–13. ASM Press, Washington, DC.

Hattoir, M. & Taylor, T.D. (2009) The human intestinal microbiome: a new frontier of human biology. *DNA Res.* **16**, 1–12.

Hauschild, A.H.W., Hilsheimer, R., Jarvis, G. & Raymond, D.P. (1982) Contribution of nitrite to the control of *Clostridium botulinum* in liver sausage. *J. Food Protect.* **45**, 500–506.

Havenaar, R. & Huis in't Veld, J.H. (1992) Probiotics: a general view. In: *The Lactic Acid Bacteria in Health and Disease, The Lactic Acid Bacteria*, Vol. **1** (ed. B.J. Wood), pp. 209–224. Chapman and Hall, New York, London.

Heitzier, A., Kohler, H.E., Reichert, P. & Hamer, G. (1991) Utility of phenomenological models for describing temperature dependence of bacterial growth. *Appl. Environ. Microbiol.* **57**, 2656.

Helms, M., Vastrup, P., Gerner-Smidt, P. & Molbak, K. (2003) Short and long term mortality associated with foodborne bacterial gastrointestinal infections: registry based study. *BMJ* **326**, 357–362.

Henderson, B., Wilson, W., McNab, R. & Lax, A. (1999) *Cellular Microbiology. Bacteria-Host Interactions in Health and Disease.* John Wiley & Sons, Chichester.

Hendrickx, M., Ludikhuyze, L., Vanden Broeck, I. & Weemaes, C. (1998) Effects of high pressure on enzymes related to food quality. *Trends Food Sci. Technol.* **9**, 197–203.

Hennesy, T.W., Hedberg, C.W., Slutsker, L., *et al.* (1996) A national outbreak of *Salmonella enteritidis* infections from ice cream. *New Engl. J. Med.* **334**, 1281–1286.

Hermon-Taylor, J., Bull, T.J., Sheridan, J.M., Cheng, J., Stellakis, M.L. & Sumar, N. (2000) The causation of Crohn's disease *Mycobacterium avium* subspecies *paratuberculosis. Can. J. Gastroenterol.* **14**, 521–539.

Hermon-Taylor, J. (2001) *Mycobacterium avium* subspecies *paratuberculosis*: the nature of the problem. *Food Control* **12**, 331–334.

Hermon-Taylor, J. (2009) *Mycobacterium avium* subspecies *paratuberculosis,* Crohn's disease and the Doomsday Scenario. *Gut Pathog.* **1**, 15.

Hill, A.F., Desbruslais, M., Joiner, S., *et al.* (1997) The same prion strain causes vCJD and BSE. *Nature* **389**, 448–450.

Hirasa, K. & Takemasa, M. (1998) Antimicrobial and antioxidant properties of spices. In: *Spice Science and Technology*, pp. 163–200. Marcel Dekker, New York.

Hitchins, A.D., Feng, P., Watkins, W.D., Rippey, S.R. & Chandler, L.A. (1998) *Escherichia coli* and the coliform bacteria. Chapter 4. In: *Food and Drug Administration Bacteriological Analytical Manual* (ed. R.L. Merker), 8th edn, revision A, Chapter 4. AOAC International, Gaithersburg, MD.

Hjelle, J.T., Miller-Hjelle, M.A., Poxton, I.R., *et al.* (2000) Endotoxin and nanobacteria in polycystic kidney disease. *Kidney Int.* **57**, 2360–2374.

Hjertqvist, M., Johansson, A., Svensson, N., Abom, P.E., Magnusson, C., Olsson, M., Hedlund, K.O. & Andersson, Y. (2006) Four outbreaks of norovirus gastroenteritis after consuming raspberries, Sweden, June–August 2006. *Euro Surveill.* **11**(9), E060907.1.

Holah, J.T. & Thorpe, R.H. (1990) Cleanability in relation to bacterial retention on unused and abraded domestic sink materials. *J. Appl. Bacteriol.* **69**, 599–608.

Holcomb, D.L., Smith, M.A., Ware, G.O., Hung, Y.C., Brackett, R.E. & Doyle, M.P. (1999) Comparison of six dose–response models for use with food-borne pathogens. *Risk Anal.* **19**, 1091–1100.

Holmes, C.J. & Evans, R.C. (1989) Resistance of bacterial biofilms to antibiotics. *J. Antimicrobiol. Chemother.* **24**, 84.

Holyoak, C.D., Stratford, M., McMullin, Z., *et al.* (1996) Activity of the plasma-membrane H+-ATPase and optimal glycolytic flux are required for rapid adaption and growth in the presence of the weak acid preservative sorbic acid. *Appl. Environ. Microbiol.* **62**, 3158–3164.

Hoornstra, E. & Notermans, S. (2001) Quantitative microbiological risk assessment. *Int. J. Food Microbiol.* **66**, 21–29.

Houf, K., Devriese, L.A., De Zutter, L., Van Hoof, J. & Vandamme, P. (2001) Susceptibility of *Arcobacter butzleri*, *Arcobacter cryaerophilus,* and *Arcobacter skirrowii* to antimicrobial agents used in selective media. *J. Clin. Microbiol.* **39**, 1654–1656.

Hugenholtz, J. & Kleerebezem, M. (1999) Metabolic engineering of lactic acid bacteria: overview of the approaches and results of pathway rerouting involved in food fermentations. *Curr. Opin. Biotechnol.* **10**, 492–497.

Huis in't Veld, J.H.J. (1996) Microbial and biochemical spoilage of foods: an overview. *Int. J. Food Microbiol.* **33**, 1–18.

Huisman, G.W. & Kolter, R. (1994) Sensing starvation: a homoserine lactone-dependent signalling pathway in *Escherichia coli. Science* **265**, 537–539.

Humphrey, T., O'Brien, S. & Madsen, M. (2007) Campylobacters as zoonotic pathogens: a food production perspective. *Int. J. Food Microbiol.* **117**, 237–257.

Hutin, Y.J., Pool, V., Cramer, E.H., *et al.* (1999) A multistate, foodborne outbreak of hepatitis A. National hepatitis A investigation team. *New Engl. J. Med.* **340**, 595–602.

Hyytiä-Trees, E.K., Cooper, K., Ribot, E.M. & Gerner-Smidt, P. (2007) Recent developments and future prospects in subtyping of foodborne bacterial pathogens. *Future Microbiol.* **2**, 175–185.

Ibarra-Sanchez, L.S., Alvarado-Casillas, S., Rodriguez-Garcia, M.O., Martinez-Gonzalez, N.E. & Castillo, A. (2004) Internalization of bacterial pathogens in tomatoes and their control by selected chemicals. *J. Food Prot.* **67**, 1353–1358.

IFBC (1990) Biotechnologies and food: assuring the safety of foods produced by genetic modification. *Regul. Toxicol. Pharmacol.* **12**, 3.

ILSI (2000) Revised framework for microbial risk assessment. An ILSI Risk Science Institute workshop report. International Life Sciences. Available from: http://www.ilsi.org/file/mrabook.pdf.

Inouye, S., Yamashita, K., Yamadera, S., Yoshokawa, M., Kato, N. & Okabe, N. (2000) Surveillance of viral gastroenteritis in Japan: pediatric cases and outbreak incidencts. *J. Infect. Dis.* **181**, S270–S274.

International Commission on Microbiological Specifications for Foods (ICMSF) (1968, revised 1978, 1982, 1988) *Microorganisms in Foods: Their Significance and Methods of Enumeration*, 2nd edn. (1978) *Microorganisms in Foods, Book 1.* University of Toronto Press, Toronto.

International Commission on Microbiological Specifications for Foods (ICMSF) (1974, revised 1986) *Microorganisms in Foods. Book 2. Sampling for Microbiological Analysis: Principles and Specific Applications.* University of Toronto Press, Toronto.

International Commission on Microbiological Specifications for Foods (ICMSF) (1980a) *Microorganisms in Foods. Book 3. Factors Affecting the Life and Death of Microorganisms. Vol. 1. Microbial Ecology of Foods.* Academic Press, New York.

International Commission on Microbiological Specifications for Foods (ICMSF) (1980b) *Food Commodities. Microbial Ecology of Foods. Vol. 2.* Academic Press, New York.

International Commission on Microbiological Specifications for Foods (ICMSF) (1988) *Microorganisms in Foods. Book 4. Application of the Hazard Analysis Critical Control Point (HACCP) System to Ensure Microbiological Safety and Quality.* Blackwell Scientific Publications, Oxford.

International Commission on Microbiological Specifications for Foods (ICMSF) (1996a) *Micro-organisms in Foods. Book. 5. Characteristics of Microbial Pathogens.* Blackie Academic & Professional, London.

International Commission on Microbiological Specification for Foods (ICMSF) (1996b) The International Commission on Microbiological Specifications for Foods: update. *Food Control* **7**, 99–101.

International Commission on Microbiological Specification for Foods (ICMSF) (1997) Establishment of microbiological safety criteria for foods in international trade. *World Health Stat. Q.* **50**, 119–123.

International Commission on Microbiological Specification for Foods (ICMSF) (1998a) *Microorganisms in Foods. Book 6. Microbial Ecology of Food Commodities.* Blackie Academic & Professional, London.

International Commission on Microbiological Specification for Foods (ICMSF) (1998b) Potential application of risk assessment techniques to microbiological issues related to international trade in food and food products. *J. Food Prot.* **61**, 1075–1086.

International Commission on Microbiological Specification for Foods (ICMSF) (1998c) Principles for the establishment of microbiological food safety objectives and related control measures. *Food Control* **9**, 379–384.

International Commission on Microbiological Specification for Foods (ICMSF) (2002) *Microorganisms in Foods. Book 7. Microbiological Testing in Food Safety Management.* Kluwers Academic/Plenum Publishers, New York.

International Health Regulations (IHR) (2005) Available from: http://www.who.int/csr/ihr/IHRWHA58_3-en.pdf.

Institute of Food Science and Technology (IFST) (1999) *Development and Use of Microbiological Criteria for Foods.* Institute of Food Science & Technology (UK), London.

International Life Science Institute (ILSI) (1998) *Food Safety Management Tools* (eds J.L. Jouve, M.F. Stringer & A.C. Baird-Parker). Report prepared under the responsibility of ILSI Europe Risk Analysis in Microbiology task force. International Life Sciences Institute.

International Life Science Institute (ILSI) (2000) *Revised Framework for Microbial Risk Assessment.* An ILSI Risk Science Institute workshop report. International Life Sciences. Available from: http://www.ilsi.org/file/mrabook.pdf.

ISO (International Standardisation Organisation) (1994) ISO 9000 Series of Standards: ISO 9000: Quality Management and Quality Assurance Standards, Part 1: Guidelines for selection and use. ISO 9001: Quality Systems – Model for Quality Assurance in design/development, production, installation and servicing. ISO 9002: Quality Systems – Model for Quality Assurance in production and installation. ISO 9004–1: Quality Management and Quality System elements, Part 1: guidelines. ISO 8402 Standard: Quality Management and Quality Assurance Standards – Guidelines for selection and use – vocabulary.

ISO 10272-1:2006. Microbiology of food and animal feeding stuffs – horizontal method for detection and enumeration of *Campylobacter* spp. Part 1: Detection method.

ISO 11290–1:1996 and 2:1998 (1996, 1998) Microbiology of food and animal feeding stuffs – Horizontal method for the detection and enumeration of *Listeria monocytogenes*. Part 1 – Detection method. Part 2: Enumeration method.

ISO 16654:2001 (2001) Microbiology of food and animal feeding stuffs – Horizontal method for the detection of *Escherichia coli* O157.

ISO 21567:2004 (2004) Microbiology of food and animal feeding stuffs – Horizontal method for the detection of *Shigella* spp.

Isolauri, E., Juntunen, M., Rautanen, T., Sillanaukee, P. & Koivula, T. (1991) A *Lactobacillus* strain (*Lactobacillus* GG) promotes recovery from acute diarrhoea in children. *Paediatrics.* **88**, 90–97.

Iversen, C., Druggan, P. & Forsythe, S.J. (2004) A selective differential medium for *Enterobacter sakazakii. Int. J. Food Microbiol.* **96**, 133–139.

Iversen, C. & Forsythe, S. (2007) Comparison of media for the isolation of *Enterobacter sakazakii. Appl. Environ. Microbiol.* **73**, 48–52.

Iversen, C., Mullane, N., McCardell, B., Tall, B.D., Lehner, A., Fanning, S., Stephan, R. & Joosten, H. (2008) *Cronobacter* gen. nov., a new genus to accommodate the biogroups of *Enterobacter sakazakii*, and proposal of *Cronobacter sakazakii* gen. nov., comb. nov., *Cronobacter malonaticus* sp. nov., *Cronobacter turicensis* sp. nov., *Cronobacter muytjensii* sp. nov., *Cronobacter dublinensis* sp. nov., *Cronobacter genomospecies* 1, and of three subspecies, *Cronobacter dublinensis* subsp. *dublinensis* subsp. nov., *Cronobacter dublinensis* subsp. *lausannensis* subsp. nov. and *Cronobacter dublinensis* subsp. *lactaridi* subsp. nov. *Intl. J. System. Evol. Microbiol.* **58**, 1442–1447.

Jacobs-Reitsma, W., Kan, C. & Bolder, N. (1994) The induction of quinolone resistance in *Campylobacter* bacteria in broilers by quinolone treatment. *Lett. Appl. Microbiol.* **19**, 228–231.

JEMRA (2000) Preliminary document: WHO/FAO guidelines on hazard characterization for pathogens in food and water. Available from: http://www.fao.org/docrep/006/y4666e00.htm.

JEMRA (2003) *Risk Assessment of Salmonella in Eggs and Broiler Chickens. MRA Series, Number 2.* WHO, Geneva, Switzerland.

JEMRA (2008) Risk assessment for *Enterobacter sakazakii* in powdered infant formula. Available from: http://www.mramodels.org/ESAK/default.aspx.

Jessen, B. (1995) Start cultures for meat fermentation. In: *Fermented Meats* (eds G. Campbell-Platt & P.E. Cook), pp. 130–159. Blackie Academic & Professional, Glasgow.

448 **The microbiology of safe food**

Jeyamkondan, S., Jayas, D.S. & Holley, R.A. (1999) Pulsed electric field processing of foods: a review. *J. Food Prot.* **62**, 1088–1096.

Jonas, D.A., Antignac, E., Antoine, J.-M., *et al.* (1996) The safety assessment of novel foods. *Food Chem. Toxicol.* **34**, 931–940.

Jouve, J.L., Stringer, M.F. & Baird-Parker, A.C. (1998) *Food Safety Management Tools.* Report prepared under the responsibility of ILSI Europe Risk Analysis in Microbiology task force. International Life Sciences Institute.

Jouve, J.L. (2001) Reducing the microbiological food safety risk: a major challenge for the 21st century. WHO Strategic Planning Meeting, Geneva, 20–21 February 2001. Available from: http://www.who.int/fsf/mbriskassess/index.htm.

Juven, B.J. & Pierson, M.D. (1996) Antibacterial effects of hydrogen perioxide and methods for its detection and quantification. *J. Food Prot.* **59**, 1233–1241.

Kaila, M., Isolauri, E., Soppi, E., Vitanenen, V., Lane, S. & Arvilommi, H. (1992) Enhancement of circulating antibody secreting cell response in human diarrhoea by a human *Lactobacillus* strain. *Pediatr. Res.* **32**, 141–144.

Kaila, M., Isolauri, E., Saaxelin, M., Arvilommi & Vesikari, T. (1995) Viable versus inactivated *Lactobacillus* strain GG in acute rotavirus diarrhoea. *Arch. Dis. Child.* **72**, 51–53.

Kalchayanand, N., Sikes, A., Dunne, C.P. & Ray, B. (1998) Factors influencing death and injury of foodborne pathogens by hydrostatic pressure-pasteurization. *Food Microbiol.* **15**, 207–214.

Kapikan, A.Z., Wyatt, R.G., Dolin, R., Thornhill, T.S., Kalica, A.R. & Channock, R.M. (1972) Visualisation by immune electron microscopy of a 27 nm particle associated with acute infectious nonbacterial gastroenteritis. *J. Virol.* **10**, 1075–1081.

Kaplan, J.E., Feldman, R., Champbell, D.S., Lookabaugh, C. & Gary, G.W. (1982) The frequency of a Norwalk-like pattern of illness in outbreaks of acute gastro-enteritis. *Am. J. Publ. Health* **72**, 1329–1332.

Karoonuthaisiri, N., Charlermroj, R., Uawisetwathana, U., Luxananil, P., Kirtikara, K. & Gajananandana, O. (2009) Development of antibody array for simultaneous detection of foodborne pathogens. *Biosens Bioelectron.* **24**, 1641–1648.

Kelly, L., Anderson, W. & Snary, E. (2000) Preliminary report: exposure assessment of *Salmonella* spp. in broilers. JEMRA. Available from: http://www.fao.org/WAICENT/FAOINFO/ECONOMIC/ESN/pagerisk/mra005.pdf.

Kilsby, D. (1982) Sampling schemes and limits. In: *Meat Microbiology* (ed. M.H. Brown), pp. 387–421. Applied Science Publishers, London.

Kilsby, D.C., Aspinall, L.J. & Baird-Parker, A.C. (1979) A system for setting numerical microbiological specifications for foods. *J. Appl. Bacteriol.* **46**, 591–599.

Klaenhammer, T.R. (1993) Genetics of bacteriocins produced by lactic acid bacteria. *FEMS Microbiol. Rev.* **12**, 39–86.

Klaenhammer, T.R. & Kullen, M.J. (1999) Selection and design of probiotics. *Int. J. Food Microbiol.* **50**, 45–57.

Klapwijk, P.M., Jouve, J.-L. & Stringer, M.F. (2000) Microbiological risk assessment in Europe: the next decade. *Int. J. Food Microbiol.* **58**, 223–230.

Kleerebezem, M., Quadri, L.E.N., Kuipers, O.P. & De Vos, W.M. (1997) Quorum sensing by peptide pheromones and two-component signal-transduction systems in Gram-positive bacteria. *Mol. Microbiol.* **24**, 895–904.

Koopmans, M. & Duizer, E. (2002) *Foodborne Viruses: An Emerging Problem.* ILSI Press. Available from: http://www.ilsi.org/file/RPFoodbornvirus.pdf.

Koopmans, M., von Bonsdorff, C-H., Vinjé, J., de Medici, D. & Monroe, S. (2002) Foodborne viruses. *FEMS Microbiol. Rev.* **746**, 1–19.

Kothary, M.H. & Babu, U.S. (2001) Infective dose of foodborne pathogens in volunteers: a review. *J. Food Saf.* **21**, 49–73.

Kramer, J.M. & Gilbert, R.J. (1989) *Bacillus cereus* and other *Bacillus* species. In: *Foodborne Bacterial Pathogens* (ed. M.P. Doyle), pp. 21–70. Marcel Dekker, New York.

Krysinski, E.P., Brown, L.J. & Marchisello, T.J. (1992) Effect of cleaners and sanitizers of *Listeria monocytogenes* attached to product contact surfaces. *J. Food Prot.* **55**, 246–251.

Kuchenmüller, T., Hird, S., Stein, C., Kramarz, P., Nanda, A. & Havelaar, A.H. (2009) Estimating the globan burden of foodborne diseases – a collaborative effort. *Eurosurveillance* **14**(18), 19195. Available from: http://www.eurosurveillance.org/ViewArticle.aspx?Articleld = 19195.

Kuipers, O.P., de Ruyter, P.G.G.A., Kleerebezem, M. & de Vos, W.M. (1997) Controlled overporduction of proteins by lactic acid bacteria. *TIBTECH* **15**, 135–140.

Kuipers, O.P. (1999) Genomics for food biotechnology: prospects of the use of high-throughput technologies for the improvement of food microorganisms. *Curr. Opin. Microbiol.* **10**, 511–516.

Kumar, C.G. & Anand, S.K. (1998) Significance of microbial biofilms in food industry: a review. *Int. J. Food Microbiol.* **42**, 9–27.

Kwon, Y.M. & Ricke, S.C. (1998) Induction of acid resistance of *Salmonella typhimurium* by exposure to short-chain fatty acids. *Appl. Env. Microbiol.* **64**, 3458–3463.

Kyriakides, A. (1992) ATP bioluminescence applications for microbiological quality control in the dairy industry. *J. Soc. Dairy Technol.* **45**, 91–93.

Lake, R., Hudson, A., Cressey, P. & Nortje, G. (2002) Risk profile: *Listeria monocytogenes* in processed ready-to-eat meats. Report prepared for New Zealand Food Safety Authority by contract for scientific services. Institute of Environmental Science & Research Limited (ESR), Christchurch, NZ. Available from: http://www.nzfsa.govt.nz/science/risk-profiles/listeria-in-rte-meat.pdf.

Lammerding, A.M. & Paoli, G.M. (1997) Quantitative risk assessment: an emerging tool for emerging foodborne pathogens. *Emerg. Infect. Dis.* **3**, 483–487.

Langeveld, L.P.M., van Spoosen, W.A., van Beresteijn, E.C.H. & Notermans, S. (1996) Consumption by healthy adults of pasteurised milk with a high concentration of *Bacillus cereus*: a double-blind study. *J. Food Prot.* **59**, 723–726.

Lazcka, O., Del Campo, F.J. & Muñoz, F.X. (2007) Pathogen detection: a perspective of traditional methods and biosensors. *Biosens. Bioelectron.* **22**, 1205–1217.

Le Chevallier, M.W., Cawthon, C.D. & Lee, R.G. (1988) Inactivation of biofilm bacteria. *Appl. Environ. Microbiol.* **54**, 2492–2499.

Leclercq-Perlat, M.-N. & Lalande, M. (1994) Cleanability in relation to surface chemical composition and surface finishing of some materials commonly used in food industries. *J. Food Eng.* **23**, 501–517.

Le Minor, L. (1988) Typing *Salmonella* species. *Euro. J. Clin. Microbiol. Infect. Dis.* **7**, 214–218.

Lederberg, J. (1997) Infectious disease as an evolutionary paradigm. *Emerg. Infect. Dis.* **3**, 417–423.

Lees, D. (2000) Viruses and bivalue shellfish. *Int. J. Food Microbiol.* **59**, 81–116.

Li, S.Z., Marquardt, R.R. & Abramson, D. (2000) Immunochemical detection of molds: a review. *J. Food Prot.* **63**, 281–291.

Lin, Y.H., Chen, S.H., Chuang, Y.C., Lu, Y.C., Shen, T.Y., Chang, C.A., Lin, C.S. (2008) Disposable amperometric immunosensing strips fabricated by Au nanoparticles-modified screen-printed carbon electrodes for the detection of foodborne pathogen *Escherichia coli* O157:H7. *Biosens. Bioelectron.* **23**, 1832–1837.

Lindgren, S.E. & Dobrogosz, W.J. (1990) Antagonistic activities of lactic acid bacteria in food and feed fermentations. *FEMS Microbiol. Rev.* **87**, 149–163.

Lindqvist, R. & Westöö, A. (2000) Quantitative risk assessment for *Listeria monocytogenes* in smoked or gravad salmon and rainbow trout in Sweden. *Int. J. Food Microbiol.* **58**, 181–196.

Lindsay, J.A. (1997) Chronic sequelae of foodborne disease. *Emerg. Infect. Dis.* **3**, 443–452.

Linkous, D.A. & Oliver, J.D. (1999) Pathogenesis of *Vibrio vulnificus*. *FEMS Microbiol. Lett.* **174**, 207–214.

Lotong, N. (1998) Koji. In: *Microbiology of Fermented Foods* (ed. B.J.B. Wood), pp. 659–695. Blackie Academic & Professional, London.

Lund, B.M, Graham, A.F., George, S.M. & Brown, D. (1990) *The combined effect of inoculation temperature, pH and sorbic acid on the probability of growth of non-proteolytic type B Clostridium botulinum. J. Appl. Bacteriol.* **69**, 481–492.

Lundgren, O. & Svensson, L. (2001) Pathogenesis of rotavirus diarrhoea. *Microbes Infect.* **3**, 1145–1156.

Luo, Y., Han, Z., Chin, S.M. & Linn, S. (1994) Three chemically distinct types of oxidants formed by iron mediated Fenton reactions in the presence of DNA. *Proc. Natl. Acad. Sci. U. S. A.* **91**, 12438–12442.

Ly, K.T. & Casanova, J.E. (2007) Mechanisms of *Salmonella* entry into host cells. *Cell. Microbiol.* **9**, 2103–2111.

Mack, D.R., Michail, S., Wei, S., McDougall, L. & Hollingsworth, M.A. (1999) Probiotics inhibit enteropathogenc *E. coli* adherence *in vitro* by inducing intestinal mucin gene expression. *Am. J. Physiol.* **276**, G941–G950.

Majowicz, SE., Dore, K., Flint, J.A, Edge, VL., Read, S., Buffett, M.C., *et al.* (2004) Magnitude and distribution of acute, self-reported gastrointestinal illness in a Canadian community. *Epidemiol. Infect.* **132**, 607–617.

Makarova, K.S. & Koonin, E.V. (2007) Evolutionary genomics of lactic acid bacteria. *J. Bacteriol.* **189**, 1199–1208.

Manafi, M. (2000) New developments in chromogenic and fluorogenic culture media. *Int. J. Food Microbiol.* **60**, 205–218.

Mansfield, L.P. & Forsythe, S.J. (1996) Collaborative ring-trial of Dynabeads® anti-Salmonella for immunomagnetic separation of stressed *Salmonella* cells from herbs and spices. *Int. J. Food Microbiol.* **29**, 41–47.

Mansfield, L.P. & Forsythe, S.J. (2000a) Arcobacters, newly emergent human pathogens. *Rev. Med. Microbiol.* **11**, 161–170.

Mansfield, L.P. & Forsythe, S.J. (2000b) Salmonellae detection in foods. *Rev. Med. Microbiol.* **11**, 37–46.

Mansfield, L.P. & Forsythe, S.J. (2001) Demonstration of the Rb_1 lipopolysaccharide core structure in *Salmonella* strains with the monoclonal antibody M105. *J. Med. Microbiol.* **50**, 339–344.

Marie, D., Brussard, C.P.D., Thyhaug, R., Bratbak, G., Vaulot, D. (1999) Enumeration of marine viruses in culture and natural samples by flow cytometry. *Appl. Environ. Microbiol.* **65**, 45–52.

Marks, H.M., Coleman, M.E., Lin, J.C.-T. & Roberts, T. (1998) Topics in microbial risk assessment: dynamic flow tree process. *Risk Anal.* **18**, 309–328.

Marks, P., Vipond, I., Varlisle, D., Deakin, D., Fey, R. & Caul, E. (2000) Evidence for airborne transmission of NLV in a hotel restaurant. *Epidemiol. Infect.* **124**, 481–487.

Marshall, D.L. & Schmidt, R.H. (1988) Growth of *Listeria monocytogenes* at 10°C in milk preincubated with selected pseudomonads. *J. Food Protect.* **51**, 277.

Marteau, P., Flourie, B., Pochart, P., Chastang, C., Desjeux, J.F. & Rambeau, J.C. (1990) Effect of the microbial lactase activity in yogurt on the intestinal absorption of lactose: an *in vivo* study in lactase-deficient humans. *Br. J. Nutr.* **64**, 71–79.

Marteau, P., Vaerman, J.P., Dehennin, J.P., *et al.* (1997) Effects of intracellular perfusion and chronic ingestion of *Lactobacillus johnsonii* strain La1 on serum concentrations and jejunal secretions of immunoglobulins and serum proteins in healthy humans. *Gastroenterol. Clin. Biol.* **21**, 293–298.

Martin, S.A., Wallsten, T.S. & Beaulieu, N.D. (1995) Assessing the risk of microbial pathogens: application of a judgment-encoding methodology. *J. Food Prot.* **58**, 289–295.

Mast, E.E. & Alter, M.J. (1993) Epidemiology of viral hepatitis. *Semin. Virol.* **4**, 273–283.

Matsuzaki, T. (1998) Immunomodulation by treatment with *Lactobacillus casei* strain Shirota. *Int. J. Food Microbiol.* **41**, 133–140.

McClure, P.J., Boogard, E., Kelly, T.M., Baranyi, J. & Roberts, T.A. (1993) A predictive model for the combined effects of pH, sodium chloride and temperature, on the growth of *Brochothrix thermosphacta*. *Int. J. Food Microbiol.* **19**, 161–178.

McCullough, N.B. & Eisele, C.W. (1951a) Experimental human salmonellosis. I. Pathogenicity of strains of *Salmonella meleagridis* and *Salmonella annatum* obtained from spray-dried whole egg. *J. Infect. Dis.* **88**, 278–289.

McCullough, N.B. & Eisele, C.W. (1951b) Experimental human salmonellosis. II. Pathogenicity of strains of *Salmonella newport*, *Salmonella derby* and *Salmonella bareilly* obtained from spray-dried whole egg. *J. Infect. Dis.* **89**, 209–213.

McCullough, N.B. & Eisele, C.W. (1951c) Experimental human salmonellosis. III. Pathogenicity of strains of *Salmonella pullorum* obtained from spray-dried whole egg. *J. Infect. Dis.* **89**, 259–266.

McDonald, K. & Sun, D-W. (1999) Predictive food microbiology for the meat industry: a review. *Int. J. Food Microbiol.* **52**, 1–27.

McLauchlin, J. (1990a) Distribution of serovars of *Listeria monocytogenes* isolated from different categories of patients with listeriosis. *Eur. J. Clin. Microbiol. Infect. Dis.* **9**, 201–203.

McLauchlin, J. (1990b) Human listeriosis in Britain, 1967–1985: a summary of 722 cases. 1. Listeriosis during pregnancy and in the newborn. *Epidemiol. Infect.* **104**, 181–189.

McMeekin, T.A., Olley, J.N., Ross, T. & Ratkowsky, D.A. (1993) *Predictive Microbiology.* John Wiley & Sons, Chichester.

Mead, P.S., Slutsker, L., Dietz, V., *et al.* (1999) Food-related illness and death in the United States. *Emerg. Infect. Dis.* **5**, 607–625.

Medema, G.J., Teunis, P.F.M., Havelaar, A.H. & Haas, C.N. (1996) Assessment of the dose–response relationship of *Campylobacter jejuni. Int. J. Food Microbiol.* **30**, 101–111.

Meng, J. & Genigoergis, C.A. (1994) Delaying toxigenesis of *Clostridum botulinum* by sodium lactate in 'sous-vide' products. *Lett. Appl. Microbiol.* **19**, 20–23.

Meng, X.J., Purcell, R.H., Halbur, P.G., *et al.* (1997) A novel virus in swine is closely related to the human hepatitis E virus. *Proc. Natl. Acad. Sci. U. S. A.* **94**, 9860–9865.

Merican, I., Guan, R., Amarapuka, D., Alexander, M.J., Chutaputti, A., Chien, R.N., Hasnian, S.S., Leung, N., Lesmana, L., Phiet, P.H., Sjalfoellah Noer, H.M., Sollano, J., Sun, H.S. & Xu, D.Z. (2000) Chronic hepatitis B virus infection in Asian countries. *J. Gastroenterol. Hepatol.* **15**, 1356–1361.

Mermin, J.H. & Griffin, P.M. (1999) Invited commentary: public health crisis in crisis-outbreaks of *Escherichia coli* O157:H7 in Japan. *Am. J. Epidemiol.* **150**, 797–803.

Messens, W., Herman, L., De Zutter, L., Heyndrickx, M. (2009) Multiple typing for the epidemiological study of contamination of broilers with thermotolerant *Campylobacter. Vet Microbiol.* **138**, 120–131.

Mierau, I., Kunji, E.R.S., Leenhouts, K.J., *et al.* (1996) Multiple-peptidase mutants of *Lactococcus lactis* subsp. *cremoris* SK110 and its nisin-immune transconjugant in relation to flavour development in cheese. *Appl. Enviorn. Microbiol..* **64**, 1950–1953.

Miller, A.J., Whitting, R.C. & Smith, J.L. (1997) Use of risk assessment to reduce listeriosis incidence. *Food Technol.* **51**, 100–103.

Mintz, E.D., Cartter, M.L., Hadler, J.L., *et al.* (1994) Dose–response effects in an outbreak of *Salmonella enteritidis. Epidemiol. Infect.* **112**, 13–19.

Morris, J.G. Jr. & Potter, M. (1997) Emergence of new pathogens as a function of changes in host susceptibility. *Emerg. Infect. Dis.* **3**, 435–441.

Mortimore, S. & Wallace, C. (1994) *HACCP – A Practical Approach.* Practical Approaches to Food Control and Food Quality Series No.1. Chapman and Hall, London.

Moseley, B.E.B. (1999) The safety and social acceptance of novel foods. *Int. J. Food Microbiol.* **50**, 25–31.

Mossel, D.A.A., Corry, J.E.L., Struijk, C.B. & Baird, R.M. (1995) In: *Essentials of the Microbiology of Foods. A Textbook for Advanced Studies.* pp. 223. John Wiley & Sons, Chichester.

Muhammad-Tahir, Z. & Alocilja, E.C. (2003) A conductometric biosensor for biosecurity. *Biosens Bioelectron.* **18**, 813–819.

Murata, M., Legrand, A.M., Ishibashi, Y., Fukui, M. & Yasumoto, Y. (1990) Structures and configurations of ciguatoxin from the morey eel *Gymnothora javanicus* and its likely precursor from the dinoflagellate *Gambierdiscus toxicus. J. Am. Chem. Soc.* **112**, 4380–4386.

Muyzer, G. (1999) DGGE/TGGE: a method for identifying genes from natural ecosystems. *Curr. Opin. Microbiol.* **2**, 317–322.

Nanduri, V., Bhunia, A.K., Tu, S.I., Paoli, G.C. & Brewster, J.D. (2007) SPR biosensor for the detection of *L. monocytogenes* using phage-displayed antibody. *Biosens. Bioelectron.* **23**, 248–252.

National Advisory Committee on Microbiological Criteria for Foods (NACMCF) (1992) Hazard analysis and critical control point system. *Int. J. Food Microbiol.* **16**, 1–23.

National Advisory Committee on Microbiological Criteria for Foods (NACMCF) (1998a) Principles of risk assessment for illness caused by foodborne biological agents. *J Food Protect.* **16**, 1071–1074.

National Advisory Committee on Microbiological Criteria for Foods (NACMCF) (1998b) Hazard analysis and critical control point principles and application guidelines. *J. Food Protect.* **61**, 1246–1259.

National Research Council (NRC) (1993) *Risk Assessment in the Federal Government: Managing the process.* National Academy Press, Washington, DC.

National Research Council (2007) The new science of metagenomics: revealing the secrets of our planet. Available from: http://www.books.nap.edu/catalog.php?record_id = 11902.

Nauta, M.J. (2000) Separation of uncertainty and variability in quantitative microbial risk assessment models. *Int. J. Food Microbiol.* **57**, 9–18.

Nelson, K.E., Fouts, D.E., Mongodin, E.F., *et al.* (2004) Whole genome comparisons of serotype 4b and 1/2a strains of the food-borne pathogen *Listeria monocytogenes* reveal new insights into the core genome components of this species. *Nucleic Acids Research* **32**, 2386–2395.

Nes, I.F., Diep, D.B., Havarstein, L.S., Brurberg, M.B., Eijsink, V. & Holo, H. (1996) Biosynthesis of bacteriocins in lactic acid bacteria. *Ant. van Leeuw.* **70**, 113–128.

Notermans, S. & Van Der Giessen, A. (1993) Foodborne diseases in the 1980's and 1990's: the Dutch experience. *Food Contam.* **4**, 122–124.

Notermans, S., Zwietering, M.H. & Mead, G.C. (1994) The HACCP concept: Identification of potentially hazardous micro-organisms. *Food Microbiol. (London),* **11**, 203–214.

Notermans, S. & Mead, G.C. (1996) Incorporation of elements of quantitative risk analysis in the HACCP system. *Int. J. Food Microbiol.* **30**, 157–173.

Notermans, S., Mead, G.C. & Jouve, J.L. (1996) Food products and consumer protection: A conceptual approach and a glossary of terms. *Intl. J. Food Microbiol.* **30**, 175–183.

Notermans, S., Nauta, M.J., Jansen, J., Jouve, J.L & Mead, G.C. (1996) A risk assessment approach to evaluating food safety based on product surveillance. *Food Control.* **9**, 217–223.

Notermans, S., Dufreene, J., Teunis, P., *et al.* (1997) A risk assessment study of *Bacillus cereus* present in pasteurised milk. *Food Microbiol.* **14**, 143–151.

Notermans, S. & Batt, C.A. (1998) A risk assessment approach for food-borne *Bacillus cereus* and its toxins. *J. Appl. Microbiol* Suppl. **84**, 51S–61S.

Notermans, S., Dufreene, J., Teunis, P. & Chackraborty, T. (1998) Studies on the risk assessment of *Listeria monocytogenes.* *J. Food Protect.* **61**, 244–248.

Notermans, S., Hoornstra, E., Northolt, M.D. & Hofstra, H. (1999) How risk analysis can improve HACCP. *Food Sci. Technol. Today* **13**, 49–54.

O'Donnell-Maloney, M.J., Smith, C.L. & Contor, C.R.E. (1996) The development of microfabricated arrays for DNA sequencing and analysis. *Trends Biotechnol.* **14**, 401–407.

O'Hara, A.M. & Shanahan, F. (2006) The gut flora as a forgotten organ. *EMBO Rep.* **7**, 688–693.

Oberman, H. & Libudzisz, Z. (1998) Fermented milks. In: *Microbiology of Fermented Foods* (ed. B.J.B. Wood), pp. 308–350. Blackie Academic & Professional, London, Weinheim, New York, Tokyo, Melbourne, Madras.

OECD (1993) *Safety Evaluation of Foods Produced by Modern Biotechnology – Concepts and Principles.* Organisation for Economic Cooperation and Development, Paris.

Ogier, J.-C., Lafarge, V., Girard, V., *et al.* (2004) Molecular fingerprinting of dairy microbial ecosystems by use of temporal temperature and denaturing gradient gel electrophoresis. *Appl. Environ. Microbiol.* **70**, 5628–5643.

Oh, D.H. & Marshall, D.L. (1995) Destruction of *Listeria monocytogenes* biofilms on stainless steel using monolaurin and heat. *J. Food Prot.* **58**, 251–255.

Oliver, J.D. (2005) The viable but nonculturable state in bacteria. *J. Microbiol.* **43**, 93–100.

Olsvik, O., Popovic, T., Skjerve, E., *et al.* (1994) Magnetic separations techniques in diagnostic microbiology. *Clin. Microbiol. Rev.* **7**, 43–54.

On, S.L.W, Dorrell, N., Petersen, L., *et al.* (2006) Numerical analysis of DNA microarray data of *Campylobacter jejuni* strains correlated with survival, cytolethal distending toxin and haemolysin analyses. *Int. J. Med. Microbiol.* **296**, 353–363.

Oscar, T.P. (1998a) Growth kinetics of *Salmonella* isolates in a laboratory medium as affected by isolate and holding temperature. *J. Food Prot.* **61**, 964–968.

Oscar, T.P. (1998b) The development of a risk assessment model for use in the poultry industry. *J. Food Safety* **18**, 371–381.

Ouwehand, A.C., Kirjavainen, P.V., Shortt, C. & Salminen, S. (1999) Probiotics: mechanisms and established effects. *Int. Dairy J.* **9**, 43–52.

OzFoodNet Working Group (2003) Foodborne disease in Australia: incidence, notifications and outbreaks. Annual report of the OzFoodNet network, 2002. *Commun. Dis. Intell.* **27**, 209–243.

Paoli, G. (2000) Health Canada risk assessment model for *Salmonella* Enteritidis. (Quoted by Fazil et al. 2000).

Park, S.F. (2005) The physiology of *Campylobacter* species and its relevance to their role as food borne pathogens. *Int. J. Food Microbiol.* **74**, 177–188.

Parkhill, J., Wren, B.W., Mungall, *et al.* (2000) The genome sequence of the food-borne pathogen *Campylobacter jejuni* reveals hypervariable sequences. *Nature* **403**, 665–668.

Parry, R.T. (ed.) (1993) *Principles and Applications of Modified Atmosphere Packaging of Foods.* Blackie Academic & Professional, Glasgow.

Peck, M.W. (1997) *Clostridium botulinum* and the safety of refrigerated processed foods of extended durability. *Trends Food Sci. Technol.* **8**, 186–192.

Peeler, J.T. & Bunning, V.K. (1994) Hazard assessment of *Listeria monocytogenes* in the processing of bovine milk. *J. Food Prot.* **57**, 689–697.

Peterson, W.L., MacKowiak, P.A., Barnett, C.C., Marling-Cason, M. & Haley, M.L. (1989) The human gastric bactericidal barrier: mechanisms of action, relative antibacterial activity, and dietary influences. *J. Infect. Dis.* **159**, 979–983.

PHLS (1992) Provisional microbiological guidelines for some ready-to-eat foods sampled at point of sale: notes for PHLS Food Examiners. *PHLS Microbiol. Dig.* **9**, 98–99.

PHLS (1996) Microbiological guidelines for some ready-to-eat foods sampled at the point of sale: an expert opinion from the PHLS. *PHLS Microbiol. Dig.* **13**, 41–43.

PHLS (2000) Guidelines for the microbiological quality of some ready-to-eat foods sampled at the point of sale. *Commun. Dis. Public Health* **3**, 163–167.

Picket, C. & Whitehouse, D. (1999) The cytolethal distending toxin family. *Trends Microbiol.* **7**, 292–297.

Piper, P., Mahe, Y., Thompson, S., *et al.* (1998) The Pdr12 ATP-binding cassette ABC is required for the development of weak acid resistance in *Saccharomyces cerevisiae*. *EMBO J.* **17**, 4257–4265.

Ponka, A., Maunula, L., von Bonsdorff, C.H. & Lyytikainen, O. (1999) An outbreak of calicivirus associated with consumption of frozen raspberries. *Epidemiol. Infect.* **123**, 469–474.

Potter, M.E. (1996) Risk Assessment Terms and Definitions. *J. Food Prot.* Suppl., 6–9.

Power, U.F. & Collins, J.K. (1989) Differential depuration of poliovirus, *Escherichia coli*, and a coliphage by the common mussel, *Mytilus edulis*. *Appl. Environ. Microbiol.* **55**, 1386–1390.

Pridmore, R.D., Berger, B., Desiere, F., *et al.* (2004) The genome sequence of the probiotic intestinal bacterium *Lactobacillus johnsonii* NCC 533. *Proc. Natl. Acad. Sci. U. S. A.* **101**, 2512–2517.

Pruitt, K.M. & Kamau, D.N. (1993) Mathematical models of bacterial growth, inhibition and death under combined stress conditions. *J. Ind. Microbiol.* **12**, 221.

Raso, J., Pagán, R., Condón, S. & Sala, F. (1998) Influence of temperature and pressure on the lethality of ultrasound. *Appl. Environ. Microbiol.*. **64**, 465–471.

Reddy, B.S. & Riverson, A. (1993) An inhibitory effect of *Bifidobacteriumm longum* on colon, mammary and liver carcinogenesis induced by 2-amino-3-methylimidazo[4,5-f] quinolone, a food mutagen. *Cancer Res.* **53**, 3914–3918.

Reid, G., Millsap, K. & Busscher, H.J. (1994) Implantation of *Lactobacillus casei* var. *rhamnosus* into the vagina. *Lancet* **344**, 1229.

Reid, G. (1999) The scientific basis for probiotic strains of *Lactobacillus*. *Appl. Env. Microbiol.* **65**, 3763–3766.

Renouf, V., Claisse, O., Miot-Sertier, C. & Lanvaud-Funel, A. (2006) Lactic acid bacteria evolution during winemaking: use of *rpoB* gene as a target for PCR-DGGE analysis. *Food Microbiol.* **23**, 136–145.

Rhodehamel, E.J. (1992) FDA concerns with sous vide processing. *Food Technol.* **46**, 73–76.

Ribot, E.M., Fitzgerald, C., Kubota, K., Swaminathan, B. & Barrett, T.J. (2001) Rapid pulsed-field gel electrophoresis protocol for subtyping of *Campylobacter jejuni*. *J. Clin. Microbiol.* **39**, 1889–1894.

Richards, G.P. (2001) Enteric virus contamination of foods through industrial practices: a primer on intervention strategies. *J. Ind. Microbiol. Biotechnol.* **27**, 117–125.

Roberts, D. & Greenwood, M. (2003) *Practical Food Microbiology*, 3rd edn. Blackwell Publishing, Oxford.

Roberts, J.A. (1996) *Economic Evaluation of Surveillance*. Department of Public Health and Policy, London.

Roberts, T.A. & Gibson, A.M. (1986) Chemical methods for controlling *Clostridium botulinum* in processed meats. *Food Technol.* **40**, 163–171.

Robinson, A., Gibson, A.M. & Roberts, T.A. (1982) Factors controlling the growth of *Clostridium botulinum* types A and B in pasteurized, cured meats. V. Prediction of toxin production: non-linear effects of storage temperature and salt concentration. *J. Food Technol.* **17**, 727–744.

Robinson, D.A. (1981) Infective dose of *Campylobacter jejuni* in milk. *Br. Med. J.* **282**, 1584.

Rollins, D.M. & Colwell, R.R. (1986) Viable but nonculturable stage of *Campylobacter jejuni* and its role in survival in the natural aquatic environment. *Appl. Env. Microbiol.* **52**, 531–538.

Rondon, M.R., Goodman, R.M. & Handelsman, J. (1999) The Earth's bounty: assessing and accessing soil microbial diversity. *TIBTECH* **17**, 403–409.

Rose, J.B., Haas, C.N. & Regli, S. (1991) Risk assessment and control of waterborne giardiasis. *Am. J. Publ. Health* **81**, 709–713.

Rose, J.B. & Sobsey, M.D. (1993) Quantitative risk assessment for viral contamination of shellfish and coastal waters. *J. Food Prot.* **56**, 1043–1050.

Rosenquist, H., Nielsen, N.L., Sommer, H., Nørrung, B. & Christensen, B. (2003) Quantitative risk assessment of human campylobacteriosis associated with thermophilic *Campylobacter* species in chickens. *Int. J. Food Microbiol.* **83**, 87–103.

Ross, T. (1993) Belehardek-type models. *J. Ind. Microbiol.* **12**, 180.

Ross, T. & McMeekin, T.A. (1994) Review Paper. Predictive microbiology. *Int. J. Food Microbiol.* **23**, 241–264.

Ross, T., Todd, E. & Smith, M. (2000, Withdrawn) Preliminary report: exposure assessment of *Listeria monocytogenes* in ready-to-eat foods. *JEMRA*.

Ross, T. & McMeekin, T.A. (2003) Modeling microbial growth within food safety risk assessments. *Risk Anal.* **23**, 179–197.

Ross, W. (2000) From exposure to illness: building a dose–response model for risk assessment (Quoted by Fazil et al. 2000).

Rowan, N.J., Anderson, J.G. & Smith, J.E. (1998) Potential infective and toxic microbiological hazards associated with the consumption of fermented foods. In: *Microbiology of Fermented Foods* (ed. B.J.B. Wood), pp. 263–307. Blackie Academic & Professional, London.

Rowan, N.J., MacGregor, S.J., Anderson, J.G., Fouracre, R.A., McIlvaney, L. & Farish, O. (1999) Pulsed-light inactivation of food-related microorganisms. *Appl. Environ. Microbiol.* **65**, 1312–1315.

Rowland, I.R. (1990) Metabolic interactions in the gut. In: *Probiotics* (ed. R. Fuller), pp. 29–52. Chapman & Hall, New York.

Rowland, I.R. (1999) Probiotics and benefits to human health – the evidence in favour. *Environ. Microbiol.* **1**, 375–382.

Ruthven, P.K. (2000) Food and health economics in the 21st century. *Asia Pacific. J. Clin. Nutr.* **9** (Suppl.) S101–S102.

Ryan, C.A., Nickels, M.K., Hargrett-Bean, N.T., *et al.* (1987) Massive outbreak of antimicrobial-resistant salmonellosis traced to pasteurised milk. *J. Am. Med. Assoc.* **258**, 3269–3274.

Safarík, I., Safaríková, M. & Forsythe, S.J. (1995) The application of magnetic separations in applied microbiology. *J. Appl. Bacteriol.* **78**, 575–585.

Salmon, R. (2005) Outbreak of verotoxin producing *E. coli* O157 infections involving over forty schools in south Wales, September 2005. *Eurosurveillance* **10**(40), 2804. Available from: http://www.eurosurveillance. org/ViewArticle.aspx?ArticleId=2804.

Sanders, M.E. (1993) Summary of conclusions from a consensus panel of experts on health attributes of lactic cultures: significance of fluid milk products containing cultures. *J. Dairy Sci.* **76**, 1819–1828.

Sanders, M.E. (1998) Overview of functional foods: emphasis on probiotic bacteria. *Int. Dairy J.* **8**, 341–349.

SCF (1997) Commission Recommendation 97/618/EEC concerning the scientific aspects of the presentation of information necessary to support applications for the placing on the market of novel foods and novel food ingredients and the preparation of initial assessment reports under regulation (EC) No 258/97 of the European Parliament and of the Council: *Offic. J. Euro. Commun.* L253, Brussels.

Schellekens, M., Martens, T., Roberts, T.A., *et al.* (1994) Computer aided microbial safety design of food processes. *Int. J. Food Microbiol.* **24**, 1–9.

Schellekens, M. (1996) New research in sous-vide cooking. *Trends Food Sci. Technol.* **7**, 256–262.

Schena, M., Heller, R.A., Theriault, T.P., Konrad, K., Lachenmeir, E. & Davis, R.W. (1998) Microarrays: biotechnology's discovery platform for functional genomics. *TIBTECH* **16**, 301–306.

Scheu, P.M., Berghof, K. & Stahl, U. (1998) Detection of pathogenic and spoilage microorganisms in food with the polymerase chain reaction. *Food Microbiol.* **15**, 13–31.

Schiff, G.M., Stefanovic, E., Young, E.C., Sander, D.S., Pennekamo, J.K. & Ward, R.L. (1984) Studies of Echovirus 12 in volunteers: Determination of minimal infectious dose and the effect of previous infection on infectious dose. *J. Infect. Dis.* **150**, 858–866.

Schiffrin, E., Rouchat, F., Link-Amster, H., Aeschlimann, J. & Donnet-Hugues, A. (1995) Immunomodulation of blood cells following the ingestion of lactic acid bacteria. *J. Diary Sci.* **78**, 491–497.

Schleifer, K.-H., Ehrmann, M., Brockmann, E., Ludwig, W. & Ammann, R. (1995) Application of molecular methods for the classification and identification of the lactic acid bacteria. *Int. Dairy J.* **5**, 1081–1094.

Schlundt, J. (2002) New directions in foodborne disease. *Intl. J. Food Microbiol.* **78**, 3–17.

Schwartz, I. (2000) Microbial genomics: from sequence to function. *Emerg. Infect. Dis.* **6**, 493–495.

Schwiertz, A., Gruhl, B., Lobnitz, M., Michel, P., Radke, M. & Blaut, M. (2003) Development of the intestinal bacterial composition in hospitalized preterm infants in comparison with breast-fed, full-term infants. *Ped. Res.* **54**, 393–399.

SCOOP (1998) Reports on tasks for scientific co-operation, Microbiological Criteria: Collation of scientific and methodological information with a view to the assessment of microbiological risk for certain foodstuffs, Report of experts participating in Task 2.1, European Commission, EUR 17638, Office for Official Publications of the European Communities, Luxembourg.

Scott, A.E., Timms, A.R., Connerton, P.L., Loc-Carillo, C., Radzum, K.A. & Connerton, I.F. (2007) Genome dynamics of *Campylobacter jejuni* in response to bacteriophage predation. *PLoS Pathog* **3**, e119.

Seera, J.A., Domenech, E., Escriche, I. & Martorelli, S. (1999) Risk assessment and critical control points from the production perspective. *Int. J. Food Microbiol.* **46**, 9–26.

Sela, D.A., Chapman, J., Adeuya, A., *et al.* (2008) The genome sequence of *Bifidobacterium longum* subsp. *infantis* reveals adaptations for milk utilization within the infant microbiome. *PNAS* **105**, 18964–18969.

Sergeev, N., Distler, M., Coirtney, S., *et al.* (2004) Multipathogen oligonucleotide microarray for environmental and biodefense applications. *Biosens. Bioelectron.* **20**, 684–698.

Sethi, D., Wheeler, J.G., Cowden, J.M., *et al.* (1999) A study of infectious intestinal disease in England: plan and methods of data collection. *Commun. Dis. Publ. Health* **2**, 101–107.

Sethi, D., Cumberland, P., Hudson, M.J., *et al.* (2001) A study of infectious intestinal disease in England: risk factors associated with group A rotavirus in children. *Epidemiol. Infect.* **126**, 63–70.

Shapton, D.A. & Shapton, N.E. (1991) *Principles and Practises for the Safe Processing of Foods*. Butterworth-Heinemann, Oxford.

Shaw, R.D. (2000) Viral infections of the gastrointestinal tract. *Curr. Opin. Gastroenterol.* **16**, 12–17.

Shivananda, S., Lennard-Jones, J., Logan, R., *et al.* (1996) Incidence of inflammatory bowel disease across Europe: is there a difference between North and South? Results of the European collaborative study on inflammatory bowel disease (ECIBD) *Gut* **39**, 690–697.

Silley, P. & Forsythe, S. (1996) Impedance microbiology – a rapid change for microbiologists. *J. Appl. Bacteriol.* **80**, 233–243.

Sinell, H.J. (1995) Control of food-borne infections and intoxications. *Intl. J. Food. Microbiol.* **25**, 209–217.

Skirrow, M.B. (1977) *Campylobacter* enteritidis: a "new" disease. *Br. Med. J.* **2**, 9–11.

Slauch, J., Taylor, R. & Maloy, S. (1997) Survival in a cruel world: how *Vibrio cholerae* and *Salmonella* respond to an unwilling host. *Genes Dev.* **11**, 1761–1774.

Smith, K., Besser, J., Hedberg, C., *et al.* (1999) Quinolone-resistant *Campylobacter jejuni* infections in Minnesota, 1992–1998. *New Eng. J. Med.* **340**, 1525–1532.

Snyder, O.P. Jr (1995) HACCP-TQM for retail and food service operations. In: *Advances in Meat Research – Volume 10. HACCP in Meat, Poultry & Fish Processing.* (eds A.M. Pearson & T.R. Dutson). Blackie Academic & Professional, London.

Sockett, P.N. (1991) Food poisoning outbreaks associated with manufactured foods in England and Wales: 1980–89. *Commun. Dis. Rep.* **1**, Rev No. 10, R105–R109.

Soker, J.A., Eisenberg, J.N. & Olivier, A.W. (1999) Case study human infection through drinking water exposure to human infections rotavirus. ILSI Research Foundation. Risk Science Institute, Washington, DC.

Sopwith, W., Matthews, M., Fox, A., *et al.* (2006) *Campylobacter jejuni* multilocus sequence types in humans, Northwest England, 2003–2004. *Emerg. Infect. Dis.*. 2006 Oct. Available from: http://www.cdc.gov/ncidod/EID/vol12no10/06-0048.htm.

Sozer, N. & Kokini, J.L. (2009) Nanotechnology and its applications in the food sector. *Trends Biotechnol.* **27**, 82–89.

Sparks, P. & Shepherd, R. (1994) Public perceptions of the potential hazards associated with food production and food consumption: an empirical study. *Risk Anal.* **14**, 799–806.

Stabel, J.R., Steadham, E. & Bolin, C.A. (1997) Heat inactivation of *Mycobacterium paratuberculosis* in raw milk: are current pasteurisation conditions effective? *Appl. Environ. Microbiol.* **63**, 4975–4977.

Stanley, G. (1998) Cheeses. In: *Microbiology of Fermented Foods* (ed. B.J.B. Wood), pp. 263–307. Blackie Academic & Professional, London.

Stickler, D. (1999) Biofilms. *Curr. Opin. Microbiol.* **2**, 270–275.

Stringer, S.C., George, S.M. & Peck, M.W. (2000) Thermal inactivation of *Escherichia coli* O157:H7. *J. Appl. Microbiol.* **88** (Suppl.), 79S–89S.

Stutz, H.K., Silverman, G.J., Angelini, P. & Levin, R.E. (1991) Bacteria and volatile compounds associated with ground beef spoilage. *J. Food Sci.* **55**, 1147–1153.

Sugita, T. & Togawa, M. (1994) Efficacy of *Lactobacillus* preparation biolactis powder in children with rotavirus enteritis. *Jpn. J. Pediatr.* **47**, 899–907.

Surkiewicz, B.F., Johnson, R.W., Moran, A.B. & Krumm, G.W. (1969) A bacteriological survey of chicken eviscerating plants. *Food Technol.* **23**, 1066–1069.

Sutherland, J.P. & Bayliss, A.J. (1994) Predictive modeling of growth of *Yersinia enterocolitica*: The effects of temperature, pH and sodium chloride. *Int. J. Food Microbiol.* **21**, 197–215.

Sutherland, J.P., Bayliss, A.J. & Braxton, D.S. (1995) Predictive modeling of growth of *Escherichia coli* O157:H7: The effects of temperature, pH and sodium chloride. *Int. J. Food Microbiol.* **25**, 29–49.

Sweeney, R.W., Whitlock, R.H. & Rosenberger, A.E. (1992) *Mycobacterium paratuberculosis* cultured from milk and supramammary lymph nodes of infected asymptomatic cows. *J. Clin. Microbiol.* **30**, 166–171.

Tam, C.C., O'Brien, S.J., Adak, G.K., Meakins, S.M. & Frost, J.A. (2003) *Campylobacter coli* – an important foodborne pathogen. *J. Infect.* **47**, 28–32.

Tannock, G.W. (1995) *Normal Microflora*. Chapman and Hall, London.

Tannock, G.W. (1997) Probiotic properties of lactic-acid bacteria: plenty of scope for fundamental R&D. *Trends Biotechnol.* **15**, 270–274.

Tannock, G.W. (1998) Studies of the intestinal microflora: a prerequisite for the development of probiotics. *Int. Dairy J.* **8**, 527–533.

Tannock, G.W. (1999a) Identification of lactobacilli and bifidobacteria. In: *Probiotics: A Critical Review* (ed. G.W. Tannock). Horizon Scientific Press, Norfolk.

Tannock, G.W. (ed.) (1999b) *Probiotics: A Critical Review*, pp. 1–4. Horizon Scientific Press, Norfolk.

Taormina, P.J., Beuchat, L.R. & Slutsker, L. (1999) Infections associated with eating seed sprouts: an international concern. *Emerg. Infect. Dis.* **5**, 626–634.

Tatsozawa, H., Murayama, T., Misawa, N., *et al.* (1998) Inactivation of bacterial respiratory chain enzymes by singlet oxygen. *FEBS Lett.* **439**, 329–333.

Tauxe, R.V. (1997) Emerging foodborne diseases: an evolving public health challenge. *Emerg. Infect. Dis.* **3**, 425–434.

Tauxe, R.V. (2002) Emerging foodborne pathogens. *Int. J. Food Microbiol.* **78**, 31–41.

Taylor, A.D., Ladd, J., Yu, Q., Chen, S., Homola, J., Jiang, S. (2006) Quantitative and simultaneous detection of four foodborne bacterial pathogens with a multi-channel SPR sensor. *Biosens. Bioelectron.* **22**, 752–758.

te Giffel, M.C., Beumer, R.R., Granum, P.E. & Rombous, F.M. (1997) Isolation and characterisation of *Bacillus cereus* from pasteurised milk in households refrigerators in The Netherlands. *Int. J. Food Microbiol.* **34**, 307–318.

Teunis, P., Havelaar, A., Vliegenthart, J. & Roessink, C. (1997) Risk assessment of *Campylobacter* species in shellfish: identifying the unknown. *Water Sci. Technol.* **35**, 29–34.

Teunis, P.F.M., Van Der Heijden, O.G., Van Der Giessen, J.W.B. & Havelaar, A.H. (1996) The dose–response relation in human volunteers for gastro-intestinal pathogens. National Institute of Public Health and the Environment. Report 284–550–002. Bilthoven.

Teunis, P.F.M. & Havelaar, A.H. (1999) *Cryptosporidium* in drinking water: evaluation of the ILSI/RSI quantitative risk assessment framework. Report No. 284-530-006. National Institute of Public Health and the Environment, Bilthoven.

Thomas, P. & Newby, M. (1999) Estimating the size of the outbreak of new-variant CJD. *Br. Food J.* **101**, 44–57.

Thomson, G.T.D., Derubeis, D.A., Hodge, M.A., *et al.* (1995) Post-*Salmonella* reactive arthritis-late clinical sequelae in a point-source cohort. *Am. J. Med.* **98**, 13–21.

Thorns, C.J. (2000) Bacterial food-borne zoonoses. *Rev. Sci. Tech. Off. Int. Epiz.* **19**, 226–239.

Titbull, R.W., Naylor, C.E. & Basak, A.K. (1999) The *Clostridium perfringens* α-toxin. *Anaerobe* **5**, 51–64.

Todd, E.C.D. (1989a) Preliminary estimates of costs of foodborne disease in Canada and costs to reduce salmonellosis. *J. Food Prot.* **52**, 586–594.

Todd, E.C.D. (1989b) Preliminary estimates of costs of foodborne disease in the U.S. *J. Food Prot.* **52**, 595–601.

Todd, E.C.D. (1996a) Risk assessment of use of cracked eggs in Canada. *Int. J. Food Microbiol.* **30**, 125–143.

Todd, E.C.D. (1996b) Worldwide surveillance of foodborne disease: the need to improve. *J. Food Prot.* **59**, 82–92.

Todd, E.C.D. & Harwig, J. (1996) Microbial risk assessment of food in Canada. *J. Food Prot.* (Suppl.) S10–S18.

Todd, E.C.D., Greig, J.D., Bartleson, C.A. & Michaels, B.S. (2007a) Outbreaks where food workers have been implicated in the spread of foodborne disease. Part 2. Description of outbreaks by size, severity, and settings. *J. Food Prot.* **70**, 1975–1993.

Todd, E.C.D., Greig, J.D., Bartleson, C.A. & Michaels, B.S. (2007b) Outbreaks where food workers have been implicated in the spread of foodborne disease. Part 3. Factors contributing to outbreaks and description of outbreak categories. *J. Food Prot.* **70**, 2199–2217.

Tomlinson, N. (1998) Worldwide regulatory issues: legislation and labelling. In: *Genetic Modification in the Food Industry* (eds S. Roller & S. Harlander), pp. 61–68. Blackie Academic & Professional, London.

Tompkins, D.S., Hudson, M.J., Smith, H.R., *et al.* (1999) A study of infectious intestinal disease in England: microbiological findings in cases and controls. *Commun. Dis. Public Health* **2**, 108–113.

Torpdahl, M., Sorensen, G., Lindstedt, B-A. & Nielsen, E.M. (2007) Tandem repeat analysis for surveillance of human *Salmonella* Typhimurium infections. *Emerg. Infect. Dis.* **13**, 388–395.

Townsend, S.M., Hurrell, E., Gonzalez-Gomez, I., *et al.* (2007) *Enterobacter sakazakii* invades brain capillary endothelial cells, persists in human macrophages influencing cytokine secretion and induces severe brain pathology in the neonatal rat. *Microbiology* **153**, 3538–3547.

Townsend, S.M. & Forsythe, S.J. (2008) The neonatal intestinal microbial flora, immunity, and infections. In: *Emerging Issues in Food Safety: Enterobacter sakazakii* (eds J. Farber & S.J. Forsythe). ASM Press, Washington, DC.

Townsend, S.M., Hurrel, E. & Forsythe, S.J. (2008) Virulence studies of *Enterobacter sakazakii* isolates associated with a neonatal intensive care unit outbreak. *BMC Microbiol.* **8**, 64. Available from: http://www.biomedcentral.com/1471-2180/8/64.

Trucksess, M.W., Mislevec, P.B., Young, K., Bruce, V.E. & Page, S.W. (1987) Cyclopiazonic acid production by cultures of *Aspergillus* and *Penicillium* species isolated from dried beans, corn, meal, macaroni and pecans. *J. Assoc. Off. Anal. Chem.* **70**, 123–126.

US Census Bureau (1998). Available from: http://www.census.gov.

Vallejo-Cordoba, B. & Nakai, S. (1994) Keeping quality of pasteurised milk by multivariate analysis of dynamic headspace gas chromatographic data. *J. Agric. Food Chem.* **42**, 989–993 & 994–999.

Van Der Poel, W., Vinjé, J., van der Heide, R., Herrera, I., Vivo, A. & Koopmans, M. (2000) Norwalk-like calicivirus genes in farm animals. *Emerg. Infect. Dis.* **6**, 36–41.

Van de Venter, T. (1999) Prospects for the future: emerging problems – chemical/biological. Conference on International Food Trade Beyond 2000: Science-based decisions, harmonization, equivalence and mutual recognition. Melbourne, Australia. Food and Agriculture Organization of the United Nations.

Van Gerwen, S.J.C. & Zwietering, M.H. (1998) Growth and inactivation models to be used in quantitative risk assessments. *J. Food Prot.* **61**, 1541–1549.

Van Regenmortel, M.H.V., Fauquet, C.M. & Bishop, D.H.L. (2000) *Virus Taxonomy: Classification and Nomenclature of Viruses*, pp. 725–739. Academic Press, San Diego, CA.

Van Schothorst, M. (1996) Sampling plans for *Listeria monocytogenes*. *Food Control* **7**, 203–208.

Van Schothorst, M. (1997) Practical approaches to risk assessment. *J. Food Prot.* **60**, 1439–1443.

Vandenberg, O., Dediste, A., Houf, K., *et al.* (2004) *Arcobacter* species in humans. *Emerg. Infect. Dis.* [Serial on the Internet]. 2004 Oct. Available from: http://www.cdc.gov/ncidod/EID/vol10no10/04-0241.htm.

Vaughan, E.E., Mollet, B. & de Vos, W.M. (1999) Functionality of probiotics and intestinal lactobacilli: light in the intestinal tract tunnel. *Curr. Opin. Biotech.* **10**, 505–510.

Vazquez-Boland, J.A., Dominguez-Bernal, G., Gonzalez-Zorn, B., Kreft, J. & Goebel, W. (2001) Pathogenicity islands and virulence evolution in *Listeria*. *Microbes Infect.* **3**, 571–584.

Velmurugan, G.V., SU, C. & Dubey, J.P. (2009) Isolate designation and characterization of *Toxoplasma gondii* isolates from pigs in the United States. *J. Parasitol.* **95**, 95–99.

Veterinary Laboratory Agency (2005) Assessment of relative to other pathways, the contribution made by the food chain to the problem of quinolone resistance in microorganisms causing human infections. Report to Food Standards Agency. VLA, UK.

Voetsch, A.C., Van Gilder, T.J., Angulo, F.J., *et al.* (2004) FoodNet estimate of the burden of illness caused by nontyphoidal *Salmonella* infections in the United States. *Clin. Infect. Dis.* **38**, Suppl. 3: S127–S134.

Vose, D. (1996) *Quantitative Risk Analysis: A Guide to Monte Carlo Simulation Modelling.* John Wiley & Sons, New York.

Vose, D. (1997) The application of quantitative risk analysis to microbial food safety. *J. Food Prot.* **60**, 1416.

Vose, D.J. (1998) The applications of quantitative risk assessment to microbial food safety. *J. Food Prot.* **61**, 640–648.

Voysey, P.A. & Brown, M. (2000) Microbiological risk assessment: a new approach to food safety control. *Int. J. Food Microbiol.* **58**, 173–180.

Walker, S.J. (1994) The principles and practice of shelf-life prediction for microorganisms. In: *Shelf-life evaluation of foods* (eds C.M.D. Man & A.A. Jones), pp. 40–51. Blackie Academic & Professional, London.

Wan, J., Wilcock, A. & Coventry, M.J. (1998) The effect of essential oils of basil on the growth of *Aeromonas hydrophila* and *Pseudomonas fluorescens*. *J. Appl. Microbiol.* **84**, 152–158.

Wang, H. (2002) Rapid methods for detection and enumeration of *Campylobacter* spp. in foods. *J. AOAC Int.* **85**, 996–999.

Ward, R.L., Berstein, D.I. & Young, E.C. (1986) Human rotavirus studies in volunteers of infectious dose and serological response to infection. *J. Infect. Dis.* **154**, 871–877.

Wassenaar, T.M. (1997) Toxin production by *Campylobacter* spp. *Clin. Microbiol. Rev.* **10**, 466–476.

Wassenar, T.M. & Newell, D.G. (2000) Genotyping of *Campylobacter* spp. *Appl. Environ. Microbiol.* **66**, 1–9.

Wegener, H.C., Hald, T., Wong, D.L.F., *et al.* (2003) *Salmonella* control programs in Denmark. *Emerg. Infect. Dis.* [serial online]. Available from: http://www.cdc.gov/ncidod/EID/vol9no7/03-0024.html.

Wei, D., Oyarzabal, O.A., Huang, T.S., Balasubramanian, S., Sista, S. & Simoman, A.L. (2007) Development of a surface plasmon resonance biosensor for the identification of *Campylobacter jejuni*. *J. Microbiol. Methods* **69**, 78–85.

Weidman, M., Bruce, J.L, Keating, C., Johnson, A.E., McDonough, P.L. & Batt, C.A. (1997) Ribotypes and virulence gene polymorphisms suggest three distinct *Listeria monocytogenes* lineages with differences in pathogenic potential. *Infect. Immun.* **65**, 2707–2716.

Wierup, M., Engstrom, B., Engvall, A. & Wahlstrom, H. (1995) Control of *Salmonella enteritidis* in Sweden (review). *Int. J. Food Microbiol.* **25**, 219–226.

Wells, J.M., Robinson, K., Chamberlain, L.M., Schofiled, K.M. & LePage, R.W. (1996) Lactic acid bacteria as vaccine delivery vehicles. *Ant. van Leeu.* **70**, 317–330.

Wesley, I. (1997) *Helicobacter* and *Arcobacter*: potential human foodborne pathogens? *Trends Food Sci. Tech.* **8**, 293–299.

Wheeler, J.G., Sethi, D., Cowden, J.M., *et al.* (1999) Study of infectious intestinal disease in England: rates in the community, presenting to general practice, and reported to national surveillance. *Br. Med. J.* **318**, 1046–1050.

Whiting, R.C. (1995) Microbial modeling in foods. *Crit. Rev. Food Sci. Nutr.* **35**, 467–494.

Whiting, R.C. (1997) Microbial database building: what have we learned? *Food Technol.* **51**, 82–86.

Whiting, R.C. & Buchanan, R.L. (1997) Development of a quantitative risk assessment model for *Salmonella* Enteritidis in pasteurized liquid eggs. *Int. J. Food Microbiol.* **36**, 111–125.

WHO (1984) Regional Office for Europe. Toxic oil syndrome: mass food poisoning in Spain. Report of a WHO meeting, Madrid, 21–25 May 1983, Copenhagen: WHO Regional Office for Europe.

WHO (1991) *Strategies for Assessing the Safety of Foods Produced by Biotechnology.* Report of joint FAO/WHO consultation. World Health Organisation, Geneva.

WHO (1995) Report of the WHO Consultation on Selected Emerging Foodborne Diseases, Berlin, 20–24 March 1995. WHO/CDS/VPH/95.142, World Health Organisation, Geneva.

WHO (1999) Strategies for implementing HACCP in small and/or less developed businesses. Geneva, Switzerland: World Health Organisation. WHO/SDE/FOS/99.7.

WHO (2001) Available from: http://www.who.int/director-general/speeches/2001/english/20010314_foodchain2001uppsala.en.html.

WHO (2002) Available from: http://www.who.int/foodsafety/publications/general/terrorism/en/.

WHO (2007) 'Safe preparation, storage and handling of powdered infant formula guidelines', and associated specialised documents for various care situations. Available from: http://www.who.int/foodsafety/publications/micro/pif2007/en/index.html.

WHO/FAO (2000a) Preliminary document: WHO/FAO guidelines on hazard characterization for pathogens in food and water. RIVM, Bilthoven, 2000. Available from: https://apps.who.int/fsf/Micro/Scientific_documents/HC_guidelines.pdf.

WHO/FAO (2000b) The interaction between assessors and managers of microbiological hazards in foods. Kiel, Germany. WHO/SDE/PHE/FOS/007. Available from: http://www.who.int/foodsafety/publications/micro/en/march2000.pdf.

Wijtzes, T., van't Riet, K., in't Veld, J.H.J., Huis & Zwietering, M.H. (1998) A decision support system for the prediction of microbial food safety and food quality. *Int. J. Food Microbiol.* **42**, 79–90.

Will, R.G, Ironside, J.W, Zeidler, M., *et al.* (1996) A new variant of Creutzfeldt–Jakob diese in the UK. *The Lancet* **347**, 921–925.

Wimptheimar, L., Altman, N.S. & Hotchkiss, J.H. (1990) Growth of *Listeria monocytogenes* Scott A, serotype 4 and competitive spoilage organisms in raw chicken packaged under modified atmospheres and in air. *Int. J. Food Microbiol.* **11**, 205.

Wood, B.J.B. (ed.) (1998) *Microbiology of Fermented Foods.* Blackie Academic & Professional, London.

Wuytack, E.Y., Phuong, L.D.T., Aetsen, A., *et al.* (2003) Comparison of sublethal injury induced in *Salmonella enterica* serovar Typhimurium by heat and by different nonthermal treatments. *J. Food Protect.* **66**, 31–37.

Yeh, P.L., Bajpai, R.K. & Lannotti, E.L. (1991) An improved kinetic model for lactic acid fermentation. *J. Ferment. Bioeng.* **71**, 75.

Zink, D.L. (1997) The impact of consumer demands and trends on food processing. *Emerg. Infect. Dis.* **3**, 467–469.

Zoppi, G. (1998) Probiotics, prebiotics, synbiotics and eubiotics. *Pediatr. Med. Chir.* **20**, 13–17.

Zottola, E.A. & Sasahara, K.C. (1994) Microbial biofilms in the food processing industry – should they be a concern? *Int. J. Food Microbiol.* **23**, 125–148.

Zwietering, M.H., Wijtzes, T., de Wit, J.C. & van't Reit, K. (1992) A decision support system for prediction of the microbial spoilage in foods. *J. Food Prot.* **55**, 973.

Zwietering, M.H., de Wit, J.C. & Notermans, S. (1996) Application of predictive microbiology to estimate the number of *Bacillus cereus* in pasteurised milk at the point of consumption. *Int. J. Food Microbiol.* **30**, 55–70.

Index

Food Science and Technology

GENERAL FOOD SCIENCE AND TECHNOLOGY

Drying Technologies in Food Processing	Chen	9781405157636
Nutraceuticals, Glycemic Health and Type 2 Diabetes	Pasupuleti	9780813829333
Frozen Food Science and Technology	Evans	9781405154789
Biotechnology in Flavor Production	Havkin-Frenkel	9781405156493
High Pressure Processing of Foods	Doona	9780813809441
Water Activity in Foods	Barbosa-Canovas	9780813824086
Food and Agricultural Wastewater Utilization and Treatment	Liu	9780813814230
Multivariate and Probabilistic Analyses of Sensory Science Problems	Meullenet	9780813801780
Applications of Fluidisation in Food Processing	Smith	9780632064564
Encapsulation and Controlled Release Technologies in Food Systems	Lakkis	9780813828558
Accelerating New Food Product Design and Development	Beckley	9780813808093
Handbook of Meat, Poultry and Seafood Quality	Nollet	9780813824468
Chemical Physics of Food	Belton	9781405121279
Handbook of Organic and Fair Trade Food Marketing	Wright	9781405150583
Sensory and Consumer Research in Food Product Design and Development	Moskowitz	9780813816326
Sensory Discrimination Tests and Measurements	Bi	9780813811116
Food Biochemistry and Food Processing	Hui	9780813803784
Handbook of Fruits and Fruit Processing	Hui	9780813819815
Nonparametrics for Sensory Science	Rayner	9780813811123
IFIS Dictionary of Food Science and Technology	IFIS	9781405125055
Kosher Food Production	Blech	9780813825700
Managing Food Industry Waste	Zall	9780813806310
Food Processing – Principles and Applications	Smith	9780813819426
Flavor Perception	Taylor	9781405116275
Food Supply Chain Management	Bourlakis	9781405101684
Plant Food Allergens	Mills	9780632059829
Welfare of Food	Dowler	9781405112451
Food Flavour Technology	Taylor	9781841272245
Adverse Reactions to Food	Buttriss	9780632055470
Advanced Dietary Fibre Technology	McCleary	9780632056347

INGREDIENTS

Food Colours (2nd Edition)	Emerton	9781905224449
Sweeteners Handbook (3rd Edition)	Wilson	9781905224425
Sweeteners and Sugar Alternatives in Food Technology	Mitchell	9781405134347
Emulsifiers in Food Technology	Whitehurst	9781405118026
Technology of Reduced Additive Foods (2nd Edition)	Smith	9780632055326
Water-Soluble Polymer Applications in Foods	Nussinovitch	9780632054299
Food Additives Data Book	Smith	9780632063956
Enzymes in Food Technology	Whitehurst	9781841272238

FOOD SAFETY, QUALITY AND MICROBIOLOGY

Color Atlas of Postharvest Quality of Fruits and Vegetables	Nunes	9780813817521
Bioactive Compounds in Foods	Gilbert	9781405158756
Microbiological Safety of Food in Health Care Settings	Lund	9781405122207
Nondestructive Testing of Food Quality	Irudayaraj	9780813828855
Food Biodeterioration and Preservation	Tucker	9781405154178
Mycotoxins	Botana	9780813827001
Advances in Food Diagnostics	Nollet	9780813822211
Advances in Thermal and Nonthermal Food Preservation	Tewari	9780813829685
Biofilms in the Food Environment	Blaschek	9780813820583
Food Irradiation Research and Technology	Sommers	9780813808826
Preventing Foreign Material Contamination of Foods	Peariso	9780813816395
Irradiation Food Safety	Sheward	9781405115810
Food Microbiology and Laboratory Practice	Bell	9780632063819
Listeria (2nd Edition)	Bell	9781405106184
Preharvest and Postharvest Food Safety	Beier	9780813808840
Practical Food Microbiology (3rd Edition)	Roberts	9781405100755
Metal Contamination of Food (3rd Edition)	Reilly	9780632059270
The Microbiology of Safe Food (2nd Edition)	Forsythe	9781405140058
Food Safety in Shrimp Processing	Kanduri	9780852382707
Shelf Life	Man	9780632056743
HACCP	Mortimore	9780632056484
Salmonella	Bell	9780632055197
Microbiology of Safe Food	Forsythe	9780632054879
Clostridium botulinum	Bell	9780632055210
E coli	Bell	9780751404623

Food Science and Technology from Wiley-Blackwell

FOOD LAWS AND REGULATIONS

BRC Global Standard – Food: A Guide to a Successful Evaluation	Kill	9781405157964
Food Labeling Compliance Review (4th Edition)	Summers	9780813821818
Guide to Food Laws and Regulations	Curtis	9780813819464
Regulation of Functional Foods and Nutraceuticals	Hasler	9780813811772
Concept Research in Food Product Design and Development	Moskowitz	9780813824246

DAIRY FOODS

98145	Tamime	9781405145305
Whey Processing, Functionality and Health Benefits	Onwulata	9780813809038
Dairy Processing and Quality Assurance	Chandan	9780813827568
Cleaning-in-Place (3rd Edition)	Tamime	9781405155038
Advanced Dairy Science and Technology	Britz	9781405136181
Structure of Dairy Products	Tamime	9781405129756
Brined Cheeses	Tamime	9781405124607
Fermented Milks	Tamime	9780632064588
Manufacturing Yogurt and Fermented Milks	Chandan	9780813823041
Handbook of Milk of Non-Bovine Mammals	Park	9780813820514
Probiotic Dairy Products	Tamime	9781405121248
Mechanisation and Automation of Dairy Technology	Tamime	9781841271101
Technology of Cheesemaking	Law	9781841270371

BAKERY AND CEREALS

Bakery Food Manufacture and Quality: Water Control and Effects (2nd Edition)	Cauvain	9781405176132
Whole Grains and Health	Marquart	9780813807775
Baked Products – Science, Technology and Practice	Cauvain	9781405127028
Bakery Products Science and Technology	Hui	9780813801872
Pasta and Semolina Technology	Kill	9780632053490

BEVERAGES AND FERMENTED FOODS/BEVERAGES

Handbook of Fermented Meat and Poultry	Toldra	9780813814773
Microbiology and Technology of Fermented Foods	Hutkins	9780813800189
Carbonated Soft Drinks	Steen	9781405134354
Brewing Yeast and Fermentation	Boulton	9781405152686
Food, Fermentation and Micro-organisms	Bamforth	9780632059874
Wine Production	Grainger	9781405113656
Chemistry and Technology of Soft Drinks and Fruit Juices (2nd Edition)	Ashurst	9781405122863
Technology of Bottled Water (2nd Edition)	Senior	9781405120388
Wine Flavour Chemistry	Clarke	9781405105309
Beer: Health and Nutrition	Bamforth	9780632064465
Brewing Yeast Fermentation Performance (2nd Edition)	Smart	9780632064984

PACKAGING

Food Packaging Research and Consumer Response	Moskowitz	9780813812229
Packaging for Nonthermal Processing of Food	Han	9780813819440
Packaging Closures and Sealing Systems	Theobald	9781841273372
Modified Atmospheric Processing and Packaging of Fish	Otwell	9780813807683
Paper and Paperboard Packaging Technology	Kirwan	9781405125031
Food Packaging Technology	Coles	9781841272214
PET Packaging Technology	Brooks	9781841272221
Canmaking for Can Fillers	Turner	9781841272207
Design and Technology of Packaging Decoration for the Consumer Market	Giles	9781841271064
Materials and Development of Plastics Packaging for the Consumer Market	Giles	9781841271163
Technology of Plastics Packaging for the Consumer Market	Giles	9781841271170

OILS AND FATS

Oils and Fats in the Food Industry	Gunstone	9781405171212
Trans Fatty Acids	Dijkstra	9781405156912
Chemistry of Oils and Fats	Gunstone	9781405116268
Rapeseed and Canola Oil – Production, Processing, Properties and Uses	Gunstone	9781405116251
Vegetable Oils in Food Technology	Gunstone	9781841273310
Fats in Food Technology	Rajah	9781841272252
Edible Oil Processing	Hamm	9781841270388

For further details and ordering information, please visit

www.blackwellfood.com

Join our free e-mail alert service to receive journal tables of contents with links to abstracts and news of the latest books in your field.